ITALIAN PHYSICAL SOCIETY

PROCEEDINGS
OF THE
INTERNATIONAL SCHOOL OF PHYSICS « ENRICO FERMI »

Course XXVIII

edited by L. Gratton
Director of the Course

VARENNA ON LAKE COMO
VILLA MONASTERO
AUGUST 20 - SEPTEMBER 1 1962

Star Evolution

1963

ACADEMIC PRESS • NEW YORK AND LONDON

SOCIETÀ ITALIANA DI FISICA

RENDICONTI
DELLA
SCUOLA INTERNAZIONALE DI FISICA «ENRICO FERMI»

XXVIII CORSO

a cura di L. GRATTON
Direttore del Corso

VARENNA SUL LAGO DI COMO
VILLA MONASTERO
20 AGOSTO - 1 SETTEMBRE 1962

Evoluzione delle stelle

1963

ACADEMIC PRESS • NEW YORK AND LONDON

ACADEMIC PRESS INC.
111 Fifth Avenue
New York 3, N. Y.

United Kingdom Edition
Published by
ACADEMIC PRESS INC. (London) Ltd.
Berkeley Square House, London W. 1

Library of Congress Catalog Card Number: 63-20575

PRINTED IN ITALY

INDICE

Preface.

L. Gratton

Director of the Course

This volume conta'ns the full text of the Lectures and Seminars given at the XXVIII Course of the International School of Physics « Enrico Fermi ». The task of editing the book proved to be very long and difficult; this is the reason why it appears one year and a half after the time at which the lectures were actually given.

The evolution of the theories of stellar evolution proved to be very fast during this time and although the A.A. took most care in giving in their lectures the last results in the respective fields, many subjects which are not included in this book are now at the center of the researchers interest. I may mention only the problems connected with radiogalaxies, gravitational collapse and relativistic Astrophysics.

The various chapters differ widely in character. Some of them are more elementary, a few very advanced; G. Burbidge's chapter is an article-review, which is reprinted here through the kindness of the Editors of the *Annual Review of Nuclear Science*. Some lectures contain original results which are published here for the first time.

Such as it is I believe that the book presents a fair picture of the present state of the subject, and I hope that it will be found useful to many workers in this field.

I wish to thank all those who contributed to the success of these lectures; the Authors, who gladly accepted my invitation and whose work constitutes the value of the book; those, students and teachers, who contributed to the very lively and interesting discussions, which followed each lecture and which are not printed in the book for technical difficulties; the Italian Society of Physics that kindly included among the 1962 Courses the subject of Stellar Evolution; last, not least, Prof. Germanà to whom all who enjoyed their stay at Varenna are greatly indebted for his ceaseless efforts on their behalf.

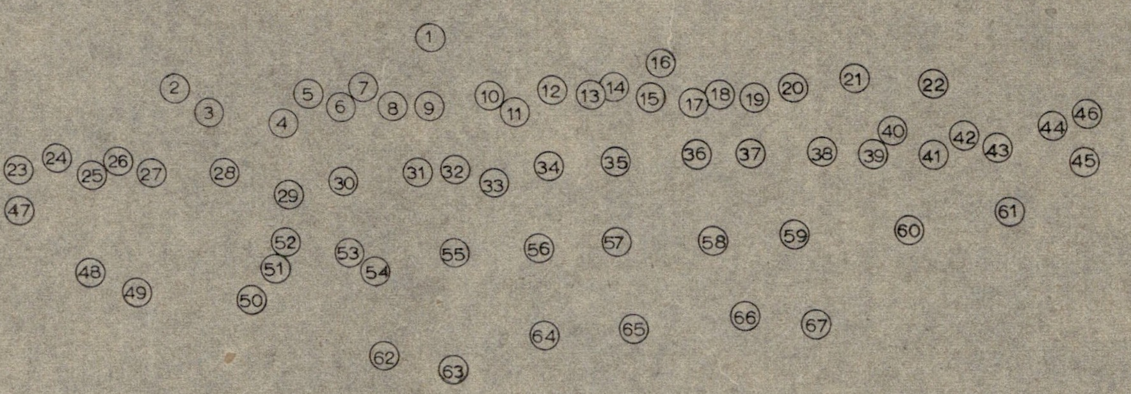

1. G. Mannino
2. A. Weigert
3. M. G. Fracastoro
4. R. Viotti
5. G. Burbidge
6. P. Broglia
7. S. Cristaldi
8. S. Delli Santi
9. D. Cattani
10. A. Sandage
11. N. Virgopia
12. R. C. Henry
13. L. Gratton
14. V. De Sabbata
15. P. Ledoux
16. F. Pacini
17. M. Hack
18. E. Hofmeister
19. P. Maffei
20. G. Pocci
21. R. Kandel
22. A. Finzi
23. A. Kranjc
24. J. Kristian
25. I. W. Roxburgh
26. M. H. Wrubel
27. P. Giannone
28. P. Bertola
29. A. M. Giannuzzi
30. U. Steinlin
31. M. Cimino
32. G. Setti
33. E. Schatzman
34. G. Chincarini
35. L. Rosino
36. N. Dallaporta
37. F. Zagar
38. M. M. Shapiro
39. R. Kippenhahn
40. Z. Zinamon
41. N. Baker
42. A. Boury
43. H. U. Schmidt
44. R. P. Fenkart
45. L. Navone
46. G. A. Tammann
47. R. Cristaldi
48. G. Germanà
49. J. Faulkner
50. K. Griffiths
51. D. Cristaldi
52. T. Chincarini
53. G. De Sabbata
54. P. De Sabbata
55. Sig.ra Maffei
56. O. Fracastoro
57. R. Rosino
58. E. M. Burbidge
59. Sig.ra Sandage
60. M. C. Bretz
61. T. Vives
62. R. Gratton
63. F. Fracastoro
64. R. Margoni
65. F. Occhionero
66. A. La Farge
67. F. Manoni

SOCIETÀ ITALIANA DI FISICA

SCUOLA INTERNAZIONALE DI FISICA « E. FERMI »

XXVIII CORSO - VARENNA SUL LAGO DI COMO - VILLA MONASTERO - 20 Agosto - 1 Settembre 1962

Stellar Evolution - Introductory Lecture.

L. GRATTON
Università - Roma
Laboratorio Gas Ionizzati - Frascati

Sed quibus ille modis conjectus materiai
Fundarit terram et caelum, pontique profunda,
Solis lunai cursus, ex ordine ponam.
(LUCRETIUS, *De Rerum Natura*, Lib. V, 415-417)

1. – The past.

At our epoch the Theory of Stellar Evolution is so closely connected with all branches of Astrophysics that a Course on Stellar Evolution is, practically, a course on Astrophysics with evolutionary concepts acting as links between the various arguments.

Indeed, if we look to the historical development of our ideas on Stellar Evolution, we see that even in the Antiquity the only rational attempt to a scientific description of the world, that of DEMOCRITUS, involved deep evolutionary ideas. The following passages from LUCRETIUS, the great Epicurus' disciple, who adopted Democritus' physical theories, are very interesting even to day.

« Sic igitur, solem, lunam, stellasque putandum
Ex alio atque alio lucem jactare subortu
Et primum quidquid flammarum perdere semper;
Inuiolabilia haec ne credas forte uigere. »
(LUCRETIUS, *De Rerum Natura*, Lib. V, 302-305)

« Haut igitur Leti praeclusa est janua caelo,
Nec soli, terraeque, nec altis aequoris undis;
Sed patet immani, et uasto respectat hiatu.
Quare etiam natiua necessumst confiteare
Haec eadem: neque enim, mortali corpore quae sunt,
Ex infinito jam tempore adhuc potuissent
Immensi ualidas aeui contemnere uireis. »
(LUCRETIUS, *De Rerum Natura*, Lib. V, 373-379)

I am sorry for not being able to reproduce in English the deep melancholy of these beautiful verses; but from the scientific point of view, we find here the two main reasons for an Evolution of the World: a *law of conservation* and the *immense extent of time*. Indeed, if we would call after the name of a philosopher the law of conservation, the name of Democritus would be the best choice. Hear again LUCRETIUS:

> « Quas ob res, ubi uiderimus, nil posse creari
> De nihilo: tum quod sequimur, jam rectius inde
> Perspiciemus: et unde queat res quaeque creari,
> Et quo quaeque modo fiant opera sine diuom. »
> . . . .
> « Haut igitur redit ad nihilum res ulla; sed omnes
> Discidio redeunt in corpora materiai. »

(LUCRETIUS, *De Rerum Natura*, Lib. I, 156-159; 248-249)

I have left purposedly undetermined the possessive of the property « con, servation ». We know at present half a dozen of conservation laws (energy-momentum, electric charge, number of baryons, etc.) so that in a sense it is immaterial which is that is conserved. Democritus' great discovery is that a world in which there are no conservation laws of any sort is a mad world in which everything may happen and there is no possibility of rational knowledge. He rightly infers that the Sun and the Stars cannot derive their light from nothing, but must produce it at the cost of some changes in their structure; however small these changes during the enormous extent of time they must cause some radical readjustment of the constitution of these bodies.

Of course, Democritus' knowledge of Physics was insufficient to go further; but he had another brilliant idea connected with his hypothesis of an atomic structure of the matter. His description of the original assembly of atoms and their subsequent grouping in macroscopic objects closely reminds us of our own picture of an uniform cloud of gas and its fractioning in nebulae and stars.

Perhaps it would be excessive to give him too much credit for this, since his ideas were in other respects very crude even for his days; it is noticeable that his successors, the Epicureans, were not aware of, or anyway did not take at all into consideration, the brilliant work of Greek astronomers, and especially ARISTARCHUS, but based their Physics upon older and rougher doctrines. But certainly we owe to DEMOCRITUS the first rational attempt to explain the formation of celestial objects.

The purpose of this rather long excursion on DEMOCRITUS was to show that even the oldest tentatives of a scientific description of the Universe, were in many principles similar to ours, because Evolution is primarily an *episthemologic* concept as well as an experimental fact; « Greek Atomism was not a mere lucky guess. Crude science though it was, it was the result of scientific thinking » (W. E. LEONARD, 1942).

But it is well known that rational scientific thinking was practically inexistent during the next two thousand years, due to the general decadence of occidental civilization and to the disastrous effects of religious prejudices and of platonic and aristotelian philosophy. Thus, until the eighteenth century it is impossible to find anything comparable with DEMOCRITUS and EPICURUS wonderful anticipations. Although some bright speculations appear here and there among the works of men like NICHOLAS CUSANUS, T. DIGGES, G. BRUNO and others, still on 1658 the Archbishop USSHER would write, quite seriously, that God created the world on Sunday, 23rd October 4004 B.C. (J. USSHER, 1658).

However the seventeenth and eighteenth centuries marked the beginning of a new epoch for Science. A frequently quoted passage from a Newton's letter to R. BENTLEY is of great interest to us: « ... if the matter were evenly distributed throughout an infinite space... some of it would convene into one mass and some into another, so as to make an infinite number of great masses, scattered great distances from one another throughout all that infinite space. And thus might the Sun and fixed Stars be formed, supposing the matter were of a lucid nature ».

This is quite modern Science; it follows only a few years after Archbishop Ussher's nonsenses. Of course, Newton's genius is largely responsible for it, but his clear way of reasoning was made possible by two major events which took place during the century before him: the great discoveries of observational Astronomy after Galileo's first use of the telescope for scientific purposes and the foundation of Newtonian Mechanics, which first gave solid grounds for theoretical developments. The Cosmologies of T. WRIGHT, I. KANT and J. H. LAMBERT are bold generalizations based on these discoveries and contain several points which remained unchanged until our days. The famous Theory which goes under the names of KANT and LAPLACE (although Laplace's version is actually considerably different from Kant's) is somewhat more limited in its scope, since it sought to explain only the origin of the Solar System. But both KANT and LAPLACE believed that Stars are entirely similar to the Sun and that their origin is due to quite identical processes. « Millions and whole myriads of millions of centuries will flow on, during which always new worlds and systems of worlds will be formed after each other in the distant regions away from the centre of Nature » (I. KANT, 1755). LAPLACE is even more explicit and surprisingly near to our own ideas in several respects; the reader is referred especially to the last « Note » of his « Exposition du Système du Monde » (P. S. LAPLACE, 1796).

With LAPLACE, Mechanics reached a practically definitive asset; to make further progresses new branches of Physics were necessary. This is why we must wait for more than half a century before finding new conceptions.

During this time enormous advances took place. The work of the HERSCHELS and of ROSSE disclosed the real structure of nebulae; accurate measurements of the motions and distances of the stars and also of their masses and luminosities became available and provided sound observational basis for theoretical reasoning. Spectral analysis was still too crude and atomic Physics was not yet born, but another branch of Physics attained great perfection which proved to be of decisive importance for the study of stellar structure and evolution: Thermodynamics.

The first Theory of stellar structure developed according to the modern methods of mathematical Physics was purely thermodynamical, as it ought to be. The main contributions are due to KELVIN, RITTER, LANE, BETTI, etc., and the best account of the Theory is contained in Emden's book « Gaskugeln » (EMDEN, 1907).

Observational Astronomers, however, were not completely satisfied even at that time and began soon to realize that things were much more complicated. The extraordinary growth of Spectral Analysis and of all other branches of observational Astrophysics due to the greatly increased power of modern telescopes (mostly through the use of celestial photography) revealed a bewildering quantity of facts which were very difficult to fit into the frame of the Theory; anyway they needed the recourses of Atomic Physics to be understood. Hale's book published one year after Emden's (G. H. HALE, 1908) clearly reflects the situation at that time. Another result which made it clear that Kelvin's theory contained at best only a part of the truth was the discovery that the age of some terrestrial rocks was some 100 times greater than that assigned by the theory to the Sun.

The history of how the growth of Atomic Physics and of Observational Astronomy led gradually to the present conceptions on Stellar structure and evolution is but yesterday's history; a very attractive account of it is contained in STRÖMGREN's two articles for Hynek's « Astrophysics » (J. A. HYNEK, 1951).

Without going into details, I will mention only some fundamental points. On the observational side, the discovery of the properties of Hertzsprung-Russel's diagram, of the mass-luminosity law, of several properties of variable stars, of the true dimensions of our Galaxy and of extragalactic Nebulae, the realization of the general uniformity of the chemical composition of stellar and interstellar matter, the final determination of the right time-scale, were only some of the most important events of this period. Concerning the theory, Saha's explanation of the spectral classes and the great importance of radiative equilibrium (as compared with the former theory of purely convective equilibrium), both in the interior and in the outer layers of a star, were probably the most fundamental discoveries; they marked the transition from a mechanical-thermodynamical model of a star to more realistic models in which atomic Physics had the prominent role. Researches on ionization, opacity, gas-degeneration, etc., permitted gradually to reach a fairly complete and detailed knowledge of the physical properties of matter in the interior of « so simple a thing as a star. »

However, the theory of stellar evolution was in a rather confusing state; Eddington's famous book (EDDINGTON, 1926) which concludes and summarizes the results obtained during the first quarter of our century, refers often to the problem of stellar evolution, but refrains carefully from any rash conclusion. « For the past 50 years stellar astronomy has been guided by successive theories of evolution. Today we have no theory of stellar evolution, pending the settlement of the laws of subatomic energy ». As a matter of fact the discovery of these laws by Bethe a little more than ten years after Eddington's writings, opened at once the way to the present conceptions.

2. – The present.

As the present state of the Theory of Stellar Evolution is the subject of this Course, I will spend only a few words on it. In doing so, I will especially stress some controversial points, which may afford a theme for discussions.

The time-scale of stellar evolution is now well established. The oldest globular clusters have ages of the order of 10^{10} years; this is also the age of some galactic clusters, but there is evidence that the formation of stars in the so-called stellar Associations is still going on in our Galaxy and presumably in other too. Therefore we know that in the spiral arms of our Galaxy there are stars of all ages from, say, zero up to 10^{10} years.

In Hertzsprung-Russell's diagram, stars spend most of their lives near the Main Sequence, which corresponds to the position in that diagram of stars that derive their energy from thermonuclear reactions in which hydrogen is transformed in helium (hydrogen-burning process); differences of initial chemical composition do not have a great influence upon the position of a star on the main sequence, which is largely determined by the mass.

Successive evolution brings the star to the right and above the main sequence; but there are some differences in the shape of the evolutionary paths in the Hertzsprung-Russell diagram according to the masses and also to the initial chemical composition of the stars. In this way the subgiant and giant branches observed in old galactic clusters and in globular clusters are formed. During this time, stars derive their energy from helium-burning thermonuclear reactions and also from gravitational contraction. At some stages of this phase, at least massive stars undergo a period in which stable radial pulsations occur.

Although there may be (and indeed there are) several questions of detail which need further investigation, there is to day a general agreement among astrophysicists upon the main lines of the above picture. What follows is much more uncertain and open to discussion.

First comes the problem of how the stars are formed. By far the large majority among the Astronomers believe that stars condense from diffuse clouds of interstellar gas (and dust), through successive fragmentation (probably due to gravitational instability) in smaller and smaller clouds and final contraction to relatively dense bodies. However AMBARTSUMIAN and his school strongly oppose this view, mainly on account of various observational difficulties relative to the properties of stellar Associations and of the time-scale of the contractional stage. They affirm that stars and interstellar matter do originate together from some « prestellar » kind of matter, whose properties are at present completely unknown, but is believed to be rather dense, probably much denser than a normal star. Of course this is too vague to be discussed, but the possibility that the formation of stars is accompanied by processes which we are not able to describe for the simple reason that they depend upon physical laws as yet undiscovered, is not to be dismissed lightly. It appears that careful studies of stellar Associations and especially of the youngest objects contained in them (for instance, *T Tauri* stars and their small nebulosities, *P Cygni* stars, some supergiants), as well as of the diffuse nebulae in which they are imbedded might throw more light upon this subject.

There is also a general agreement that the last phases of its evolution will carry a star into the low left region of the Hertzsprung-Russell diagram (the white-dwarf region); to accomplish this, massive stars must lose a large fraction of their mass through ejection of matter in space. But the whole evolutionary process connecting the white-dwarf region with the giant branch is very little known.

During this evolution important changes of chemical composition take place. This brings to the front the problem of how the present chemical composition of stars did originate; the problem of the origin of chemical elements or Nucleogenesis is closely connected with general Cosmology.

A somewhat older Theory maintains that the present abundance of the different nuclear species was determined in a not too well specified state of the Universe which preceded the present stage. In the so-called « ylem » theory proposed by ALPHER, GAMOW and HERMAN (1950), this initial stage is supposed to be characterized by rather high density and very high temperature; under these conditions matter consisted entirely of free neutrons, protons and electrons, but in a very short time progressive aggregation processes led to the synthesis of heavier nuclei in more or less the observed relative abundances. Successive chemical evolution consisted mainly in the synthesis of helium from four protons, which occurs in the interior of stars.

There exist different versions of these ideas, but at present there seems to be among astronomers a strong preference towards the rival theory which was proposed by HOYLE, FOWLER, BURBIDGE, and BURBIDGE. This theory will be discussed in this Course by one of the Authors; according to it, the synthesis of elements takes place inside the stars, the original substance being practically pure hydrogen. Several processes are involved; some of them include also catastrophic changes of stellar structure.

The attractive feature of this theory is that it does not start from any more or less artificial hypothesis concerning the initial state, but it fits quite naturally into the general frame of the above sketched evolutionary process. An interesting point is that during their life stars send back to space a considerable part of their mass (through more or less continuous ejection at various phases) or even the totality of it (through Supernova explosions). In this way interstellar matter too undergoes an important chemical evolution. As, in turn, new generations of stars are continuously born from interstellar matter in the course of time, stars of the same mass, but belonging to different « generations », have different initial chemical compositions and thus (not too) different physical properties. By this way one obtains a quite reasonable explanation of the existence of different kinds of Stellar « Populations », one of the most fundamental discoveries of the last twenty years which is due to BAADE.

There can be little doubt that the theory of the stellar origin of the elements is in a much better agreement with observed facts than any other so far proposed and gives a much clearer understanding of stellar structure and evolution. Whether it contains the whole truth and nothing more than the truth is quite a different question.

Among the points which do not fit too easily into the general frame of the theory, but need further assumptions, I will mention only those connected with stellar magnetic fields and stellar rotation. On the contrary, at least some of the types of stellar variability may be satisfactorily interpreted.

Cosmological theories are outside the scope of this Course; but obviously the ylem theory is tied with some particular cosmological models. On the contrary, the theory of the synthesis of elements in stars is compatible with almost every cosmological model, including Hoyle, Bondi and Gold's theory of the continuous creation of matter.

We will not discuss in this Course theories of the origin of the solar system, which at present seem rather disconnected from the main body of the theory of stellar evolution.

3. – The future.

After this short sketch of the past and present states of the theory of stellar evolution, it is not out of place to attempt to get a glimpse of its near future, were it only for the sake of argument. Of course, I do not wish to commit myself to rash and arbitrary profecies, but I think that there may be a general agreement upon some points at least.

One of these is doubtless the easy guess that in the near future there will be some important and perhaps fundamental developments of the theory. This may seem almost trivial, but the analysis of the reasons for such a statement may disclose some points of interest.

Our review of the past has shown that every important progress in observational Astronomy and every development of a new branch of Physics has brought up a corresponding advance of our conceptions on stellar evolution. Thus Laplace's theory of the origin of the solar System followed shortly after the great discoveries of the seventeenth and eighteenth centuries and the development of Mechanics. The progress of Stellar Astronomy and the creation of Thermodynamics during the last century were directly responsible for the evolutionary concepts contained in Lane, Kelvin and Ritter's theories. The extraordinary advances of spectral analysis and of Atomic Physics during the first quarter of this century led to Eddington's theory of stellar structure. Finally our present ideas followed at once from the further progress of observational techniques (for instance, photoelectric Photometry and Radioastronomy) as well as from the growth of Nuclear Physics.

This remark allows to make a somewhat more detailed and perhaps less trivial guess as to the future.

First we have to consider advances in observational techniques; and here the enormous possibilities in the near future of astrophysical observations from outside the Earth atmosphere are obvious. There can be little doubt that the progress which will ensue from these observations will be similar to that which followed the first use of the telescope. A manyfold increase of the range of accessible wavelengths is only one aspect of the possibilities offered by these observations, although, may be, the most important. But we must not forget the increase that we may expect of the accuracy of the measurements of stellar position, due to the increase of resolving power, and also of photometric measurements, due to the elimination of the perturbing influence of the atmosphere; these will cause decisive advances in our knowledge of stellar motions, distances, masses, etc.

Another important advance will come from observations of corpuscolar radiation. Cosmic-ray physics is becoming each day more a branch of Astrophysics. Here also observations from outside the atmosphere will be decisive. It is well known that numerous projects both national and international are in progress at various places, so that very quick progress may be expected.

Recently another interesting possibility has been suggested, which is not tied with extraterrestrial observations. It seems that a not impossible improvement of present techniques will permit to observe the flux of neutrinos coming from the Sun and other possible astrophysical sources. This offers indeed wonderful possibilities, due to the very small capture cross-section of neutrinos, which makes it possible to them to cross without absorption a body as big as a star. Thus neutrinos coming from the Sun may carry direct information upon its deep interior.

Which branch of Physics, is now the other question, is more likely to influence the future progress of our evolutionary conceptions? My guess is (and it is not a too difficult one) that, apart from important contributions due to Nuclear Physics and Magnetohydrodynamics, the most fundamental advances of the theory of evolution will be associated with the Physics of Elementary Particles.

So far this branch of Physics has plaid a negligible role in the development of theoretical Astrophysics, certainly because it is still very far from a definitive asset. But some points, connected with the Physics of Elementary Particles did already come to the Astrophysicists' attention.

BURBIDGE and HOYLE have discussed the implications of « anti-matter » consisting of antiprotons and electrons (1956). They showed that in our Galaxy the ratio of anti-matter to ordinary matter cannot exceed 10^{-7}. However with a mean density of $6 \cdot 10^{-29}$ g/cm^3 in intergalactic space, the life-time of an anti-atom against annihilation is of the order of 10^{12} years; thus the existence of anti-matter is quite admissible.

I mentioned earlier the possibilities of neutrino Astronomy from an observational point of view; theoretically, neutrinos may prove to be even more important.

The first to consider the importance of neutrinos in stellar problems was, I think, GAMOW in his theory of the so-called « Urca-process »; more recently the problem of neutrinos was considered again by B. PONTECORVO (1959), G. MARX and N. MENYHÁRD (1959, 1960, unpublished) and H. Y. CHIU (1961). Among other things it was pointed out that the cooling effect due to the escape of neutrinos might be in some cases so strong as to prevent a Supernova explosion when all nuclear energy sources become exhausted.

In general, when the temperature is high, the loss of energy by neutrinos escape greatly exceeds the radiation losses. This may have important evolutionary consequences.

Quite recently AMBARTSUMIAN and SAAKYAN (1960, 1962) improved and extended some earlier work by OPPENHEIMER and VOLKOFF (cf. LANDAU and LIFSHITZ, 1959) and considered the properties of a gas whose density is of the order of nuclear density or higher. If the temperature is low enough, so that all fermions are degenerate, hyperons will appear due to Pauli principle. (See also CAMERON, 1959.) It was found that at densities $\varrho = 10^7$ g/cm^3 neutrons are stable and at $\varrho = 10^{15}$ first μ^--mesons and then Σ^--hyperons, followed by Λ-hyperons, excited neutrons, etc., will become stable.

Of course, it is highly improbable that any known cosmic body will correspond to these densities; nevertheless it is interesting to speculate upon the properties of a sufficiently massive body in gravitational equilibrium. This will consist of a hyperon core, a neutron layer around the core and a proton-electron envelope. The total mass of such a body comes out of the order of $\frac{1}{3}$ solar mass and the radius will be of the order of a dozen km, so that only small masses can reach a superdense state.

Another branch of Physics which may become extremely important for our problem is the Theory of Gravitation. At present this theory is considered by many Physicists to be in a quite unsatisfactory state and it is not possible to foresee in which direction progress will take place and which implications, if any, it will have upon the theory of stellar structure and evolution.

But let me make a last guess. In the past, not only Astronomy has benefitted from the developments of some branches of Physics, but very often it has paid back a high reward to Physics; it is enough to mention Newtonian Mechanics, General Relativity, the development of Spectroscopy, and, recently, of Plasma Physics, which all arose or at least were greatly stimulated by astrophysical researches. Is it too risky to prophesy that it is highly probable that in the next future, precisely those branches of Physics which I mentioned in this section, namely the Physics of Elementary Particles and the Theory of Gravitation may receive some decisive contributions from the study of astronomical problems?

BIBLIOGRAPHY

ALPHER R. A. and HERMAN R. C.: 1950, *Rev. of Mod. Phys.*, **22**, 153.
AMBARTSUMIAN V. A. and SAAKYAN G. S.: 1960, *RAJ*, **37**, 187.
AMBARTSUMIAN V. A. and SAAKYAN G. S.: 1962, *RAJ*, **38**, 601.
BURBIDGE G. R. and HOYLE F.: 1956, *Il Nuovo Cimento*, **4**, 558.
CAMERON A. G. W.: 1959, *Ap. J.*, **130**, 884.
CHIU H. Y.: 1961, *Ann. of Phys.*, **15**, 1.
CHIU H. Y.: 1961, *Ann. of Phys.*, **16**, 321.
EDDINGTON A. S.: 1926, *The internal Constitution of the stars*, Cambridge.
EMDEN R.: 1907, *Gaskugeln*, Leipzig.
HALE G. E.: 1908, *The study of Stellar Evolution.*
HYNEK J. A.: 1951, *Astrophysics*, New York.
KANT I.: 1755, *Universal Natural History and the Theory of the Heavens*, Transl. in *Kant's Cosmogony*, by Hastie, 1900.
LANDAU, L. D. and LIFSHITZ E. M.: 1959, *Statistical Physics*, London.
LAPLACE P. S.: 1796, *Exposition du Système du Monde*, Transl. by *Rev. Harte H. H.*, 1830.
LEONARD W. E.: 1942, *Introduction to Lucretii, De Rerum Natura*, Madison.
MARX G. and MENYARD N.: 1959, *Science*, **131**, 299.
MARX G. and MENYARD N.: 1960, *Budapest Sternw.*, No. 48.
PONTECORVO B.: 1959, *Soviet Phys. JETP*, **9**, 1148.
USSHER J.: 1658, *The Annals of the World deduced from the origin of Time*, London.

Observational Approch to Stellar Evolution (*).

A. R. Sandage

Mt. Wilson and Palomar Observatories - Pasadena, Cal.

L. Gratton

Laboratorio di Astrofisica e Laboratorio Gas Ionizzati - Frascati (Roma)

1. – Theory of the Hertzsprung-Russell diagram.

1·1. *The Hertzsprung-Russell diagram.* – Since its first coming into use in the beginning of this century the Hertzsprung-Russell (H-R) diagram proved to be extremely useful, because it summarizes clearly and neatly a great deal of information about the stars.

An excellent account of the history of the H-R diagram and its properties can be found in a monograph by H. C. Arp (1958). There are three common forms in which the H-R diagram is ordinarily presented. In the original, historical form, which is little used today, the visual absolute magnitude M_v is plotted against the spectral type. The second observational form plots M_v against the colour index of the stars, which will be defined precisely in Sec. 2. The final form is that used by theoreticians and plots the bolometric absolute magnitude, M_{bol}, or the log of the luminosity, L, against the log of the effective temperature, T_e. Transformations from one form to others require a knowledge of the relation between spectrum and colour index, of the temperature scale and of the bolometric correction, $M_{bol} - M_v$; they are, therefore, very difficult if one wishes to obtain a fair degree of accuracy.

Fig. 1 gives a general outline of the form of the diagram and of the terminology of the various sequences and regions. It contains also a rough sketch of the regions occupied by some types of variable stars (RR Lyrae, Classical and Pop II Cepheids, Long Period variables, β Canis Majoris Stars and Novae).

(*) The lectures were given by Sandage, but this fine summary was prepared by Gratton, who improved many of the arguments and made the whole into a coherent manuscript. A. S.

The actual observations will be treated in more detail later; instead we will commence by dealing with the theory of the H-R diagram and, in parti-

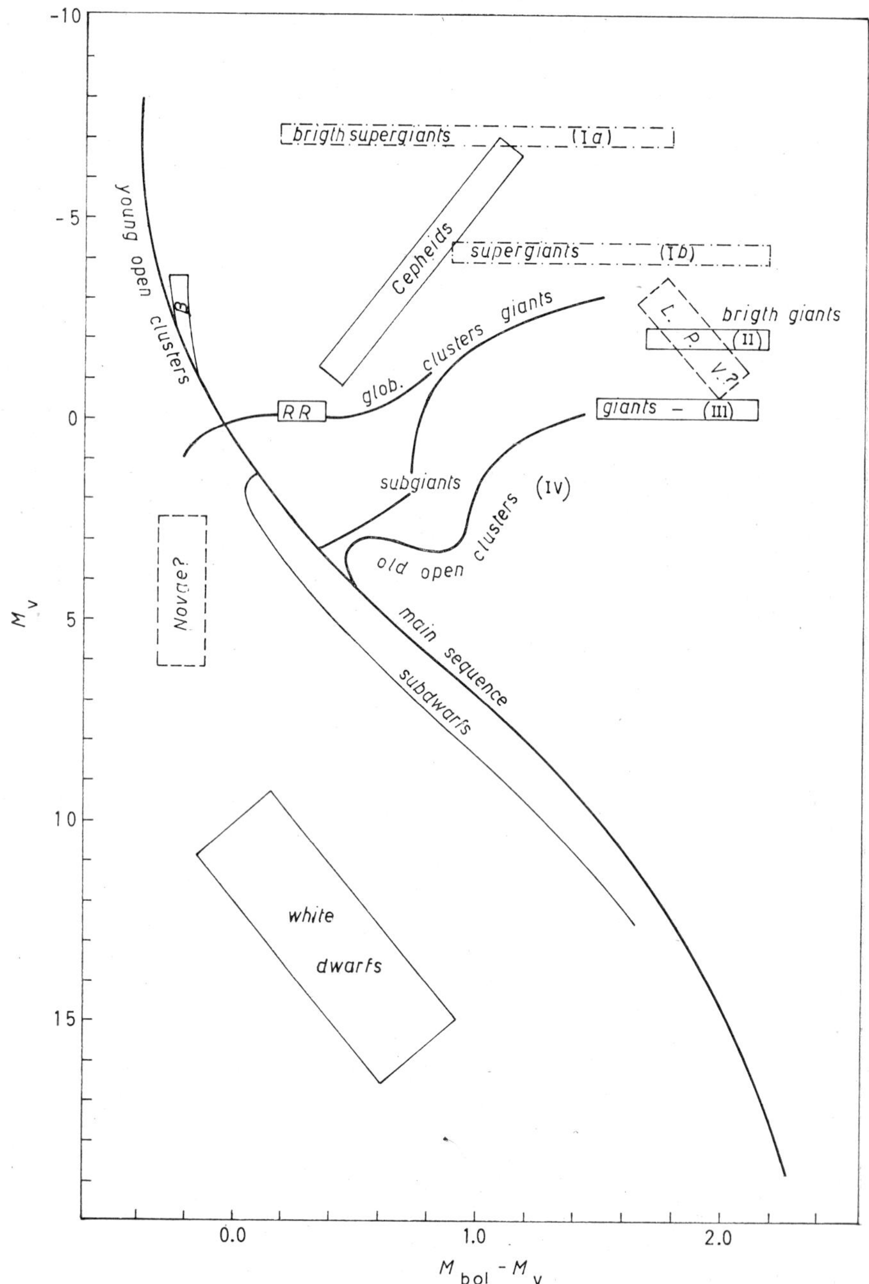

Fig. 1. – Sequences and regions of the H-R diagram.

cular, the equation of the main sequence will be derived under a set of simple (homology) assumptions.

Clearly any attempt to deal theoretically with the complex set of phenomena shown by the diagram of Fig. 1 *in toto* would be premature at this moment. Therefore we will try first to explain theoretically the dominant feature of the diagram, viz. the Main Sequence, which contains alone the large majority of all the stars. Later we will examine how evolution can at least partially account for the other features of the H-R diagram.

1.2. *The theory of main sequence.*

1.2.1. Fundamental equations; homology. The derivation of the main sequence equation will follow the method of STRÖMGREN (1952). To simplify things, we shall restrict ourselves to a group of stars satisfying the following conditions.

a) Chemical homogeneity. We choose as parameters representing the chemical composition the fractional abundances by weight of hydrogen, helium and heavy elements, which we call X, Y, Z respectively; then X, Y, Z are assumed to remain constant throughout a star. Moreover: $X+Y+Z=1$.

b) Hydrostatic equilibrium. i.e. the following equations must be satisfied, assuming spherical symmetry,

$$\mathrm{d}P = -G\frac{\mathcal{M}_r}{r^2}\varrho\,\mathrm{d}r\,, \tag{1.1}$$

$$\mathrm{d}\mathcal{M}_r = 4\pi r^2\varrho\,\mathrm{d}r\,, \tag{1.2}$$

here P and ϱ are the pressure and density at radius r, $\mathcal{M}_r$ is the total mass inside a sphere of radius r and $G=6.67\cdot 10^{-8}$ (c.g.s.) is the gravity constant. From (1.1) and (1.2) we derive

$$\mathrm{d}P = -\frac{G}{4\pi}\frac{\mathcal{M}_r\,\mathrm{d}\mathcal{M}_r}{r^4}\,. \tag{1.3}$$

c) Nondegeneracy. i.e. the perfect gas law holds throughout any star of the group, therefore the pressure will be given by

$$P = NkT\,, \tag{1.4}$$

where N is the number of particles of all kinds per cm^3, T the temperature and $k=1.3805\cdot 10^{-16}$ (erg deg^{-1}) the Boltzmann's constant. If we call μ the mean molecular weight, N is given by

$$N = \frac{1}{H}\frac{\varrho}{\mu}\,, \tag{1.5}$$

H being the atomic mass unit ($1.6598 \cdot 10^{-24}$ g on the physical scale). For a complete ionized mixture μ is given with great approximation by

$$(1.6) \qquad \mu = \frac{1}{2X + \frac{3}{4}Y + \frac{1}{2}Z} = \frac{4}{3 + 5X - Z},$$

since each atom of hydrogen will give 2 particles (one proton and one electron), each atom of He, with atomic weight 4, will give 3 particles and the heavy elements $Z+1$ particles for atomic weight $A \approx 2Z$.

d) *Radiative equilibrium. i.e.* the transport of energy from the interior to the surface occurs by radiation only. It can be shown (see, for instance, WRUBEL 1958) that this condition gives

$$(1.7) \qquad \mathrm{d}\left(\frac{1}{3}\, aT^4\right) = -\frac{1}{c}\,\frac{L_r(r)\varkappa\varrho\,\mathrm{d}r}{4\pi r^2}\,.$$

Here $\varkappa$ is the opacity in $\mathrm{cm}^2\,\mathrm{g}^{-1}$ and $L_r(r)$ the net flux through the surface of a sphere of radius r which equal the total amount of energy liberated per second inside the sphere if the star radiates all the energy it generates; a and c are, as it is usual, the Stefan-Boltzmann's constant ($7.565 \cdot 10^{-15}$ erg $\mathrm{cm}^{-3}\,\mathrm{deg}^{-4}$) and the velocity of light.

e) *Opacity.* Is due only to bound-free (b-f) transitions (see later and Wrubel, p. 50.

f) *Energy.* Is produced by the thermonuclear reactions of the C-N cycle; we shall however consider briefly also the case in which energy is produced through the p-p chain.

g) The structures of the stars of our group are *homologous.*

If we now assume for the opacity and for the law of energy generatione a very simple (power) law, we will be able to derive by a dimensional analysis some important relations between the masses, radii, luminosities and effective temperature of the stars of the group. It is however clear that this treatment can never give the numerical values of the constants of the various relations, nor can the fundamental details of the deviation of the opacity from a simple power law, etc. be considered. The results we derive will only give a feeling for the physics involved, but can never replace stellar models properly computed. However the answers we shall obtain do predict the correct sense of the changes of the M-L relation, the M-R relation and the $\log L - \log T_e$ relation for arbitrary changes of the chemical parameters X, Y, Z.

Consider a particular solution of our equations corresponding to a certain star S of the group, *i.e.* a given set of functions $P(r)$, $\varrho(r)$, $\mathcal{M}_r(r)$ etc. which satisfy the fundamental equations of stellar structure. Then the homology assumption means that if we multiply all the quantities corresponding to S by suitable constants we must obtain the quantities corresponding to another star $\bar{S}$ of the group; in other words, if we call $c_1, c_2, \ldots$ a set of constants (not all of them arbitrary) and put

$$\begin{aligned} \bar{r} &= c_1 r\,, & \bar{\varrho} &= c_4 \varrho\,, \\ \bar{P} &= c_2 P\,, & \bar{T} &= c_5 T\,, \\ \bar{\mathscr{M}}_r &= c_3 \mathscr{M}_r\,, & \bar{L}_r &= c_6 L_r\,, \end{aligned}$$

then the solution of the equations corresponding to $\bar{S}$ will be given by

$$\begin{aligned} \bar{P} &= c_2 P(\bar{r}/c_1)\,, \\ \bar{\mathscr{M}}_r &= c_3 \mathscr{M}_r(\bar{r}/c_1)\,, \\ \bar{\varrho} &= c_4 \varrho(\bar{r}/c_1)\,, \\ \bar{T} &= c_5 T(\bar{r}/c_1)\,, \\ \bar{L}_r &= c_6 L_r(\bar{r}/c_1)\,, \end{aligned}$$

Note in particular that

$$\bar{R} = c_1 R\,, \qquad \bar{\mathscr{M}} = c_3 \mathscr{M}\,, \qquad \bar{L} = c_6 L\,, \tag{1.8}$$

where R, $\mathscr{M}$, L are respectively the radius, total mass and luminosity of S and $\bar{R}, \bar{\mathscr{M}}, \bar{L}$ those of $\bar{S}$.

Of course in order that this be true, the constants cannot be chosen arbitrarly, but must satisfy some conditions; for instance from (1.1) we derive easily

$$c_2 = \frac{c_3 c_4}{c_1}\,.$$

Expressing c_2, c_4 and c_5 by means of c_1, c_3 and c_6 one thus obtains from eq. (1.1) to (1.5)

$$\left|\begin{aligned} c_2 &= \frac{c_3^2}{c_1^4}\,, \\ c_4 &= \frac{c_3}{c_1^3}\,, \\ c_5 &= \frac{\bar{\mu}}{\mu}\frac{c_3}{c_1}\,, \end{aligned}\right. \tag{1.9}$$

and from eq. (1.7)

$$c_6 = \left(\frac{\bar{\mu}}{\mu}\right)^4 \frac{\varkappa}{\bar{\varkappa}}\, c_3^3 \,. \tag{1.10}$$

Substituting from (1.8) we get the *mass-luminosity* law for our homologous group of stars,

$$\frac{\bar{L}}{L} = \left(\frac{\bar{\mu}}{\mu}\right)^4 \left(\frac{\varkappa}{\bar{\varkappa}}\right) \left(\frac{\bar{\mathscr{M}}}{\mathscr{M}}\right)^3 ; \tag{1.11}$$

here $\varkappa$ and $\bar{\varkappa}$ are the opacity in two homologous points of S and $\bar{S}$, respectively that is, in two points such that $\bar{r} = c_1 r$.

1.2.2. Opacity. – Among the various possible sources of opacity it is found that the following three are the most important under stellar conditions (see, for instance WRUBEL 1958 and this vol. p. 63).

a) Bound-free transitions. i.e. photoionization of bound electrons; the opacity (Rosseland's mean over all wave lengths) due to this source can be written approximately

$$\varkappa_{\mathrm{b-f}} = 3 \cdot 10^{25} Z(1+X) \varrho T^{-3.5} \left(\frac{g}{t}\right) \mathrm{cm^2\, g^{-1}} \,. \tag{1.12}$$

The « guillotine factor » (g/t) is a rather slowly varying function of ϱ and T, which may be taken equal to a constant to a first approximation.

b) Free-free transitions. i.e. transitions of an electron from a free state to another free state of higher energy with absorption of a light quantum; practically only H and He contribute to this source of opacity whose approximate expression is

$$\varkappa_{\mathrm{f-f}} = 4 \cdot 10^{22} (1+X)(X+Y) \varrho T^{-3.5} \left(\frac{g'}{t'}\right) \mathrm{cm^2\, g^{-1}} \,. \tag{1.13}$$

Here g'/t' is again a slowly varying function of ϱ and T.

c) Scattering by free electrons. This is given by

$$\varkappa_{\mathrm{el}} = 0.2(1+X)\ \mathrm{cm^2\, g^{-1}} \,. \tag{1.14}$$

As we said before, only the first of these will be considered here. Neglecting the variation of the guillotine factor, we can now eliminate $\bar{\varkappa}/\varkappa$ from eq. (1.11) using eq. (1.12); we obtain easily

$$L = \text{const}\,\frac{\mu^{7.5}\mathcal{M}^{5.5}}{Z(1+X)}\,R^{-0.5}\,. \tag{1.15}$$

This is a completely general result if: (*a*) the star is in hydrostatic equilibrium, (*b*) energy is transported by radiation only and (*c*) eq. (1.11) for the opacity (Kramer's formula) applies, with a constant guillotine factor. It does not depend on the mode of energy generation.

Equation (1.15) applies over a very large domain of the H-R diagram; by introducing the mode of energy generation we can find over what part of the H-R diagram real stars will be spread which satisfy our conditions.

1·2.3. Energy generation. If ε is the energy production in erg per gram per second, then by definition

$$L_r(r) = \int_0^r 4\pi r^2 \varrho(r)\varepsilon(r)\,\mathrm{d}r\,. \tag{1.16}$$

Of course, this equation gives a further relation between the constants $c_1, c_2, \ldots,$ *i.e.* between $\mathcal{M}$, R and L. For our discussion we are assuming that the energy production is due only to the carbon-nitrogen (C-N) cycle; in this case ε can be given approximately by

$$\varepsilon = \text{const}\,Z\times\varrho T^n\,, \tag{1.17}$$

where $16 \leqslant n \leqslant 21$. From eqs. (1.16) and (1.17) we derive

$$c_6 = \frac{\bar{Z}\bar{X}}{ZX}\,c_1^3 c_4^2 c_5^n\,,$$

from which eliminating c_4 and c_5 and using (1.8), we get as before

$$L = \text{const}\,ZX\mu^n\mathcal{M}^{n+2}R^{-(n+3)}\,. \tag{1.18}$$

We now eliminate L from eq. (1.15) by means of (1.18); solving for R, we find

$$R = \text{const}\,[Z^2X(1+X)]^{2/(2n+5)}\mu^{(2n-15)/(2n+5)}\mathcal{M}^{(2n-7)/(2n+5)}\,. \tag{1.19}$$

Choosing $n = 20$ as a representative value, eq. (1.19) becomes

$$R = \text{const}\,[Z^2X(1+X)]^{2/45}\mu^{5/9}\mathcal{M}^{11/15}\,. \tag{1.20}$$

This result can be tested observationally; from eclipsing binaries brighter than $M_v = 2$, KOPAL (1959) finds

$$R \sim \mathcal{M}^{0.64},$$

while eq. (1.19) gives for constant chemical composition

$$R \sim \mathcal{M}^{0.733}.$$

The agreement is not too bad and might be improved by choosing a somewhat smaller value for n.

The result for L, found by introducing (1.19) into (1.15) is

$$L = \text{const}\,\frac{\mu^{(14n+45)/(2n+5)}}{[Z^{2n+7}X(1+X)^{2n+6}]^{1/(2n+5)}}\,\mathcal{M}^{(10n+31)/(2n+5)}, \tag{1.21}$$

which, for $n = 20$, becomes

$$L = \text{const}\,\frac{\mu^{7.22}}{Z^{1.044}X^{0.022}(1+X)^{1.022}}\,\mathcal{M}^{5.13}. \tag{1.22}$$

The effective temperature, T_e, is related to L and R by the equation

$$L = 4\pi R^2 \sigma T_e^4,$$

σ being Stefan's constant $(=5.670 \cdot 10^{-5}$, c.g.s.); solving for T_e and inserting eqs. (1.19) and (1.21) we get

$$T_e = \text{const}\left[\frac{\mu^{10n+75}}{Z^{2n+15}X^5(1+X)^{2n+10}}\right]^{1/(8n+20)}\mathcal{M}^{(6n+45)/(8n+20)}. \tag{1.23}$$

For $n = 20$ eq. (1.23) becomes

$$T_e = \text{const}\,\frac{\mu^{1.528}}{Z^{0.306}X^{0.028}(1+X)^{0.278}}\,\mathcal{M}^{0.917}. \tag{1.24}$$

Finally eliminating $\mathcal{M}$ between (1.21) and (1.23) we obtain

$$L = \text{const}\,Z^{2/3}X^{22/(6n+45)}(1+X)^{(4/3)(n+2)/(2n+15)}\mu^{-4/3}T_e^{4(10n+31)/(6n+45)} \tag{1.25}$$

or

$$L = \text{const}\,Z^{2/3}X^{2/15}(1+X)^{8/15}\mu^{-4/3}T_e^{5.6}, \tag{1.26}$$

for $n = 20$. This is the equation for the main sequence in the L, T_e-plane.

The observed slope of the main sequence is not constant throughout (see Fig. 2). Its general average is near to $L \sim T_e^{6.5}$; for stars brighter than $M_b = 2$ the slope is about $L \sim T_e^5$ and it becomes $L \sim T_e^8$ for fainter stars. These dif-

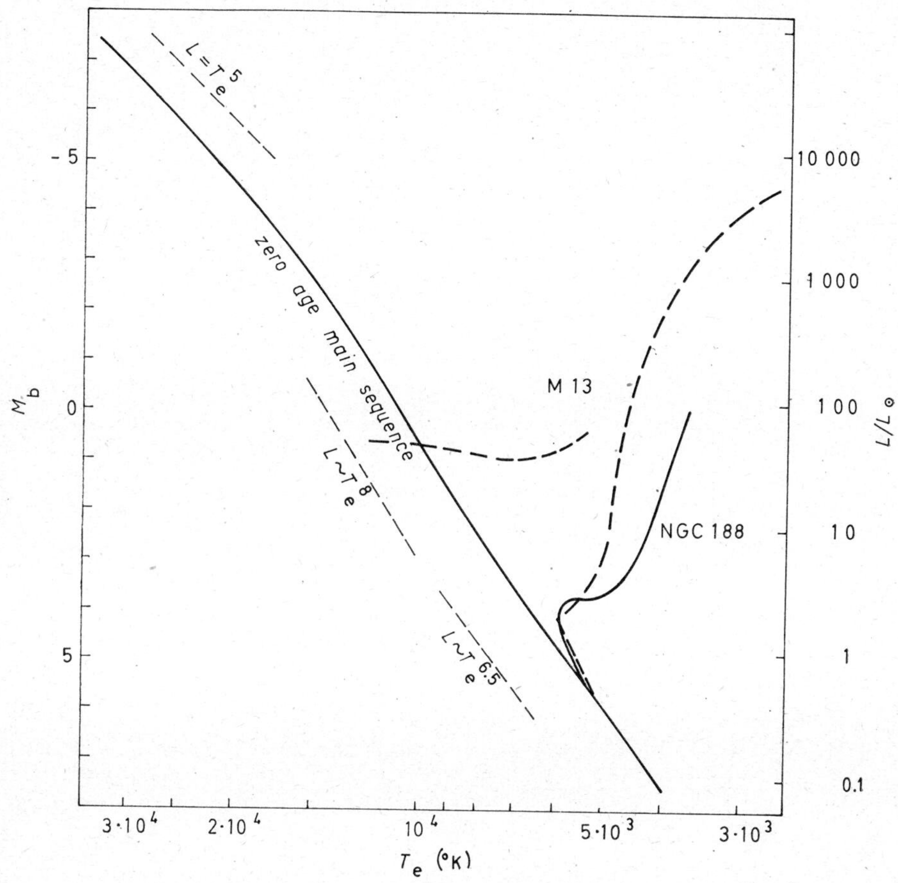

Fig. 2. – The T_e, M_b-diagram.

ferences from the constant predicted slope are not surprising in view of the crudeness of our models and of the various simplification introduced; for instance, the guillotine factor and the exponent of the law of energy generation are certainly not constant through the main sequence. However the equations are useful in showing the effect of the difference in chemical composition upon the position of the main sequence.

The same analysis could be carried through for stars in complete radiative equilibrium but where energy is produced by the p-p chain of reactions, the opacity being again due to bound-free transitions. The approximate equation for ε may be written in this case

$$\varepsilon = \text{const}\, X^2 \varrho T^n \,, \tag{1.27}$$

with n nearly equal to 4.

Going through the same computations as before, we find, for $n = 4$ and a constant guillotine-factor,

$$R = \text{const}\, Z^{0\cdot154}\, X^{0\cdot308}\, (1+X)^{0\cdot154} \mu^{-0\cdot538} \mathscr{M}^{0\cdot077} \,, \tag{1.28}$$

$$L = \text{const}\, Z^{-1\cdot077} X^{-0\cdot154} (1+X)^{-1\cdot077} \mu^{7\cdot77} \mathscr{M}^{5\cdot46} \,, \tag{1.29}$$

$$T_e = \text{const}\, Z^{-0\cdot346} X^{-0\cdot192} (1+X)^{-0\cdot346} \mu^{2\cdot21} \mathscr{M}^{1\cdot327} \,, \tag{1.30}$$

and

$$L = \text{const}\, Z^{0\cdot348}\, X^{0\cdot637}\, (1+X)^{0\cdot347} \mu^{-1\cdot333} T_e^{4.12} \,. \tag{1.31}$$

We see, thus, that the results are not very different from those of eqs. (1.20), (1.22), (1.24) and (1.26).

1·2.4. **Effect of the variation of chemical composition; subdwarfs.** To study the effect of a variation of the chemical composition it is convenient to write the chemical factors in the form $Z^\alpha X^\beta$ with constant α and β.

Using eq. (1.6) for μ it is found by numerical experiment that the equations for the C-N cycle may be written with sufficient approximation

$$R = \text{const}\, Z^{0\cdot09}\, X^{-0\cdot19} \mathscr{M}^{0\cdot73} \,, \tag{1.32}$$

$$L = \text{const}\, Z^{-1\cdot04} X^{-4\cdot33} \mathscr{M}^{5\cdot13} \,, \tag{1.33}$$

$$T_e = \text{const}\, Z^{-0\cdot31} X^{-0\cdot96} \mathscr{M}^{0\cdot92} \,, \tag{1.34}$$

and

$$L = \text{const}\, Z^{0\cdot67}\, X^{1\cdot07}\, T_e^{5.6} \,, \tag{1.35}$$

if $Z \ll X$ as one can safely assume. The corresponding equations for the p-p chain are

$$R = \text{const}\, Z^{0\cdot15}\, X^{0\cdot66} \mathscr{M}^{0\cdot08} \,, \tag{1.36}$$

$$L = \text{const}\, Z^{-1\cdot08} X^{-4\cdot78} \mathscr{M}^{5\cdot46} \,, \tag{1.37}$$

$$T_e = \text{const}\, Z^{-0\cdot35} X^{-1\cdot52} \mathscr{M}^{1\cdot33} \,, \tag{1.38}$$

and

(1.39) $$L = \text{const}\, Z^{0.35}\; X^{1.50}\, T_e^{4.12}\,.$$

Let us now first consider the effect of a decrease in the abundance of heavy elements on a star's position in the H-R diagram. Equations (1.33) and (1.34) or (1.37) and (1.38) show that both the luminosity and the effective temperature for constant $\mathcal{M}$, will increase in such a way that the star falls bellow the « normal » main sequence (Fig. 3).

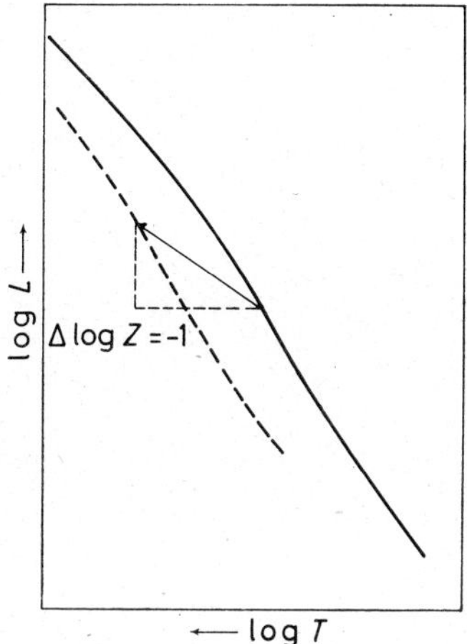

Fig. 3. – Effect of a change of the heavy-element content in the H-R diagram.

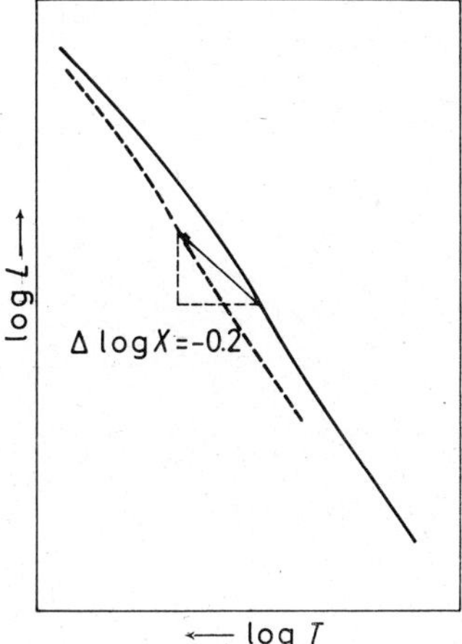

Fig. 4. – Effect of a change of the hydrogen content in the H-R diagram.

Variation of X have the same general effect (Fig. 4). By increasing X whilst decreasing Z one can get a compensation in the L-T_e plane, but of course not simultaneously in the $\mathcal{M}$-L plane, since the exponents of Z and X in eq. (1.33) are both negative. This will be further discussed when observational data on masses, luminosities and temperatures will be presented (Sect. 3·2.2), for stars with different X and Z.

Exact calculations using correct stellar models have been done by DEMARQUE (1960) and earlier by REIZ (1954). Hoyle and his students (see the Seminar by FAULKNER and GRIFFITH in these Lectures) have recently redone these calculations with modern opacity tables. They all give the same sense for the changes in L and T_e relative to various of X and Z as that given by the dimensional analysis.

1·2.5. *Evolution without chemical inhomogeneities.* Let us now consider how a star will change with time, as its hydrogen is burnt. As X

decreases, $\mathscr{M}$ remaining effectively constant, L and T_e will increase (eqs. (1.32) and (1.34)).

The equation of the evolutionary line in the H-R diagram is given by

$$T_e = \text{const}\, X^{-0.96}\,, \qquad L = \text{const}\, X^{-4.33}\,,$$

or

$$L = \text{const}\, T_e^{4.5}\,. \tag{1.40}$$

Comparing the slope of this line, 4.5, with the theoretical slope of the main sequence, 5.6, we see that the stars will evolve toward brighter luminosities, but will stay in the main sequence. Thus we cannot possibly account for the existence of giant stars. For example, in the case of the Sun, if the hydrogen abundance were to decrease by a factor of 2 in the course of its evolution, then

$$\frac{L_{\text{final}}}{L_{\text{initial}}} = 2^{4.33} (= -\,3.3 \text{ magnitudes})\,,$$

and

$$\frac{T_{\text{final}}}{T_{\text{initial}}} = 2^{0.96} = 1.94\,.$$

The final visual absolute magnitude and the final effective temperature will be $1^{\text{M}}.5$ and 11 000 °K. Thus the Sun will become much brighter and move slightly below the main sequence in this type of evolution.

Since giant stars do in fact exist one of our assumptions must be wrong. In later sections we will see how the assumption of nonmixed models leads to correct evolutionary paths.

2. – Observational evidence for evolution.

The H-R diagram that emerges from observations plots absolute magnitudes against either spectral types or colour indices (colour-magnitude diagram). To-day the most commonly used system of colours is Johnson and Morgan *U-B-V*-system, with its three colours at effective wavelengths of

$$\lambda_U = 3600\ \text{Å}\,,$$

$$\lambda_B = 4430\ \text{Å}\,,$$

$$\lambda_V = 5540\ \text{Å}\,,$$

which provides the two colour indices *B-V* and *U-B*.

Figure 5 shows a colour-magnitude diagram produced by such observations for the stars of the solar neighbourhood with parallaxes $\pi \geqslant 0''.067$, that is stars within 15 pc from the Sun, using the material of the Yale 1952 Parallax Catalogue and Eggen's photometric data (1955). Besides some white dwarfs with M_v between +11 and +16 and B-V between 0.1 and +1.0, there are almost no stars outside the main sequence and no stars with absolute magnitude brighter than zero. The dispersion of the main sequence is largely due to differences in metal content (as we shall show later).

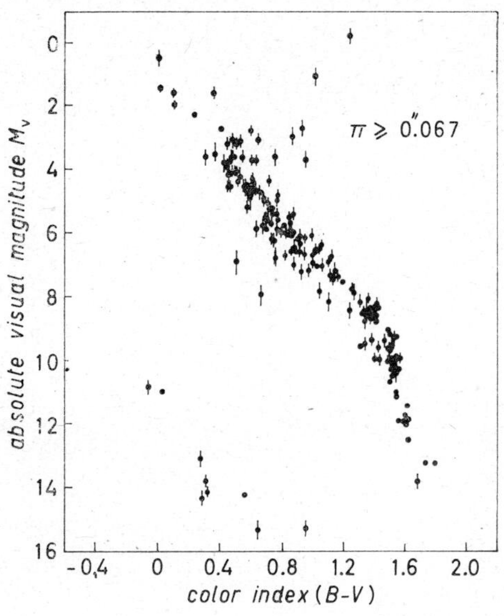

Fig. 5. – Colour-magnitude diagram for stars with trigonometric parallaxes larger than $0''.067$ (SANDAGE: *Vatican Symposium on Stellar Populations*).

A very different colour-magnitude diagram is observed in the globular clusters of the galactic halo (Fig. 6) with its remarkably extended branch through the region of red giants and the horizontal branch crossing the main sequence, with the characteristic gap in which fall the *RR Lyrae* variables.

As the three diagrams for the globular clusters M3, M5 and M13 in Fig. 7 show quite well, the turn-off point for these three (and also for all well-observed globular clusters) is virtually the same, at least to within the accuracy of our present data.

Figure 8 shows the colour-magnitude diagram for a characteristic sample of open (galactic) clusters. For these the turn-off point from the main sequence is different from one cluster to another; as we shall see later the position of the turn-off point was shown to be a function of the age. NGC 2362 is the youngest cluster of the sample and the clusters with increasing age have turn-off points

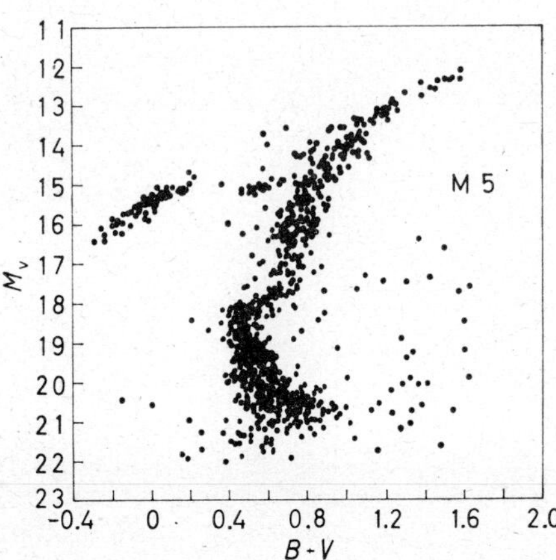

Fig. 6. – The colour-magnitude diagram for M5 (Arp, 1962).

farther and farther down along the main sequence. M67 has an age comparable to that of globular clusters, but as Fig. 8 shows, the shape of its giant branch

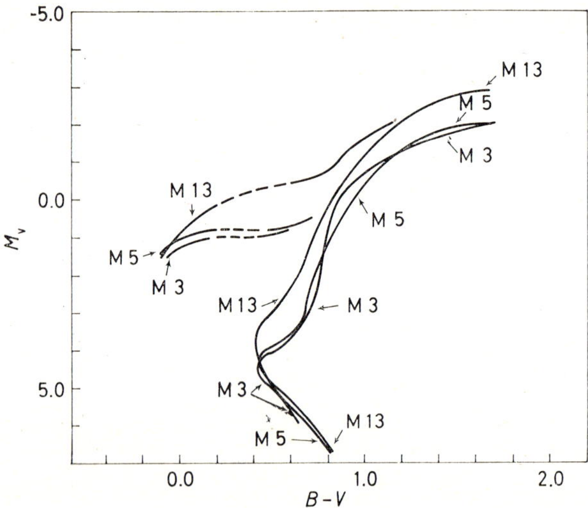

Fig. 7. – Sequences for three globular clusters in the colour-magnitude diagram (magnitudes and colours are corrected for reddening, but not for blanketing).

is somewhat different from that of the globular clusters due to different metal content. The evolutionary changes have also brought the brightest stars of the younger clusters to the right side of the diagram (red supergiants in h and χ *Per*, and giants in M41 etc.).

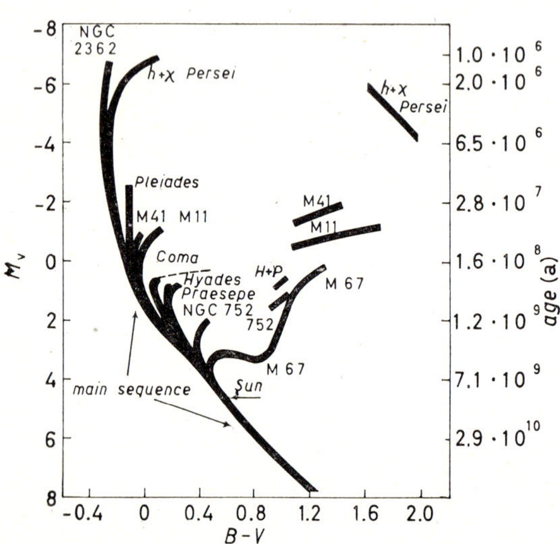

Fig. 8. – Schematic sequences for galactic clusters in the colour-magnitude (diagram adapted from SANDAGE: *Vatican Symposium on Stellar Populations*).

The determination of luminosities for field stars needs a more elaborate method. O. WILSON (1959) found that the width W (in Å) of the emission core of the H and K lines shows a linear correlation with the visual absolute magnitudes. By use of this method the absolute magnitudes of field stars in the solar neighbourhood can be determined.

The colour-magnitude diagram based on these absolute magnitudes is given in Fig. 9.

Only stars of spectral type F0 and later, which show H and K in emission could be measured. There is a definite enveloping curve towards the right which coincides rather well with the diagram of the open cluster NGC 188 (the heavy line to the right in Fig. 9). This means that field stars scatter over a large range of ages up to the limit given by the age of NGC 188, with no field star evolved beyond the evolution stage of the stars of this cluster.

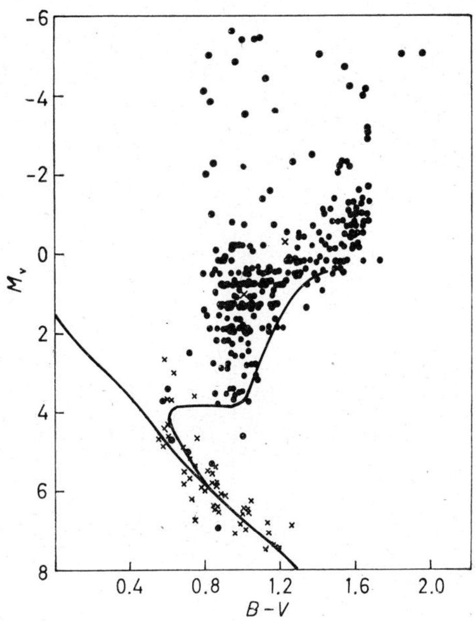

Fig. 9. – Comparison of the NGC 188 colour-magnitude diagram with that of the nearby field stars after O. C. WILSON. The NGC sequences have been corrected for absorption and reddening. The crosses represent stars with trigonometric parallaxes (SANDAGE, 1962).

The scale of the ordinates shown on the right side of the diagram of Fig. 8 gives the correlation between the position of the turn-off point and the age; it is calibrated only to within a factor 2. This correlation exists for the following reason. Due to the slowness of diffusion in stellar conditions and the absence of strong convective motions over the largest part of stellar radius, a complete mixing of the stellar material cannot take place; instead a central core of helium is gradually formed, because the hydrogen-burning reactions possess a very strong temperature-dependence. In the absence of convection hydrogen is exhausted in the core, whose boundary slowly moves outwards as hydrogen burning continues. In this condition the star builds up an isothermal hydrogen-exhausted core, surrounded by a thin shell in which thermonuclear reactions take place. It was shown, however, by CHANDRASEKHAR and SCHÖNBERG (1942) that the isothermal core cannot exceed a certain size, roughly corresponding to the one tenth of the total mass of the star (Schönberg-Chandrasekhar limit). The physical reason for this is that a star of finite mass is not able to build in its outer parts a temperature gradient steep enough to maintain more than a certain fraction of its mass at the highest temperature.

Once the helium core reaches the Chandrasekhar-Schönberg limit, the star moves off the main sequence. The critical core size is reached of course more rapidly in the heavier stars. Naturally what is observed in the colour-magnitude diagrams are not evolutionary tracks, but isochronical lines for stars of different masses, *i.e.* the loci of points corresponding to a number of stars having the same age and (initial) chemical composition but different masses. The older a cluster, the smaller is the mass of the stars whose core has reached the critical size and therefore the lower is the position of the turn-off point in the H-R diagram.

The time scale of this evolution follows from energy requirements. The energy available equals

$$(2.1) \qquad E_n = 0.007\,\mathcal{M}Xc^2\,,$$

E_n being the energy from the nuclear process involved and X the fraction of the star mass, $\mathcal{M}$, available in the form of hydrogen.

From (2.1) follows the equation

$$(2.2) \qquad \mathrm{d}E_{ni} = 0.007\,\mathcal{M}Xc^2\,\mathrm{d}q_i = L_i\,\mathrm{d}t_i\,,$$

for the fraction $\mathrm{d}q_i$ of the mass burnt during the time $\mathrm{d}t_i$, L_i being the mean luminosity of the star during this time. By integration from the time the star begins to burn its hydrogen (*i.e.* the instant it reaches the zero-age main sequence after its initial stage of contraction) to the time it reaches the Chandrasekhar-Schönberg (S-C) limit we obtain

$$(2.3) \qquad T = 0.007\mathcal{M}c^2\int_{q=0}^{q=q_L} X\,\frac{\mathrm{d}q_i}{L_i}\,.$$

Stellar models must be available to give the rise in luminosity as a given fraction of fuel is consumed. The fraction $\mathrm{d}q_i/L_i$ also depends on chemical abundances.

The theoretical tracks which have so far been developed cannot completely explain the further evolution of the stars as suggested by the diagrams for globular clusters. It has proved so far too difficult to account exactly for the evolution subsequent the turn-off point from a purely theoretical standpoint. Thus semiempirical methods have to be adopted. The following method was developed by SANDAGE (1957) and applied to the old open cluster M67 and to the globular cluster M3.

The central idea of the method is to compare the observed luminosity function for a given cluster with a universal initial luminosity function supposed to be the same for all clusters. The luminosity function is defined as the number of stars having their absolute magnitude in an interval of 1 magnitude around a value M_i. Let us call $\Phi(M_v)$ the observed luminosity function and $\Psi(M_v)$ the initial luminosity function; then $\Phi(M_v)$ is known from observations and $\Psi(M_v)$ has been determined by SALPETER (1955). Of course all the stars from which $\Phi(M_v)$ has been obtained are distributed along the present observed C-M diagram of the cluster; they are supposed to have been initially distributed along the zero-age main sequence, so that $\Psi(M_v)$ refers actually to the density distribution of the stars along the zero-age main sequence.

To determine the evolutionary tracks on the H-R diagram one needs further a theory of the early stages of the evolution on or close to the main sequence. The Schönberg-Chandrasekhar models provide the initial evolutionary history of main sequence stars until a limit configuration where the star has burnt all its hydrogen inside a core containing a fraction, q_L, of the total mass (the SC limit). For a jump of mean molecular weight of 2 between the core and the envelope (which corresponds to an initial H-content of about 0.6) this limiting fraction is $q_L = 0.12$. The evolutive tracks from the main sequence to the SC limit are shown in Fig. 10 and 11 extending from the line labelled « original main sequence » to the light line marked « SC ». The heavy line cutting across these tracks represents the observed C-M diagram for the cluster, converted to the M-log T_e plane.

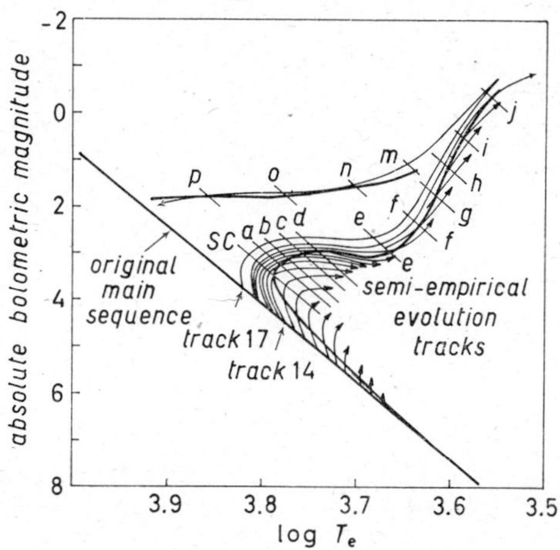

Fig. 10. – Semiempirical evolutionary tracks for stars in M67; see the text for details (SANDAGE, 1957).

In the Fig. 10 and 11, the first ten tracks from below reach the cluster line before they cross the SC line; this means that the theory is sufficient to represent the whole past history of these stars from the original main sequence to their present state. To obtain the tracks for brighter stars the procedure is the following. Take as an example the case

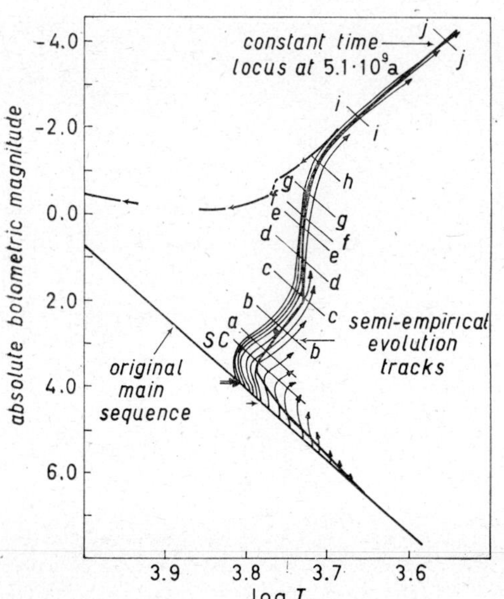

Fig. 11. – Semiempirical evolutionary tracks for stars in M3 (SANDAGE, 1957).

of M67; it is found in this case that the cluster M_v-log T_e line intersects the SC line at $M_b = +3.46$, and the track through this point leaves the original main sequence at $M_b = 4.4$. Now, counts in the two functions $\Phi(M_v)$ and $\Psi(M_v)$ show that there were as many stars between $M_b = 4.4$ and $M_b = 4.3$ along the original main sequence as there are now from point SC to a along the observed M_v-log T_e line of the cluster. It is therefore concluded that a star now at point a left the original main sequence at $M_b = 4.3$. Similarly, there were equal numbers of stars between $M_b = 4.3$ and $M_b = 4.2$ on the main sequence as there are now between points a and b; consequently the end point of the evolutionary track leaving the main sequence at $M_b = 4.2$ is at point b. By this stepwise counting procedure the end points of the various tracks of evolution are located on the observed cluster M_b-log T_e line.

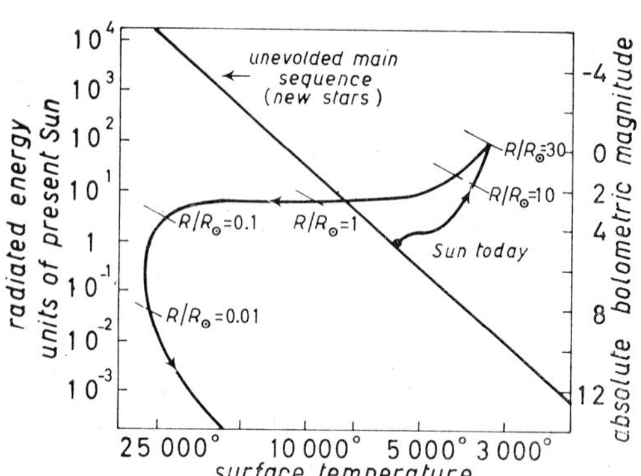

Fig. 12. – Evolutionary track of the Sun.

Note that the actual counts are not made on Salpeter's original function $\Psi(M_v)$ but on a modified function to account for (a) the escape of stars of small mass from the cluster and (b) an initial evolution along the main sequence before the stars leave it. For details we refer to Sandage's original paper.

After the end points of the various tracks have been located, the complete tracks are found by constructing a parallel line (this assumes that the stars are homologous) to the track found in the immediately preceding step.

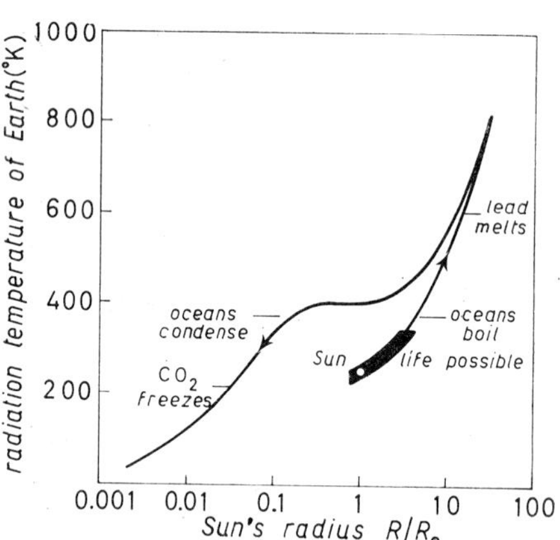

Fig. 13. – Temperature of the Earth surface as a function of the Solar radius.

From the results obtained for M67 we may try to make a tentative prediction of the future evolution of the Sun. According to this, the Sun is now expanding and will continue to expand to a radius of about 30 times the present value, the surface temperature decreasing to about 3500 °K. Then it will contract and heat up untill eventually it reaches the white dwarf stage (see Fig. 12). From the radius and the temperature of the Sun, the temperature of the Earth will follow, assuming that the latter is in radiative equilibrium and that the radius of the Earth's orbit is constant; we have, writing R_E and T_E for the radius and temperature of the Earth and r for the distance Sun-Earth,

$$\frac{L}{4\pi r^2}\,\pi R_E^2 = 4\pi R_E^2 \sigma T_E^4\,.$$

In Fig. 13 the temperature of the Earth is given as a function of the radius of the evolving Sun. It will rise to about 800 °K and then rapidly fall towards very low temperatures.

As to the time scale of this evolution the rise of the Earth temperature in the present stage is only $3\cdot10^{-9}$ °K/year for the next $6\cdot10^9$ years. Then with the quick evolution of the Sun, the temperature rises at a rate of $1\cdot10^{-6}$ °K/year to a maximum of 800 °K. The present rate of increase holds of course also for the evolution in the past up to the present value.

The rather small number of available stars in open clusters prevents from obtaining a satisfactory luminosity function $\Phi(M_v)$; on this respect the situation is much better in globular clusters.

In at least one globular clusters, M3, a number of stars are found on the main sequence beyond the turn-off point. These stars pose a difficult problem for the theory.

3. – Problems of fitting observed main sequences.

3·1. *Interstellar reddening.* – The determination of the main sequences of clusters rests on the following method. First a standard main sequence is determined in the (M_v, B-V)-diagram. To this purpose the stars of the Hyades can be used for which one obtains very accurate parallaxes by means of the moving cluster method.

The observational data on the U, B, V magnitudes for the Hyades, including new observations, have been recently collected and discussed by Johnson, Mitchell and Iriarte (1962); they provide the most accurate colour-magnitude and two-colour diagrams available at present for an open cluster.

For clusters whose distances are unknown, the observations provide only the main sequence in a (m_v, B-V)-diagram. Vertical shift of this diagram until a coincidence is obtained with the standard (Hyades) main sequence for not too bright stars yields the distance modulus m-M. This implies the assumption that all galactic clusters have the same main sequence (below the turn-off point) regardless of differences in chemical composition; we shall come later to justify this assumption.

However this method assumes that there is no interstellar reddening present. If there is any absorbing matter, the distance modulus has to be corrected for the amount of interstellar absorption which causes a considerable colour excess. If E_{B-V} is the colour excess, that is the difference A_B-A_V of the absorption in the two wavelength regions B and V, it is found that

$$A = 3.0(\pm 0.2)E_{B-V} \,. \tag{3.1}$$

This results comes from observational determinations of the absorption law.

To obtain E_{B-V} one uses a *two-colour diagram*, which plots U-B againts B-V. On such a diagram stars which radiate like black bodies would lie approximately upon a straight line; actual stars lie on an S-shaped curve, due to the Balmer absorption in the U-region, which is very strong in A-stars and fainter both in earlier and later spectral types. Any three-colour system including one colour in a spectral region which deviates from the blackbody emissivity would show a similar deviation of the two-colour curve from a straight line. Figure 14 shows the two-colour diagram of the Hyades.

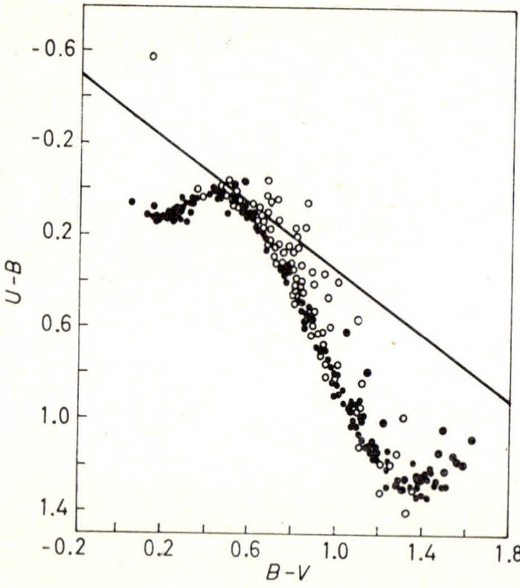

Fig. 14. – Two-colo r diagram for the Hyades (JOHNSON *et al.*, 1962). The filled circles represent the main sequence cluster members; the open circles represent probable field stars. The diagonal line is the reddening line.

Stars reddened by interstellar absorption are found displaced with respect to unreddened stars of the same intrinsic colour parallel to the so called reddening path, the slope of which is given by the ratio

$$\frac{E_{U-B}}{E_{B-V}} = 0.78 \,, \tag{3.2}$$

between the colour excesses. This value is based upon the (observed) absorption law; direct observations give a somewhat smaller ratio of 0.72.

Comparing the two-colour diagram for a reddened cluster with that for unreddened stars, we will therefore find it displaced parallel to the reddening path by an amount equal to the colour excesses both in abscissae (E_{B-V}) and in ordinates (E_{B-V}), as it is shown schematically in Fig. 15. The colour excess can thus be determined.

In many cases — especially if there are stars in a cluster which have evolved appreciably off the main sequence — the two-colour diagram may not give a clear picture of the absorption because evolved stars settle down into the horizontal branch of the curve. In this case it is advisable to consult the two-

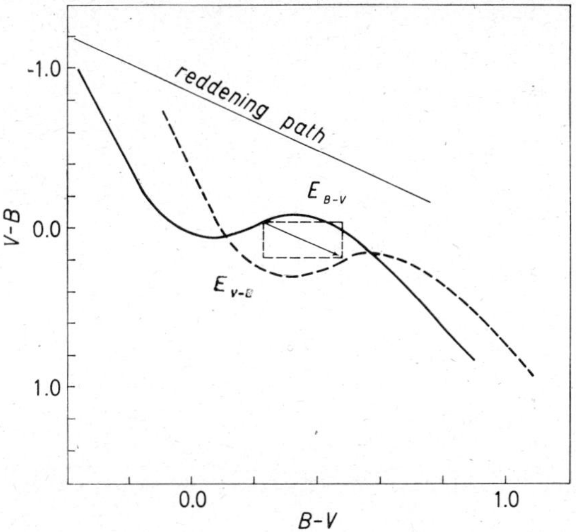

Fig. 15. – Reddening effect on the two-colour diagram for open clusters: ——— standard main sequence, – – – reddened main sequence.

colour-magnitude diagrams, (M, B-V) and (M, U-B), which give both a value for the absorption and for the distance and therefore allow to check each other.

3·2. *Observed evidence of the heavy element abundances.*

3·2.1. The blanketing effect. There are certain groups of stars which in the two-colour diagram are found to lie to the right and above the lower part of the curve. These stars include among others the red giants in the globular clusters NGC 4147, M3, M5, M13 and M92, subdwarfs and the high-velocity stars.

This fact cannot be explained by interstellar reddening; we will see that it is due to the effect of the Fraunhofer lines, which in the above-mentioned cases causes an ultraviolet excess larger than that of the assumed standards (the Hyades).

The general effect of the absorption lines upon the observed magnitudes and colours is loosely termed the *blanketing effect*, although the effect is twofold and only one part of it is properly an effect of blanketing.

The first effect is obviuos; if a large number of strong absorption lines are present in a certain spectral region they will block part of the energy radiated in the continuum at the same wavelengths. If the band pass corresponding to a given magnitude system happens to fall in that spectral region, clearly the corresponding magnitude will be found fainter than otherwise. The characteristic shape of the $(B - V, U - B)$-diagram is partly due to this effect because the depression of the continuum due to the crowding of the Balmer lines near the series limit adds to the continuous absorption at the violet of the limit itself; the resulting effect is expecially strong in A-stars.

The second effect, which is somewhat less obvious, depends on a general distortion of the temperature distribution throughout the star's atmosphere, caused by the presence of absorption lines, relative to the distribution one would find if the absorption lines were absent. In short, since nearly all the energy removed from the continuum by the Fraunhofer lines — which are formed in the outer atmospheric layers — is radiated back, the temperature gradient in the outer layers is much steeper than without absorption lines; hence the true effective temperature of the star is lower than the effective temperature we need to describe the flux in the continuum.

From this it is now clear that if we use observed magnitudes and colours to get bolometric magnitudes and effective temperatures, the results from stars having lines of different intensities will not be consistent. We must therefore correct the magnitudes and colours for the effect of blanketing before we apply the fundamental relations to obtain M_b and T_e.

To find the corrections for blanketing let us call $F(\lambda)$ the flux actually observed at wavelength λ (thus including absorption lines if present), $F_c(\lambda)$ the flux in the continuum obtained by means of extrapolating over the regions occupied by absorption lines the curve drawn through wavelengths free from lines and, finally, $F_0(\lambda)$ the flux corresponding to an ideal atmosphere having the same total flux (or effective temperature) as the real one, but whose opacity is due to continuous absorption only, line absorption being neglected. Let, further, $\varepsilon(\lambda)$ be the fractional energy absorbed by Fraunhofer lines at wave length λ.

Thus

$$F(\lambda) = [1 - \varepsilon(\lambda)] F_c(\lambda) \,, \tag{3.3}$$

and $\varepsilon(\lambda) = 0$ in a line-free spectral region, $\varepsilon(\lambda) = 1$ in the core of a very strong line.

Then, if we call $\delta_1 m$ the *blocking* effect of absorption lines, that is the difference between the magnitudes corresponding to flux F_c and that corresponding to F — the latter is therefore the magnitude actually observed — we have

$$\delta_1 m = -2.5 \lg \frac{\int_0^\infty S(\lambda) F_c(\lambda) \, d\lambda}{\int_0^\infty S(\lambda) F_c(\lambda) [1 - \varepsilon(\lambda)] \, d\lambda} \,, \tag{3.4}$$

where $S(\lambda)$ is the intrumental sensitivity function of the photometric system to which eq. (3.4) refers. Ordinarily $S(\lambda)$ differs from zero only in a finite range of wavelengths, say from λ_1 to λ_2 and hence the integrations extend in practice only from λ_1 to λ_2.

Now the effective temperature corresponding to F is

$$\sigma T_e^4 = \int_0^\infty F(\lambda)\,d\lambda = \int_0^\infty [1-\varepsilon(\lambda)]\,F_c(\lambda)\,d\lambda\,, \tag{3.5}$$

which we can write

$$\sigma T_e^4 = (1-\eta)\int_0^\infty F_c(\lambda)\,d\lambda\,, \tag{3.6}$$

where

$$\eta = \frac{\int_0^\infty \varepsilon(\lambda) F_c(\lambda)\,d\lambda\,,}{\int_0^\infty F_c(\lambda)\,d\lambda}\,, \tag{3.7}$$

is the mean blocking coefficient.

For definition, $F_0(\lambda)$ has the same effective temperature as $F(\lambda)$; then

$$(1-\eta)\int_0^\infty F_c(\lambda)\,d\lambda = \int_0^\infty F_0(\lambda)\,d\lambda\,. \tag{3.8}$$

Equation (3.8) is sufficient to define F_0 if we have a model atmosphere for the required effective temperature with continuous absorption only.

The magnitude difference between a star whose flux is F_c and another whose flux is F_0 is, of course,

$$\delta_2 m = -2.5 \log \frac{\int_0^\infty S(\lambda) F_0(\lambda)\,d\lambda}{\int_0^\infty S(\lambda) F_c(\lambda)\,d\lambda}\,. \tag{3.9}$$

The sum of (3.4) and (3.9),

$$\delta m = \delta_1 m + \delta_2 m\,, \tag{3.10}$$

may be taken as the total correction for blanketing to be applied to the observed magnitudes to make consistent the results for stars with different line intensities.

This method has been recently employed by WILDEY, BURBIDGE, SANDAGE and BURBIDGE (1962) to obtain the corrections for blanketing on U, B, V measurements necessary to make consistent the radiation parameters for stars with different metals-to-hydrogen ratio. $\varepsilon(\lambda)$ was determined by means of high dispersion spectra of 4 Stars; to these measurements previous results obtained by MELBOURNE (1960) on 8 more stars were added. To compute the fluxes F_0, a set of atmospheric models constructed by DE JAGER and NEVEN (1957) was used.

Now, consider a number of stars of the same effective temperature and surface gravity, but with different chemical composition; if we limit ourselves to stars later than F0 practically all the lines are due to metals, while the continuous absorption is due to the negative hydrogen ion. We may, then, put

$$\varepsilon(\lambda) = \eta\varphi(\lambda) ,$$

where $\varphi(\lambda)$ is the same function of λ for all the stars of the group; η, the mean blocking coefficient, will be different from one star to another and is roughly proportional to Z, the heavy element abundance, or, better to Z/X, the metals-to-hydrogen ratio.

Clearly, then, in the two-colour diagram all these stars will lie upon a definite line, since the magnitudes and colours are functions of the parameter η only. This *blanketing line* can be drawn by means of the theory sketched above if the function $\varphi(\lambda)$ is known from high-dispersion spectra of a few selected stars.

As for the stars of the main sequence of a cluster there is a definite relation between temperature and surface gravity, starting from the points of the two-colour diagram of the Hyades, we can map a region of the (B-V, U-B)-plane by means of a single family of blanketing lines. This region will be limited on the upper side by a *blanketing envelope* corresponding to stars with zero metal content. Keeping constant the effective temperature and the surface gravity, the influence of the blanketing will move a star down along the blanketing line from a starting point on the blanketing envelope until a point corresponding to the maximum content of heavy elements, crossing the Hyades line at a point B corresponding to the chemical composition of the Hyades.

Of course, starting from a cluster with a very different relation between effective temperature and surface gravity, strictly speaking we would obtain a different set of blanketing lines. But the effect of surface gravity cannot be large on this connection, as it is proved by the fact that the giants belonging to the Hyades fall exactly on the same two-colour line obtained from the main sequence stars. Therefore we can neglect possible small differences between, the relations connecting surface gravity and effective temperature corresponding to the main sequences of different clusters and use a unique set of blanketing lines.

For practical purposes one can construct a table giving for each point P of a certain blanketing line the differences Δm_V, $\Delta(U-B)$, $\Delta(B-V)$ between the values of the magnitude and colours corresponding to P and those corresponding to the point B where the blanketing line through P crosses the standard two-colour line of the Hyades, as functions of the vertical displacement of P (see Fig. 16) relative to the Hyades line, and B—V.

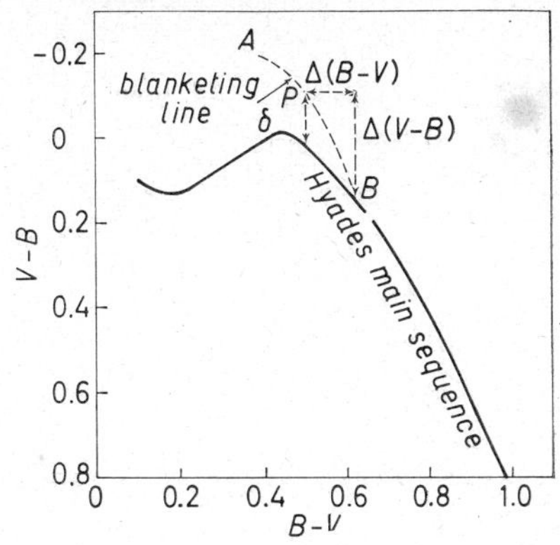

Fig. 16. – Effect of blanketing upon the position of a star in the two-colour-diagram. See the text for the meaning of lines and symbols.

The high-velocity stars scatter over an area of the two-colour diagram which lies between the Hyades line and the blanketing envelope (see Fig. 17); this latter is rather well represented by the line corresponding to the globular clusters M13, M15 and M92 (see SANDAGE and EGGEN, 1959, EGGEN and SANDAGE, 1959 and 1962 and also ARP, 1962). This means of course that the heavy-element content of the high-velocity stars ranges from that of the Hyades to almost zero; the globular clusters have a very low metal content.

There is a sharp limit to the left of the area occupied by the high-velocity stars at $B-V=+0.36$. This means that there are no (main sequence) high-velocity stars with $B-V$ less than $+0.35$; therefore all high-velocity stars are at least as old as a globular cluster with its turn-off point at $B-V=+0.35$. We shall come later to this point.

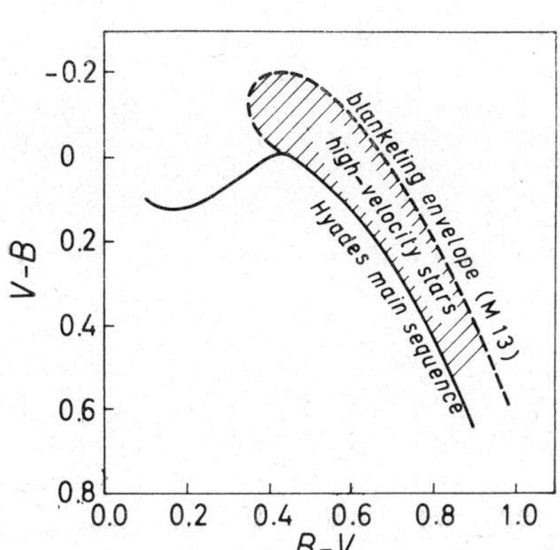

Fig. 17. – Position of high-velocity stars in the two-colour diagram due to blanketing.

3·2.2. Subdwarfs in the L, T_e-plane and the mass-luminosity relation. The subdwarfs are a rather extreme case of high-velocity stars; according to the results of the previous section one would then expect that their heavy-element content is remarkably low. Accordingly, one may suspect that if the magnitudes and colours of the subdwarfs are corrected for the blanketing their position on the L, T_e-plane might considerably change. Indeed it was found (SANDAGE and EGGEN, 1959, EGGEN and SANDAGE 1962) that if the magnitudes and colours of subdwarfs are corrected for blanketing, they fall in the temperature-luminosity diagram exactly on the Hyades line. As a matter of fact it was found that this holds even in the most extreme cases of low metal content. If corrections for blanketing are applied, the dispersion of individual stars around the main sequence is substantially reduced; the residual scatter from stars redder than the turn-off point of the globular clusters sequence can be entirely explained by observational errors (mainly of the parallaxes).

This very remarkable result (which in no way could have been foreseen) can mean only one thing, if our theory of the main sequence is correct: for some reason in all groups of stars a lower metal abundance is compensated by a higher hydrogen abundance, in such a way that the product $Z^{0.67}X^{1.07}$ is constant (see eq. (1.35)). Of course the actual value of the exponents might be somewhat different — for instance, if the pp-chain rather than the CN cycle, is the source of stellar energy in these star the product to remain constant is $Z^{0.35}X^{1.50}$ — but the physical fact remains.

This conclusion can be checked by means of the mass-luminosity law. Since the dependence of the mass-luminosity relation on X and Z [eq. (1.33)] is different from that of the T_e-L relation, the compensation will not in general occur in the former if it occurs in the T_e-L relation (the H-R diagram) Various binaries are known to belong to the Hyades and to Eggen's Sirius moving group (see EGGEN, 1960 for a general summary on stellar groups). From our previous simple considerations, the abundances of hydrogen and metals in the Hyades and in the Sirius group are related by

$$Z_H^{0.67}X_H^{1.07} = Z_S^{0.67}X_S^{1.07}, \tag{3.11}$$

where the subscripts H and S refer to the Hyades and to the Sirius group, respectively. Hence

$$\frac{X_S}{X_H} = \left(\frac{Z_H}{Z_S}\right)^{0.40}. \tag{3.12}$$

The WALLERSTEIN and CARLSON (1960) calibration yields, for a value of $\delta(U-B) = +0^m.04$ corresponding to the Sirius group, a relative metal abundance of

$$\frac{Z_H}{Z_S} = 1.3\ ;$$

we find, therefore,

$$\frac{X_S}{X_H} = (1.3)^{0.4} = = 1.18\,. \tag{3.13}$$

Assuming that the hydrogen content of the Sun, $X = 0.67$, is the same as that of the Sirius group, the hydrogen content of the Hyades turns out to be

$$X_{\text{H}} = 0.57 .$$

Now Z and X appear in eq. (1.33) under the product $Z^{-1.04}X^{-4.33}$; we will expect, then, that in the mass-luminosity diagram, stars belonging to the Hyades and to the Sirius group lie upon different lines. Actually it is found that in such a diagram the stars of the Sirius group are about $0^{\text{m}}.75$ fainter than those of the Hyades having the same mass; hence, from (1.33)

$$\left(\frac{Z_{\text{S}}}{Z_{\text{H}}}\right)^{1.04} \left(\frac{X_{\text{H}}}{X_{\text{S}}}\right)^{4.33} = 2.0 ,$$

and from this equation, since $Z_{\text{H}}/Z_{\text{S}} = 1.3$, we get

$$\frac{X_{\text{S}}}{X_{\text{H}}} = 1.25 . \tag{3.14}$$

Owing to the crudeness of our analysis, the results expressed by (3.13) and (3.14) are to be considered in good agreement; the latter values give 0.54 as the hydrogen content of the Hyades. Incidentally it may be mentioned that recently PARKER, GREENSTEIN, HELFER and WALLERSTEIN (1961) from an analysis of high-dispersion spectra, by the curve-of-growth method, obtained a 20% higher metals-to-hydrogen ratio for the Hyades with respect to the Sun.

A rather interesting consequence is that in principle this argument may permit an evaluation of the helium content of the various groups relative to the Hyades. The simple homology argument can only predict the direction of the expected changes; to obtain the actual values one has to use more detailed stellar models. For a preliminary estimate EGGEN and SANDAGE (1962) have used some stellar model calculations by DEMARQUE (1960) together with their observed data on the blanketing. They found that if for the Hyades $X = 0.6$, $Z = 0.04$, then for the extreme subdwarfs (corresponding to $Z \to 0$), $X \simeq 1$ and therefore $Y \simeq 0$.

Since estreme subdwarfs are the oldest stars known, the primeval helium abundance in the Galaxy must have been very nearly zero.

4. – Correlation between kinematical properties, age and chemical composition.

Investigations of the space motions of the stars show a variation of the mean velocity relative to the Sun between different groups of stars from the low velocity disk population stars up to subdwarfs with the highest speeds around 300 km/s. There exists a unique relation between the space velocities of the various groups and their chemical composition. This fact was revealed for the first time by the analysis of the stars with known U, B, V, magnitudes in the Catalogue of high velocity stars of N. ROMAN (1955), although, of course, it was known since several years that the high-velocity stars possessed many physical properties (including chemical composition) sharply different from those of low-velocity stars.

A new discussion based on more complete data on magnitudes and space motions has been recently published by EGGEN, LYNDEN-BELL and SANDAGE (1962); their results have disclosed facts connected with the early history of the galactic system and the process of star formation.

The material upon which their discussion is based is contained in two catalogues (EGGEN, 1961, 1962) which give the computed values of the (U, V, W)-velocity vectors for a number of well-observed stars all of which lie within about 300 parsecs from the Sun. Here U and V are the components of the star space velocity in the galactic plane, U being directed towards the galactic center and V perpendicular to it in the direction of the general rotation of the galaxy; W is the component perpendicular to the galactic plane. The first catalogue contains approximately 4000 stars with well-determined proper motions and radial velocities; the second catalogue contains all stars for which the present available data indicate a total motion in excess of 100 km/s. Although the data from the second catalogue are generally of lower weight, the resulting uncertainty in the (U, B, W)-vectors is not excessively great.

As it is well known, the galactic rotational velocity of the Sun is of the order of 250 km/s, which is less than 100 km/s smaller than the velocity of escape; therefore all the so called high-velocity stars must have smaller velocities relative to the galactic center than that of the Sun (otherwise their resulting velocity would be larger than the escape velocity). This is the well-known explanation of the asymmetry of the motions of the high-velocity stars. In other words the stars of high velocity must possess highly elliptical orbits which bring them into the central regions of the Galaxy.

For these orbits around the galactic center, which are of course no closed orbits, but to a first approximation may be described in the galactic plane as rotating ellipses, one can define an eccentricity according to the equation

$$e = \frac{R_{\max} - R_{\min}}{R_{\max} + R_{\min}}, \tag{4.1}$$

where $R_{\max}$ and $R_{\min}$ are, respectively, the maximum and minimum distance from galactic center (see Fig. 17).

Consider, first, the two-dimensional motion in the galactic plane, for the case in which the potential is not a function of time (*i.e.* the Galaxy is in dynamical equilibrium). Assuming cylindrical symmetry, we have the two integrals of energy and rotational momentum about the galactic axis, which can be written

$$\tfrac{1}{2}(\dot{R}^2 + R^2\dot{\Phi}^2) - \psi(R) = E\,, \tag{4.2}$$

and

$$R^2\dot{\Phi} = h\,, \tag{4.3}$$

where E and h are constants, $\psi(R)$ is the gravitational potential in the galactic plane and Φ an azimuthal co-ordinate.

For a given form of $\psi(R)$ these two equations can be integrated to obtain the orbit of the stars for any value of E and h. The following expression is assumed for $\psi(R)$.

$$\psi = \frac{GM}{b + (R^2 + b^2)^{\frac{1}{2}}}\,, \tag{4.4}$$

on the ground that it is apparently the most general form for which eq. (4.2) and (4.3) may be integrated in a simple way and which has the limiting conditions

$$\psi \to \frac{GM}{R}\,, \quad \text{as } R \to \infty\,,$$

$$R^{-1}\frac{\partial\psi}{\partial R} \to \text{const}\,, \quad \text{as } R \to 0\,,$$

expressing the facts that the galaxy has a finite mass M and the density at $R=0$ is not infinite. b is a constant of the dimension of a length which is zero for a simple Newtonian potential. Of course eq. (4.4) is only a rough approximation but it may be considered sufficient for our purpose.

To determine the constants M and b the conditions are used that the circular velocity at $R=10$ kpc is 250 km/s and Oort's constants of differential rotation have the values $A = 15$ km/s kpc and $B = -10$ km/s kpc. It is found in this way

$$M = 2.4\cdot 10^{11} \text{ solar masses}\,,$$

$$b = 2.74 \text{ kpc}\,.$$

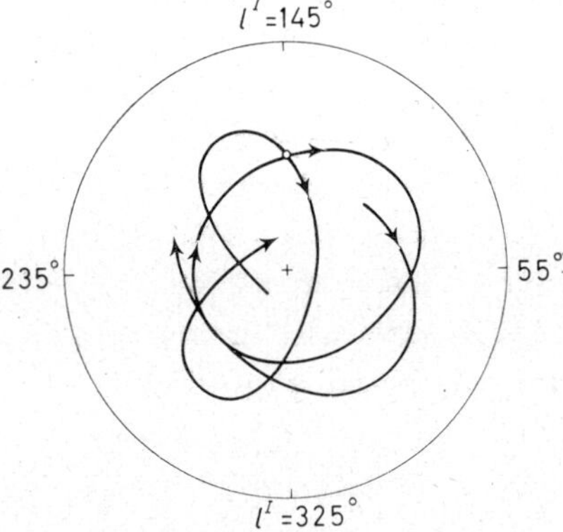

Fig. 18. – Segment of the galactic orbits for two stars. The more circular orbit corresponds to HD 117635 with an ultraviolet excess of $\delta = +0^m.05$; the more eccentric orbit is for HD 11980 with $\delta = +0^m.17$. Both orbits pass through the solar neighbourhood, which is designated by a circle at and a distance or 10 kpc from the galactic center (shown by a cross). The outer circle has a radius of 20 kpc (Eggen *et al.*, 1962).

The orbits of two stars are shown in Fig. 18.

The effect of the motion perpendicular to the galactic plane is next discussed. It is found that the effect of this is to increase slightly both apogalacticum and perigalacticum distance; however the effect is small and can be neglected. For each star with known (U, V)-vectors one can thus compute a parameter e, the « planar » eccentricity, defined by eq. (4.1) and the integrated equations of motion in the galactic plane; even when the neglect of the motion perpendicular to the galactic plane is not valid and the actual eccentricity differs considerably from e, this parameter still classifies orbits into « near circular » and « high eccentric » groups.

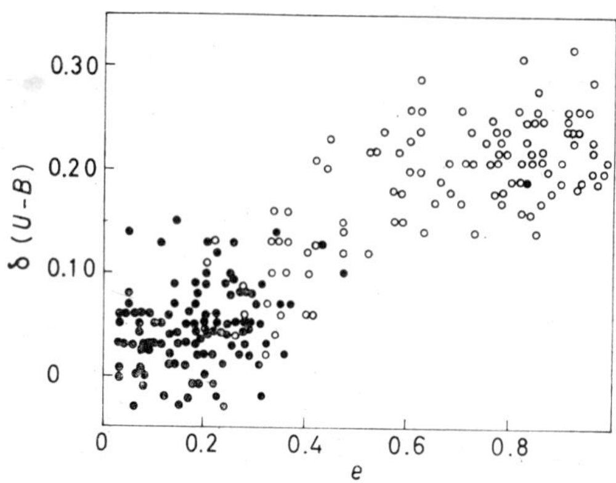

Fig. 19. – The correlation between the ultraviolet excess, $\delta(U-B)$ and the orbital eccentricity, e, for a sample of 221 stars; the filled and open circles represent stars from Eggen's two catalogues, respectively (EGGEN *et al.*, 1962).

Figure 19 plots the planar eccentricity e against the ultraviolet excess $\delta(U-B)$. While observational selection may account for the scarcity of stars with e near to 0.5, there is no doubt that the general increase of the ultraviolet excess with increasing e is real. Stars now in the vicinity of the Sun with large values of the ultraviolet excess have very eccentric orbits, whereas those with little or no excess have nearly circular orbits.

Of course this means that stars with high metal content move in nearly circular orbits, while those with low metal content move in highly eccentric orbits (in the neighbourhood of the Sun). It is suggested that this is also a correlation with age. Indeed the subdwarfs in the general field, with $\delta(U-B)$ greater than about $0^m.2$ appears to be identical with the main sequence stars of globular clusters; this indicates then that these subdwarfs are as old as the globular clusters themselves—near 10^{10} years. On the other hand, a variety of well-known arguments makes it clear that stars with little or no ultraviolet excess are relatively young. While it is very probable that the gradual enrichment by heavy elements of the prestellar medium was not uniform in space, one can then assume that stars with greater ultraviolet excess and therefore lower metal abundance are on the average older than those with smaller ultraviolet excess.

Then the correlation shown in Fig. 19 gives the remarkable result that older stars (at least in the vicinity of the Sun) have larger eccentricities than younger stars.

Figure 20 plots in a similar way the ultraviolet excess against the absolute value of W. It is seen, that the low-velocity stars with small ultraviolet excess have values of $|W|$ of not more than some 40 km/s, while some high-velocity stars of large $\delta(U-B)$ possess values of $|W|$ as high as 200 km/s. Of course these large velocities perpendicular to the galactic plane bring these stars at great heights above it; the scale on the right of Fig. 20 shows that stars with the highest values of $|W|$ can travel to heights larger than 10 000 pc, putting them into the realm of the halo globular clusters. This is convincing evidence that the extreme subdwarfs with ultraviolet excesses greater than $0^m.2$ can be identified with the main sequence stars of globular clusters.

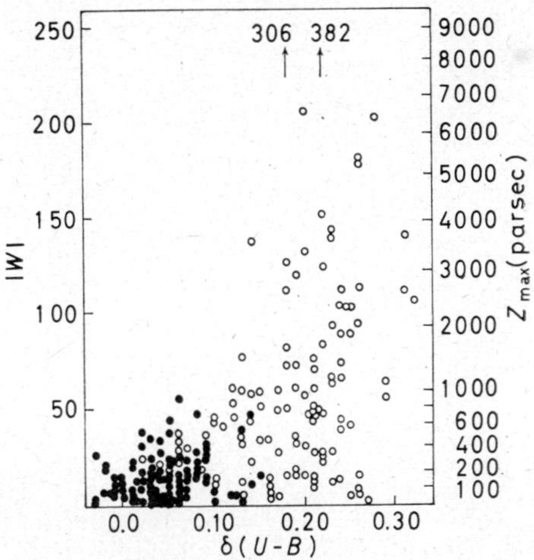

Fig. 20. – The correlation between the velocity perpendicular to the galactic plane W and the ultraviolet excess $\delta(U-B)$ of the stars in Fig. 19; the scale to the right shows the height over the galactic plane reached by stars whose W value is on the left scale.

If the metal abundance is a function of the age of the stars then it is clear that the mean height reached by stars is correlated with the epoch of formation. Young stars with high metal abundance show W-components of their motion which confine their orbits within the galactic disk with a half-thickness of about 400 pc. This result is in agreement with that found by KINMAN (1959) for globular clusters. Following earlier work by MAYALL and by MORGAN he found that clusters with weak metal lines form an extended halo reaching heights of at least 10 000 pc; clusters of intermediate line strength extend until about 4 000 pc and strong-line clusters are located near the galactic plane. Similar results were found by PRESTON (1959) for the *RR Lyrae* in the general field.

This leads us to the notion of a galactic collapse. That is these facts can be understood if we adopt the view that the conglomeration of stars and interstellar matter which forms to-day the galactic disk has condensed from a rather large spherical mass of gas. Streams of gas have come mainly down to the galactic plane from the direction of the z-co-ordinate. Stars which have condensed in these streams at these early stages have more elliptical orbits and high W-velocities, which were not much changed later by subsequent encounters; while the gas clouds have considerably modified their initial motions through collisions, thus assuming nearly circular motions close to the galactic plane, corresponding to the present state of galactic rotation. Stars which were born in this later stage from these clouds show more circular orbits with hardly any W-component in their motion.

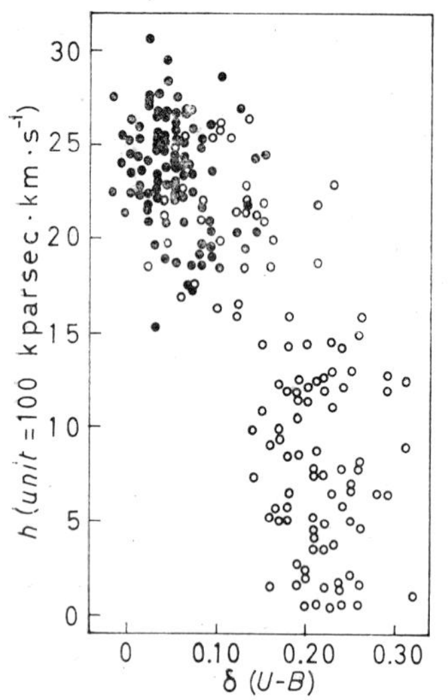

Fig. 21. – The correlation between the angular momentum, h, in units of 10^2 kpc km/s and the ultraviolet excess of the stars in Fig. 19 and 20.

Figure 21 shows the distribution of the angular momentum about the galactic axis as a function of the ultraviolet excess $\delta(U-B)$ and therefore as a function of age. All lately formed stars show large momenta according to their circular orbits whereas the older stars have randomly distributed momenta in relation to their more or less elliptical orbits.

Stars in these elliptical orbits were formed at a time when the Galaxy had not yet reached the present state of equilibrium, that is, gas was still streaming towards the galactic plane; the eccentricity of their orbits was later substantially preserved, while the remnants of the gas streams, out of which they were formed, came after collision with other masses of gas to an equilibrium state of circular motion.

The rate of collapse may be estimated in the following way. If the collapse was slow compared with the period of galactic rotation, the velocity of the prestellar gas in the radial direction was small compared with that in the direction of rotation; but then the initial orbital eccentricity of the stars formed during this early stage would have been small and we would not now observe the present eccentric orbits since the orbital eccentricity remains practically constant during a slow collapse in quasi equilibrium. Hence we conclude that the collapse of gas out of the halo was so rapid that only about $2 \cdot 10^8$ years (that is a time of the order of one galactic rotation) elapsed between the time when star formation started and the time when no material remained from which subsequent generations of halo stars could be formed.

Incidentally this explains why all the halo globular clusters for which accurate photometry is available — M3, M5, M13, M15 and M92 — are so nearly of the same age.

Coming to the linear scale of the collapse, we can argue as follows. The apogalactic distances of the stars with very high orbital eccentricities are in several cases as large as 50 kpc. But Fig. 21 shows that the angular momenta of the oldest stars with high eccentricity is less than $1.5 \cdot 10^3$ kpc km/s, which is the momentum of a circular orbit of some 5 kpc radius in the present state of the Galaxy. From our point of view second generation stars moving in circular orbits and first generation stars with the same angular momentum but moving in highly eccentric orbits, must have been produced by the same parent gas. Because the ratio of the apogalactic distance of the old stars to the radius of the circular orbit of the young stars with the same angular momentum is of the order of 10 to 1, we conclude that the present Galaxy has a radius in the galactic plane which is about one tenth that of the original mass of gas which formed the protogalaxy. The reason why the collapse factor perpendicular to the galactic plane is much larger (25 to 1) is understandable because the latter is not impeded by centrifugal acceleration.

Evidence from the old galactic cluster NGC 188 (SANDAGE 1962) shows that in the disk of the Galaxy near the Sun there has been no appreciable enrichment of the heavy-element content since the oldest galactic clusters were formed. During the very short time of $2\cdot10^8$ years which the collapse of the galaxy lasted, the entire halo population, including the globular clusters and the high-velocity stars, was created. That means that during this time the rate of star formation was very high. The increase in metal abundances during the same time requires a rate of formation of heavy elements of 5 solar masses for year. If the enrichment of metal content of the interstellar gas is due to Supernovae, one Supernova explosion per year is required or, roughly, hundred times the present rate.

5. – The intrinsic variables.

We will now discuss very briefly some properties of intrinsic variable stars, mainly connected with the period-luminosity relation in Cepheids.

5·1. *Intrinsic variables in the* H-R *diagram.* – It is well known that intrinsic variables are found only in well-defined areas of the H-R diagram (see Fig. 1); we can refer to Arp's monograph (ARP 1958) for a detailed account.

RR Lyrae stars are known to lie in a very small area of the orizontal branch of globular clusters, the so called *RR Lyrae* gap. The horizontal width of the gap is of the order of $0^{\text{m}}.25$ in $B-V$, with the center at about $B-V=0.30$. The mean absolute magnitude is less well determined; since it is often used to obtain distance moduli for globular clusters, it is important to fix its value with good accuracy. According to ARP (1962) the best absolute magnitude for the *RR Lyrae* stars of M3 and M5 is $M_{\text{v}}=+0.6\pm0.2$ in agreement with the value found by EGGEN and SANDAGE (1959) for *RR Lyrae* itself.

W Virginis stars are found in the H-R diagrams of globular clusters in a small strip about $0^{m}.2$ wide in $B - V$, vertically above the *RR Lyrae* gap with a rather large scatter in luminosities and periods. For example in M2, M_v ranges from -2.1 to -3.2 and $\log P$ from 1.19 to 1.82 (P in days).

The period-luminosity relation for these stars is

$$M_v = -1.72 \log P - 0.08 \,, \tag{5.1}$$

if one adopts $+0.6$ as the mean M_v for *RR Lyrae* stars in globular clusters.

Classical Cepheids occupy a small and almost vertical strip in the upper part of the H-R diagram whose width is, roughly, $0^{m}.20$ in $B - V$. Due to the well-known period-luminosity relation the brightest Cepheids with $M_v \sim -6^{M}$ have the longest periods (about 100 d), the faintest ones, with $M_v \sim -1^{M}$ have the shortest periods (about 2 days).

The absolute magnitude of Cepheids has always been a difficult observational problem. Since they are used as distance indicators (through the period-luminosity law) it is very important to have good values of the mean absolute magnitude of Cepheids of a given period. The question was recently examined by KRAFT (1961) following a line of analysis developed by SANDAGE (1958). We will now briefly consider this problem.

5·2. *The period-luminosity relation for Cepheids.* – The justification for the use of the period luminosity (*P-L*) relation for Cepheids has come mainly from empirical data, especially by Shapley and others, on the Cepheids in the two Magellanic clouds; although these data show considerable scatter, it was assumed that this was due to various observational uncertainties, including internal absorption inside the clouds. When however, more accurate data on Cepheids in clusters became available, it was clear that there was an intrinsic dispersion in the period-luminosity relation, and that a third parameter was involved. The best data show that this parameter may be the (*B-V*)-colour, in the sense that bluer stars have a slightly shorter period than redder stars of the same luminosity, so that the lines of constant period are tilted down to the right in the (M_v, *B-V*)-diagram.

Classical Cepheids belong to population I and from the observations it seems probable that a star moving horizontally through the H-R diagram from the higher main sequence to the red giants region passes through this stage of variability. Several Cepheids in galactic clusters are known which fall in the gap between the brightest stars of the main sequence and the red giants. An interesting example is the very rich cluster NGC 1866 in the large Magellanic cloud, which contains abount 10 Cepheids, all of the same luminosity and with periods ranging from 2.6 to 3.5 days, wich fall exactly on the connecting line between the cluster main sequence and evolved red giants (SANDAGE unpublished and also HODGE 1961).

From a simple dimensional argument, we can assume for homologous pulsating stars that

$$P\sqrt{\varrho} = \text{const}\,, \tag{5.2}$$

the well-known period-density law. Using

$$\varrho = \frac{\mathscr{M}}{\frac{4}{3}\pi R^3}\,,$$

and

$$L = 4\pi R^2 \sigma T_e^4\,,$$

we obtain from (5.2)

$$\log P = \frac{1}{2}\log\frac{\mathscr{M}}{\mathscr{M}_\odot} + 0.3(M_b - M_{b\odot}) + 3\log\frac{T_e}{T_{e\odot}} = \lg Q\,, \tag{5.3}$$

where Q is a constant.

We eliminate, first, the mass through the mass-luminosity relation. For the bright stars ($-8 < M_{bol} < +1$) the observational data are well represented by

$$M_b = 3.96 - 8.22\log\frac{\mathscr{M}}{\mathscr{M}_\odot}\,,$$

adopting $M_{b,\odot} = 4.77$; this refers, however, to main sequence stars. If the evolutionary tracks for Cepheids are similar to those of other stars in the giant region, we may assume that their absolute bolometric magnitudes are 1^M brighter than that of the main sequence stars of the same mass; hence we may adopt for the Cepheids

$$M_b = 2.96 - 8.22\log\frac{\mathscr{M}}{\mathscr{M}_\odot}\,. \tag{5.4}$$

Consider, next, the bolometric corrections and effective temperatures. A rediscussion of these relations for the stars in the supergiant region gives approximately

$$M_b = M_v + 0.145 - 0.322(B-V)\,, \qquad (1.0 > B-V > 0.4)\,, \tag{5.5}$$

$$\log T_e = 3.886 - 0.175(B-V)\,, \qquad (1.0 > B-V > 0.4)\,. \tag{5.6}$$

Using eqs. (5.4), (5.5) and (5.6) to eliminate $\mathscr{M}$, M_b and T_e, we obtain from (5.2)

$$\log P + 0.239M_v - 0.602(B-V) = 0.838 + \log Q\,. \tag{5.7}$$

From eq. (5.7) it is seen that the period-luminosity relation must show a scatter which is not due as it was assumed for a long time, to observational difficulties, but which is an intrinsic scatter depending on the colour.

The amplitudes of the variation of light are largest for stars near the central line of the area of the Cepheids in the H-R diagram (dashed line in Fig. 22)

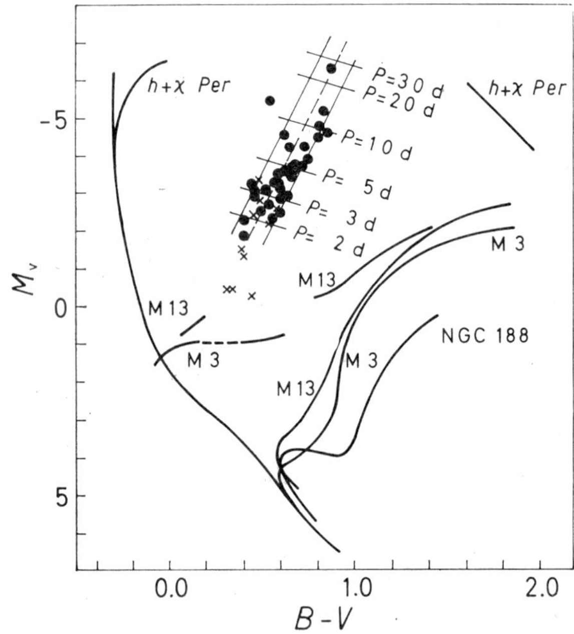

Fig. 22. – Cepheids on the color-magnitude diagram. The filled circles represent 32 Cepheids discussed by KRAFT (1961*b*), the crosses represent Cepheids in globular clusters; the M_v of the latter were derived assuming an absolute magnitude of $+0.6$ for the *RR Lyrae* stars. The sequences for the globular clusters differ from those of Fig. 7, because a correction for blanketing has been applied.

and get smaller for stars farther away from it. If one assumes that the stars are crossing the area of variation during their evolution, then the amplitude of the light variation slowly increases untill the star reaches this central line in the instability strip and then decreases again, as the star passes to the right-hand edge of the strip.

If we adopt as the period-luminosity relation that corresponding to the central strip, we can use eq. (5.7) to predict this relation.

For this purpose the constant Q can be computed by means of 6 Cepheids in galactic clusters, whose magnitudes and colours have been recently discussed by KRAFT (1961*a*, *b*); giving the same weight to all the stars, we find

$$\log Q = -1.294 \, .$$

With this value we can now compute the absolute magnitude for all the Cepheids whose periods and colours are known. The values for 32 Cepheids discussed by Kraft are plotted as dots in Fig. 22. These values are slightly different from those computed by Kraft, because he adopted two equations somewhat different from (5.5) and (5.6) for the bolometric corrections and the temperatures. As the discussion by Kraft shows, the result is very little affected by a different choice of the constants in eqs. (5.5) and (5.6).

The central line (drawn by free hand) through the Cepheids domain in the H-R diagram, has an equation

$$M_V = 2.70 - 10.0(B-V) \, . \tag{5.8}$$

The predicted *P-L* relation (corresponding to a zero width of the Cepheid domain) is

$$\log P + 0.299 M_V = 1.000 + \log Q = -0.294 \, , \tag{5.9}$$

when eq. (5.8) is substitued in eq. (5.7). Substituting the above value for Q in (5.9) predicts $M_V = -0.98 - 3.34 \log P$ for the center of the strip on the basis of constant Q. This should be compared with the observed relation $M_V = -1.67 - 2.54 \log P$ (KRAFT 1961*b*). The agreement is not particularly good. If exact agreement is required, then $\log Q = -1.50 + 0.241 \log P$, which follows from eq. (5.9) and the observed *P-L* relation. If this equation for Q is correct, then the Cepheid variables are not homologous, which is of course a reasonable result. A check on the variation of Q with P can be obtained by considering the *P-L* relation in blue light. From

$$M_B = M_V + B - V \, ,$$

equations (5.8) and (5.7) become

$$M_B = 2.70 - 9.0(B-V) \, ,$$

and

$$\log P + 0.333 M_B = 1.090 + \log Q \, . \tag{5.10}$$

Substitution of $\log Q = -1.50 + 0.241 \log P$ into eq. (5.10) gives

$$M_B = -1.23 - 2.28 \log P \, , \tag{5.11}$$

for the center of the instability strip, which is in excellent agreement with the observed relation

$$M_B = -1.33 - 2.25 \log P. \qquad (5.12)$$

The agreement gives some confidence that the derived variation of Q with period is approximately correct, but it must be remembered that the exact numerical value of Q depends rather critically on the adopted relations (5.4), (5.5) and (5.6) which are subject to improvement.

REFERENCES

ARP, H. C.: 1958, *Hand. d. Phys.* Bd. 51, 75.

ARP, H. C.: 1960, *A. J.*, **65**, 404.

ARP, H. C.: 1962, *Ap. J.*, **135**, 311.

ARP, H. C. and KRAFT, R. P.: 1961, *Ap. J.*, **133**, 420.

CHANDRASEKHAR, S. and SCHÖNBERG, M.: 1942, *Ap. J.*, **96**, 161.

CODE, A. D.: 1960, *Stellar Atmospheres* (ed. by J. L. GREENSTEIN).

DEMARQUE, P.: 1960, *Ap. J.*, **132**, 366.

EGGEN, O. J.: 1955, *A. J.*, **60**, 401.

EGGEN, O. J.: 1960, *La Plata Symposium on Stellar Evolution* (ed. by J. SAHADE).

EGGEN, O. J.: 1961, *R. Obs. Bull.*, n. 51.

EGGEN, O. J.: 1962, *R. Obs. Bull.* (in preparation).

EGGEN, O. J. and SANDAGE, A. R.: 1959, *M. N.*, **119**, 255.

EGGEN, O. J. and SANDAGE, A. R.: 1962, *Ap. J.*, **136**, 735.

EGGEN, O. J., LYNDEN-BELL, D. and SANDAGE, A. R.: 1962, *Ap. J.*, **136**, 748.

HODGE, P. W.: 1961, *Ap. J.*, **133**, 413.

DE JAGER, C. and NÉVEN, L.: 1957, *Rech. Astr. Utrecht*, **13**, 4.

JOHNSON, H. L., MITCHELL, R. I. and IRIARTE, B.: 1962, *Ap. J.*, **136**, 75.

KINMAN, T. D.: 1959, *M. N.*, **119**, 538.

KOPAL, Z.: 1959, *Close Binary Systems*.

KRAFT, R. P.: 1961*a*, *Ap. J.*, **133**, 39.

KRAFT, R. P.: 1961*b*, *Ap. J.*, **134**, 616.

MELBOURNE, W. G.: 1960, *Ap. J.*, **132**, 101.

PARKER, R., GREENSTEIN, J. L., HELFER, H. L. and WALLERSTEIN, G.: 1961, *Ap. J.*, **133**, 101.

PRESTON, G. W.: 1959, *Ap. J.*, **130**, 507.

REIZ, A.: 1954, *Ap. J.*, **120**, 342.

ROMAN, N.: 1955, *Ap. J. Suppl.*, n. 18, 195.

SALPETER, E. E.: 1955, *Ap. J.*, **121**, 161.

SANDAGE, A. R.: 1957, *Ap. J.*, **126**, 326.

SANDAGE, A. R.: 1958, *Ap. J.*, **127**, 537.

SANDAGE, A. R.: 1962, *Ap. J.*, **135**, 333.

SANDAGE, A. R. and EGGEN, O. J.: 1959, *M. N.*, **119**, 278.

STRÖMGREN, B.: 1952: *A. J.*, **57**, 65.
WALLERSTEIN, G. and CARLSON, M.: 1960, *Ap. J.*, **132**, 276.
WILDEY, R. L., BURBIDGE, E. M., SANDAGE, A. R. and BURBIDGE, G.: 1962, *Ap. J.*, **135**, 94.
WILSON, O. C.: 1959, *Ap. J.*, **130**, 496.
WOOLLEY, R. v. D. R., SANDAGE, A. R., EGGEN, O. J., ALEXANDER, J. B., MATHER, L., EPPS, E. and JONES, S.: 1962, *R. Obs. Bull.*, n. 58.
WRUBEL, M.: 1958, *Handb. d. Phys.* Bd. **51**, 1.

Construction of Stellar Models.

M. H. WRUBEL

Indiana University - Bloomington, Ind.

1. – Introduction.

Several summaries of the theory of stellar interiors have recently appeared (SCHWARZSCHILD (1958), WRUBEL (1958)), and others are in press or in progress. It is quite unnecessary to duplicate these articles on this occasion. Therefore, rather than devoting this chapter to a comprehensive survey of the subject from its foundations, I have chosen instead to concentrate on a few aspects selected largely because of their particular personal interest.

In contrast with Sandage's observational approach to stellar evolution, which proceeds from the color-magnitude diagram to evolutionary tracks, the theoretical approach, based upon stellar models, predicts the tracks to be compared with observations. In the end, both methods must, of course, agree. The ultimate aim of the stellar model approach is to describe the internal structure of the star, beginning with its formation from the interstellar medium, through all its changes in luminosity and radius, until its ultimate gradual extinction, or its cataclismic annihilation as a supernova. Unfortunately, this goal is still quite remote.

At this juncture questions remain in every region of the color-magnitude diagram. In the pre-main sequence contraction phases, Hayashi's (1961) results are at variance with earlier work of HENYEY *et al.* (1955) and COX and BROWNLEE (1961). The main sequence of HOYLE and HASELGROVE (1959) has predicted questionable ages for NGC 188 and M3 (SANDAGE (1961), WOOLF (1962)). SCHWARZSCHILD has experienced difficulties carrying models through the « helium flash » in red giants. And there are virtually no models for those stars in which element synthesis beyond carbon is going on.

The majority of this chapter will be devoted to consideration of two types of problems: those of physical questions and those of numerical methods. In each case recent results foreshadow changes in the models to be computed in the future.

2. – Differential equations of stellar structure.

The derivation of the differential equations of stellar structure has appeared in the literature so often that there is no point in rederiving them here (see CHANDRASEKHAR (1939), SCHWARZSCHILD (1958), and WRUBEL (1958)). We will merely define the symbols and state the equations. They describe a spherical star in equilibrium.

The variables are:

r	radial co-ordinate, chosen here as the independent variable,
G	gravitational constant,
P	total pressure,
ϱ	density,
$M(r)$	mass interior to a sphere of radius r,
T	temperature,
a	radiative constant,
c	velocity of light,
$\varkappa$	Rosseland mean opacity,
$L(r)$	energy crossing surface of radius r per second,
ε	energy sources in erg $g^{-1}\, s^{-1}$,
Γ	adiabatic exponent.

ϱ, T, P, $\varkappa$, ε, and Γ, are further related through « constitutive equations » which describe the properties of the material out of which the star is made and which therefore depend upon the detailed composition.

The differential equations of a spherical star in equilibrium are:

a) Hydrostatic equilibrium:

$$\frac{\mathrm{d}P}{\mathrm{d}r} = -\frac{GM(r)}{r^2}\,\varrho\,; \tag{1}$$

b) Conservation of mass:

$$\frac{\mathrm{d}M(r)}{\mathrm{d}r} = 4\pi r^2 \varrho\,; \tag{2}$$

c) Energy transport:

i) Radiation; diffusion approximation:

$$\frac{\mathrm{d}T}{\mathrm{d}r} = -\frac{3}{4ac}\,\frac{\varkappa\varrho}{T^3}\,\frac{L(r)}{4\pi r^2}\,; \tag{3}$$

ii) Convection; adiabatic approximation:

$$\frac{dT}{dr} = \frac{\Gamma - 1}{\Gamma} \frac{T}{P} \frac{dP}{dr} ; \tag{4}$$

(When the adiabatic approximation is not adequate to describe convective transport, an iterative procedure, described in Dr. KIPPENHAHN's chapter, is used.)

d) Energy production:

$$\frac{dL(r)}{dr} = 4\pi r^2 \varrho \varepsilon . \tag{5}$$

To these we add equations describing the time rate of change of the abundance of elements that are altered by the nuclear reactions. For example, for reactions that convert hydrogen to helium:

$$\frac{dX}{dt} = -\frac{dY}{dt} = -\frac{4H}{Q} \varepsilon(X) , \tag{6}$$

where: X fractional abundance of hydrogen by mass,
Y fractional abundance of helium by mass,
H mass of hydrogen atom,
Q energy released per conversion of four hydrogen atoms to helium,
$\varepsilon(X)$ energy production due to these reactions, in erg gm^{-1} s^{-1}.

Various transformations of the above equations can be selected for practical computation (see WRUBEL (1958)).

The equations must be solved subject to physical boundary conditions. At the center of the star both $M(r)$ and $L(r)$ must tend to zero. The condition at the « surface » has given a good deal of trouble recently. Traditionally the study of the outer layers, the stellar « atmosphere », has been a discipline distinct from the study of stellar interiors. It is more concerned with the region in which the observed stellar spectrum originates. There, $M(r)$ is practically M, the total stellar mass; and $L(r)$ is equal to L, the observed luminosity of the star. The determination of the pressure-temperature relation in a stellar atmosphere has been studied independently of, and concurrently with, studies of stellar structure. In recent years it has become clear that simplified boundary conditions for stellar models (such as $P \to 0$ and $T \to 0$ as $r \to R$) which were once thought adequate for interiors studies are not. Rather, the outer boundary condition a stellar model should fulfill is ideally to be determined by model stellar atmospheres. The physical variables at the top of the model interior must be continuous with the values of the same variable at the base of the atmosphere. Unfortunately, just these layers, deep in the atmosphere, are influenced by convection where current methods are not completely satisfactory. A further discussion of the outer boundary condition will be found in the chapters by Dr. KIPPENHAHN.

3. – Statistical description of material in stellar interios.

We can lay the ground work for further discussions by studying the nature of the material with which we must deal. We will be interested in a range of density from about 10^{-7} to 10^{+6} g/cm^{-3} and temperatures from several thousand degrees to above 10^{10} °K. (Problems of higher-density material are discussed in the chapter by Dr. SCHATZMAN.) The chemical composition of the material will also be quite varied. For a study of « first generation » stars, in the language of the nucleogeneticists, we must consider almost pure hydrogen; but for main-sequence stars of later generations, about $\frac{1}{4}$ of the mass would be helium, and a fraction ranging from a few tenths of a per cent to a few per cent would be heavier elements. In the core of a star that is evolving away from the main sequence the hydrogen content is negligible and helium is the predominant element. Within the heavy-element fraction in each case, the distribution of individual elements might vary depending upon the nuclear processes to which the material has been exposed.

We therefore begin with a very general approach. Consider a mixture of various positive nuclei, each identified by its atomic number, Z. Let the number of atoms of a particular nuclear species Z be N_Z. To this mixture we add enough electrons so that the result is macroscopically neutral; that is

$$N_e = \sum_Z Z N_Z . \tag{7}$$

If the mixture is in thermodynamic equilibrium at temperature T, some of the electrons will be captured by the nuclei to form ions. If we use r to designate the state of ionization, we have, in obvious notation:

$$N_Z = \sum_r N_{Z,r} . \tag{8}$$

Each ion, r, is capable of many levels of excitation, which we can designate by l. Thus.

$$N_{Z,r} = \sum_l N_{Z,r,l} . \tag{9}$$

A complete description of the system would specify the distribution of electrons among all the available free states and all the bound states associated with the positive ions. In any event the total number of electrons would always be given by

$$N_e = N_e^{(f)} + N_e^{(b)} = \sum_Z ZN_Z \,. \tag{10}$$

For many purposes the *ionic* representation is too explicit. We would be satisfied if only the principal quantum numbers were given for the attached electrons. That is, it would be adequate to know how many atoms of species Z have respectively 0, 1, or 2 K-electrons, for example; or, equivalently, the average number of electrons in the various shells of the atoms. We will denote the number of electrons bound to the n-th shell of nuclei of species Z by

$$N_e^{(b,Z,n)} \,. \tag{11}$$

We can then define « occupation numbers » as follows

$$\bar{N}_e^{(b,Z,n)} \equiv \frac{N_e^{(b,Z,n)}}{N_Z} \equiv \text{average number of electrons bound to the } n\text{-th shell of nuclei of type } Z \,. \tag{12}$$

3·1. *Free electrons.* – The electrons are statistically described as a Fermi gas, since they obey the Pauli exclusion principle, which states that no two electrons can simultaneously occupy the same state.

Consider the free electrons: if the number of available states with energies between ε_f and $\varepsilon_f + \mathrm{d}\varepsilon_f$ is given by c_f, the number of electrons to be found in the range of energies $\mathrm{d}\varepsilon_f$ at temperature T is

$$\mathrm{d}N_e^{(f)} = \frac{c_f\,\mathrm{d}\varepsilon_f}{\exp[\alpha + \beta\varepsilon_f] + 1} \,, \tag{13}$$

where $\beta = 1/kT$. The constant, α, introduced in this expression is called the « degeneracy parameter » and is evaluated by recognizing that the total number of free electrons must be

$$N_e^{(f)} = \int_0^\infty \frac{c_f\,\mathrm{d}\varepsilon_f}{\exp[\alpha + \beta\varepsilon_f] + 1} \,. \tag{14}$$

The « statistical weight » of the band of energies $\mathrm{d}\varepsilon_f$ can be evaluated as follows

$$C_f\,\mathrm{d}\varepsilon_f = \frac{8\pi p_f^2\,\mathrm{d}p_f}{h^3}\,V \,, \tag{15}$$

where p is the momentum of the electron V is the volume, and h is Planck's constant. In the nonrelativistic approximation this becomes

$$c_f \, d\varepsilon_f = \frac{8\sqrt{2}\,\pi m_e^{\frac{3}{2}} V}{h^3} \varepsilon_f^{\frac{1}{2}} \, d\varepsilon_f \,. \tag{16}$$

At low densities and high temperatures, $\exp[\alpha+\beta\varepsilon_f]$ is much larger than 1 and eq. (13) goes over to the (nonrelativistic) form

$$\begin{cases} dN_e^{(f)} = \exp[-\alpha] \exp[-\varepsilon_f/kT] c_f \, d\varepsilon_f \,, \\ \exp[+\alpha] = \dfrac{V}{N_e^{(f)}} \dfrac{2(2\pi m_e kT)^{\frac{3}{2}}}{h^3} \,. \end{cases} \tag{17}$$

Under these circumstances the gas is Maxwellian and degeneracy effects are negligible.

At high densities and relatively low temperatures, degeneracy effects are quite important. In that case the parameter α is not so easily evaluated. Equation (4) can be rewritten as

$$N_e^{(f)} = \frac{V}{h^3} 2^{\frac{7}{2}} \pi (m_e kT)^{\frac{3}{2}} F_{\frac{1}{2}}(-\alpha) \,, \tag{18}$$

where the function

$$F_W(\psi) = \int_0^\infty \frac{u}{\exp[u-\psi]+1} du \,, \tag{19}$$

is tabulated by McDougall and Stoner (1939).

We can regard

$$\tilde{\omega}_f = \frac{1}{\exp[\alpha+\beta\varepsilon_f]+1} \,, \tag{20}$$

as the probability that the band of states $d\varepsilon_f$ is fully occupied. Therefore

$$q_f = 1 - \varpi_f = 1 - \frac{1}{\exp[\alpha+\beta\varepsilon_f]+1} \,, \tag{21}$$

is the probability that there is still room for an electron in the range $d\varepsilon_f$. This quantity is of considerable importance in any transition in which the final electron state is in the range $d\varepsilon_f$ (as in photo-ionization, for example): for unless there is a state available to it, the electron cannot make the transition and the process will be suppressed.

3·2. *Bound electrons. Occupation numbers.* – Bound states are, of course, available to the electrons just as free states are, and the electrons that occupy them must obey the same statistics. We can therefore write

$$N_e^{(b,Z,n)} = \frac{N_Z C_{Zn}}{\exp[\alpha + \beta\varepsilon_{Zn}] + 1} . \tag{22}$$

The numerator indicates that for each of the N_Z nuclei, shell n can accommodate C_{Zn} electrons. In this notation β is $1/kT$ as before; C_{Zn} is the statistical weight of a bound state and equals $2n^2$. The same degeneracy parameter, α, holds for all the electrons in the medium and therefore is the same for the bound electrons as the free.

As a first approximation we may ignore interactions between the electrons so that the energy levels are unperturbed by the presence of bound electrons, although the limitations of the Pauli principle still apply. If the free electron gas is Maxwellian, we have, as above

$$e^{\alpha} = \frac{V}{N_e^{(f)}} \frac{2(2\pi m_e kT)^{\frac{3}{2}}}{h^3} , \tag{23}$$

and the occupation numbers, eq. (22), become

$$\bar{N}_e^{(b,Z,n)} = \frac{2n^2}{1 + (2(2\pi m_e kT)^{\frac{3}{2}}/h^3)(V/N_e^{(f)}) \exp[-\chi_n/kT]} , \tag{24}$$

where the energy of an electron in shell n is given by the hydrogenic approximation:

$$\varepsilon_{Zn} \equiv \chi_n \simeq -\frac{cR_\infty hZ'^2}{n^2} , \tag{25}$$

(see for example CHANDRASEKHAR (1939)) where Z' is the effective nuclear charge.

This approximation is suitable only for the roughest estimates. Effects that we have ignored must be allowed for. A detailed discussion of these effects has been given by H. MEYER (1947) in a report available to qualified research workers through the Los Alamos Scientific Laboratory.

In point of fact the electrons are not independent. When interactions are taken into account the problem becomes much more complicated, because the energies, ε_{Zn}, which appear in the equations for the occupation numbers, themselves depend upon the occupation numbers and the number of free electrons.

The detailed determination of ε_{Zn} would carry us very far afield so we restrict our discussion only to an enumeration of those effects that must be included. They are:

1) Screening of the nucleus by other bound electrons;
2) Interaction of the bound electrons with the free electrons;
3) Interaction of the free electrons with themselves;
4) Interaction with neighboring ions.

It should be remarked that the continuum is depressed as a result of interactions: that is, the highest states which would be discrete in the case of an isolated atom are broadened and merge into a continuum in the presence of other particles. Electrons in these states are considered free and the zero of energy is adjusted so as to coincide with the energy of a free electron with zero kinetic energy.

The calculation of occupation numbers can be carried out as an iteration, which is schematically described as follows (MAYER (1947)).

Trial values of α, $\bar{N}_e^{(b,Z,n)}$, and $N_e^{(f)}$ are selected, satisfying the requirement that the total number of electrons is a specified constant:

$$N_e = N_e^{(b)} + N_e^{(f)} . \tag{26}$$

The energies, ε_{Zn}, can be determined once the numbers of bound and free electrons are specified. Using these values of ε_{Zn} and the trial value of α, a new approximation to $\bar{N}_e^{(b,Z,n)}$ can be obtained from eq. (22). The total number of bound electrons is given by

$$N_e^{(b)} = \sum_Z \sum_n N_e^{(b,Z,n)} , \tag{27}$$

and the corresponding number of free electrons follows from eq. (26).

The number of free electrons can be used to determine a new approximation for α by using the relation (valid in the nonrelativistic case)

$$N_e^{(f)} = \frac{V}{h^3} 2^{\frac{7}{2}}\pi(mkT)^{\frac{3}{2}} F_{\frac{1}{2}}(-\alpha) . \tag{28}$$

The new values of α, $\bar{N}_e^{(b,Z,n)}$, and $N_e^{(f)}$ can now be used to repeat the process until it converges.

An alternate approach, described by A. N. COX (1962), which performs the equivalent computation using individual Saha ionization equations for each ion of each element, has been found especially useful in the range of conditions in the boundary region between the interior and the atmosphere.

In mixtures containing a variety of elements, either method of computation is so lengthy that only a large electronic computer can handle them in a reasonable time. A very extensive program for the IBM 7090 has been developed by A. N. Cox with the assistance of D. EILERS at Los Alamos Scientific Laboratory. Dr. Cox has generously expressed his willingness to perform calculations upon request for astrophysicists who wish to investigate particular mixtures.

3·3. *Pressure and entropy.* – Thus far we have concentrated on occupation numbers, but it is only a short step to the equation of state. We will consider next the pressure exerted by the gas.

The total pressure is the sum of radiation pressure and gas pressure due to the particles,

$$P = P_r + P_g \,, \tag{29}$$

where the radiation pressure is

$$P_r = \tfrac{1}{3} a T^4 \,. \tag{30}$$

The gas pressure can conveniently be regarded as a sum of the partial pressures of the ions and the free electrons

$$P_g = P_i + P_e \,. \tag{31}$$

The ions are so massive that they are not appreciably degenerate even at white dwarf densities; therefore the perfect gas law applies and

$$P_i = \frac{\sum N_Z}{V} kT \,. \tag{32}$$

If conditions are such that the free electrons are also not degenerate the free electron pressure is given by

$$P_e = \frac{N_e^{(f)}}{V} kT \,. \tag{33}$$

The total gas pressure is then

$$P_g = \frac{kT}{V} \sum_Z \{N_Z(1 + Z - \sum_n \bar{N}_e^{(b,Z,n)})\} \,, \tag{34}$$

where we have explicitly indicated the number of free electrons contributed by each element as equal to the nuclear charge less the number of bound electrons.

If A_z is the atomic weight of element Z, and H is the mass of unit atomic weight. The density of the material is

$$\varrho = \frac{H}{V} \sum_{Z} N_Z A_Z . \tag{35}$$

The gas pressure can then be rewritten in the form

$$P_g = \frac{k}{\mu H} \varrho T , \tag{36}$$

where the mean molecular weight, μ, is identified as

$$\mu^{-1} = \frac{\sum_{Z} \{N_Z(1 + Z - \sum_{n} \bar{N}_e^{(b,Z,n)})\}}{\sum_{Z} N_Z A_Z} . \tag{37}$$

If the material is predominantly hydrogen and helium, it is convenient to adopt special symbols for the fractional mass of hydrogen and the fractional mass of helium. Thus if X refers to the fraction of each gram of stellar material made up of hydrogen, and Y refers to a similar quantity for helium, we have

$$\left\{ \begin{aligned} & X = \frac{N_H}{\sum N_Z A_Z}, \qquad Y = \frac{4N_{He}}{\sum N_Z A_Z}, \\ & 1 - X - Y = \frac{\sum_{Z>2} N_Z A_Z}{\sum_{Z} N_Z A_Z} . \end{aligned} \right. \tag{38}$$

If we approximate the number of free particles per unit atomic weight for all elements heavier than helium as $\frac{1}{2}$, we get the usual approximation for the mean molecular weight

$$\mu = \frac{4}{2 + 6X + Y} . \tag{39}$$

Of more interest is the case when the electron gas is degenerate. Then the partial pressure due to the electrons must be evaluated from the expression

$$P_e = \frac{8\pi}{3h^3} \int_0^\infty \frac{p_f^3}{\exp[\alpha + \beta\varepsilon_f] + 1} \frac{\partial \varepsilon_f}{\partial p_f} \, dp_f , \tag{40}$$

where the energy and momentum of the free electron are related by

$$\varepsilon_f = m_0 c^2 \left[\left(1 + \frac{p_f^2}{m_0^2 c^2}\right)^{\frac{1}{2}} - 1\right], \tag{41}$$

if interactions between free electrons are neglected.

When relativistic effects are negligible

$$P_e = \frac{N_e^{(f)}}{V} kT \left[\frac{\frac{2}{3} F_{\frac{3}{2}}(-\alpha)}{F_{\frac{1}{2}}(-\alpha)}\right]. \tag{42}$$

(The ratio in brackets has been tabulated by MacDougall and Stoner (1939).)

For densities above the order of 10^6 and temperatures above 10^9 relativistic effects are not negligible, but unfortunately a general tabulation of the integrals appropriate in this case is not available.

Under the same approximations as in eq. (42), the energy of the electron gas is given by

$$E_e^{(f)} = \frac{8\pi V}{h^3} \int_0^\infty \frac{\varepsilon_f p_f^2 \, \mathrm{d}p_f}{\exp[\alpha + \beta\varepsilon_f] + 1}. \tag{43}$$

The degeneracy parameter, α, is related to the Helmholtz free energy, Ψ, by the relation

$$\Psi = -\alpha kT. \tag{44}$$

Since Ψ is related to the entropy, S, according to

$$\Psi = E_e^{(f)} - TS, \tag{45}$$

we can find the entropy of the electron gas using previously determined quantities:

$$S = \frac{E_e^{(f)} - \Psi}{T}. \tag{46}$$

3·4. *Pair degeneracy.* – In gases at temperatures exceeding 10^9, electron-positron pairs can be created. As pointed out by Temesvary (1935), on the basis of earlier work by Koppe (1948), the pressure exerted by these pairs is not negligible. Its value can approach $\frac{7}{4}$ times the radiation pressure.

The study of pairs by the methods of statistical mechanics leads to two very similar expressions for the distribution of electrons and positrons among the available energy states. As before, the number of free electrons in a range of momentum is given by

$$\mathrm{d}N_e = V\,\frac{8\pi p^2\,\mathrm{d}p}{h^3}\,\frac{1}{\exp[\alpha+\beta\varepsilon]+1}\,. \tag{47}$$

The corresponding expression for the positron is identical except that the α is replaced by $(-\alpha)$

$$\mathrm{d}N_p = V\,\frac{8\pi p^2\,\mathrm{d}p}{h^3}\,\frac{1}{\exp[-\alpha+\beta\varepsilon]+1}\,. \tag{48}$$

The total number of free electrons is now due in part to ionization of heavier elements and in part to pair creation; and the degeneracy parameter, α, is evaluated by subtracting the number of positrons from the total number of electrons and equating the result, which represents unpaired electrons, to the number of electrons available through ionization

$$N_e - N_p = \sum_Z N_Z Z\,. \tag{49}$$

The pressure of the positrons can be calculated by an expression similar to eq. (40).

4. – Opacity.

We now turn to the interaction between radiation and the material described in the preceding section. An authoritative review article on the subject of opacity calculations has recently been prepared by A. N. Cox (1962) of the Los Alamos Scientific Laboratory where the most extensive recent work in this subject has been done. The treatise by HARRIS MAYER (1947) referred to earlier, also gives explicit details of opacity calculation procedures. Anyone intending to do detailed opacity calculations of his own is referred to these papers and the bibliography contained in them. For our purposes we will concentrate on the main features of the problem and some of the recent methods of attack.

The term « opacity » refers to those physical phenomena that remove energy from an incident beam of radiation. A distinction must be made between *absorptive* processes, which truly remove photons from the beam, and *scattering* processes that merely redirect the photons. The distinction is significant because absorptive processes must be corrected for stimulated emission, which decreases the effectiveness of the absorption by a factor $(1-\exp[-h\nu/kT])$.

On the other hand, as pointed out by MAYER (1947) and independently by RUDKJØBING (1947), this correction is not appropriate for scattering processes because stimulated scattering out of the beam is exactly compensated by stimulated scattering into beam, unless there is a change in photon frequency by scattering (SAMPSON (1959)).

The Rosseland mean which is the basis for all opacity calculations in stellar interiors can be written

$$\frac{1}{\varkappa} = \frac{\displaystyle\int_0^\infty \frac{(\mathrm{d}B_\nu/\mathrm{d}T)\,\mathrm{d}\nu}{\varkappa_{\nu a}(1-\exp[-h\nu/kT])+\varkappa_{\nu s}}}{\displaystyle\int_0^\infty \frac{\mathrm{d}B_\nu}{\mathrm{d}T}\,\mathrm{d}\nu}\,. \tag{50}$$

$B_\nu(T)$ is the Planck function at frequency ν and temperature T; $\varkappa_{\nu a}$ is the monochromatic absorptive opacity and $\varkappa_{\nu s}$ the monochromatic scattering opacity. For a derivation of Rosseland's formula, see for example, WRUBEL (1959, Section **8**).

For the moment let us ignore scattering. Then the above equation can be written in the following form convenient for calculation

$$\frac{1}{\varkappa} = \int \frac{W(u)}{D(u)}\,\mathrm{d}u\,, \tag{51}$$

where

$$\begin{cases} u = h\nu/kT\,, \\[2mm] W(u) = \dfrac{15}{4\pi^4}\,u^7\exp[2u](\exp[u]-1)^{-3}\,, \\[2mm] D(u) = \varkappa_\nu u^3\,. \end{cases} \tag{52}$$

The weighting function, $W(u)$, has a sharp maximum at 6.98 and beyond $u=30$ it is negligible.

The Rosseland mean is really a transparency mean since it involves the reciprocal of the opacity. This should be kept in mind. In particular, if the opacity in some region of the spectrum is already high and it is doubled, the effect on the Rosseland mean will be negligible because the transparency in either case would be negligible. Effects which produce only weak absorption are effective only in « windows » where other effects are weak.

The three absorptive processes we will consider are

1) bound-free (photoionization);
2) free-free;
3) bound-bound (line absorption).

4·1. *Bound-free absorption.* – The physical basis of bound-free absorption is the photoelectric effect. A photon of energy $h\nu$ strikes an atom; part of the energy goes into overcoming the attraction of the nucleus and the remainder goes into the kinetic energy of the electron that is freed.

Consider a gas consisting of a single element at high temperature. The energy of an electron in shell n can be estimated by again using the hydrogenic approximation:

$$\chi_n \simeq -cR_\infty h \frac{Z'^2}{n^2} . \tag{53}$$

A photon must have the energy $h\nu_n = |\chi_n|$ to liberate such an electron. The opacity increases discontinuously going from $\nu < \nu_n$ to $\nu > \nu_n$ because a photon with slightly less than the ionization energy cannot remove an electron from shell n and will therefore not be absorbed, while the reverse is true for a photon with slightly more than the ionization energy. This jump in absorption is called an « absorption edge ». In opacity computations, we are primarily concerned with those absorption edges that lie in the region where the weighting function, $W(u)$, is appreciable.

If we use the notation

$$u_n = \frac{h\nu_n}{kT} , \tag{54}$$

we note that as the temperature is increased, first the M, then the L, and finally the K shell will come under the maximum of the weighting function.

In a gas containing a mixture of elements, the edges due to different elements will be intermingled.

In practical computations, the absorption edges must be located more precisely by using the energies of the bound electrons, ε_{Zn}, described in Section **3**·2. The absorption edge corresponding to the n-th shell of element Z is located at

$$u_{Zn} = \frac{|\varepsilon_{Zn}|}{kT} . \tag{55}$$

This is also an approximation since there is actually an edge for every excited level of every ion. The computer program of Cox and Eilers takes these ionic configurations into account.

Let $a_{bf}^{(Z,n)}(\nu)$ refer to the contribution to the bound-free absorption coefficient due to shell n of element Z. To the red of the edge, there is no contribution; to the violet, the contribution can be written as follows:

$$a_{bf}^{(Z,n)}(\nu) = \left[\text{const}\,\frac{Z^4}{n^5}\,\frac{1}{\nu^3}\,\bar{N}_e^{(b,Z,n)}\right] g_{bf}\,q_f\,, \qquad (\nu > \nu_{Zn})\,. \tag{56}$$

The expression in square brackets was derived by KRAMERS and exhibits the characteristic behavior as a function of Z, n, and ν. Note the proportionality to the occupation numbers, $\bar{N}_e^{(b,Z,n)}$, since a shell is effective only to the extent that it is populated.

The Gaunt factor, g_{bf}, is a correction for quantum-mechanical effects and is a slowly varying function of frequency. These factors were recently tabulated by KARZAS and LATTER (1961) on the assumption that the nucleus exerts a Coulomb force.

The correction, q_f, takes into account the fact that the band of free states into which the electron is to go when liberated from the atom may already be occupied. Its value is given by eq. (21).

The contribution of each edge in a mixture of element to the bound-free absorption must be weighted by the abundance of the element. The result can be written in the form:

$$\varkappa_{bf}^{(Z,n)}(u) = \frac{N_Z}{\varrho V}\,a_{bf}^{(Z,n)}(u) = \frac{D_{bf}^{(Z,n)}(u)}{u^3}\,, \tag{57}$$

where

$$D_{bf}^{(Z,n)} = \begin{cases} 0\,, & u < u_n\,, \\ \dfrac{N_Z}{\varrho V}\,\dfrac{64\pi^4}{3\sqrt{3}}\,\dfrac{Z'^4}{n^5}\,\dfrac{m_e e^{10}}{c(hkT)^3}\,\bar{N}^{(b,Z,n}\,g_{bf}\,q_f\,, & u \geqslant u_n\,, \end{cases} \tag{58}$$

$D_{bf}^{(Z,n)}(u)$ is a discontinuous function, jumping from the value zero for u less than u_n to the value given above.

This is the basis of Strömgren's method for calculating absorption coefficients. Consider the locations of all the absorption edges that contribute to the opacity of the mixture, and number them in order of increasing frequency: $u^{(1)}$, $u^{(2)}$, etc. Two successive edges, $u^{(i)}$ and $u^{(i+1)}$, define a wavelength interval. Although the absorption coefficient changes discontinuously at the boundaries of this interval, it is continuous in the interior.

The integral, eq. (51), is evaluated by summing the contributions of the strips between absorption edges. Within the strip bounded by the edges $u^{(i)}$ and $u^{(i+1)}$, the value of D_{bf} is a sum of contributions from all edges to the red of and including $u^{(i)}$ but none of the edges to the violet

$$D_{bf}(i) = \sum_{u_n < u^{(i)}} D_{bf}^{(Z,n)} . \tag{59}$$

The contribution of this strip to the integral is

$$\int_{u^{(i)}}^{u^{(i+1)}} \frac{W(u)}{D_{bf}(i)} \, du . \tag{60}$$

The D and W functions have been defined in such a way as to throw the u^3 frequency-dependence of the bound-free absorption coefficient into the weighting function, $W(u)$. Thus the only frequency-dependent part of D is the product $g_{bf} q_f$. It this is negligible, the evaluation of eq. (60) is simple, because D will then be constant in the interval from $u^{(i)}$ to $u^{(i+1)}$, and

$$\int_{u^{(i)}}^{u^{(i+1)}} \frac{W(u)}{D_{bf}(i)} \, du = \frac{1}{D_{bf}(i)} \left[S(u^{(i+1)}) - S(u^{(i)}) \right] , \tag{61}$$

where

$$S(u) = \int_0^u W(u) \, du , \tag{62}$$

is a tabulated function.

4.2. *Free-free absorption.* – Free-free absorption is due to the absorption of a photon by a free electron while in the vicinity of a heavy nucleus of effective charge Z', with the result that the electron is raised to a free state of greater energy. Unlike the bound-free case, there are no « edges ». The free-free absorption coefficient is a continuous function of frequency, again varying principally as ν^{-3}. The corresponding $D_{ff}(u)$ is therefore.

$$D_{ff} = \frac{N_e^{(f)}}{\varrho V} \frac{16\pi^2}{3\sqrt{3}} \frac{e^6 h^2}{c(2\pi m_e)^{\frac{3}{2}}} \frac{Z'^2}{(kT)^{\frac{7}{2}}} g_{ff} q_f . \tag{63}$$

KARZAS and LATTER (1961) have also tabulated free-free Gaunt factors as a function of the incident electron energy and have integrated them over a Maxwellian distribution in the free electron gas. More recent calculations have extended these computations to a degenerate gas and the latter computations are used in the program of Cox and Eilers.

The correction q_f is also necessary in this case, to account for the fact that the range of states to which the electron is to be raised may already be occupied.

The D factor for free-free absorption is added to the D factor for bound-free absorption in Strömgren's method. If the abundance of heavy elements is appreciable and the occupation numbers are high, the transparency is practically unaffected by free-free absorption since it is already very small due to the bound-free absorption. On the other hand, when the metal abundance is low, free-free transitions in the vicinity of hydrogen and helium ions can predominate over bound-free absorption.

4·3. *Bound-bound absorption.* – When an electron is in a bound state, it can be raised to another bound state by the absorption of a photon, but the range of energy over which this process is possible is quite narrow. In stellar atmospheres this effect produces a narrow absorption line, but in stellar interiors the effect has been neglected because a narrow line has virtually no effect on the Rosseland mean opacity.

MAYER analyzed this question and showed that lines could produce an appreciable effect under some circumstances. Nevertheless, no published tables of stellar opacities included this process until the recent work of COX and EILERS.

At high temperatures, lines will not be important because of the small number of bound electrons; but at low temperatures two effects contribute to their importance: one is the large number of lines that can be produced by all the bound energy levels of the material; the second is the extreme broadening produced by interactions.

Only the extreme wings of the lines are important in Rosseland mean calculations, and in most cases these can be represented by the so-called « damping form » of the absorption coefficient, where the line center is at u_c:

$$\varkappa_u \propto \frac{\gamma/4\pi}{(u-u_c)^2+(\gamma/4\pi)^2}\,. \tag{64}$$

COX and EILERS found that the opacities were not very sensitive to the choice of γ. The values they used were computed from a formula due to MEYEROTT and MOSZKOWSKI (1951) in all cases except for lines of hydrogen and ionized helium. In the latter case, absorption coefficients due to Griem based upon Stark broadening were used.

Lines become more and more important contributors to opacity as the temperature decreases, and at temperature below 10^6 °K the lines can increase the opacity by more than a factor of two (COX, private communication). Even with a large computer, the number of lines that can be included is limited; therefore, the evaluation of the contribution to the opacity due to lines is still somewhat uncertain.

If line absorption is to be included in the Rosseland mean calculation as a D function it should be noted that it does not have the inverse cube frequency-dependence that makes the D function so convenient for bound-bound or bound-free calculations; however, an appropriate D function can be defined by multiplying the strong-frequency-dependent line shape by u^3.

4·4. *Conduction.* – When the degeneracy of the electrons is appreciable, conductivity effects become important. The effect can be included as an « opacity » because the conductive flux is proportional to the temperature gradient just as the radiative flux is.

The details of conductivity calculations can be found in SCHATZMAN (1958) so they will be omitted here. The net effect of conductive opacity on the Rosseland mean is to decrease the opacity at high densities.

4·5. *Scattering.* – Thomson scattering by electrons, although intrinsically a small effect, can be important when the transparency due to other effects is high. The Thomson scattering coefficient is frequency-independent; therefore, in contrast to D_{bf} and D_{ff}, D' is not constant but goes as u^3.

The stimulated-emission correction is already included in the weighting function, $W(u)$; but for Thomson scattering this correction is not appropriate. It must therefore be removed by multiplying the scattering coefficient by $(1-e^{-u})^{-1}$.

At very high frequencies, the Thomson coefficient must be modified according to the Klein-Nhisina formula. Since under these circumstances there is a change in the frequency of the scattered photon due to Compton effect, scattering out of the beam at frequency ν is not compensated by scattering into the beam at the same frequency. SAMPSON (1959) has studied this effect and tabulated appropriate correction factors, $G(u)$.

SAMPSON has also pointed out that above the threshold for pair production, the electrons and positrons created contribute to the scattering process.

We get, finally, the appropriate D_s for the scattering process:

$$D_s = \frac{u^3}{1-\exp[-u]} \frac{8\pi e^4}{3m_e^2 c^4} \frac{N_e}{\varrho V} G(u) . \tag{65}$$

4·6. *Concluding remarks.* – From the complications of this section it should seem clear that it is utterly impractical to calculate opacities *ab initio* as required in the course of a stellar model integration. Instead, opacity calculations are a problem in themselves, to be solved for the particular densities, temperatures, and mixtures anticipated, and the results alone should be used in the model integrations.

In the past, simple interpolation formulae have been used to approximate the detailed results. These have the advantage of providing expressions that are relatively easy to compute as needed, but they are of very limited accuracy.

For the best results, the method advocated by COX should probably be used. In preparation for the computation of an evolutionary track, opacity tables should be computed for the anticipated range of density, temperature, and composition. These tables should then be stored in the computer for use during the computation of the models. With the large computers now available, sufficiently detailed tables can be stored so that only linear interpolation is necessary.

5. – Nuclear energy sources.

5·1. *Introduction.* – There is an adequate introduction to nuclear energy sources in WRUBEL (1958). Two types of processes are generally involved: processes in which two nuclei interact, and β-decays.

For the first type of process, it can be shown that the energy produced per gram per second depends upon the following kinds of quantities:

a) those that depend upon local condition alone:

ϱ density,
T temperature;

b) those that depend upon the reacting nuclei and local conditions:

x_1, x_2 mass fractions of reacting nuclei,
f screening factor;

c) nuclear constants:

Z_1, Z_2 nuclear charges,
A_1, A_2 nuclear masses;

d) functions that depend upon the specific reaction:

S_0 cross-section (keV barns),
Q energy released per reaction.

A discussion of screening factors would take us too far afield. We will merely note that SALPETER (1954) has given a technique for computing them.

Two quantities are of concern in the β-decay processes: the energy produced, and the lifetime of the parent nucleus. As is well known, part of the energy released during β-decay is carried off by neutrinos which do not interact with the stellar material and therefore do not contribute to the energy available to the star.

β-decay lifetimes were once considered to be the same in stellar interiors and in the laboratory. In an interesting series of papers, BAHCALL (1961, 1962) has shown that stellar β-decay lifetimes are influenced by the high ionization and degeneracy of the stellar gas. In the terrestrial laboratory, the ionization is low and bound states are fully occupied. Under these circumstances, β-decay occurs to continuum states. But when bound states are poorly populated, as is the case in stellar interiors, β-decay to bound orbits must be considered. Furthermore, when the continuum states are filled, as in the case of a degenerate gas, decay to continuum states may be blocked.

These differences between stellar and terrestrial rates play a small role in energy production calculations but they may be important in blocking and opening different paths in problems of nucleogenesis (BAHCALL and PETERSON (1962)).

5'2. *Reactions in the proton-proton chain.* – Below temperatures of about $18\cdot 10^6$ °K, the predominant energy source is the proton-proton chain. The energy production is proportional to the square of the mass fraction of hydrogen, and goes approximately as the fourth power of the temperature.

The first two reactions of the chain are

$$\begin{cases} {}^1\mathrm{H}(\mathrm{p},\beta^+\nu)^2\mathrm{D}\,, \\ {}^2\mathrm{D}(\mathrm{p},\gamma)^3\mathrm{He}\,. \end{cases} \tag{66}$$

The first is the slowest reaction in the entire chain and determines the rate at which the chain proceeds.

It was thought for several years that the chain is completed through

$$ {}^3\mathrm{He}({}^3\mathrm{He},2\mathrm{p})^4\mathrm{He}\,. \tag{67}$$

This step requires two ^{3}He nuclei, which means that the steps from ^{1}H to ^{3}He must be repeated twice before the last step can be carried through once. Experiments by HOLMGREN and JOHNSTON (1958) showed, however, that in the presence of a sufficient amount of ^{4}He, the process

$$ {}^3\mathrm{He}(\alpha,\gamma)^7\mathrm{Be} \tag{68}$$

is competitive. The ^{7}Be thus produced may go to ^{4}He either through reactions beginning with electron capture:

$$ {}^7\mathrm{Be}(\beta^-,\nu)\,{}^7\mathrm{Li}(\mathrm{p},\alpha)^4\mathrm{He} \tag{69}$$

or beginning with proton capture:

(70) $$^7\mathrm{Be}(p,\gamma)^8\mathrm{Be}(\beta^+\nu)^8\mathrm{Be}^*(\alpha)^4\mathrm{He}\,.$$

The fact that ^{3}He can thus react with ^{4}He implies a much more rapid ^{4}He production rate, since in this case the steps from ^{1}H to ^{3}He need only take place once to produce ^{4}He.

Since the $^3\mathrm{He}(\alpha,\gamma)$ reaction is more likely at higher ^{4}He concentrations, it is more effective in the present sun than the « primeval sun ». This reaction is also favored at high temperatures. FOWLER (1958) has represented the change in helium production due to the new path, by a factor φ which is 1 at low temperatures and goes to 2 at high temperatures.

The neutrino loss when the chain is completed through electron capture by ^{7}Be is markedly different from the alternate case of proton capture. In the first case the loss is only 4%; in the second it is 29%. Thus the actual amount of energy produced per ^{4}He nucleus depends upon which path is followed. Recent measurements by KAVANAGH (1960) indicate that the proton capture branch is important only at high temperatures; but at these temperatures the carbon cycle predominates unless the heavy element content is low. It therefore seems that the proton capture process is only important in Population II stars (low metal content) of high luminosity (FOWLER (1959)). For a solar model in which the possible branches are taken into account see SEARS (1959); and for explicit expressions for reactions rates see FOWLER (1959), IBEN and EHRMAN (1962) and BAHCALL (1962*c*).

5·3. *Carbon-nitrogen cycle reactions.* – Up-to-date values of the cross-sections, energies, and lifetimes of nuclei involved in these reactions may be found in FOWLER (1959).

It is worth noting that it seems reasonably well established that there is no resonance at stellar energies in the $^{14}\mathrm{N}(p,\gamma)^{15}\mathrm{O}$ reaction (HONSAKER (1960)) so that this reaction is the slowest in the cycle.

5·4. *Three α reaction.* – When the abundance of helium is large and the temperature is high, three α-particles can form carbon via the series of reactions (SALPETER (1957)):

(71) $$\begin{cases} 2\alpha + 94\ \mathrm{keV} \to {}^8\mathrm{Be}\,, \\ {}^8\mathrm{Be} + \alpha + 278\ \mathrm{keV} \to {}^{12}\mathrm{C}^*\,, \\ {}^{12}\mathrm{C}^* \to {}^{12}\mathrm{C} + 2\gamma + 7.282\ \mathrm{MeV}\,. \end{cases}$$

The first reaction is a result of the fact that two α-particles with 94 keV relative kinetic energy can form a ^{8}Be nucleus in the ground state. The inverse of the second reaction has been observed by COOK, FOWLER, LAURITSEN, and LAURITSEN, who established the energy of the excited state of ^{12}C as (7.653 ± 0.008) MeV. (It had been predicted by HOYLE to lie at about 7.70 MeV.) The excited state decays according to the last equation.

The net result is

$$3\alpha \to {}^{12}\mathrm{C} + 2\gamma + 7.282\ \mathrm{MeV}\,. \tag{72}$$

SALPETER has discussed the reactions in detail and obtained the following equation for the energy production per gram per second:

$$\varepsilon_{3\alpha} = A Y^{3} \varrho^{2} T^{-3} \exp[-B/T]\,, \tag{73}$$

where Y is the relative mass abundance of helium, eq. (38),

$$\left\{\begin{aligned} A &= 1.38\cdot 10^{35}\,\frac{\Gamma_{\pm e} + \Gamma_{\gamma}}{10^{-3}}\,,\\ B &= 43.2\cdot 10^{8}\ {}^{\circ}\mathrm{K}\,. \end{aligned}\right. \tag{74}$$

$\Gamma_{\pm e}$ is the partial width for decay to the ground state by emitting an electron-positron pair. Γ_{γ} is the partial width for γ emission. SALPETER suggests

$$\Gamma_{\pm e} + \Gamma_{\gamma} = 1\cdot 10^{-3}\ \mathrm{eV}\,. \tag{75}$$

SALPETER estimates that the overall rate of this reaction is uncertain by a factor of 30 either way.

These rates have been used by J. P. COX and R. T. GIULI (1961), to study pure helium stars.

6. – Numerical methods.

6·1. *Stellar evolution as a problem for an electronic computer.* – Since much of the current work in stellar evolution is done on electronic computers, it is appropriate to begin a survey of numerical methods with a brief introduction to the computer approach. From the point of view of a computing machine, a stellar model describing a star at a particular time, is a set of tables stored in the memory, giving the physical variables and chemical composition at discrete points in the stellar interior. The problem of stellar evolution is to compute, on the basis of tables describing an initial configuration, corresponding tables for future times.

The procedure for solving a problem on a computer must be precisely specified as a « program », giving the steps that must be followed to reach the answer, in the proper logical sequence. Large programs are frequently constructed of small programs, called « subroutines ». Each subroutine has a limited function: for example, one subroutine might compute the pressure, given information about the density, temperature and composition; another might compute the energy production by the carbon cycle, etc.

Recipes for solving problems of various general types are called « algorithms » and they must be chosen with care. Procedures that work in theory may be quite impractical because they require an excessive number of operations. The poor choice of an algorithm may increase the time necessary to compute a solution beyond practical limits.

A rapidly converging iterative procedure is one of the best kinds of algorithms for a computer. In methods of this general type, an approximate solution is improved by a definite sequence of operations. The program is executed repeatedly until two successive approximations differ by a negligible amount.

We will discuss two algorithms for solving the problem of stellar structure and evolution. The first, used for example by HASELGROVE and HOYLE (1959), SCHWARZSCHILD (1958), and many others, we will call the « stepwise method ». The second, developed by HENYEY (1961) and his collaborators, and independently in a different form by COX and BROWNLEE (1961), we will call the « shell method ». Each method is concerned with finding the solution to the problem at discrete space points and at discrete times.

6·2. *The stepwise method.* – We will discuss this method under the following limited condition especially suitable for this approach. We will assume that the nuclear source is the only significant term in the energy equation and that, consequently, time derivatives do not explicitly appear in the equations of equilibrium.

At the beginning of our computation of an equilibrium model, we will assume that we are provided with a table of composition as a function of position in the star. (In practice this will be the result of previous model computations in the evolutionary sequence.) This table, together with the boundary conditions, is all that is needed for the computation.

It will be somewhat easier if we describe the basic algorithm in terms of a simpler problem. Consider the differential equation

$$y' = f(x, y) . \tag{76}$$

Given the value of y at some point x_i (for brevity, we will write $y_i \equiv y(x_i)$), several classical methods of numerical analysis can be used to predict the value of y at the next point, x_{i+1}. The simplest of these is Euler's method in which the true curve $y(x)$ is replaced by a line segment having the same slope as the curve at the point x_i

$$\frac{y_{i+1} - y_i}{x_{i+1} - x_i} \equiv \frac{\Delta y}{\Delta x} \simeq \left(\frac{\mathrm{d}y}{\mathrm{d}x}\right)_{x=x_i} = f(x_i, y_i) , \tag{77}$$

or explicitly

$$y_{i+1} = y_i + hf(x_i, y_i) , \qquad h = x_{i+1} - x_i . \tag{78}$$

This procedure is easily programmed if a subroutine is available for computing $f(x, y)$.

Once we have y_{i+1}, we can apply the same procedure to obtain y_{i+2}. Successive applications of this algorithm permit us to proceed from the value of the function at some initial point x_0 stepwise to its value at any subsequent point. This method can therefore solve the classical « initial value problem »: given y_0, find y_j, $j = 1, 2, ..., n$.

Every technique for the numerical solution of differential equations suffers from two sources of error. The first is due to the fact that the differential equation itself is not solved, but only an approximation to it. In this case, the differential equation has been replaced by a difference equation. The inaccuracy, due to the fact that the differential equation is approximated to some order, is called « truncation error ». The second type of error is due to the fact that the computations can only be carried out with a finite number of significant figures. The last digit is usually « rounded », depending upon the magnitude of the first figure dropped and any actual calculation is subject to this « rounding error ».

Truncation error is proportional to a power of the step size, h, and is decreased by making the step size smaller. Rounding error accumulates with each step and gradually decreases the number of significant digits; therefore, it is wise to begin the computation with many more significant figures than are eventually to be used.

The truncation error of Euler's method is proportional to h^1. To proceed forward one step requires that we enter the subroutine for $f(x, y)$ once. Other methods are more efficient in the sense that N evaluations of the subroutine can achieve an error proportional to h^p where p is greater than 1.

Most widely used of these more effective methods is the Runge-Kutta method. To cross an interval, h, requires four evaluations of the subroutine;

but the truncation error of this process goes as h^4.

$$(79)\qquad \begin{cases} k_1 = f(x, y)h\,, \\ k_2 = f\left(x + \dfrac{h}{2},\quad y + \dfrac{k_1}{2}\right)h\,, \\ k_3 = f\left(x + \dfrac{h}{2},\quad y + \dfrac{k_2}{2}\right)h\,, \\ k_4 = f(x + h\,,\quad y + k_3)h\,, \\ \Delta y = \frac{1}{6}[k_1 + 2k_2 + 2k_3 + k_4]\,. \end{cases}$$

The extension of this technique to a system of first order equations is straightforward (SCARBOROUGH (1955)). A variation of this method with an even smaller error has been given by GILL (1951).

Now consider its application to the computation of an equilibrium stellar model, which is in reality a two-point boundary value problem—one pair of conditions at the center and another pair at the surface. To use the method outlined above, the problem is split into two initial-value problems: one beginning at the center and integrating outward, the other beginning at the surface and integrating inward.

For the outward integration, the center boundary conditions are imposed, together with trial values of P_c and T_c; for the inward integration the surface conditions are imposed together with trial values of L and R. (The actual choice of independent variables may vary from one region to another so that none of the derivatives become excessively large.)

The integrations are continued in both directions up to a selected meeting point where the physical variables are to be « fitted ». P, T, $M(r)$ and $L(r)$ must be continuous at this point, and the trial values of P_c, T_c, L and R must be adjusted to achieve a continuous fit. The details of this procedure have been given by SEARS (1958). « Deep » inward integrations are computed with trial values of L and R, and the behavior is watched in the $(U, n+1)$ plane; U and $n+1$ are dimensionless parameters defined as

$$(80)\qquad \begin{cases} U = \dfrac{\mathrm{d}\log M(r)}{\mathrm{d}\log r}\,, \\ n + 1 = \dfrac{\mathrm{d}\log P}{\mathrm{d}\log T}\,. \end{cases}$$

Most trial runs develop instabilities before reaching $U = 3$, which is the value it must have at the center. The nature of the instability indicates whether L or R needs adjustment (SEARS (1959)).

When an L and R are found with the proper behavior, six trial integrations—three inward and three outward—are run with the following trial values

$$(81)\qquad \begin{cases} P_c,\ T_c & L,\ R \\ P_c+\delta P_c,\ T_c & L+\delta L,\ R \\ P_c,\ T_c+\delta T_c & L,\ R+\delta R \end{cases}$$

where δ indicates a small variation.

At the interface between the inward and outward integrations the magnitudes of the jumps in the physics variables are noted. It is assumed that these jumps are linear functions of the variations δP_c, δT_c, δL, and δR; and the magnitudes of the corrections are determined such that the jumps will be reduced to zero. Two new integrations with the revised values are then run as a check. The entire fitting procedure can be made completely automatic on a sufficiently large computer.

Once a stable configuration is achieved for a particular time, t, it may be evolved to $t+\Delta t$. If we assume that the conditions in the interior remain sufficiently constant over the time interval, we can compute the rate of change of hydrogen to helium and, by solving eq. (6), we can compute the new hydrogen content at the end of the time interval. The new composition then replaces the old in storage and the process can be repeated.

The stepwise method has been very successful in model integrations and it is still widely used. Boundary conditions are easily applied and the interval between steps can be easily halved or doubled in regions where a finer mesh is needed or a coarser one is tolerable. Difficulties are encountered, however, when there are large and rapid readjustments between models. Problems of that kind seem to be handled more easily by the shell method.

6·3. *Numerical instabilities.* – Before proceeding to the shell method, we will discuss a numerical instability to which time-dependent integrations are subject.

The classic example of the kind of pathological condition that can develop is exhibited by the heat equation (Richtmyer (1957).) It is well known that the stability of numerical solutions of the heat equation depends upon the scheme used for replacing the differential equation by the difference equation.

Consider a grid of points in one space co-ordinate, x, and one time co-ordinate, t. We will use the notation y_j^n to denote the value of y at $x_0+j\,\Delta x$ at time $t_0+n\,\Delta t$, and we will speak of point j at time n.

If one considers the simple heat equation

$$\frac{\partial y}{\partial t} = \sigma \frac{\partial^2 y}{\partial t^2}\,, \tag{82}$$

with the initial condition

$$y(x, t_0)\,, \tag{83}$$

and boundary conditions

$$y(x_0, t) = C_1\,, \qquad y(x_J, t) = C_2\,, \tag{84}$$

and make the naive approximation

$$\frac{y_j^{n+1} - y_j^n}{\Delta t} = \frac{\sigma}{(\Delta x)^2}(y_{j-1}^n - 2y_j^n + y_{j+1}^n)\,, \tag{85}$$

one has a system that can be explicitly solved for y^{n+1} as a function of known values at time n. Unfortunately, however, this scheme is unstable and can lead to violently fluctuating errors. The initial tendency to reduce these errors by using a smaller interval in x only makes the situation worse.

This difference scheme was studied by Courant, Friedrichs, and Lewy (1928), who showed that stability can be achieved only if the mesh size in t and x are related such that

$$\frac{\sigma \Delta t}{(\Delta x)^2} \leqslant 1\,. \tag{86}$$

Other difference schemes are unconditionally stable, and thus present obvious advantages. One method which generally leads to stability is the use of an *implicit* formulation in that the space derivative is evaluated at the time $n+1$,

$$\frac{y_j^{n+1} - y_j^n}{\Delta t} = \sigma \frac{y_{j+1}^{n+1} - 2y_j^{n+1} + y_{j-1}^{n+1}}{(\Delta x)^2}\,. \tag{87}$$

These equations form an implicit linear system for the unknowns, y_j^{n+1}, and this is therefore called an « implicit method ». An alternative description is to note that the time derivative goes « backward » from the time at which the space derivative is evaluated. The properties of various difference schemes are described in detail by Richtmyer (1957).

6·4. *The shell method.* – We will consider the form of the shell method as developed and described in detail by HENYEY *et al.* (1959). Although HENYEY no longer uses the precise form described in that article, the general principles are retained. A related method is described in BROWNLEE and SEARS (1962).

In this approach the star is considered to be made up of a finite number of shells, J, numbered so that the center is at $j \simeq 0$ and the surface is at $j = J$. The locations of the interfaces between the shells are denoted by r_j. Each shell is assigned a mass which remains invariable throughout the evolution, but shells are not assigned equal masses, since in some regions a finer subdivision is needed than in others. Although in the course of evolution the positions of the interfaces change, interfaces can obviously never cross.

The mass interior to a sphere of radius r_j is denoted by $4\pi m_j$ and the energy flowing across interface j is denoted by $4\pi l_j$. The other variables (pressure, temperature, density, and composition) represent average values for a shell and it is convenient to evaluate them at some point intermediate between j and $j+1$ designated as $j+\frac{1}{2}$.

It is instructive to examine the transformation of at least one of the equations; for example, the pressure equation. If for the moment j is regarded as some suitably chosen parameter, and we assume that the mass distribution is given in terms of this parameter alone, then the differential equation for the pressure can be written in the form

$$\frac{\partial}{\partial j}\left(\frac{P}{T^3}\right) + \frac{3}{T}\left(\frac{P}{T^3}\right)\frac{\partial T}{\partial j} + \frac{4\pi G m}{r^4 T^3}\frac{\mathrm{d}m}{\mathrm{d}j} = 0\,, \tag{88}$$

where P/T^3 is used in preference to pressure alone because it is a more linear function of the subdivision variable.

The above equation can be transformed to a difference equation in the following way. The specific function $m(j)$ is assigned and the derivative $\partial/\partial j$ is replaced by the difference of the function at $j+\frac{1}{2}$ and $j-\frac{1}{2}$, where j adopts integral values only. Values of variables defined at half-integral points, but needed at integral points, are evaluated by simple averaging. Denoting $\mathrm{d}m/\mathrm{d}j$ by m'_j the result becomes

$$\left(\frac{P}{T^3}\right)_{j+\frac{1}{2}} - \left(\frac{P}{T^3}\right)_{j-\frac{1}{2}} + 3\,\frac{(P/T^3)_{j+\frac{1}{2}} + (P/T^3)_{j-\frac{1}{2}}}{T_{j+\frac{1}{2}} + T_{j-\frac{1}{2}}}\,(T_{j+\frac{1}{2}} - T_{j-\frac{1}{2}}) + \frac{32\pi G m_j m'_j}{r_j^4 (T_{j+\frac{1}{2}} + T_{j-\frac{1}{2}})^3} = 0\,. \tag{89}$$

One such equation must be satisfied at every interface j within the star. Note that at an interface, j, the equation involves physical variables at $j+\frac{1}{2}$ and $j-\frac{1}{2}$ but no points more distant from the point j.

Similar transformations can be carried out on all the equations of stellar structure. We will assume that a model is given at time n; the problem is to find a solution at time $n+1$.

Special note should be taken of the equations of energy production and composition change which contain explicit time derivations. The difference scheme chosen to replace these differential equations must be stable in the sense of the previous section. The particular choice made by HENYEY contains backward derivatives as follows for the gravitational terms in the energy equation:

$$-\left(\frac{\partial E}{\partial t} + P\frac{\partial V}{\partial t}\right), \tag{90}$$

becomes

$$-\frac{1}{\Delta t}\left[E^{n+1}_{j+\frac{1}{2}} - E^{n}_{j+\frac{1}{2}} + \tfrac{1}{2}\left(P^{n+1}_{j+\frac{1}{2}} + P^{n}_{j+\frac{1}{2}}\right)\left(V^{n+1}_{j+\frac{1}{2}} - V^{n}_{j+\frac{1}{2}}\right)\right], \tag{91}$$

where E denotes the internal energy of the gas.

The dependent variables of the problem are now

$$\left|\begin{array}{lllll} T_{\frac{1}{2}}, & T_{\frac{3}{2}}, & T_{\frac{5}{2}}, & \ldots, & T_{J-\frac{1}{2}} \\ V_{\frac{1}{2}}, & V_{\frac{3}{2}}, & V_{\frac{5}{2}}, & \ldots, & V_{J-\frac{1}{2}} \\ r_1, & r_2, & r_3, & \ldots, & r_J \\ l_1, & l_2, & l_3, & \ldots, & l_J \end{array}\right. \tag{92}$$

and the differential equations have been transformed into a system of *nonlinear* algebraic equations inter-relating this large number of unknowns. Fortunately, because the derivatives at a point have been approximated by using only closely neighboring values of the variables, no single equations contain more than a few of the unknowns. The task of finding a solution is, therefore, not as formidable as it first appears.

The problem is solved by an iterative procedure which is a generalization of the Newton-Raphson method; this is a familiar scheme for converging to the root of a function if the function can be differentiated. A brief outline of the procedure follows.

Suppose we want to find the value of x for which a known function, $y(x)$, vanishes. Newton's method is based upon the following geometrical concept. Select a test value of $x = x^{(0)}$, say. Evaluate $y(x^{(0)}) = y^{(0)}$ which in general will not be zero. (Of course, if $y^{(0)} = 0$, the problem has been solved: $x^{(0)}$ is the value we are seeking.)

If $y^{(0)}$ does not equal zero, consider the line tangent to the function at $x^{(0)}$, and passing through the point $(x^{(0)}, y^{(0)})$. This line intercepts the x-axis at

$$x = x^{(0)} - \frac{y(x^{(0)})}{y'(x^{(0)})} . \tag{93}$$

This intercept is chosen as the new trial value, $x^{(1)}$, and the process continues:

$$x^{(i+1)} = x^{(i)} - \frac{y(x^{(i)})}{y'(x^{(i)})} . \tag{94}$$

The root is thus approached along a saw-toothed path.

The most familiar use of this is the problem of finding the square root of a number, c. In the case we seek the solution of the equation

$$y = x^2 - c = 0 , \tag{95}$$

and the iteration scheme reduces to

$$x^{(i+1)} = \tfrac{1}{2}\left(x^{(i)} + \frac{c}{x^{(i)}}\right) . \tag{96}$$

This is a useful algorithm for an electronic computer because it can be programmed very simply and the process can be terminated when a pre-assigned accuracy is reached.

Alternatively, the Newton-Raphson technique can be regarded in the following way. Given a trial value, $x^{(0)}$; what correction, $\delta x^{(0)}$, whould be applied so that $f(x^{(0)} + \delta x^{(0)})$ will vanish. Expanding as the first two terms of a Taylor series, we find

$$f(x^{(0)} + \delta x^{(0)}) \simeq f(x^{(0)}) + \delta x^{(0)} f'(x^{(0)}) = 0 . \tag{97}$$

This procedure can be generalized to K nonlinear functions of K variables, $x_1, \ldots, x_K$, that must vanish simultaneously. Assume trial values of the variables, $x_1^{(0)}, x_2^{(0)}, \ldots, x_K^{(0)}$, and let $f_k^{(0)}$ and $(\partial f_k/\partial x_j)_0$ refer to evaluation of the functions and their derivatives at the trial point. Applying the Newton-Raphson scheme, we get a linear system to be solved for the corrections $\delta x_j^{(0)}$:

$$f_k^{(0)} + \sum_{j=1}^{K} \left(\frac{\partial f_k}{\partial x_j}\right)_0 \delta x_j^{(0)} = 0 \qquad k = 1, \ldots, K . \tag{98}$$

The next trial values of the variables are

$$\begin{cases} x_1^{(1)} = x_1^{(0)} + \delta x_1^{(0)} , \\ x_2^{(1)} = x_2^{(0)} + \delta x_2^{(0)} , \\ \text{etc.} . \end{cases} \tag{99}$$

Equation (98) can be rewritten as

$$AY = B \,, \tag{100}$$

where the vectors Y and B and the matrix A have the components

$$\begin{cases} y_j = \delta x_j \,, \\ b_k = -f_k \,, \\ a_k = \dfrac{\partial f_k}{\partial x_j} \,. \end{cases} \tag{101}$$

The application to the problem at hand is straightforward. The equations of the shell method may be regarded as a system of nonlinear relations that must vanish simultaneously. Using the procedure of the preceding paragraphs, therefore, these can be transformed into a linear system to be solved for the corrections to the physical variables δT, δV, δr, and δl, at integral or half-integral mesh points.

At the end point of the mesh, $j = 0$ and $j = J$, the variations are restricted by the boundary conditions of the problem. For example, at $j = 0$:

$$\begin{cases} \delta r_j = 0 \,, \\ \delta l_j = 0 \,. \end{cases} \tag{102}$$

The surface boundary condition is more difficult to apply but the procedure is described in detail in HENYEY *et al.* (1959).

As a consequence of the form of the nonlinear equations described above, each of the linear equations contains corrections for only a few of the variables. The resulting matrix of coefficients is zero except near the diagonal, where a regular pattern is repeated.

(103)

$\delta T_{j-\frac{1}{2}}$	$\delta V_{j-\frac{1}{2}}$	δr_j	δl_j	$\delta T_{j+\frac{1}{2}}$	$\delta V_{j+\frac{1}{2}}$	δr_{j+1}	δl_{j+1}	
—	—	—	0	—	—	0	0	
—	—	—	—	—	—	0	0	
0	0	—	0	0	—	—	0	
0	0	—	—	—	—	—	—	
				—	—	—	0	
				—	—	—	—	
				0	0	—	0	
				0	0	—	—	etc.

The problem has thus been reduced to solving the equation

$$AY = B \tag{104}$$

when the elements of the matrix A cluster about the diagonal.

Instead of discussing the solution of these equations in full generality we will discuss the simpler case of a matrix that arises in another boundary value problem. The generalization of this method is given in HENYEY *et al.* (1959).

Consider the following system of linear equations:

$$\gamma_{-1}y_{j-1} + \gamma_0 y_j + \gamma_{+1}y_{j+1} = f(x_j) \qquad j = 1, \ldots, J, \tag{105}$$

where γ_{-1}, γ_0, γ_{+1} are specified constants, and y_0 and y_J are given by boundary conditions. This can be expressed as the equation

$$AY = B \tag{106}$$

where Y is the $(J-1)$-dimensional vector

$$Y = \begin{pmatrix} Y_1 \\ \vdots \\ Y_{J-1} \end{pmatrix}, \tag{107}$$

B is the vector

$$B = \begin{pmatrix} f(x_1) - \gamma_{-1}y_0 \\ f(x_2) \\ \vdots \\ f(x_{J-2}) \\ f(x_{J-1}) - \gamma_{+1}y_J \end{pmatrix}, \tag{108}$$

and the matrix A is tridiagonal; that is, its elements are zero everywhere except for three elements on each row (except the first and last), symmetrical about the diagonal:

$$A = \begin{pmatrix} \gamma_0 & \gamma_{+1} & & & & & \\ \gamma_{-1} & \gamma_0 & \gamma_{+1} & & & & \\ & \gamma_{-1} & \gamma_0 & \gamma_{+1} & & & \\ & & & \cdot & & & \\ & & & & \cdot & & \\ & & & & & \cdot & \\ & & & \gamma_{-1} & \gamma_0 & \gamma_{+1} & \\ & & & & \gamma_{-1} & \gamma_0 & \gamma_{+1} \\ & & & & & \gamma_{-1} & \gamma_0 \end{pmatrix} \tag{109}$$

(A typical problem in which such a matrix arises is given in HENRICI (1962) from which this solution has been adapted.)

A practical procedure for solving this problem is a variation of the Gaussian elimination method. It is successful in this case because each row of the matrix has only a few nonzero elements.

Consider the first two equations:

$$(110)\qquad \begin{cases} \gamma_0\, y_1 + \gamma_{+1} y_2 = b_1 \\ \gamma_{-1} y_1 + \gamma_0\, y_2 + \gamma_{+1} y_3 = b_2 \end{cases}$$

The variable y_1 can be eliminated to give y_2 in terms of y_3. This relation can be used to eliminate y_2 in the next equation

$$(111)\qquad \gamma_{-1} y_2 + \gamma_0 y_3 + \gamma_{+1} y_4 = b_3$$

which can be solved for y_3 as a function of y_4. This can be used to eliminate y_3 in the next equation; and so on. But since each equation introduces a new variable, the best that can be done is to derive a relation for y_{J-1}, in terms of y_J until we come to the last equation.

No new variable is introduced in the last equation (because y_J is given by the boundary conditions). Therefore, at that point y_{J-1} can be explicitly determined.

Once y_{J-1} is known, y_{J-2} can be determined from the relation between them previously derived. Similarly, y_{J-3} can be found from y_{J-2}; and so on.

The practical computation can be carried out systematically by calculating the elements of auxiliary matrices and vectors, based on information obtained by moving downward through matrix A and vector B. When the last row is reached, the value of y_{J-1} can be determined, and the vector Y can be filled in backward.

This amounts to beginning at x_1, having applied the boundary condition at x_0, and proceeding to x_{J-1}, calculating the auxiliary functions along the way. At this point, the application of the boundary condition at x_J permits y_{J-1} to be determined, and subsequent values of Y can be constructed.

In the tridiagonal case the algorithm is as follows. The auxiliary functions are two matrices L and U and a vector Z (HENRICI (1962), p. 354).

$$u_{11} = a_{11}\,,$$

$$z_1 = b_1\,,$$

$$
(112) \quad \begin{cases} l_{j,j-1} = \dfrac{a_{j,j-1}}{u_{j-1,j-1}} \\ u_{jj} = a_{jj} - l_{j,j-1} a_{j-1} , \\ z_j \ = b_j \ - l_{j,j-1} z_{j-1} , \end{cases} \qquad j = 2, 3, \ldots, J-1 ,
$$

$$
y_{J-1} = \frac{z_{J-1}}{u_{J-1,J-1}} ,
$$

$$
y_j = \frac{z_j - a_{j,j+1} y_{j+1}}{u_{j,j}} , \qquad j = J-2, J-3, \ldots, 1 .
$$

The generalization of this technique to the problem involving the matrix in eq. (103) can be found in HENYEY *et al.* (1959).

Perhaps we should review the procedure since it is rather complex. The steps can be schematically outlined as follows:

1) replace the differential equations by a system of nonlinear difference equations;
2) transform to a set of linear equations for variations in the unknowns;
3) solve the linear system by the elimination technique.

The shell method has been used by both Henyey's group and COX and BROWNLEE (1961) to study the pre-main sequence contraction phases. It seems to run into difficulties when dealing with situations such as are presented by boundaries between convective and radiative zones since these boundaries are not likely to fall precisely at an interface between shells. Special procedures must be used to accommodate these conditions. Despite the difficulties, this type of procedure is very promising.

* * *

My thanks are due to Drs. A. N. COX, R. R. BROWNLEE, and R. L. SEARS for an opportunity to see their recent review articles prior to publication.

BIBLIOGRAPHY

BAHCALL, J. N.: 1961, *Phys. Rev.*, **124**, 495.
BAHCALL, J. N.: 1962*a*, *Phys. Rev.*, **126**, 1143.
BAHCALL, J. N.: 1962*b*, *Ap. J.* **136**, 445.
BAHCALL, J. N.: 1962*c*, *Phys. Rev.*, **128**, 1297.
BAHCALL, J. N. and PETERSON, V.: 1962, *A. J.* (in press).
BROWNLEE, R. R. and SEARS, R. L.: in *Stars and Stellar Systems*, Ed. G. P. KUIPER and B. M. MIDDLEHURST (Chicago), vol. VIII (in press).

CHANDRASEKHAR, S.: 1939, *An Introduction to the Study of Stellar Structure* (Chicago), (Reprinted New York, 1957).
COOK, C. C., FOWLER, W. A., LAURITSEN, C. C. and LAURITSEN, T.: 1957, *Phys. Rev.*, **107**, 508.
COURANT, R., FRIEDRICHS, K. O. and LEWY, H.: 1928, *Math. Ann.*, **100**, 32.
COX, A. N.: 1962, in *Stars and Stellar Systems*, Ed. G. P. KUIPER and B. M. MIDDLEHURST (Chicago), vol. VIII (in press).
COX, A. N. and BROWNLEE, R. R.: 1961, *Sky and Telescope*, **5**, 2.
COX, J. P. and GIULI, R. T., 1961, *Ap. J.*, **133**, 755.
FOWLER, W. A.: 1958, *Ap. J.*, **127**, 551.
FOWLER, W. A.: 1959, *Mem. Soc. R. Sci. Liege*, **16**, 207.
GILL, S.: 1951, *Proc Cam. Phil. Soc.*, **47**, 96.
HAYASHI, C.: 1961, *Pub. Astron. Soc. Japan*, **13**, 450.
HENRICI, P. : 1962, *Discrete Variable Methods in Ordinary Differential Equations* New York.
HENYEY, L. G., LELEVIER, R. and LEVEE, R. D.: 1955, *P.A.S.P.*, **67**, 154.
HENYEY, L. G., WILETS, L., BOHM, K. H., LELEVIER, R. and LEVEE, R. D.: 1959, *Ap. J.*, **129**, 628.
HOLMGREN, H. D. and JOHNSTON, R. L.: 1958, *Bull. Am. Phys. Soc.*, Ser. II, **2**, 377.
HONSAKER, J. L.: 1960, *Ap J.*, **132**, 517.
HOYLE, F. and HASELGROVE, C. B.: 1959, *M. N.*, **119**, 112.
IBEN, I., Jr. and EHRMAN, J. R.: 1962, *Ap. J.*, **135**, 770.
KARZAS, W. J. and LATTER, R.: 1961, *Ap. J. Suppl.*, **6**, 167.
KAVANAGH, R. W.: 1960, *Nuclear Phys.*, **15**, 411.
KOPPE, H.: 1948, *Ann. d. Phys.*, Series 6, **3**, 103.
MACDOUGALL, J. and STONER, E. C.: 1939, *Phil. Trans. Roy. Soc. London*, **237**, 67.
MAYER, H: 1947, Los Alamos Scientific Laboratory, report LA-647.
MEYEROTT, R. E. and MOSZKOWSKI, S. A.: 1951, Argonne National Laboratory, report ANL-4594
RICHTMYER, R. D.: 1957, *Difference Methods for Initial Value Problems* (New York).
RUDKJØBING, M.: 1947, *Pub. Copenhagen Obs.*, No. 145.
SALPETER, E. E.: 1954, *Aust. J. of Phys.*, **7**, 373.
SALPETER, E. E.: 1957, *Phys. Rev.*, **107**, 516.
SAMPSON, D. H.: 1959, *Ap. J.*, **129**, 734.
SANDAGE, A.: 1962, *Ap. J.* **135**, 349.
SCARBOROUGH, J. B.: 1955, *Numerical Mathematical Analysis* (Baltimore).
SCHATZMAN, E.: 1958, *Hdb. d. Phys.* (Berlin), **51**, 729.
SCHWARZSCHILD, M.: 1946, *Ap. J.*, **104**, 203.
SCHWARZSCHILD, M.: 1958, *Structure and Evolution of the Stars* (Princeton).
SEARS, R. L.: 1958, *Ph. D. Thesis*, Indiana University (Available through University Microfilms, Ann Arbor, Mich.).
SEARS, R. L.: 1960, *Mem. Soc. R. Sc. Liege*, **16**, 479.
TEMESVARY, S.: 1953, *Mem. Soc. Roy. des Sci. de Liege*, **13**, 122.
WOOLF, N. J.: 1962, *Ap. J.*, **135**, 644.
WRUBEL, M. H.: 1958, *Hdb. d. Phys.* (Berlin), **51**, 1.

Stellar Models.

J. Faulkner and K. Griffiths

Department of Applied Mathematics and Physics, Cambridge University - Cambridge

1. – Method.

During this seminar we shall discuss work recently carried out by Professor Hoyle, and ourselves at Cambridge. The computations referred to have been performed on an IBM 7090 electronic computer in London.

1·1. *The surface routine.* – Considerable uncertainty has come to light in recent years concerning the surface zones and their important effects on evolutionary tracks in the H-R diagram. Many workers have tried to use a mixing-length theory, but have found it necessary to introduce a parameter, the « efficiency factor », in order to obtain good models. There is also another uncertainty, the mixing length itself, often chosen to be a simple multiple of the scale height. This can be unsatisfactory in those ionization regions where the fluid elements, in travelling « a mixing length », lose their identity in a region where the scale height may have changed by a very large factor.

We have recently been working on a more dynamical, non mixing-length theory which, when fitted to complete stellar interiors, has given an excellent main sequence with the observed gradients, and close to that reported by Dr. Sandage. The theory has also been applied to the Hayashi problem and to sunspot phenomena, with interesting results in both situations.

The function of the surface routine is to determine the pressure, P, and radius $R_s - h$ at $T = 10^5$ °K. The photosphere is obtained by iterative solution of the equation

$$\left(\Theta + X + \frac{Y}{4}\right)\varkappa P_e = g\Theta\,,$$

where Θ/m_H = no. of free electrons per gram, and

$$\Theta = \frac{Xx}{1+x} + \frac{Y}{4}\,\frac{y(1+2z)}{1+y+yz} + A\sum_{i=1}^{4}\frac{c_i u_i}{1+u_i}\,.$$

The last term takes account of five metals with relatively low ionization potentials, A being the relative abundance by no. of Mg+Si to H+4He. Contributions of Al, Na and K are included with Al/(Mg+Si) = 0.1, etc., these ratios being the c_i, $i = 2, 3, 4$. x, y and the u_i refer to the first stages of ionization of the respective elements (*e.g.* $x =$ HII/HI), and $z =$ HeIII/HeII. The Saha equation, written in a slightly different form to that normally employed, then gives

$$x, y, z, u_i \sim \frac{T^{\frac{5}{2}}}{P_e} \exp\left[B - \frac{A}{T}\right].$$

The absorption coefficient, $\varkappa$, includes contributions from free electrons, H$^-$-ions, proton-electron recombination, electron-HeII and electron-HeIII. Care is taken both in this and Θ to use only the free hydrogen, the amount tied up in molecules being calculated in the usual way. However, this reduces $\varkappa$ some times considerably, without putting anything back to include effects of molecular absorption and Rayleigh scattering. A small correction term is added to $\varkappa$ to prevent the opacity from becoming too small in such cases. The resulting opacities have been checked against the tables of Cox and show good agreement.

To integrate inwards from the photosphere, differential equations are constructed with T as independent variable; the step in T is chosen so that $\Delta P_e/P_e \leqslant 0.4$, to prevent « skipping » of ionization zones. The variables integrated are

h, depth below photosphere.

δT, temperature difference between rising and falling elements,

V^2, where V is the velocity of the moving elements, and

P_e, the electron pressure.

P_e and T give Θ, and these three give ϱ; P_e and Θ give P.

The equations used have the following form:

$$\frac{dP_e}{dT} = \left(\frac{\alpha}{\beta}\right)_{rad} \frac{P_e}{T}, \qquad \text{where} \qquad \alpha_{rad} \sim \frac{1}{F - C} + \text{terms from } \frac{d\Theta}{dT},$$

α_{rad}, β_{rad}, α_{conv} and β_{conv} (see below) are large expressions, only α_{rad} containing C, the convective flux, obtained from certain considerations. The convective energy consists of the thermal and ionization differences between rising and falling material, so that $C \sim V\varrho\{E(T) - E(T - \delta T)\}$. However, one can set an upper limit, C_{II} to this by constructing an equation for dP_e/dT at constant entropy, which gives

$$\frac{dP_e}{dT} = \left(\frac{\alpha}{\beta}\right)_{conv} \frac{P_e}{T}.$$

This equation is not used for P_e directly, but sets an upper limit to C by putting

$$\left(\frac{\alpha}{\beta}\right)_{\text{rad}} = \left(\frac{\alpha}{\beta}\right)_{\text{conv}}.$$

As only α_{rad} contains C, this equation is easily solved to give C_{II}.

The main difference between this and former approaches now enters in the equations for δT and V^2. If the material were in violent convective motion, the temperature in descending material would remain essentially constant. When

$$\left(\frac{\alpha}{\beta}\right)_{\text{rad}} = \left(\frac{\alpha}{\beta}\right)_{\text{conv}},$$

we expect $\mathrm{d}\,\delta T/\mathrm{d}T$ to be zero. We therefore empirically take

$$\frac{\mathrm{d}\,\delta T}{\mathrm{d}T} = 1 - \frac{(\alpha/\beta)_{\text{rad}}}{(\alpha/\beta)_{\text{conv}}}.$$

Similar considerations suggest that when material is in violent convection conditions approach those of free fall. But, for conservation of mass, upward velocities must approximately equal the downward velocities, the upward velocities being driven by the pressure gradient which must operate to reduce the downward velocities. Thus when conditions are superadiabatic we are led to the equation $\mathrm{d}V^2/\mathrm{d}h = g$. Once again, we add a term to take account of the strength of the convection, so that

$$\frac{\mathrm{d}V^2}{\mathrm{d}h} = g\left(1 - \frac{(\alpha/\beta)_{\text{rad}}}{(\alpha/\beta)_{\text{conv}}}\right) = g\,\frac{\mathrm{d}\,\delta T}{\mathrm{d}T}.$$

Thus

$$\frac{\mathrm{d}V^2}{\mathrm{d}T} = g\,\frac{\mathrm{d}\,\delta T}{\mathrm{d}T}\,\frac{\mathrm{d}h}{\mathrm{d}T},$$

and we use essentially the usual equation for $\mathrm{d}h/\mathrm{d}T$, the reduced flux taking the place of the total flux.

The above differential equations for δT and V^2 are used when the right-hand sides are positive, until one reaches the maximum of $C(C_{\text{II}})$ from entropy requirements. Thereafter, the δT and V^2 are reduced by an iterative procedure which requires that the two ways of computing the convective flux should give the same result, *i.e.* by putting

$$C_{\text{II}} \sim V\varrho\left[E(T) - E(T - \delta T)\right],$$

and the reductions in V^2 and δT are made so that the equation

$$\frac{\mathrm{d}V^2}{\mathrm{d}T} = g\,\frac{\mathrm{d}\,\delta T}{\mathrm{d}T}\,\frac{\mathrm{d}h}{\mathrm{d}T}\,,$$

remains satisfied.

1·2. *The central stellar regions.* – The method used is essentially that described by Prof. WRUBEL as the step-wise method. The main difference is that a pressure grid is used with $\ln P$ as the independent variable, steps of 0.1 in $\ln P$ being taken. As at present the program provides for 450 such steps from the centre to $T_6 = 0.1$, but in the main sequence, only some 150 to 200 are needed. The interface selected in initial trial integrations is 10 steps from the centre.

One consequence of using $\ln P$ as the independent variable is that we must do seven integrations rather than six, for when a δP_c is made, the interface changes its position and thus the values produced at the interface by an inward integration will be functions of P_c.

One may wonder why the extra complications of this method should be thought worthwhile. Experience of evolutionary models has shown that in giants, a « gas bubble » effect is likely to arise, in which the pressure may change by a factor 2 or more for a 10^{-4} change in mass. All detail when integrating with respect to mass is then lost. It is hoped that the current way will enable observation of the dynamical passage of hydrogen through this region to join the helium core, as several steps will be inside the region.

New variables have also been used to obtain the initial main sequence. From the central developments given in Schwarzschild's book, and from an examination of his main-sequence models, the following variables were found to have an almost linear dependence:

$$\ln P,\quad \ln T,\quad m,\quad r\quad \text{and}\quad q\,,$$

where:

$$m = \mu M^{\frac{2}{3}},\qquad r = \lambda R^2,\qquad q = \nu Q^{\frac{1}{6}} R^{\frac{3}{2}}.$$

In terms of these, the equations take the simple form:

$$\frac{\mathrm{d}m}{\mathrm{d}\ln P} = -P\left(\frac{r}{m}\right)^2, \qquad \left(-\frac{P}{\varrho^{\frac{4}{3}}}\right)$$

$$\frac{\mathrm{d}r}{\mathrm{d}\ln P} = -\frac{P}{\varrho}\left(\frac{r}{m}\right)^{\frac{3}{2}}, \qquad \left(-\frac{P}{\varrho^2}\right)$$

$$\frac{\mathrm{d}q}{\mathrm{d}\ln P} = -\frac{3}{4}\,\frac{P}{\varrho}\left(\frac{r}{m}\right)^{\frac{3}{2}}\left(\frac{q}{r}\right)\left[\varrho\varepsilon\left(\frac{r}{q}\right)^6 + 1\right], \qquad \left(-\frac{P}{\varrho^2}\,(3\varrho\varepsilon)^{\frac{1}{6}}\right),$$

$$\frac{\mathrm{d}\ln T}{\mathrm{d}\ln P} = C'\frac{P\varkappa}{T^4}\left(\frac{q}{r}\right)\left(\frac{r}{m}\right)^{\frac{3}{2}}, \qquad \left(\frac{3C'P\varkappa\varepsilon}{T^4}\right),$$

or

$$= \frac{\Gamma - 1}{\Gamma} \text{ (smallest being chosen)},$$

λ, μ, ν have been suitably chosen, Γ is the adiabatic exponent of Chandrasekhar, and C' is a numerical constant.

The quantities in brackets on the right are the central gradients, expressible exactly in terms of the central values of the physical variables.

In the physical routines, ϱ and $\varkappa$ are calculated by formulae slightly different to those of HOYLE and HASELGROVE, and giving somewhat better values. The nuclear reaction rates are taken from Prof. Fowler's latest data.

In using the surface routines, three sample surfaces are calculated with the basic r and q, and with changes in these quantities of about 1%. A further set of 3 is calculated if subsequent attempts should fall outside this range, interpolation being made for h and $\ln P$ within the range.

The resulting stars appear to be very good, and in particular a model for the zero-age Sun gives indications of obtaining the present radius $\pm 1\%$ at the present luminosity.

We have also begun to fit empirical formulae of the homology type suggested by Dr. SANDAGE, with the intention of determining the «pseudo-subdwarf» abundances. Such relationships are based on extremely simplified assumptions and it is not surprising that detailed integrations do not fit precise homology formulae. Where such a formula can be constructed, it does give similar trends in the variables, however, and the homology method may be considered useful for exhibiting the correct behaviour in small regions of the H-R diagram.

2. – Summary of results.

2·1. *Comparison with observed zero-age main sequence.* – Let us see how well our computed models agree with observation in the H-R Diagram ($\log L$, $\log T_e$). We may compare Sandage's standard zero-age sequence (SCHWARZSCHILD, 1958), represented by a continuous line l_0 in Fig. 1, with the theoretical track computed for homogeneous stars of varying mass, with composition $X = 0.675$, $Z = 0.015$, represented in the same figure by a broken line, l_t. X is the amount by weight of hydrogen and Z the amount by weight of elements heavier than helium. Although the agreement is not perfect, it must be considered very satisfactory, since nowhere does the separation between the curves exceed 0.5 mag. In addition, as we shall see in **2·2** one would not expect complete agreement, because of composition differences between observed stars and theoretical ones. We might note here that the composition $X = 0.675$, $Z = 0.015$ was taken as being representative of the solar composition.

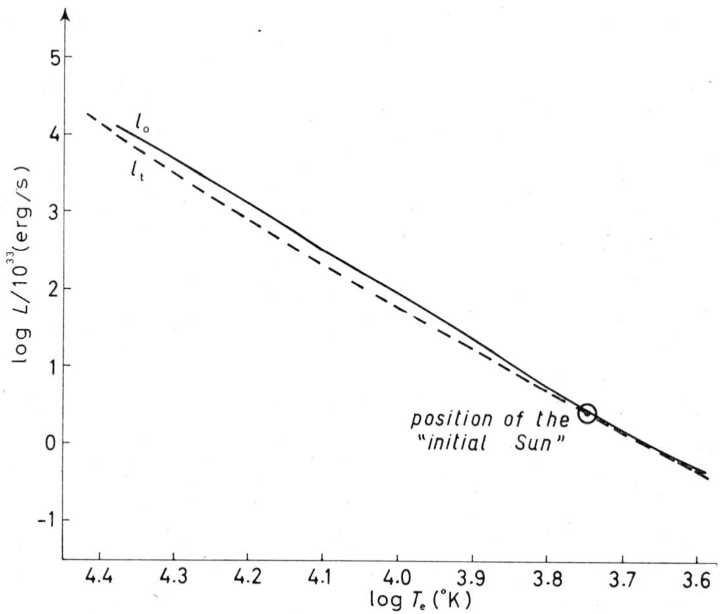

Fig. 1. – Comparison between observed (———) and theoretical (- - -) zero-age Main Sequence.

It is very interesting to compare the gradients of l_0 and l_t. l_0 is very near to a straight line of gradient 5.7 as we go down to log $T_e = 3.75$. At this point, which is near the «initial Sun», the slope falls to become about 5.3, at which it stays till the sequence ends at log $T_e = 3.6$. The theoretical track, l_t, starts high on the main-sequence with a slope of about 7, which decreases as the observed track is approached, and levels out at 5.5 between log $T_e = 4.3$ and log $T_e = 3.7$. Below this temperature, the slope decreases slightly to about 5.3. Thus we see that there is good agreement with l_0 below the position of the « initial Sun », but immediately above this region there is a slight difference in the slope. Once again, 2·2 will show that there need be no contradiction here. In passing, it is worth noting that homology arguments, using Kramer's opacity and the C-N cycle energy generation law, which would be applicable in this region just above the initial Sun, predict a slope of 5.6.

2·2. *Effect of varying composition.* – Besides the track at $X = 0.675$, $Z = 0.015$, several other sequences have been computed for different X and Z. Whilst computational difficulties have prevented the completion of some of the tracks, it has still been possible to derive some information on how a composition change might alter a star's position in the H-R diagram. It is hoped that the difficulties referred to, which seem to be of a type met before and overcome, will soon be dealt with and the sequences extended to lower luminosities (*).

The effect of changing X is shown in Fig. 2, where sequences are drawn for $X = 0.325$, 0.675, 0.985, with $Z = 0.015$ in each case. Clearly, an increase in X raises the curve, and a decrease lowers it, in agreement with the simple homology requirements:

$$(1) \qquad L \propto X^{1.15} Z^{0.55} T_e^{5.6} ,$$

for Kramer's opacity and C-N cycle energy generation.

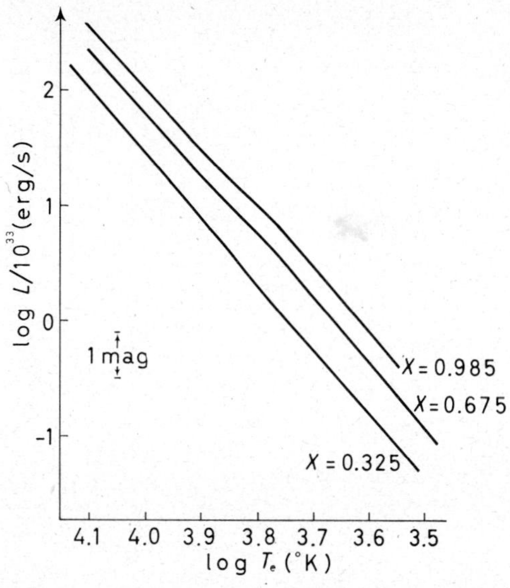

Fig. 2. – Sequences at constant Z (= 0.015).

The displacement is more or less a parallel shift and is considerable. It may be up to 1.5 mag at some temperatures, which is far greater than the observed spread of the main sequence, when line-blanketing corrections have been made. Clearly, reasoning from equation (1) if stars exist with varying hydrogen content, their heavy-element content must vary too in some compensatory fashion, if they are still to lie on the main sequence.

In fact, if a star wishes to have a greater X whilst keeping on the same sequence, it must decrease its Z. Figure 3 shows the results of runs made at $X = 0.985$, with $Z = 0.015$ and 0.0015. The decrease of Z does in fact lower the curve, in agreement with equation (1) but not however by a constant amount. Whereas the slope of $X = 0.985$, $Z = 0.015$ was about 5.5, the slope of $X = 0.985$, $Z = 0.0015$ is about 4.4. At $\log T_e = 3.8$, there is an indication that the $Z = 0.0015$ curve is about to become parallel to the $Z = 0.015$ one. Unfortunately at this point the computer refused to find the next solution down the sequence. On a run at $X = 0.985$, $Z = 0.00015$, an exactly similar situation was encountered; once again, the track seemed to be on the point

(*) *Note added in proofs.* – The difficulties referred to were caused by cases of marginal surface convection in trial integrations. These have been overcome, and a full discussion of methods and results for main sequences of some ten compositions will appear in M.N.R.A.S.

of bending towards the track of $X = 0.985$, $Z = 0.015$, when the run stopped. This track was just below $Z = 0.0015$ and parallel to it. However if one compares the tracks of HASELGROVE and HOYLE (1959) for their type IIa stars which are very similar in abundance to our $X = 0.985$, $Z = 0.0015$ stars, one sees that it does exhibit this parallel effect below $\log T_e = 3.8$. We may therefore very tentatively extrapolate our track as shown in Fig. 3.

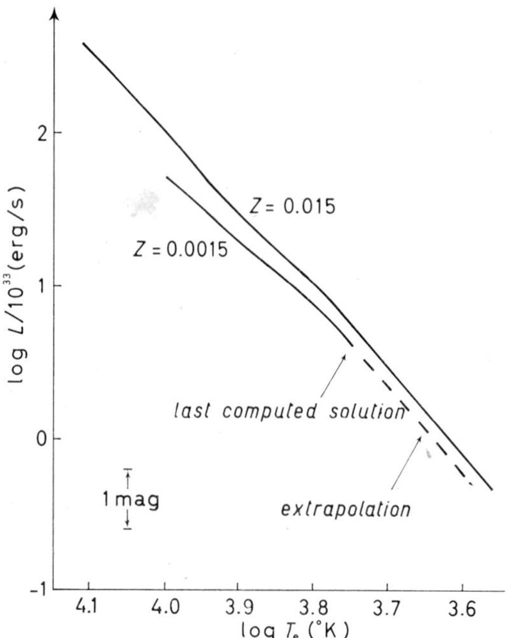

Fig. 3. – Sequences at constant $X(=0.985)$.

Less extensive tracks were obtained for $X = 0.675$, $Z = 0.0015$; $X = 0.675$, $Z = 0.00015$; $X = 0.325$, $Z = 0.0015$. Each of these has a slope of about 4.4 but none was carried to a low enough luminosity for the bend to appear. Once again, the $Z = 0.00015$ curve was everywhere close to and below the corresponding $Z = 0.0015$ curve.

We would thus deduce that for any X, decreasing Z from 0.015 will lower the sequence, and for $\log T_e \geqslant 3.8$ the slope is decreased too. Homology shows one the effect of changing composition on the position of the main sequence but as yet we are not able to understand why a decrease of the heavy-element content should decrease its slope.

Now we can see that the disagreement between l_0 and l_t (2·1) is not unexpected. Composition differences between the observed stars which make up l_0 — the Hyades, the Pleiades and NGC 2362 — and our theoretical ones would produce the sideways displacement which is evident in Fig. 1, and even a difference in slope in parts. Moreover, the composition differences between the clusters must make one wary of assigning complete reality to l_0 as the unique zero-age main sequence.

Following a suggestion of Dr. SANDAGE, we have in the past week been attempting to fit our curves to a homology relation of the type

$$L \propto X^a (Z + \alpha)^b T_e^c . \tag{2}$$

If such a relation could be established it would enable one for instance to deduce X for a given sequence which fell on the observed main sequence but had a Z of, say, one hundredth of the normal metal abundance. However for different T_e, the quantities a, b, α and c vary so widely that it is impossible to give values satisfactory in all situations. It should be possible, though, to obtain values good in restricted ranges of T_e. Work on this is proceeding.

2'3. *Mixed stars.* – Two tracks have been obtained for stars of constant mass and Z, but with varying X. Such tracks may be interpreted as giving the evolution of stars which, for some reason or other, mix quickly as their hydrogen is depleted. For a star of four solar masses, the evolution was found to be upwards along a line of gradient 4.2 — *i.e.*, close to the main sequence but below it. As X decreases from 0.675 to 0.325, the star increases by 2 mag and is then 0.75 mag below the main sequence. By extrapolation of our present results we may say that a further increase of 2 mag is suggested before the hydrogen is completely depleted, when the star will be 1.4 mag below the main sequence. Now, when the observed main sequence has been corrected for line-blanketing, its width is about 0.25 mag. Of course, in the later stages of hydrogen depletion, the star many not behave in the way we have assumed in making our extrapolation, but nevertheless for most of its life it will lie below the main sequence and be easily differentiated from a main-sequence star.

2'4. *Mass-luminosity relation.* – Plots have been drawn of the mass-luminosity relation for the sequences $X = 0.675$, $Z = 0.015$ and $X = 0.985$,

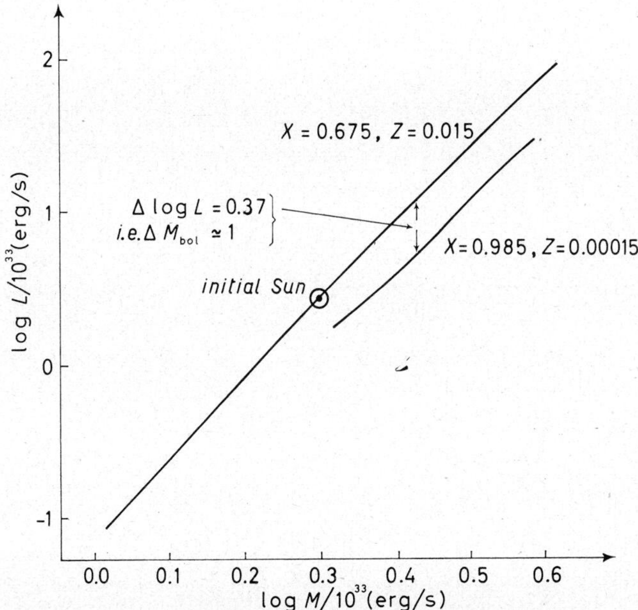

Fig. 4. – Comparison of mass-luminosity curves.

$Z = 0.00015$. These sequences, because of the compensatory effect of increasing X and decreasing Z, are approximately coincident in the H-R diagram. However, since the (M, L) homology relation has different exponents of X and Z from the exponents of X and Z in the (T_e, L) relation, one would not expect the curves to coincide in the (M, L) diagram. Indeed, they do not, differing by 1 mag on average (see Fig. 4).

2'5. *Comparison with present-day Sun.* – Since we have used what are believed to be the relative abundances of X, Y and Z in the atmosphere of the present-day Sun, and therefore throughout the « initial Sun » if no appreciable mixing has taken place, a good check on our computations should be a comparison with observed values for the Sun. Of course, we must take into account the effect of evolution on the Sun. Using Hoyle's evolutionary track, we may compute the time the Sun must have taken to reach its present luminosity of $3.86 \cdot 10^{33}$ erg/s, starting at $2.79 \cdot 10^{33}$ erg/s, our computed value for a star of mass $1.982 \cdot 10^{33}$ g, very close to the mass of the Sun. This age comes out at $4.5 \cdot 10^9$ years, in remarkably close agreement with the geologically determined age of the earth, $(4.55 \pm 0.05) \cdot 10^9$ years.

Once again, using Hoyle's tracks, one is able to compute the factor by which the radius of his solar model changed in $4.5 \cdot 10^9$ years. This factor is 1.088. Now, our « initial Sun » has a radius of $.6388 \cdot 10^{11}$ cm. If one assumes that the effect of evolution would be to cause an increase by this factor of 1.088 one obtains a value for the present-day radius of $0.6388 \cdot 1.088 = 0.6950 \cdot 10^{11}$ cm. The observed value is $0.6951 \cdot 10^{11}$ cm! This very close agreement must be considered a little coincidental, however; to obtain a stricter value of the present radius, we hope to carry out an evolutionary run for the Sun, starting with our new values of radius and luminosity. However, it seems clear that the effect of introducing the new routine to discover the boundary condition at the surface of a star has been to improve the accuracy of our models by a considerable amount.

REFERENCES

HASELGROVE, C. B. and HOYLE, F.: 1959, *M.N.R.A.S.*, **119**, 112.

SCHWARZSCHILD, M.: 1958, *Structure and Evolution of the Stars*, Princeton.

Nuclear Astrophysics (*).

G. BURBIDGE

University of California - San Diego, Cal.

1. – Introduction.

Applications of nuclear physics in astronomy fall under five headings:

a) Energy generation in stars.

b) Nucleosynthesis in stars.

c) Age determinations in astronomy using radioactive nuclei.

d) Matter at high densities.

e) High-energy nuclear physics in astrophysical circumstances.

Topics *a*) and *b*) are intimately related, but *a*) has been the subject of review separately in this series by SALPETER (1953) and we shall only summarize rather briefly this aspect of the subject at the beginning of *b*). The major developments in the field come under heading *b*). Apart from the pioneering paper of RUTHERFORD (1929), all of the modern developments under *c*) are closely related to *b*), but there has been such an outgrowth of ideas and speculation in this field that we have thought it worth-while to discuss it separately. There has also been a considerable amount of interest in recent years in topics which come under heading *d*). The necessity for some discussion of the topics in *e*) has come because of modern developments in the theories of the origin of cosmic rays and of nonthermal radio emission. Some of the most baffling problems of modern astrophysics are encountered here.

(*) This paper was prepared by Dr. G. BURBIDGE for the *Annual Review of Nuclear Science* for 1962. As he based his lectures on it, the full paper is printed here without additions or suppressions through the kindness of the Editors of the *Annual Review of Nuclear Science*.

2. – Synthesis of the elements.

2·1. *Historical summary.* – The last review of nuclear astrophysics by CAMERON (1958 *a*) gave some account of the early work and described the major developments up to 1958. In early attempts to describe the conditions in which the elements could be built, their place of origin was assumed to be either at a very early stage in the expansion of the universe, or in stars of unknown types (GAMOW 1953, ALPHER and HERMAN 1950, CHANDRASEKHAR and HENRICH 1942, KLEIN 1947, BESKOV and TREFFENBURG 1947, ATKINSON and HOUTERMANS 1928, ATKINSON 1931).

It is of interest to note that after the discovery that the reaction $3\alpha \rightarrow {}^{12}C$ would enable the difficulties in element building at mass 5 and mass 8 to be overcome, HAYASHI and NISHIDA (1956) attempted to modify the Ylem theory in an arbitrary fashion so that this reaction could be used in the formation of the elements in the first stage of evolution of the universe. They concluded that the building up of light nuclei to A in the range 16 to 20 out of protons and neutrons was possible for densities of 10^6 to 10^7 g/cm³ and $T = 10^{10}$ degrees. By the time that this synthesis had been completed very few initial neutrons would be left and it was necessary to appeal to reactions such as (α, n) to build further.

The modern developments began with the fundamental work of BETHE and CRITCHFIELD (BETHE 1939, BETHE and CRITCHFIELD 1938) and VON WEIZSÄCKER (1938) on hydrogen-burning in stars and the later work of SALPETER (1952) which showed that helium-burning could take place under the conditions found in stellar interiors. Extensive work on nucleosynthesis in stars was then carried out in the period 1954-1957 culminating in a number of research papers and an article attempting to delineate all aspects of the nucleosynthesis theory by Burbidge, Burbidge, Fowler, and Hoyle (HOYLE 1954, FOWLER, BURBIDGE and BURBIDGE 1955 *a* and *b*, BURBIDGE, BURBIDGE, FOWLER and HOYLE 1957), and a number of research papers and reports by CAMERON (1954, 1955, 1957, 1958 *b*). In describing more recent work it will be convenient to use the method of presentation adopted by BURBIDGE, BURBIDGE, FOWLER and HOYLE (1957) which we shall refer to as B²FH. We begin by summarizing briefly the different types of nuclear processes which are required to build all of the elements starting with hydrogen in stars. Then, after considering the present status of modern results on stellar evolution as they affect nucleosynthesis, we shall describe the most recent developments relating to each of the processes described in B²FH. At the end of this section we shall discuss the evidence for element synthesis in stars obtained from astronomical observations.

Eight separate types of nuclear process are required to build all of the elements in stars. They were described by B²FH as follows:

1) *Hydrogen-burning* either through the p-p chain, or through the CNO cycle, which convert hydrogen to helium. If nuclei produced in helium-burning are already present this will produce nuclei in the range $A=12$ to $A=23$.

2) *Helium-burning* through the Salpeter reaction $3\alpha \rightarrow {}^{12}C$, which converts helium to carbon, and then the successive addition of α-particles to ${}^{12}C$, which was thought to produce ${}^{16}O$ and ${}^{20}Ne$ under similar conditions of temperature and density as the 3α reaction.

3) The *α-process*, which by releasing α-particles by reactions of the type ${}^{20}Ne(\gamma,\alpha){}^{16}O$ would provide a source of α-particles which could be captured by ${}^{20}Ne$ to form successively ${}^{24}Mg$, ${}^{28}Si$ etc., up to nuclei in the range $A = 40$ to 48

4) The *equilibrium process* which builds the elements in the iron peak of the abundance curve; *i.e.*, this accounts for the range in A from 50 to 60.

The processes all occur through a temperature sequence ranging from the lowest temperature at which the p-p chain goes only to ${}^{3}He$, $8\cdot10^{6}$ degrees, up to about $4\cdot10^{9}$ degrees where the equilibrium process is effective. In addition to these, two processes involving chains of neutron-capture are required. These are:

5) The *s-process* which is a neutron-capture chain taking place on a time scale long enough so that the β-active nuclei in the capture chain are able to decay before further neutron-capture occurs. This chain will build many of the nuclei up to ${}^{209}Bi$.

6) The *r-process* which is a neutron-capture chain on a short time scale less than one hundred seconds, in which the neutron-rich heavy elements uranium, thorium, and nuclei with atomic weights up to ~ 270 will be built.

Finally two other processes are required. They are:

7) The *p-process* which is required to produce some of the low-abundance heavy isotopes which are proton-rich. This was thought to be due to reactions mainly of the (p,γ) and (γ,n) types on already existing heavy isotopes in a high-temperature ($\sim 10^{9}$ degrees) medium.

8) The *x-process* which is required to produce deuterium, Li, Be, and B which are highly perishable in stellar interiors.

In discussing the places of origin of these processes it is necessary to establish just how far our understanding of the evolutionary tracks of stars has gone, since many different phases of evolution of stars over a wide mass spectrum are probably required. These questions were discussed in detail by B^2FH and also in a review of stellar evolution written a little later than this by

BURBIDGE and BURBIDGE (1958 *b*). Some developments have been made since then and we describe next the present situation in this field.

2·2. *Stellar evolution as it relates to nucleosynthesis.* – After stars reach the main sequence, they spend most of their lives there, the conversion of hydrogen to helium taking place through the p-p chain on the lower part of the main sequence or the CNO cycle on the upper part. After a certain fraction of mass has been converted to helium in the core (this fraction varies from about 10% for stars with masses slightly greater than that of the sun to about 60% for massive stars) the stars will tend to move to the right of the main sequence in the Hertzsprung-Russell diagram, their cores contracting and the envelopes expanding until they reach the vicinity of the giant branch in the H-R diagram. Such an evolutionary track for stars of masses slightly greater than that of the sun is well established on theoretical gounds [cf. SCHWARZSCHILD (1958)]. The calculations of HOYLE and HASELGROVE (HOYLE 1953, HASELGROVE and HOYLE 1959) have been carried out with compositions thought appropriate for globular cluster and galactic cluster stars, for stellar masses 1.09, 1.16, and 1.35 $M_\odot$. They have been used to fit to the observed color-magnitude diagrams with a view to determining ages and other parameters. For stars with masses very much greater than that of the sun the beginning stages of evolution have been computed by SCHWARZSCHILD and HÄRM (1958) and these calculations show that such stars move off the main sequence to the right after a large fraction of the hydrogen has been burnt.

For stars with masses less than the sun no theoretical evolutionary tracks have yet been computed. For stars with masses 0.5 $M_\odot$ or less very deep convective zones exist (OSTERBROCK 1953, LIMBER 1958). Consequently it appears that a large fraction of the star is mixed, and a combination of this fact together with the very slow rate of conversion of hydrogen to helium suggests that the star will evolve very slowly as far as its chemical composition is concerned [for a completely mixed star the evolutionary path will presumably follow the track computed long ago by GAMOW (1939)], and within the time scale commonly associated with our Galaxy ($\sim$1 to $2\cdot10^{10}$ years) such stars will not play a significant role in nucleosynthesis.

It is clear, from the models of stars of 1.1 to 1.2 $M_\odot$ computed by SCHWARZSCHILD, HASELGROVE and HOYLE, that as the stars move into the red-giant region of the H-R diagram the conditions for helium-burning, *i.e.* a temperature between 1 and $2\cdot10^8$ degrees and densities of the order of 10^5 g/cm^3 are reached. In some work carried out since 1958 attempts have been made to take into account the helium-burning phase and relate the models to observed stars. SCHWARZSCHILD and HÄRM (1961 and in press) have followed the onset of helium-burning at a temperature of about $8\cdot10^7$ degrees in a red giant of mass 1.2 $M_\odot$. Because the zone is highly degenerate the new energy source causes heating and a thermal runaway ensues. This is terminated when the core becomes nondegenerate and subsequent helium-burning causes expansion and cooling. During this helium flash the temperature reaches about $3 \cdot 10^8$ degrees. The overall luminosity of the star hardly increases, because the bulk of the energy generated is absorbed in the deep interior. While this phase of the evolution is taking place the star may move back down the giant branch. It is also possible that mixing of the helium core and outer envelope will occur so that the star will move rapidly to the left-hand end of the H-R diagram.

Some work on evolution of globular cluster stars ($1.2\,M_{\odot}$) in the helium-burning phase by NISHIDA and SUGIMOTO (NISHIDA 1960, NISHIDA and SUGIMOTO (1962) shows that during this phase the stars stay in the region of the horizontal branch well within the span of the *RR Lyrae* gap and in a time scale of the order of $3 \cdot 10^7$ years. This latter time is a time of helium-burn-out with an assumed value for the helium fraction in the range 0.45 to 0.54. Such calculations which shift the models from the region of the giant branch to places near, or to the left, of the *RR Lyrae* gap are exceedingly sensitive to the rates of both hydrogen- and helium-burning. A similar study of the helium-burning phase for a star of $4\,M_{\odot}$ up to the point at which the helium core is burned out has also been carried out by HAYASHI, NISHIDA and SUGIMOTO (1961), and in this paper some attempt is made to correlate the evolutionary tracks with a globular cluster (NGC 458) in the Small Magellanic Cloud.

HAYASHI and his collaborators (SAKASHITA, ONO and HAYASHI 1959, HAYASHI, JUGAKU and NISHIDA 1959, HAYASHI and CAMERON 1962) have considered the evolution of stars of mass $15.6\,M_{\odot}$ starting on the main sequence where the star is burning hydrogen in a convective core. The star then exhausts hydrogen in the core which passes through a stage of gravitational contraction and then begins to burn helium in the core while hydrogen burns in the surrounding shell. Attempts were made to relate the computed evolutionary track with the positions of the stars in the color-magnitude diagram for *h* and *χ Persei*. However, the results are uncertain because of the earlier uncertainty about the energy generation rates in helium-burning and also because the assumed He/H ratio which they took is incompatible with the observations of stars of this mass existing at this time. Furthermore, an important factor which has been omitted is mass loss, which probably occurs and may well determine the rate of evolution in massive stars after they leave the main sequence.

In all of the investigations which have been described so far it has been supposed that the stars originate on a normal main sequence and have only moved into the helium-burning phase after the hydrogen in their cores has been exhausted. However, some work has been attempted in which it is supposed that the stars are composed of pure helium and that this phase may be related to the ejection of outer envelopes which contain hydrogen (BUR-

BIDGE and BURBIDGE 1958 *b*). The models can be divided into two groups:

a) Those in which the star is composed of pure helium (CRAWFORD 1953, COX and GIULI 1961).

b) Stars with helium cores which give rise to the energy generation by helium burning together with outer hydrogen envelopes which contribute negligibly to the energy production (COX and SALPETER 1961).

The models considered in *a*) give rise to a sequence which lies roughly parallel to the Population I main sequence, but is displaced towards higher effective temperatures for a given luminosity by a factor of about 4 relative to the hydrogen-burning main sequence. Stars which might approximate to such models are the hot subdwarfs, but there is little evidence that such a sequence exists in nature. However, the grafting on of an outer envelope as described in the models of type *b*) has the effect of increasing the radius by a factor of about 10. The models then lie in the H-R diagram in a place which is compatible with their representing in a qualitative fashion the nuclei of planetary nebulae, Also such models may be applied to hot subdwarfs such as HZ 44 (cf. section on observed compositions).

It is clear from this summary of the recent theoretical investigations that our detailed understanding of evolutionary tracks holds only as far as the hydrogen-burning and certain aspects of the helium-burning phases are concerned. Beyond this, when carbon-burning begins, and when processes subsequent to this become important, we can make arguments based only on very qualitative ideas concerning the sequence of evolution. As was described by B^2HF the successive stages of evolution in which the central temperatures steadily increase have characteristic time scales which speed up very rapidly. Consequently, in relating the processes of chemical evolution to the theoretical models and to the observations, we are faced with the difficulty that at the present time theoretical models are not available, and on the observational side there are very few stars which can be certainly identified with these later phases. Bearing these limitations in mind, we shall now consider the eight different nuclear processes which are required to build all of the elements from hydrogen in the stars.

2·3. *Hydrogen-burning and helium-burning.* – The situation as far as the reactions and energy generation rates in hydrogen- and helium-burning are concerned has been reviewed and continuously brought up to date by FOWLER and his collaborators (FOWLER 1954, 1958 *a* and *b*, 1959 *a* and *b*, BOSMAN-CRESPIN, FOWLER and HUMBLET 1954, and also B^2HF) and by SALPETER (1953, 1955, 1957). In the most recent papers of these authors (particularly B^2HF and FOWLER 1959 *a*), experimental and theoretical data for energy generation rates in the p-p chain and CNO cycle have been given. We mention here only a few of the most recent developments.

For the p-p chain the relatively large cross-section for the $^{3}He(\alpha,\gamma)^{7}Be$ reaction found by HOLMGREN and JOHNSON (1959) indicates that this reaction will lead to completion of the chain through $^{3}He(\alpha,\gamma)^{7}Be(e^{-},\nu)^{7}Li(p,\alpha)^{4}He$ or $^{3}He(\alpha,\gamma)^{7}Be(p,\gamma)^{8}B(\beta,\nu)^{8}Be^{*}(\alpha)^{4}He$ in stars which contain helium in comparable amounts with hydrogen and with $T>13\cdot10^{6}$ degrees. Since this mode of completion requires only one initial $^{1}H(p,\beta^{+}\nu)^{2}D$ reaction rather than the two required for completion by $^{3}He(^{3}He,2p)^{4}He$, the rate of production of ^{4}He is doubled at these higher temperatures. Following the experimental determination by KAVANAGH (1958) of the cross-section for $^{7}Be(p,\gamma)^{8}B$, FOWLER (1959) showed that $^{7}Be(p,\gamma)^{8}B$ can prevail over $^{7}Be(e^{-},\nu)^{7}Li$ only for $T\geqslant20\cdot10^{6}$ degrees. However, above about $20\cdot10^{6}$ degrees the CNO cycle will dominate over the p-p chain in all stars other than those with no carbon, nitrogen, or oxygen, so that this method of completion of the p-p chain can be neglected except for such stars.

REEVES (1959) has reconsidered the possibility that the p-p chain could be completed through $^{3}He(p,\gamma)^{4}Li(\beta^{+},\nu)^{4}He$. BETHE (1939) originally discussed this method of completion and stressed that it depended on ^{4}Li being stable against decay into $^{3}He+p$. There are strong theoretical arguments against the stability of ^{4}Li which have been summarized by FOWLER (1959 *a*), and an experiment by BASHKIN, KAVANAGH and PARKER (1959) to detect the production of ^{4}Li in $^{3}He(p,\gamma)^{4}Li$ failed by a factor 25 times lower than the minimum required if ^{4}Li were particle stable. Consequently it appears that this reaction still has no significance in the p-p chain.

The CNO cycle rates remain essentially as they were described by B^{2}FH and FOWLER (1959 *a*). Further experimental results have been discussed by FOWLER in this latter paper. From the experimental investigations of the compound nuclei of the reactions involving carbon, nitrogen, and oxygen with protons, it is now reasonably certain, except for $^{17}O(p,\alpha)^{14}N$, that resonances do not occur in the CNO cycle in the range 0 to 100 keV. CAUGHLAN and FOWLER (1962 preprint) have now derived effective cross-section factors $S(E)$ (which are defined by B^{2}FH) for the interaction of the CNO-cycle nuclei with protons and have calculated the mean lifetimes for ^{12}C, ^{13}C, ^{14}N, ^{15}N, ^{16}O and ^{17}O for the range of temperatures and densities of significance for main-sequence stars. An experimental study of the reaction $^{17}O(p,\alpha)^{14}N$ has been made by BROWN (1962). He has then calculated the $^{17}O/^{16}O$ ratio formed at equilibrium at various temperatures and has concluded that the terrestrial material was processed at a temperature of about $17\cdot10^{6}$ degrees.

The reaction rates for helium burning have been discussed by SALPETER (1957), REEVES and SALPETER (1959), and B^{2}FH. It was originally supposed that the helium-burning reactions $3\,^{4}He\rightarrow{}^{12}C$, $^{12}C(\alpha,\gamma)^{16}O$ and $^{16}O(\alpha,\gamma)^{20}Ne$

would all take place at the lowest temperatures at which helium burning occurs. In the case of $^{16}O(\alpha,\gamma)^{20}Ne$ the two levels which are of importance are the 4.969 and 5.631 MeV levels. However, it has now been established by Gove, Litherland, Clark and Ferguson (Gove, Litherland and Clark 1961, Gove, Litherland and Ferguson 1961) that the 4.969 MeV level of ^{20}Ne is inactive for low energy α-particle reactions because of its (2^-) character. Thus $^{16}O(\alpha,\gamma)^{20}Ne$ is not effective at the helium-burning stage. The next level at 5.631 MeV is (3^-) and hence it is active and ^{20}Ne can be formed through this level at temperatures $\gtrsim 5\cdot 10^8$ degrees.

2.4. *Carbon burning, oxygen burning and the α-process.* – It was originally pointed out that at a temperature of between 1 and $2\cdot 10^8$ degrees helium burning together with (α,γ) reactions would lead to the build-up of ^{12}C, ^{16}O, ^{20}Ne and perhaps a little ^{24}Mg. Following a further increase in temperature to about 10^9 degrees the first (γ,α) reaction to occur, $^{20}Ne(\gamma,\alpha)^{16}O$ would start to provide a source of α-particles and further build-up of nuclei into the region of ^{40}Ca would occur (B^2FH).

Recent calculations by Reeves and Salpeter (1959), Cameron (1959 *b*) and Hayashi, Nishida, Ohyama and Tsuda (1958 and 1959) and experimental results now lead to the following conclusions.

When the temperature has risen to $6\cdot 10^8$ degrees or greater, at a density of about 10^4 g/cm³ reactions between two ^{12}C nuclei will begin to occur at an appreciable rate. These reactions are $^{12}C(^{12}C,\gamma)^{24}Mg$, $^{12}C(^{12}C,p)^{23}Na$, $^{12}C(^{12}C,n)^{23}Mg$, $^{12}C(^{12}C,\alpha)^{20}Ne$, $^{12}C(^{12}C,^{8}Be)^{16}O$ and $^{12}C(^{12}C,2\alpha)^{16}O$ with Q values of 13.846, 2.230, -2.603, 4.619, -0.206, and 0.111 MeV respectively. After the bulk of the ^{12}C has burned up in this way, the remainder will react with nitrogen and oxygen. Thus the result so far will be the production of neon, sodium, magnesium, and oxygen, together with large fluxes of γ-rays, protons, neutrons, and α-particles. The neon, sodium, magnesium and oxygen will react with the nucleons and α-particles thus building many of the intermediate nuclei. The reaction times which are set by the temperature and density conditions lie between 1 and 10^6 years.

When the temperature rises to about $12\cdot 10^8$ degrees the $^{20}Ne(\gamma,\alpha)^{16}O$ reaction will become significant, and the α-particles released will be absorbed by ^{20}Ne, ^{23}Na, and ^{24}Mg, so that the net result will be the building up of ^{16}O, Mg, Al, and Si. At a temperature of about $14\cdot 10^8$ degrees the reactions $^{16}O(^{16}O,\alpha)^{28}Si$, $^{16}O(^{16}O,n)^{31}S$ and $^{16}O(^{16}O,p)^{31}P$ will begin to be appreciable. They all have approximately equal cross-sections. The α-particles, protons, and neutrons released in these reactions will undergo a complicated network of reactions similar to those following the $^{12}C+^{12}C$ reactions.

At about $16\cdot 10^8$ degrees the reactions $^{32}S(\gamma,\alpha)^{28}Si$ will occur. At temperatures of the order of $2\cdot 10^9$ degrees ^{24}Mg and ^{28}Si can be photodisintegrated to

^{16}O, followed by $^{16}O+^{16}O$ and further photodisintegrations. The net effect of this is to release α-particles and nucleons in the presence of heavier and heavier nuclei so that the approach to the build-up of the elements in the Fe peak is made. The reactions at this point are so profuse and the reaction times are so short that statistical methods can be used to calculate relative abundances that will result under specified conditions of temperature and density.

As was clear from the earlier discussion, stellar models in which this sequence of nuclear evolution occurs have not been computed. However, it seems probable that this sequence will occur in stars of masses slightly greater than the sun after they have left the giant region of the H-R diagram, whereas for more massive stars these phases will occur as the stars move off to the right from the main sequence. Since extensive measurements of the nuclear reactions between ^{12}C, ^{14}N, and ^{16}O and heavier nuclei have recently been carried out with the Chalk River tandem accelerator an attempt has been made by REEVES (1962) to estimate from this data energy production rates involving these nuclei. The rates for $^{12}C+^{12}C$ and $^{16}O+^{16}O$ for a range of temperatures covered in the work of REEVES and SALPETER (1959) are tabulated by REEVES, but they are uncertain by about an order of magnitude.

2·5. *The place of occurrence of the s-process.* – In the early work (CAMERON 1951, FOWLER, BURBIDGE and BURBIDGE 1955 *a*, B^2FH), it was argued that at the helium-burning stage in red-giant stars the neutron-producing reactions $^{13}C(\alpha,n)^{16}O$ and/or $^{21}Ne(\alpha,n)^{24}Mg$ which follow the products of helium-burning mixed in a hydrogen-burning zone or protons mixed in a helium-burning zone would occur, thus giving a source of neutrons which are required to build the *s*-process elements in the neutron capture chain starting with one of the abundant elements produced in another nuclear process. Difficulties associated with both of these neutron-producing reactions have been described. In the case of $^{13}C(\alpha,n)^{16}O$ these are the comparatively small amount of ^{13}C at equilibrium in a hydrogen-burning zone (and a ratio of $^{13}C/^{12}C$ considerably less than the equilibrium CNO-cycle value is now indicated in many carbon stars; cf. the discussion on observed compositions), the $^{13}C(p,\gamma)^{14}N$ reaction which removes ^{13}C, and the $^{14}N(n,p)^{14}C$ reaction which has a large cross-section and removes neutrons rapidly. It is possible to minimize some of these difficulties by supposing that the hydrogen from the envelope mixed into the helium-burning core and reacts with ^{12}C at such a rate that ^{14}N is produced, while at the same time ^{12}C mixed into the outer envelope captures the protons produced in the ^{14}N neutron-poisoning reaction $^{14}N(n,p)$ forming more ^{13}C. However, a full understanding of the mixing condition can only come from detailed stellar models and a physical theory of convective layers.

The production of neutrons by the $^{21}Ne(\alpha,n)$ reaction also requires that the production of ^{21}Ne by (p,γ) on ^{20}Ne takes place faster than its destruction by $^{21}Ne(p,\gamma)^{22}Na$ and that the neutron poison ^{14}N is depleted by $^{14}N(\alpha,\gamma)^{18}F$ before $^{21}Ne(\alpha,n)$ becomes effective. It was argued by B²FH that these conditions, together with the restrictions that the ^{21}Ne be produced before the hydrogen mixed with the ^{20}Ne was exhausted, and then be consumed by $^{21}Ne(\alpha,n)$ before helium is burned in the core, might well be fulfilled. This argument has now been further strengthened by the experimental investigation by TANNER (1962) of $^{20}Ne(p,\gamma)^{21}Na(\beta^+,\nu)^{21}Ne$. As was mentioned earlier, it now appears that ^{20}Ne will not be produced by $^{16}O(\alpha,\gamma)^{20}Ne$ in the lowest-temperature phase of helium burning. This is not a difficulty however, provided that ^{20}Ne was produced in a previous generation of stars, a view that has generally been taken by B²FH.

However, other neutron sources, and particularly those which produce neutrons at higher temperatures are under discussion. Thus, CAMERON (1959 *b*) has argued that in the midst of carbon-burning at a temperature near $7 \cdot 10^8$ degrees $^{12}C(p,\gamma)^{13}N(\beta^+,\nu)^{13}C(\alpha,n)^{16}O$ will give neutrons provided that the ^{13}N is not photodisintegrated in a time short compared with its β-decay time, which condition will be fulfilled for $T < (7.5 \text{ to } 8) \cdot 10^8$ degrees (REEVES and SALPETER 1959, CAMERON 1959 *b*). Alternatively REEVES and SALPETER (1959) have shown that the reactions in carbon-burning, and later in oxygen-burning, such as $^{12}C(^{12}C,n)^{23}Mg$ and $^{16}O(^{16}O,n)^{31}S$ will provide neutrons which in the case of carbon-burning, gives about 30 neutrons per iron nucleus. This will account for some synthesis of *s*-process elements beyond iron but with $A < 90$. Finally CAMERON (1962) has recently suggested that in a star having initially solar composition hydrogen-burning will first convert most of the carbon, nitrogen and oxygen to ^{14}N. This in helium-burning will be converted to ^{18}O and ^{22}Ne, and the principal source of neutrons will then be from $^{22}Ne(\alpha,n)^{25}Mg$ which is only slightly endoergic ($Q = -0.48$ MeV). This mechanism has not been considered in detail, but such neutron-producing reactions which might occur in the supernova stages of evolution were earlier described by B²FH.

To sum up, nuclei in the range $A = 12$ to $A = 19$ will be produced in helium- and hydrogen-burning, and it is probable that nuclei in the range from $A = 20$ to $A = 27$ will be produced in carbon-burning. Neon-burning will probably give rise to much ^{24}Mg, while oxygen-burning will produce the isotopes in the range $A = 25$ to $A = 32$ with a strong peak at ^{28}Si. The α-process will probably take over to provide the isotopes from $A = 36$ to $A = 46$. It is also possible that the neutron-producing reactions responsible for building the *s*-process elements may take place, in part, at later stages beyond the helium-burning stage.

The types of star in which these processes occur are not all known for certain. In the earlier work (FOWLER, BURBIDGE and BURBIDGE 1955 *a*, CAMERON 1955, B²FH) it was argued that those stars in which over-abundances of the elements resulting from helium-burning and the *s*-process are seen on the surface—carbon stars, S-type stars, and Ba stars—were at the helium-burning stage. If the neutron-producing reactions which are required for the *s*-process often occur at a later stage in a star's evolution, then the time-scale in such cases is very much shorter. For example, the reaction times in carbon-burning are 10^5 years at $6\cdot10^8$ degrees and 1 year at $8\cdot10^8$ degrees for a density of 10^4 g/cm^3. The energy release in carbon-burning only amounts to about 0.5 MeV per nucleon, as against 1 MeV per nucleon in helium-burning and about 7 MeV in hydrogen-burning. The main energy sources in such stars will therefore be the outer hydrogen- and helium-burning shells. However, it appears that the central regions of such stars will rapidly proceed to their final evolutionary states and the possibility of catastrophic changes will be present, in times which will be comparable with, or less than, the mixing time for material to reach the surface. Thus, if the view is taken that some of the stars in which *s*-process nuclei are seen in great abundance on the surface, are stars in which the later stages of evolution have taken place in their cores, the short time-scales in these phases may be in conflict with the relative frequencies of the observed stars as compared with the total number of giant stars. Although it may be argued that much of the element synthesis took place in the first burst of star formation and evolution when the galaxy condensed (this question will be discussed later), this would mean that little direct evidence for element synthesis in these phases is now available. Furthermore, it is still necessary to account for the stars with these *s*-process abundance anomalies which we now see. This latter observational evidence still makes it probable that a neutron source in the helium-burning phase of stellar evolution is of significance.

2·6. *Details of the s-process.* – Many years ago it was shown that the cross-sections of nuclei for neutrons in the MeV range indicated a relationship between the cross-sections and isotopic abundances N of the form $N\propto 1/\sigma$ (HUGHES, SPATZ and GOLDSTEIN 1949). This is to be expected in general in a chain of successive neutron captures, since nuclei with small cross-sections will build up to large abundances and vice-versa so that the number of captures per unit time will be uniform over considerable ranges of the chain. This means that much information is obtained by correlating empirical capture cross-sections for *s*-process nuclei with abundances. This has been done by forming the product $N\sigma$ which ought to be a smoothly varying function $f(A)$ (FOWLER, BURBIDGE and BURBIDGE 1955 *a*, B^2FH). $f(A)$ over a wide range of A is a measure of the total number of neutron captures to which the lighter beginning nuclei (*e.g.* ^{56}Fe) have been subjected. This product $f(A)$ is basic data which can be obtained from experiment and from isotopic analysis of solar system material. Thus the attempt made by CAMERON (1959 *a*, 1960) to *calculate* neutron capture cross-sections and then use these to deduce abundances and hence to adjust experimental abundance determinations is misleading, cf. the comments of BURBIDGE (1960) and CLAYTON, FOWLER, HULL and ZIMMERMAN (1961).

The neutrons are produced in the interiors of stars through (α,n) reactions, or in carbon or oxygen burning and the discussions of the previous section suggest that after being thermalized they will have energies in the range 10 to 100 keV. Cross-sections for many isotopes which were not available in the early work have now been given by the Oak Ridge and Livermore groups (MACKLIN, LAZAR and LYON 1957, GIBBONS, MACKLIN, MILLER and NEILER in press, MACKLIN, INADA and GIBBONS 1962, BOOTH, BALL and McGREGOR 1958) and some revisions of the abundances of some isotopes have been made by neutron activation techniques (REED, KIGOSHI and TURKEVICH in press, SCHMITT, SHARP, MOSEN, DUFFIELD and ZUMWALT 1962). An extensive anal-

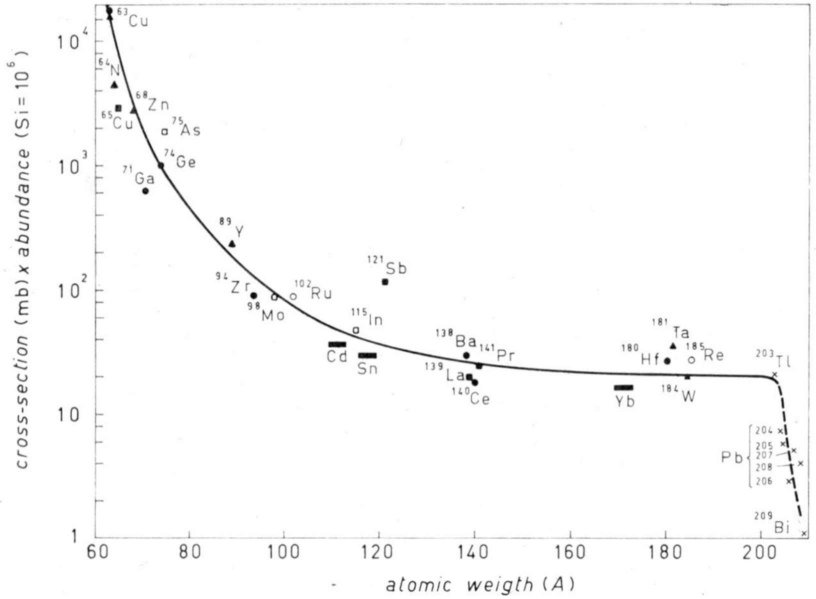

Fig. 1. – The product σN for s-process nuclei as a function of atomic weight. The cross-sections are those measured near 25 keV neutron energies by MACKLIN *et al.* at Oak Ridge, and by BOOTH *et al.* at Livermore. The abundances are those given by SUESS and UREY, SCHMITT *et al.*, REED *et al.* $T \sim 3 \cdot 10^8$ °K. The curve is drawn smoothly for visual aid through the points. The dashed tail represents an uncertain guess as to how a s dden decrease may occur for $A > 200$.

ysis of the s-process chain incorporating these data has recently been made by CLAYTON, FOWLER, HULL and ZIMMERMAN (1961) and we reproduce a plot of $f(A)$ from their paper (Fig. 1). They have solved for the abundance distribution with a chain of neutron captures starting from ^{56}Fe, by an approximate method. The assignment of s-process isotopes is completely demonstrated in this way. It is clear from the form of the σN curve that the production of the s-process elements in the abundances which are seen in the solar system has not occurred through the exposure of seed nuclei such as ^{56}Fe to a uniform integrated neutron flux. The form of the σN curve which has a large negative slope in the range $A < 90$ and a fairly constant value beyond this point, can be interpreted by supposing that during one phase the seed nuclei were subjected to a relatively large but limited neutron flux in which between three and seven neutrons were captured per seed nucleus, while in another phase, or phases, the heavier elements were built and more neutrons (30 to 140) were provided to be captured by the seed nuclei. This latter phase is required to account for the abundances of isotopes in this chain in the range $90 < A < 200$. The situation at the end of the s-process chain where both the s-process and the r-process contribute to the lead and bismuth abundances is complex. It is clear that some cycling at the end of the s-process chain has occurred in the solar system material. This was discussed in detail by B²HF and later by CLAYTON *et al.* (1961). The final abundance of Pb calculated by CLAYTON *et al.* is 6.5 ± 3 on the $Si = 10^6$ scale. The most probable value is consistent with the meteoritic value (taken from the carbonaceous chondrites). Moreover the Pb abundance in the sun was thought until recently to be an order of magnitude smaller than that calculated using these arguments. However, calculation and experimental determination of certain f-values for lead by HELLIWELL (1961) and BELL and KING (1960) give a revised value for the Pb abundance in the sun of $5.0^{+2.4}_{-1.7}$ which is in good agreement with the nucleosynthesis calculations.

Undoubtedly the s-process has played an important role in the nucleosynthesis of solar system material. A more detailed understanding will come with the measurement of more neutron capture cross-sections, the development of more sophisticated nuclear models, and perhaps determination of critical isotopes ratios outside the solar system.

2·7. *The equilibrium process.* – The peak in the abundance curve centered on ^{56}Fe has been explained by B²FH as being due to a very short-lived phase in a star in which the temperature and density conditions are so high that nuclear reactions take place in great profusion and the bulk of the material is converted to the elements at the bottom of the packing fraction curve. The only material for which both relative abundances of elements and isotopes are available is solar system material, and thus initially an attempt was made to determine the temperature and density conditions under which these elements were produced by fitting calculated abundances to the observed values. In this way a temperature $T = 3.8 \cdot 10^9$ degrees and $\log (n_n/n_p) = 2.5$ was determined. These temperature and density conditions are those in which the material was finally frozen in these abundance ratios. That is, following a very short time at this temperature and density, the material must have been removed to regions in which no further reactions could occur.

The preliminary results described in the discussion of the observations which suggest that there may be real abundance differences among elements in the iron peak between different stars, would mean that the conditions under which nuclei in this region of the abundance table were synthesized were slightly different (this would not be surprising) and that such material was ejected in the further evolution of the stars in which it was made, and condensed in new stars, before widespread mixing occurred.

2·8. *Supernova explosions.* – It was originally proposed by B^2FH that the place in which the equilibrium process nuclei were synthesized was in a star at an advanced evolutionary state just prior to a supernova explosion. Following this, in the explosion a flux of neutrons is produced on a very short time-scale and these are captured by some of the elements in the iron peak to build many of the isotopes up to $A \approx 270$. Apart from the fact that the only known astrophysical situation in which a rapid nuclear process can occur appears to be in a supernova explosion, the observational evidence for the r-process occurring in a supernova outburst has been taken to be that the light curve in a well observed supernova of Type I (that in IC 4182) decayed with a half-life of 55 days in agreement with the half-life of ^{254}Cf which decays by spontaneous fission. A fundamental question which still remains unanswered in the californium hypothesis is how the exponential decline in the visible light is reproduced without modification, from the exponential decline from the energy input of a radio-active nucleus.

Other radio-active nuclei which might possibly contribute to the light curves of Type I supernova (with half-lives in the range 40 to 70 days) were originally investigated by B^2FH. One of these which was considered and rejected was ^{59}Fe which has a half-life of 45 days. ANDERS (1959) has further considered this hypothesis. However, the basic difficulty which still remains is that approximately $0.1\,M_\odot$ of ^{59}Fe would be required to explain the observed light curve. The most recent calculations by HOYLE and FOWLER (1960) suggest that the amount of ^{59}Fe which can be made in a Type I outburst is only about $0.005\,M_\odot$.

However, theoretical collective model calculations for heavy nuclei which have been made by JOHANSSON (1959) have led to the conclusion that several heavy nuclei which will be built in the r-process will decay by spontaneous fission with half-lives in the range 10 days to one year, and may thus make significant contributions to the light curve. Thus, ^{256}Cf has an estimated lifetime of 30 days to one year, while ^{260}Fm may have a half-life in the range

10 to 100 days. Also ^{258}Cf, ^{258}Fm and isotopes with $Z = 102$ and greater may also contribute. Consequently the number of nuclei contributing fission energy is limited by the maximum atomic weight produced in the r-process, and if this limit varied from one star to another, corresponding variations in the detailed forms of the light curve could arise; this would mean that some variations in the apparent half-lives from the light curves would be seen. It has been argued that neutron-induced fission finally terminates the r-process and B²FH applied the enhancement in spontaneous fission above $N = 152$ to neutron-induced fission thus finding that the r-process ends near $A = 260$. However, Johansson's calculations now cast doubt on such an enhancement. In addition to this another important effect has been described by WHEELER (1958). This is that it is necessary to correct the nuclear surface energy for neutron-proton asymmetry. If this is done, then from the expression for the critical fission parameter ($E_{\text{Coulomb}}/E_{\text{surface}}$), it is found that the neutron-rich nuclei with large $I = N - Z$ are more susceptible to fission than was originally thought. Thus, after recovering from the steep drop in fission life-times at $N = 152$ as indicated by JOHANSSON this effect will finally be of importance, and HOYLE and FOWLER (1960) have estimated that the r-process will terminate in the range $270 < A < 275$ just short of the next closed neutron shell at $N = 184$ for which $A \approx 280$.

On this basis the number of even A nuclei produced in the r-process whose β-stable daughters decay by spontaneous fission is of the order of 10, and if four of these have half-lives in the range 30 to 100 days it is possible to recalculate the energetics of the light curves. If the total energy emitted under the exponential part of the light curve is of the order of 10^{48} erg, then since $1.75 \cdot 10^{51}$ erg is produced in $1\,M_\odot$ of fissionable material the amount required is $6 \cdot 10^{-4}\,M_\odot$. If four nuclei including ^{254}Cf contribute equally, this means that $1.5 \cdot 10^{-4}\,M_\odot$ of ^{254}Cf is built in the r-process. Since ^{254}Cf comprises about 1% of all of the r-process nuclei (B²FH) about $1.5 \cdot 10^{-2}\,M_\odot$ must be processed in a Type I supernova outburst.

The mechanism of a supernova outburst originally proposed by HOYLE (1946) and described in more detail by B²FH is that of implosion followed by explosion. It is supposed that the star has evolved to a situation in which the central temperature is greater than about $5 \cdot 10^9$ degrees and that the material is nondegenerate. Now the mechanism of the implosion hangs on the fact that under conditions of statistical equilibrium, the equilibrium at sufficiently high temperatures shifts decisively from the iron-group elements to helium plus neutrons. It is easily shown from the equations of statistical equilibrium that the material will be one-half ^{56}Fe by mass and one-half α-particles and neutrons when

$$\log \varrho = 11.62 + 1.5 \log T_9 - 38.17/T_9\,,$$

where ϱ is the density and T_9 is the temperature in units of 10^9 degrees. A plot of this function taken from the paper of HOYLE and FOWLER (1960) is shown in Fig. 2. To the left of the curve the abundance is concentrated into the iron peak nuclei while to the right of the curve it shifts rapidly towards α-particles and neutrons. The transition is very sudden so that a range $\Delta T_6 = \pm 0.5$ about this curve determines the transition strip. Now the mean binding energy of the nucleons in ^{56}Fe is 8.79 MeV, while the average binding in a mixture of 13 α-particles and 4 neutrons is only 6.57 MeV. Energy must therefore be supplied in such a transition at a rate of 2.22 MeV per nucleon or $2 \cdot 10^{18}$ erg per gram. Now this is greater than the thermal energy of the material, which, if it is ^{56}Fe only amounts to about 0.8 MeV per nucleon. Consequently gravitational energy is required to effect the transition.

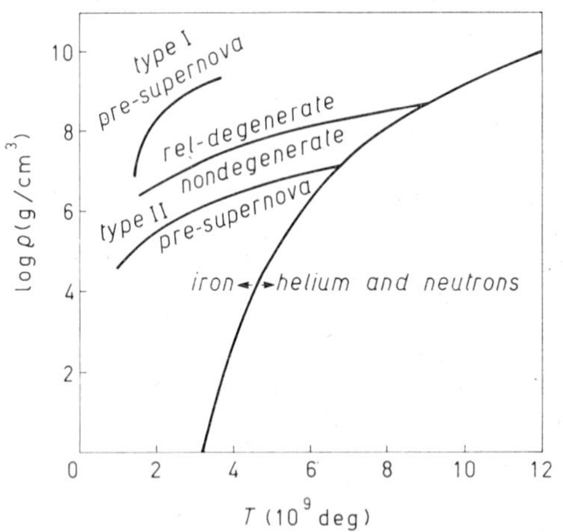

Fig. 2. – The ight-hand curve shows the density as a function of temperature when the material of a supernova core is half iron-group elements by mass and half helium plus neutrons. The upper left-hand curve shows ϱ *vs.* T throughout the interior of a star with $M \approx 1.28\ M_\odot$ prior to exploding as a Type I supernova. The lower left-hand curve shows ϱ *vs.* T throughtout the interior of a star with $M \approx 30\ M_\odot$ prior to exploding as a Type II supernova. The central left-hand curve separates relativistically degenerate matter from nondegenerate matter.

Let us consider the evolution in more detail. For material with $T \gtrsim 3 \cdot 10^9$ degrees energy loss by neutrino emission speeds up the evolution so that the central material moves to the right in the $\log \varrho$-T_9 plane in a time scale $\leqslant 10^7$ seconds. It reaches the transition strip, and since energy is not available to make the transition it moves upward along the edge of the strip. More and more of the material moves in this way, and if the star could maintain an equilibrium structure this would imply a sequence of models in which the density steadily increases. However, for a star in mechanical equilibrium the virial theorem must remain satisfied, *i.e.*

$$|\text{gravitational energy}| = n\ (\text{thermal energy}),$$

where the gravitational energy and thermal energy are integrals taken throughout the star and n $(1 < n < 2)$ measures the relative importance of gas and radiation pressures. Now it is clear that as the star contracts the gravitational energy increases rapidly while the thermal energy only increases slowly as the temperature changes very little. Consequently the virial theorem cannot continue to be satisfied, *i.e.* the star's structure no longer satisfies this condition for mechanical equilibrium. Its departure implies a catactrophic implosion in which ϱ increases without a corresponding increase in T. This implosion takes place over a time determined by the time for free fall. The outer material also falls inward in this time scale and the dynamical energy acquired is converted to heat. This sudden heating leads to a nuclear explosion in the envelope and the supernova outburst results.

An attempt to elucidate the dynamics of such an implosion by making an explicit hydrodynamical calculation has been made by COLGATE and his collaborators (COLGATE and JOHNSON 1960, GRASBERGER 1960, COLGATE, GRASBERGER and WHITE 1962). Following B^2FH they have started with a star of 10 $M_{\odot}$ and polytropic index 3 which is highly evolved, and have calculated the dynamical implosion using an equation of state which includes the iron to helium transition followed by the transition of helium to protons and neutrons. The collapse phase is reached by introducing an artificial heat sink which removes internal energy from the inner regions over a time scale of about 40 seconds. The inner regions fall inward and approach the free-fall rate. They show no sign of bouncing before a density of $3 \cdot 10^{11}$ g/cm^3 is reached. At this point their equation of state is no longer valid, because the rate of inverse β-decays of protons to neutrons becomes so fast that the core is evolving into a neutron configuration in a time of the same order as the dynamic collapse time. Just how far beyond this a physical bounce will occur is not clear at present. It seems probable that factors which have not been taken into account so far, such as the angular momentum and the stellar magnetic field, will eventually determine the point at which a bounce will occur. If these are the important factors, then the previous history of the star may well determine the critical point at which the bounce will occur. Thus the initial angular momentum and the role of the magnetic field in redistributing the angular momentum in the core as the star evolved will be of importance, provided that much of the initial angular momentum has not been removed during the slow mass-loss phases of evolution. The magnetic field may not have a direct effect since it compresses like a gas with $\gamma = \frac{4}{3}$. Finally it is of interest to consider the role of the magnetic field in a neutron configuration. Before the transition to nuclear densities the material consisted of a plasma containing a strong magnetic field. In a neutron configuration, however, the magnetic energy can be described in terms of the nuclear magnetic moment of the neutron star.

In order to investigate the explosive phases of the supernova outbursts and because the physics of the bounce are not yet understood, COLGATE *et al.* have introduced an artificial hard core which bounces the infalling material after a maximum compression by a factor of 10^4 has been reached. The outward moving shock then gains in strength as it moves through regions of decreasing density, and although the details have not yet been calculated, it will eventually reach a level at which it is of sufficient strength to blow off the remaining envelope of the star. There is some indication that for the initial conditions chosen here this cut will occur at a level such that about $1\ M_\odot$ is ejected (in the first calculations of the outgoing shock alone, Colgate *et al.* *arranged* the conditions so that $1\ M_\odot$ would be ejected). Immediately behind the shock a detonation wave will propagate. The energy generated through the burning of the light elements in this detonation wave will contribute to the energy output of the supernova, and it is in this phase that some element synthesis, originally thought to be the *r*-process and probably the *p*-process occurs. Though some neutrons may be generated here, it seems probable that the major neutron source will be the neutron core. In the calculations made by the Livermore group the energy density due to hydrodynamics in the shock is at least ten times the nuclear energy released in the detonation wave immediately behind the shock. Thus in this model, the gravitational energy is the dominant factor in the ejection, and it is the implosion which is directly responsible for the explosion.

The shock strength at the level at which the shell is ejected is so great that much material is ejected at relativistic velocities and it is argued that about 10^{26} g is ejected with energies of about 10 GeV per nucleon. Moreover the calculation of the energy distribution of the ejected material gives a distribution compatible with the primary cosmic-ray spectrum. This mechanism for the generation of cosmic rays will be discussed further in the final section.

The first investigation of the dynamics of a supernova outburst initiated by an implosion is of importance because it demonstrates that the qualitative picture developed by Hoyle and B^2FH is correct, and also because it does show that it is distinctly probable that neutron cores can be formed in this phase of stellar evolution. However, it is necessary that a considerable range of initial stellar models and masses be investigated in this way. It is probable, that such calculations will lead to changes in the estimates of the masses and initial conditions in exploding stars which become Type II supernovae.

We turn now to a consideration of another mechanism by which a nuclear explosion may be detonated in a star at an advanced stage of stellar evolution. This has been proposed by Hoyle and Fowler (1960). Again the starting point is a star in which the central material has evolved so that it lies to the left of the curve in Fig. 2 but outside the transition strip in which the conversion of iron to helium will take place very rapidly. The evolution is again taking place very rapidly because of neutrino processes though the time scale is much longer than that of free-fall. The star remains in mechanical equilibrium. The rate of evolution ($\sim 10^7$ s) is so rapid that the inner parts evolve before the radiation has had time to diffuse to the surface. To balance the energy loss by neutrino emission the inner parts of the star must contract. The effect is to increase the gravitational field that acts on the outer parts of the star, tending to cause compression, with a consequent release of thermal energy. Moreover, because of the rapid rate of evolution, radiation is trapped in the star causing further heating. Thus the net result will be a steady heating of the light nuclei and this must lead to an explosive condition unless the star can expand and cool. Since the time scale is long enough for such expanded configurations to develop this simply depends on whether such expanded configurations exist. Now calculation of models in stellar evolution shows that expanding configurations always exist if the material is nondegenerate. Thus for nondegenerate stars the implosion mechanism is required for an explosion. On the other hand if the star is degenerate the situation is quite different. A consideration of the structure of degenerate stars with polytropic index 3 (CHANDRASEKHAR 1938) has led HOYLE and FOWLER to the conclusion that explosions will only occur in stars in the narrow mass range given by

$$1 \leqslant M/M_C \leqslant 1.1 \,,$$

where M_C is the Chandrasekhar limit for degenerate stars. In our later discussion of stars with high densities it will be seen that more sophisticated models suggest that a slightly wider range of masses may be effective in this respect. Since the bulk of the material must be degenerate this means that the central temperature must be appreciably greater than the explosion temperature of the nuclear fuels. This means, for the model discussed by HOYLE and FOWLER, that the central temperatures must be near $3.5 \cdot 10^9$ degrees, for a star of $1.1\,M_C = 1.28\,M_\odot$. At such an explosion point the central density is near 10^9 g/cm³. Thus the star lies far to the left of the helium transition zone. It is estimated that such an explosion will generate in a star of this mass about $7 \cdot 10^{50}$ erg.

HOYLE and FOWLER have therefore concluded that these two explosion mechanisms occur in different astrophysical circumstances. The implosion-explosion mechanism is invoked to explain outbursts of massive stars (they have assumed a mass of $30\,M_\odot$ in the case discussed) and the explosion in degenerate matter to explain outbursts in stars in the mass range given above. If the assignment of Type II supernovae to Population I and Type I to Population II is taken at its face value then it is natural to associate the implosion-explosion mechanism with Type II supernovae and the degenerate star explosion with Type I supernovae. However, it should be borne in mind that assignment of population characteristics with masses, comes from indirect arguments since the observational features which are to be correlated with the theoretical predictions about such outbursts are all obtained from stars in external galaxies.

In the earlier work by B^2FH the mass of the pre-supernova star was assumed to be $10\ M_{\odot}$, although theoretical predictions were made with this model to account for the light curve of the supernova in IC 4182 which was a Type I supernova. The new assignment avoids this apparent inconsistency. Since the *e*-process, the *r*-process and the *p*-process were all assigned either to the immediate pre-supernova stage or to the explosion itself the places in which these processes occur must be re-examined. HOYLE and FOWLER now conclude that the *e*-process takes place in massive stars which implode. The temperatures and densities in such stars in the layers immediately below the seat of the explosion and above the imploding core are shown to be approximately those required to produce the abundances of the elements in the iron peak as determined in the solar system. As was described in the previous section a temperature near $4\cdot10^9$ degrees and a ratio of free neutrons to protons of about 300 corresponding to a mass density of $4.5\cdot10^5$ g/cm³ is required. Values near this are obtained for the appropriate layers in the massive star but, as has been discussed earlier, in the case of a small-mass degenerate stellar explosion the density is far higher than this, about 10^9 g/cm³ at the time of the explosion.

The production of the iron-peak elements in massive stars also means that the synthesis starts to take place very early in the evolution of a galaxy because of the exceedingly short time scale (10^6 to 10^7 years) for stars of these masses.

In the case of the explosion of a degenerate star the material is still concentrated in the iron peak but the degeneracy condition and the high density have led HOYLE and FOWLER to the conclusion that at a temperature of about $3\cdot10^9$ degrees the equilibrium peak now occurs at ^{58}Fe instead of ^{56}Fe, and since the abundance ratio $^{58}Fe/^{56}Fe < 0.003$ in the solar system this indicates that this material cannot contain very much material from such supernova outbursts. However, they have argued that it is these supernovae which are responsible for the *r*-process nuclei. Since (α,n) reactions are thought now not to be able to provide sufficient neutrons at the high temperatures achieved in such outbursts (at energies $\geqslant 100$ keV the bulk of the neutrons that are produced are probably captured by light nuclei) a more prolific neutron source is required. It is argued that the explosion is so violent that the inward-moving shock which originates in the exploding layer will cause sufficient heating in the central region for the iron peak elements to be disintegrated into helium and neutrons. Then if sufficient energy is focused in the central region to maintain the temperature as the star expands, *e.g.* if the density falls to a value of 10^5 to 10^6 g/cm³ before the temperature drops so that the material moves out of the helium zone, then a small fraction of the helium undergoes normal helium-burning to form ^{12}C, which by further reactions rapidly builds into the range $A = 60$ to 70. If this phase does occur, then these nuclei are the seed nuclei onto which the neutrons are captured to produce the elements up to $A=270$ in the r-process chain. The numerical estimates by HOYLE and FOWLER, based on this complex chain of events, indicate that the process will be adequate to account for the number of r-process nuclei, as indicated by the amount of ^{254}Cf and other nuclei needed to give the energy output in the exponential part of a Type I supernova light curve.

2·9. *Details of the* r-*process.* – B²FH gave a detailed account of the way in which the neutron-rich heavy elements must be built in this rapid neutron-capture chain. In the r-process there is a competition between (n,γ) and (γ,n) reactions in the region of a supernova where the temperature is about 10^9 degrees. Under these conditions successive neutron captures produce neutron-rich nuclei far to the electron-unstable side of the stable isobars. As the neutron flux dies down, the neutron-rich nuclei decay back to the stability region through repeated negative β-decays. Since the first calculations were made in 1957 accurate neutron binding energies have become available for the stable nuclei in the range $140 \leqslant A \leqslant 200$, mainly through the mass-spectroscopic work of JOHNSON and BHANOT (1957). BECKER and FOWLER (1959) have therefore made further calculations of the abundances of isotopes built in the r-process chain using this data. In B²FH the critical neutron-binding energy in MeV above which neutron capture will occur on the average was given as

$$Q_n \geqslant \frac{T_9}{5.04}\left(34.07 + \frac{3}{2}\log T_9 - \log n_n\right),$$

where T_9 is the temperature in units of 10^9 degrees and n_n is the density of neutrons. Originally the values taken to satisfy this equation were $Q_n \geqslant 2.0$ MeV corresponding to $T_9=1$, and $n_n=10^{24}/\text{cm}^3$. Since the supernova conditions under which the neutron flux is generated are not specified in detail, the temperature and density can be treated as adjustable parameters. In the recent work a better fit with the Suess-Urey abundances is obtained for $Q_n \geqslant 1.6$ MeV, which value is obtained by keeping the neutron density $=10^{24}/\text{cm}^3$ but adjusting the temperature T_9 to 0.8. The r-process capture path is changed to some extent by this adjustment in Q_n. In Fig. 3 we show this capture path together with the original one calculated by B²FH and also the path for $Q_n=3.0$ MeV corresponding to a value of $T_9=1.5$ if n_n is still kept at $10^{24}/\text{cm}^3$. On the average the path given by $Q_n=2.0$ MeV passes through nuclei for which $Z-Z_A$ (where Z_A is the charge of the hypothetically most stable nucleus in the isobaric sequence) is less than for those along the path $Q_n=1.6$ MeV. Since the decay energy W_β is linearly proportional to $Z-Z_A$ and because the mean lifetime varies as W_β^{-5}, an appreciable difference in abundances can be expected between the two cases. In Fig. 4 we show the calculated abundances in the region between $A = 140$ and $A = 200$.

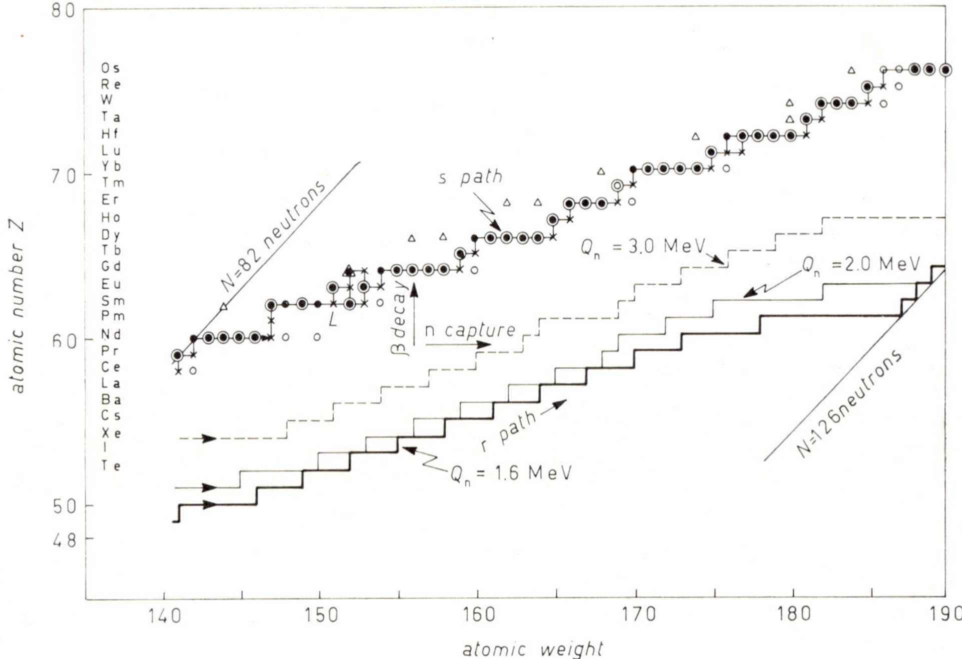

Fig. 3. – Capture paths in the r-process for different choices of critical neutron-binding energy. The symbol L denotes a life-time longer than the time scale of the s-process: △ nuclei produced in p-process; ● nuclei shielded from r-process; □ nuclei produced on s- and r-process; ○ nuclei produced only in r-process.

The neutron density obtained in the model of a degenerate stellar explosion (HOYLE and FOWLER 1960) is about $10^{26}/\text{cm}^3$, which is considerably higher than the number used here, but it is consistent with the results of B^2FH and BECKER and FOWLER for the conditions under which the r-process abundance peaks at $A = 130$, $N = 82$ and $A = 196$, $N = 126$ were synthesized. The higher neutron density does not alter these values significantly. The r-process peak at $A = 80$, $N = 50$ appears to have been produced under different conditions than the two heavier peaks, in that the total integrated neutron flux was smaller than when cycling was obtained for the heavier elements.

In B^2FH a table was given in which assignment of all of the isotopes in the abundance table to one or more nuclear processes was made. The further work on the s-process, the investigations described above, and changes in abundances from the Suess-Urey values have necessitated a re-examination of this table for the s, r, and p-processes. This analysis has been made by CLAYTON and FOWLER (1961). The assignments of B²FH are in general quite consistent with the numerical values given in their article, but they have now

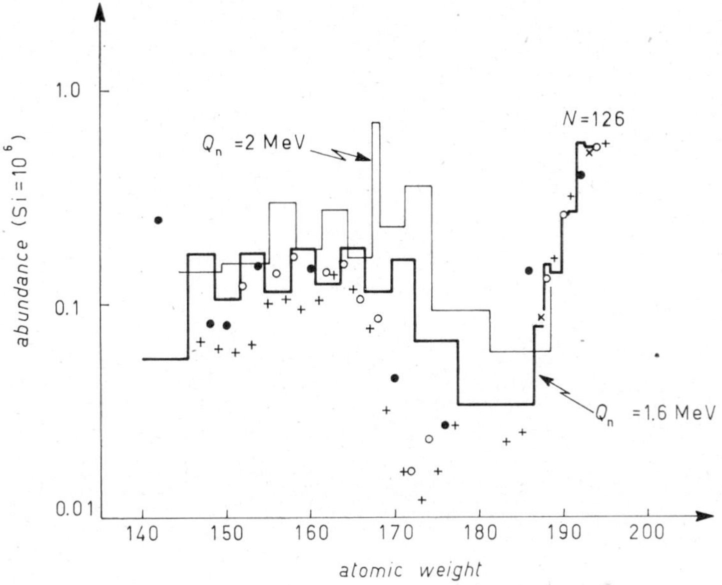

Fig. 4. – Abundances of nuclei in the region of the rare earths which are produced in the r-process: × ● rapid chain only; + ⊙ rapid chain predominantly. The empirical points are taken from SU. The histogram (—) is a curve calculated by the method more fully described in the work of B²FH. The abundances are normalized to silicon taken as 10^6 atoms. The heavy line depicts the conditions believed to be responsible for the solar-system abundances of these nuclei (Q_n critical $= 1.6$ MeV). The light line shows the effect of increasing the temperature to $1.0 \cdot 10^9$ degrees (Q_n critical $= 2.0$ MeV). Both situations are normalized to show a common abundance for an element like tellurium (not included in the diagram).

been able to give quantitative values for the ratio of contributions of *r*- and *s*-processes. The method of normalization for the *r*-process is to adjust the parameters of the model until the *r*-process abundance peaks near $A = 80$, 130, and 196 are fitted. The recent abundance determinations of the *r*-process peak elements by neutron activation analyses have led to reductions in the abundances in all of these peaks. Consequently CLAYTON and FOWLER have reduced these values as compared with those used by B²FH, which are the Suess-Urey abundances, by an overall factor of 2.5. It appears now that the calculated *r*-process abundances are in good agreement with the observed abundances for Os and Te, not quite as satisfactory for Pt and Xe, but there is a considerable discrepancy for Kr, as well as difficulties in explaining the isotope ratios of this element. As was mentioned above this lightest *r*-process peak may have been produced under different conditions from the others.

2.10. *Details of the* p-*process*. – B^2FH pointed out that the proton-rich isotopes of a large number of heavy elements cannot be built either in the *s*- or the *r*-processes. To account for these the *p*-process was introduced. This was thought to involve (p,γ) and possible (γ,n) reactions which take place on material which has already been synthesized by the *r*- and *s*-processes. Such processes will go on in hydrogen at densities $\geqslant 10^2$ g/cm³ and temperatures of $(2 \text{ to } 3)\cdot 10^9$ degrees. Arguments of a rather qualitative nature were given to show that such a mechanism could probably account for all of the *p*-process isotopes. This question has now been considered in more detail by ITO (1961) and also by FRANK-KAMENETSKII (1961).

ITO has made calculations of the (p,γ) and (γ,n) reactions and has concluded that two temperatures and several densities are required to produce the *p*-process isotopes in approximately their observed abundances. Thus, at $T = 3\cdot 10^9$ degrees and proton densities 10 to 10^2 g/cm³ the proton-rich isotopes ^{74}Se, ^{78}Kr, ^{84}Sr, ^{92}Mo, ^{94}Mo, ^{96}Ru, ^{98}Ru, ^{102}Pd, ^{106}Cd, ^{108}Cd, ^{112}Sn, ^{113}In, ^{114}Sn, ^{115}Sn, ^{120}Te, ^{124}Xe, ^{126}Xe, ^{130}Ba, ^{132}Ba, ^{136}Ce, ^{138}Ce, ^{138}La, ^{144}Sm, ^{152}Gd, ^{156}Dy, ^{158}Dy, ^{162}Er, and ^{164}Er can be synthesized by (p,γ) reactions. They come from between 1 and 6 progenitors. ^{113}In, ^{115}Sn, and ^{138}La come from single progenitors ^{111}Cd, ^{113}Cd, and ^{137}Ba respectively (^{113}In arises through (p,γ) followed by β^+ decay) while ^{144}Sm has 6 progenitors ^{136}Xe, ^{138}Ba, ^{139}La, ^{140}Ce, ^{141}Pr, and ^{142}Nd. ^{136}Ce, ^{138}Ce, ^{138}La, ^{144}Sm, ^{152}Gd, ^{156}Dy, ^{158}Dy, ^{162}Er, ^{164}Er, ^{168}Yb, ^{174}Hf, ^{180}W, ^{184}Os, ^{190}Pt, and ^{196}Hg are produced by (γ,n) are reactions on from 1 to 3 progenitors at a temperature of $2.5\cdot 10^9$ degrees and neutron densities in the range 10^{-4} to 10^{-5} g/cm³ (or $6\cdot 10^{18}$ to $6\cdot 10^{19}$ neutrons/cm³). Above about $3\cdot 10^9$ degrees the (γ,n) reaction time becomes much shorter than the (p,γ) reaction time in the very-heavy element region. It is for this reason that ITO has concluded that the very heaviest proton-rich isotopes can only be built by (γ,n) reactions.

These results show that the proton-rich isotopes can be produced in some phase of a supernova outburst. The time scales which have been taken by ITO lie in the range $1 \div 10^3$ s; these are reasonable if a supernova is responsible. However, no very clear picture has yet emerged of at what stage in the outburst such a synthesis process may occur. (p,γ) processes at these high temperatures are most probable in the hydrogen-rich envelopes of massive supernovae, and since only about 1 % of the heavy elements need to be produced in this way such events may be rather rare. (γ,n) reactions at a somewhat lower temperature may occur in another layer at a subsequent phase of the outburst. A more detailed discussion of the outburst as described in the model of COLGATE and his collaborators may lead to a more detailed description of this phase of element synthesis.

FRANK-KAMENETSKII (1961) has taken the view that this stage in element synthesis probably does not take place through (p, γ) and (γ, n) reactions. He argues that if the rate of (p, γ) reactions is sufficient for the formation of the by-passed nuclei in two stages via an unstable nucleus, for example, then the reverse photo-nuclear reactions are also rapid, so that the *p*-process is a sequence of reversible rapid thermal reactions. Accordingly he considers that a state of thermodynamic equilibrium between the by-passed nuclei and their parent nuclei must apply. He has tested this by plotting the ratio of the observed abundances against binding-energy differences of these nuclei, and has concluded that the results are in disagreement with this hypothesis of partial thermodynamic equilibrium. He also argues that these results cast doubt on the validity of the *e*-process. For this and other reasons he has rejected the idea that (p, γ) and (γ, n) reactions are responsible for the production of these isotopes and has proposed instead that reactions of the type (p, n), and (p, 2n) are more likely to account for these features of the abundance curve. Such reactions only occur under conditions in which the protons are accelerated in a nonthermal manner. Thus the argument appears to lead him to the conclusion that the *p*-process isotopes are produced in stellar atmospheres, in flares, and in other regions where acceleration can occur by nonthermal processes. Such processes have been described earlier (FOWLER, BURBIDGE and BURBIDGE 1955 *b*).

2·11. *The x-process.* – As was stated in the introduction, the synthesis of D, Li, Be, and B must take place under conditions such that these elements never mix into the regions in which hydrogen-burning will occur, since if this happens they will then be very rapidly destroyed. This leaves as possible places of origin for these elements the atmospheres of stars, or gaseous nebulae, in which the nuclear reactions are initiated by particles which have been accelerated by nonthermal processes, and also possibly in the outer parts of supernova envelopes. These various possibilities were first explored by FOWLER, BURBIDGE and BURBIDGE (1955 *b*) and by B^2FH, but the process was designated the *x*-process in this latter paper because the details remained unknown and the estimated abundances based on production in stellar atmospheres were found to be less than those in terrestrial matter by a factor of 100. Among the types of star described in these papers as possible regions in which the elements could be synthesized were the *T Tauri* stars, which are stars in the process of gravitational contraction. More recently this problem has been discussed by BASHKIN and PEASLEE (1961).

The determination of the abundance of lithium in a number of *T Tauri* stars by BONSACK and GREENSTEIN (BONSACK and GREENSTEIN 1960, BON-

SACK 1961 *a*) lends strong observational support to this idea. This was pointed out by GOLD (1960) and BONSACK and GREENSTEIN (1960) argued that high-energy nonthermal processes which derive their energy from the magnetic field in the contracting star might be able to make the required amounts of lithium. In the case of *T Tauri*, the prototype star of the class, BONSACK and GREENSTEIN concluded that in the nebula surrounding it the abundance of lithium was significantly lower than it is in the star. From this it appears that this element has been built in the star during its contraction phase and it is not lithium which was already present in the material out of which the star condensed. The high abundance of lithium in the earth can then be looked upon as being due to its production in the contraction phase of the evolution of the sun, when the planetary and meteoritic material formed part of an extensive solar nebula. On the other hand, the low abundance of lithium in the sun must be due to its destruction in a subsequent stage in solar evolution. A similar argument for the production of deuterium in such a contracting phase in a star has been made by HAYAKAWA (1960).

It was clear in the early work that the most probable nuclear process for production of these light elements is spallation on elements already built in thermal nucleosynthesis processes. The data now available on spallation cross-section (DOSTROVSKY, FRAENKEL and RABINOWITZ 1960, MILLER and HUDIS 1959) and the spallation fraction observed in the primary cosmic rays make it a fairly straightforward matter to estimate the amount of spallation activity which must have gone on to produce the elements in the observed abundances. However, to develop a more detailed picture of the circumstances which have led to these events occurring, it is necessary to consider the physics of the contraction phases of a star and its surrounding nebula. Thus, it is natural to consider this phase of nucleosynthesis for a specific model for the origin of the solar system, and use the more complete observational data available here; *i.e.*, for the solar system we can use the terrestrial and meteorite abundances of D, Li, Be, and B and isotope ratios $^{6}Li/^{7}Li$ and $^{10}B/^{11}B$, whereas if we consider the production of the light elements in *T Tauri* stars we have only fragmentary data on the atmospheric structures of such stars, and only the abundance determination for lithium.

An extensive investigation along these lines has been made by FOWLER, GREENSTEIN and HOYLE (1962). The model they have used for the solar nebula is that developed by HOYLE (1960) in which it is supposed that an equatorial disk of mass about 1% that of the sun was detached from the contracting sun when its radius was about $40\,R_\odot$. The sun was rotating rapidly at this point and was magnetically coupled to the disk to which it transferred angular momentum through this torque coupling so that the disk moved outward to a distance of about $3 \cdot 10^{14}$ cm. At this point solid particles began to condense, first iron and other metals, and later ice, ammonia, and perhaps methane.

Small planetesimals, particles which grew to a size of a meter or so, were left behind as the gas streamed outward. Such chemical separations are required to explain the origin of the terrestrial planets. The hydrogen and helium in the outer parts of this disk evaporated into the interstellar medium. With this model to work with they have considered the nuclear physics as follows.

They have pointed out that although spallation reactions are required to produce these light elements, the relative abundances do not agree with those to be expected by such reactions. Thus the Suess-Urey abundances show that the even A isotopes ^{6}Li and ^{10}B are very low in abundance relative to the odd A isotopes ^{7}Li, ^{9}Be and ^{11}B (the values relative to $\mathrm{Si} = 10^6$ are $^6\mathrm{Li} = 7.4$, $^7\mathrm{Li} = 92.6$, $^9\mathrm{Be} = 20$, $^{10}\mathrm{B} = 4.5$, $^{11}\mathrm{B} = 19.5$). Moreover $^7\mathrm{Li} \approx 5\,^9\mathrm{Be} \approx 5\,^{11}\mathrm{B}$. From these ratios they have concluded that, in addition to the production of the light elements by spallation processes, these nuclei have further been irradiated by a slow neutron flux which has given rise to the reactions $^6\mathrm{Li}(\mathrm{n},\alpha)^3\mathrm{T}(\beta^-,\nu)^3\mathrm{He}$ and $^{10}\mathrm{B}(\mathrm{n},\alpha)^7\mathrm{Li}$ which, because they have large low-energy cross-sections, will deplete ^{6}Li and ^{10}B and build up ^{7}Li and ^{3}He. On the other hand (n,α) reactions on ^{7}Li, ^{9}Be and ^{11}B are endothermic and will not occur at low energies. The effect of other neutron-induced reactions has been considered in some detail, and it is concluded that none of these contributes seriously to the depletion of neutrons, or to the building up of anomalies which would modify the abundance of other elements built in the thermal nucleosynthesis processes, provided that only a fraction of the material was irradiated by the neutron flux. The proton-induced reactions $^6\mathrm{Li}(\mathrm{p},\alpha)^3\mathrm{He}$ and $^7\mathrm{Li}(\mathrm{p},\alpha)^4\mathrm{He}$ have cross-sections such that at a temperature of about $5 \cdot 10^6$ degrees ^{6}Li could be depleted relative to ^{7}Li, but under similar conditions (p,α) reactions on ^{10}B and ^{11}B would have the opposite effect, *i.e.* they would reduce ^{11}B with respect to ^{10}B. Thus the occurrence of reactions of this type is ruled out.

Consideration in detail of the role of the neutron-induced reactions has led FOWLER *et al.* to conclude that the neutron flux amounted to about 10^7 neutrons/cm^2 s for a time of about 10^7 years. This flux also has the effect of producing radioactive ^{107}Pd and ^{129}I giving rise to the radiogenic ^{107}Ag and ^{129}Xe anomalies in the meteorites. This will be discussed further in the later section on radioactive nuclei.

The synthesis of deuterium will take place both by direct spallation reactions and by the capture of neutrons by protons. The latter process, $\mathrm{p}(\mathrm{n},\gamma)^2\mathrm{D}$, is mot effective, while the conversion of deuterium to tritium by $^2\mathrm{D}(\mathrm{n},\gamma)^3\mathrm{T}$ has a very small cross-section and can be neglected. FOWLER *et al.* have concluded that the neutron flux will therefore build up the deuterium so that the ratio $\mathrm{D/H} \approx 1.5 \cdot 10^{-3}$ in the material which is irradiated. This ratio of D/H is about 10 times greater than that which is found in the oceans and the carbonaceous chondrites. This conclusion together with the requirement that this process has not significantly changed the abundances of the heavy elements produced in earlier nucleosynthesis processes, has led these authors to the conclusion that only a small fraction of the material was subjected to such a neutron irradiation. Though there are several ways in which this can be thought to have come about, they argue that most probably it was because the material, when it was irradiated, had already condensed into bodies with radii which were large in comparison with the ranges of the particles producing the neutrons.

In order for this mechanism of nucleosynthesis to occur, a large flux of high-energy protons is required. These give rise to spallation reactions in which Li, Be, B, and neutrons are produced. The neutrons are then rapidly thermalized in hydrogen in the outer parts of the planetesimals. It is natural to suppose that the protons were accelerated in the sun when it was in a highly active state as compared with its present-day activity. No detailed theory of the generation of particles at this stage has been given. However, if it is assumed that the Li, Be, and B have the same abundances in all of the terrestrial planets, Earth, Venus, Mercury, and Mars, the total energy which was required in the form of high-energy protons is estimated to be about $6.5 \cdot 10^{43}$ erg, the average energy per proton being about 540 MeV. If only about 10% of all of the particles emitted collided with the planetesimals, this requires that the flare activity at this stage was of the order of 10^7 times as great as it was during the last solar maximum, *i.e.* it is required that about $5 \cdot 10^{47}$ high-energy particles with a total energy of about $5 \cdot 10^{44}$ erg were emitted in about 10^7 years.

The production of the light elements which are found in the earth and the meteorites are explained in this way. However, in order to account for the production of these elements in the *T Tauri* stars, it is necessary to consider these reactions going on, not in the outer parts of condensed planetesimals, but in the stellar atmospheres. In this case the great preponderance of hydrogen in these regions means that nearly all of the neutrons will be captured to form deuterium, and neutron irradiation will not significantly alter the ratios of the abundances produced in spallation. Moreover, the existence of large amounts of helium means that the dominating spallation reactions will be those due to protons on ^{4}He. These have the effect of producing ^{2}D and ^{3}He and more neutrons, which in turn will be captured to form deuterium. The final abundances which are seen, either in the surface of the sun, or in the *T Tauri* stars, are then determined by the amount of mixing into the interior and hence destruction which has occurred in the subsequent evolution of the star.

From many points of view this model for the production of the light elements is highly satisfactory. However, since it was worked out results have been obtained which may lead to some modifications in the details. Recent analyses by SHIMA (1962) of the abundances of Li, Be, and B, in a number of meteoritic samples have led to the conclusion that the abundances given by

SUESS and UREY are too high by a factor of about 10. Moreover, there is some indication that the isotope ratios ^{6}Li/^{7}Li and ^{10}B/^{11}B differ significantly between the earth and the meteorites. Clearly it is possible to revise the numbers obtained in the model to take account of these changes, and, in particular, the differences in isotope ratios may be an indication that different parts of the material were subjected to different neutron fluxes. A second point concerns the contraction phases of stars. HAYASHI (1961) has argued from rather general considerations that in their contraction phases stars will be, for significant periods, totally convective. More detailed calculations (SCHWARZSCHILD and UPTON 1962) tend to confirm this result although the situation is not completely clear (cf. COX and BROWNLEE 1961). Thus, in the case of the sun, it appears that in contracting to the main sequence it will become very much more luminous than was originally assumed, *i.e.* its track will be such that it will reach a maximum luminosity before it arrives on the main sequence, and the luminosity will then decrease as the star begins to burn hydrogen. Further, this means that its gravitational contraction time-scale will only be in the range $(5 \text{ to } 10) \cdot 10^{6}$ years, as against the Kelvin-Helmholtz time of about $50 \cdot 10^{6}$ years. This means that the phase of evolution in which the light elements were synthesized may be considerably shorter than the 10^{7} years envisaged here. If the convective phase and the element-synthesis phase take place together, or if the convective phase occurs subsequently to the light-element synthesis, this will have the effect of mixing in and destroying all of the light elements made in the star's atmosphere. Finally, the higher temperature of the solar nebula due to the higher luminosity of the proto-sun at this phase will also have to be taken into account.

An entirely different model for the formation of the solar system has recently been proposed by CAMERON (1962). In this it is supposed that the light elements D, Li, Be, and B are already present in the material out of which the solar system was condensed, since it is supposed that they have been produced in an earlier nucleosynthesis process. In this model some neutrons are then produced in the deuterium-burning phase in the contracting star through the reactions ^{2}D(d, n)^{3}He which competes with ^{2}D(p, γ)^{3}He and ^{2}D(d, p)^{3}T. In this model the time scale for the formation of the constituents of the solar system is very short, in accordance with the preliminary results of HAYASHI.

For the reasons discussed earlier it seems most probable that D, Li, Be, and B are produced in early phases of stellar evolution. This means that these elements are produced *in situ*, and as a star evolves it is probable that the bulk of these elements will eventually be destroyed *in situ*. Thus these elements should be very rare in the interstellar medium. Such a conclusion is completely at variance with the ideas of CAMERON (1962) who believes that these elements in the interstellar medium should be comparable in abundances or only a little less abundant than they are on the earth. For the elements

Li, Be, and B the attempts to determine the abundances in the interstellar medium are inconclusive (a review of this problem was given by B^2FH). However, in the case of deuterium an observational test can be used to settle this question. Attempts to detect neutral atomic deuterium by looking for the radio spectral line at 327 MHz have so far failed, leading to the conclusion that the D/H ratio does not exceed $2.5 \cdot 10^{-4}$ (STANLEY and PRICE 1956, ADGIE and HEY 1957), and in a further experiment at Green Park this limit has been lowered so that $D/H < 7.5 \cdot 10^{-5}$ (WEINREB 1962) as compared with the terrestrial value $D/H = 1.5 \cdot 10^{-4}$. If techniques are developed so that a gain in sensitivity by a factor of about 100 can be achieved even a null result may decide this critical point.

This completes the discussion of the processes by which all of the elements can be synthesized in stars. We next describe the observational evidence for element synthesis in stars. Finally in this general description of nucleosynthesis and stellar evolution it is necessary to discuss the role of neutrino processes in stellar evolution, and also the properties of matter at high densities which relate to initial conditions or to the last stages of stellar evolution.

3. – The observed compositions of stars.

If the galaxy originally condensed out of a cloud of either pure hydrogen, or hydrogen with a very small admixture of other elements, then the evolution of the stars up to the present should have left its mark in several forms which can be tested observationally. Three separate avenues of investigation are open:

1) Since the stars in the galaxy have been converting hydrogen to helium over the age of the galaxy (*i.e.* since the first stars formed), there is a direct relation between the initial H/He ratio, the ratio at present, and the total luminosity of the galaxy integrated over its age, together with any other energy-dissipating processes.

2) Several generations of stars are probably required to build all of the elements starting with hydrogen. The sense of the evolution is that the abundance of heavy elements relative to hydrogen will build up throughout the galaxy, if the mixing process is efficient, as the galaxy ages. Thus if we can sample material which was isolated from chemical evolution at different stages, evidence can be gathered of the composition at earlier times and of the rate at which material is processed. This means that we must analyse the compositions of stars and gaseous nebulae whose ages can be estimated.

3) Some stars are expected to show evidence that extensive nuclear processes are going on currently or have gone on, either in their deep interiors

or on their surfaces. Such stars will give the most direct evidence that nucleosynthesis takes place in stars.

3·1. *The* He/H *ratio.* – As far as (1) is concerned, if we suppose that the galaxy is currently emitting about $1.5 \cdot 10^{44}$ erg/s, then it is now converting hydrogen to helium at a rate of about $2.5 \cdot 10^{25}$ g/s. Thus over a time scale of $6 \cdot 10^{17}$ s ($= 20 \cdot 10^{9}$ years) it will have converted about $1.5 \cdot 10^{43}$ g of hydrogen to helium. If the total mass of the galaxy is $10^{11} M_{\odot} = 2 \cdot 10^{44}$ g then the total mass converted to helium will be about 7.5% by mass. Thus the overall He/H ratio by number should be about 0.019. This question has been discussed by BURBIDGE (1958). The He/H ratio observed in the planetary nebulae [cf. MATHIS (1957)] is about 0.16 while in the Orion nebula it is about 0.13. Now the planetary nebulae are considered on the basis of their population characteristics to have ages of about $5 \cdot 10^{9}$ years, while the Orion nebula is a good sample of present day interstellar matter. Since the gas in planetary nebulae has probably been ejected from the central star, it is possible that nuclear reactions in those particular stars are responsible for this abundance ratio so that the agreement between the two values may be fortuitous. However, an estimate of the He/H ratio for the sun has been made by an indirect method by OSTERBROCK and ROGERSON (1961), and it is in fairly good agreement with the value for the Orion nebula. Since the material of the sun condensed about $5 \cdot 10^{9}$ years ago we need to explain both the comparatively high value of the He/H ratio as compared with the estimates made earlier, and the fact that there has been little change in the relative abundance in the intervening $5 \cdot 10^{9}$ years.

Although the numbers we have used earlier to derive the theoretical He/H ratio are all uncertain, it is very difficult to see how the ratio can be significantly increased within the present framework of ideas. To do this would require either a time-scale of the order of $(50 \text{ to } 100) \cdot 10^{9}$ years, or a mass of only about $10^{10} M_{\odot}$ for the Galaxy, or a luminosity considerably higher than almost all other galaxies. Possible explanations are as follows:

a) The result may suggest that the bulk of the mass of the galaxy has not been ejected from stars and re-condensed into new stars; *i.e.* after the first burst of star formation the bulk of the material was condensed into comparatively small-mass stars in which it still remains. In this case the samples which give the ratio He/H approximately equal to 0.15 contain gas which has been processed through stars several times. This would suggest that the low-mass stars have a very small He/H ratio.

b) It is possible that in the first stages of its evolution the Galaxy became exceedingly luminous for a comparatively short period. Within the framework of an evolutionary cosmology this is possible, but in a steady-state universe such an hypothesis is not tenable, since we do not see any galaxies in this form today.

c) Nuclear energy has been released in stars which radiate strongly in the infra-red or give rise to very large-scale motions. The existence of black giant stars (large stars lying deep inside optically opaque clouds) was first suggested by BONDI, GOLD and HOYLE (1955) and energy dissipation by other processes was discussed by BURBIDGE (1958).

d) The bulk of the transmutation of hydrogen to helium took place in the first few minutes in the expansion of the universe (GAMOW 1953, ALPHER and HERMAN 1950).

The fact that there has been little or no change in the He/H ratio over the last $5 \cdot 10^9$ years is not surprising. The rate of energy conversion estimated earlier would lead to a conversion of only about 2 % by mass of the hydrogen to helium in this time, and the uncertainties and differences between the abundance determinations in different objects are of this order.

Preliminary determinations of the He/H ratio in nearby galaxies by SCHMIDT (1961) and JOHNSON (1959) give values very similar to those in our own Galaxy; *i.e.* JOHNSON obtained a value of 0.13 for the He/H ratio in a gaseous nebula in the Large Magellanic Cloud. Recently, values of 0.11 for the Small Magellanic Cloud and 0.10 for M 33 have been found by ALLEN and FAULKNER (1962) and by MATHIS, respectively.

3'2. *Stars older than the Sun.* – We turn now to the determination of the compositions of stars of different ages. In general, abundance analyses are made by comparing stars with the sun, and an extensive analysis of the abundances in the sun based on modern data has recently been made by GOLDBERG, MÜLLER and ALLER (1960). Since there are very few *f*-values available for many of the elements of interest, the analysis of abundances of most stars has been made by comparative methods.

Since the sun is used as a standard of composition and it has an age of about $5 \cdot 10^9$ years it is convenient to discuss the chemical composition of the material of the Galaxy first by considering the composition, relative to the sun, of material which condensed into stars prior to the condensation of the sun, and second, of material which condensed at approximately the same time as the sun, or at a later epoch. Age estimates for stars can be made only by comparing their positions in the Hertzsprung-Russell diagram with theoretically computed evolutionary tracks, and this is normally done by using the H-R diagrams for star clusters. In general age determinations remain uncertain, because the theoretical tracks taking into account all the complex conditions in evolution are not yet available for a range of initial masses and compositions covering the assumed masses and compositions of the stars in the observed color-magnitude diagrams. Consequently methods involving interpolation or extrapolation with assumption of homology have been used, while uncertainties in the initial composition remain. Thus, although ages as high as $(20 \text{ to } 25) \cdot 10^9$ years have been quoted for halo globular clusters and $(15 \text{ to } 20) \cdot 10^9$ years for the oldest stars so far analysed in the disk, with an uncertainty of 4 or $5 \cdot 10^9$ years (HOYLE 1959, HASELGROVE and HOYLE 1959, ARP 1962, SANDAGE 1962), it is recognized that these values are very uncertain. In the following discussion we therefore give only very rough estimates of age when discussing stars older than the sun.

The first quantitative analyses of stars which appeared to show deficiencies of the heavy elements relative to hydrogen were carried out on stars which, although they are comparatively nearby, have high velocities and form part of the halo population containing stars of ages $\gtrsim 10 \cdot 10^9$ years (ALLER and CHAMBERLAIN 1951, BURBIDGE and BURBIDGE 1956 *a*). They showed deficiencies, relative to the sun, when compared with hydrogen, of the elements such as Mg, Al, Ca, Sc, Ti, Cr, Mn, Fe, Sr, Y, Zr, Ba, by factors of up to 10 to 20. A recent discussion of four stars HD 19445, HD 140283, HD 161817, and HD 219617 by ALLER and GREENSTEIN (1960) gives the results shown in Table I. A similar analysis for one of these stars has been carried out by BASCHEK (1959). In these papers some of the stars analysed in the earlier work were re-examined. As can be seen from Table I the general conclusion is that in the extreme

TABLE I. – *Abundance in a typical extreme low metal-content star.*

Element	Solar abundance / Stellar abundance	Element	Solar abundance / Stellar abundance
C	590	Mn	51
Mg	17	Fe	65
Al	130	Co	13
Si	40	Ni	28
Ca	49	Sr	100
Sc	130	Y	250 (very uncertain)
Ti	34	Zr	120
V	68	Ba	170
Cr	42		

cases the metals are deficient by factors of from 40 to 100 as compared with the sun, while in the intermediate cases the abundances are down by factors of about 20. It also appears that the carbon abundance is reduced by a factor of about 100. Also the elements beyond the iron peak appear to be more underabundant than the others. An analysis of a hotter halo population star by TRAVING (1962) shows that it has a normal He/H ratio but that under-abundances of C, N, O, Mg, and Si by factors of from 5 to 10 are present. In some cases the underabundance factors are quite small but significant. For example, in the high velocity (and hence halo population) star *85 Pegasi*, HELFER and WALLERSTEIN (1959 *b*) have found that Na, Mg, Si, Ca, Sc, Ti, Cr, Fe, and Ni are underabundant by factors near 3.

In three other stars which are red giants, extreme metal deficiencies have been found in an analysis by WALLERSTEIN, GREENSTEIN, PARKER, HELFER and ALLER (1962). The stars, HD 122563, HD 165195, and HD 221170 are similar to giants in some globular clusters, but their ages are not known with certainty. The average metal-hydrogen ratios are decreased by factors of 800, 500 and 500 respectively when compared with the sun. The abundance ratios of other elements to iron resemble those in the sun with important exceptions. Manganese and vanadium are deficient with respect to iron, by a factor of 3. In HD 122563 all elements heavier than zinc are deficient when compared with iron by a factor of 50, so that the total deficiency factor of the heavy elements in this star is about 40 000. The authors therefore believe that these stars were formed very early in the history of the Galaxy and that in the case of HD 122563 the extreme deficiency of the very heavy elements is to be explained by supposing that when a flux of neutrons was produced the progenitors of the light s-process elements were present, but the iron peak elements were rare or absent.

Analyses have also been carried out by HELFER, WALLERSTEIN, and GREENSTEIN (1959) of three stars which lie in old star clusters, two in M 13 and one in M 92. The ages of the clusters are determined by comparison of the observed color-magnitude diagrams with theoretical evolutionary tracks. In these stars it has been found that the metals are deficient relative to hydrogen by factors of 100 or greater. In one star in one of the old clusters, the He/H ratio alone can be determined and it has been found by TRAVING (1962) to be reduced by a factor of 2 compared with a normal star. However, the abundances of elements relative to iron are the same, to within factors of about 5, in these stars as in the sun.

Curve-of-growth analyses made by WALLERSTEIN (1962) for 21 stars of very nearly solar type which have a range of ages from the age of the sun up to $(10 \text{ to } 15) \cdot 10^9$ years, show that the abundances of metals relative to hydrogen in the oldest stars in the disk are not significantly different from the abundances in the sun and the other younger stars. However, a correlation is found between the abundance parameter and the velocity characteristics, *i.e.* all of the high-velocity stars which therefore originated in the halo regions of the galaxy show a deficiency in the metals.

Since abundances derived by curve-of-growth methods are now available for about 30 field stars, WALLERSTEIN (1962) has attempted to draw conclusions about the relative abundances of different elements and relate them to the nuclear processes which produced them. He finds that among the elements in the Fe peak manganese is often deficient when compared with the other iron-peak elements. Some stars also show an excess of Mg, Si, Ca, Ti with respect to the iron-peak elements as well as with Na and Sc.

The investigations described in the last few paragraphs are the only abundance determinations which have been carried out so far by a curve-of-growth technique (*i.e.* by a crude analysis of a stellar atmosphere) for stars for which some age estimates can be made either from their dynamical characteristics, or from their membership in clusters. However, some attempts to determine compositions in a more indirect way for stars in clusters which can be dated have been made. In a stellar spectrum the absorption lines due to the metals crowd together more and more and get stronger in the ultra-violet. Now the wavelength regions over which colors are measured are U (covering the wavelength region $\lambda\, 2950 \div \lambda\, 4150$ with a maximum response set by the filter near $\lambda\, 3500$), B ($\lambda\, 3600 \div \lambda\, 5500$ with maximum near $\lambda\, 4150$) and V ($\lambda\, 4900 \div \lambda\, 7400$ with maximum near $\lambda\, 5300$). The radiation emitted is blanketed off by the absorption lines, so that if we measure for example, ultra-violet (U) and blue (B) magnitudes and compute ($U - B$), then the difference in the values of ($U - B$) for two stars, one of solar composition and one having small amounts of the metals as compared with the sun, will reflect this composition difference (cf. BURBIDGE and BURBIDGE 1961). Such color measurements have been calibrated by using stars whose compositions have been determined by curve-of-growth abundance analyses by WALLERSTEIN and CARLSON (1960). Having done this, the compositions measured by the parameter (metals taken together) to hydrogen, for stars which are too faint to be investigated in detail spectroscopically, have been made. In this way it has been found that in the old globular clusters M 5 and M 13 the metals are reduced by a factor of about 20, while for M 2 the underabundance factor is of the order of 200 (WALLERSTEIN and CARLSON 1960, ARP 1959).

Rough estimates of the compositions can also be made by examination of low-dispersion spectra. MORGAN (1959) has shown that the metallic-line intensity in the integrated spectra of globular clusters continuously varies, ranging from strong lines to very weak lines. In the weak-line category are included such clusters as M 92. There appears to be a correlation between metal deficiency and position of the cluster in the sense that the globular clusters near to the center of the galaxy have larger metal abundances than those further out in the halo. Thus clusters such as *47 Tucanae* and NGC 6356 have metal deficiencies relative to the sun of less than a factor of 10 (KINMAN 1959, SANDAGE and WALLERSTEIN 1960) while in clusters such as M 92 the factors are greater than 100.

For a few of the old clusters in the galactic disk similar qualitative discussions of the compositions from low-dispersion spectra have been made. In the case of NGC 752, which is only a few billion years old, there is some evidence that there is a small underabundance of the metals. However, low-dispersion spectra of stars in the two oldest known galactic clusters, M 67 (BURBIDGE and BURBIDGE 1959) which has an age in the range $(6 \text{ to } 10) \cdot 10^9$ years, and NGC 188 (SANDAGE 1962) with an age of $(10 \text{ to } 20) \cdot 10^9$ years, the spectra suggest that the compositions do not vary by a large factor from some of the youngest stars in the galaxy. Similar arguments suggest that the field stars in the vicinity of the sun in the galactic disk which have comparable ages, have compositions similar to those of the sun (cf. WILSON 1959).

Thus, on the basis of the preliminary results described above, the chemical compositions of stars which have ages greater than that of the sun can be summarized as follows. The halo population apparently condensed out of material containing significantly smaller amounts of the heavy elements relative to hydrogen than the solar material. The deficiencies range between factors of 10^2 or greater to factors less than 10. There is some evidence that the gas in the central bulge of the galaxy was enriched in the heavy elements to a greater extent than that further out in the halo by the time that the clusters condensed. However, in the disk population the situation is rather different. In this case it appears that there has been very little enrichment in heavy elements between the time that the oldest galactic clusters of which we have knowledge, and the sun, condensed. Although age determinations are uncertain at present this means that there has been little enrichment over a time interval of $(5 \text{ to } 15) \cdot 10^9$ years prior to the condensation of the sun. The evidence is as yet insufficient to determine how much mixing took place in the gas throughout this period.

3·3. *Stars of the same age or younger than the Sun.* – We next discuss the chemical compositions of stars which are of ages comparable to, or younger than, the sun; *i.e.* we consider the evidence bearing on composition changes over approximately the last $5 \cdot 10^9$ years.

Curve-of-growth analyses have been carried out for 3 stars in the Hyades —a cluster with an age of about 10^9 years—by HELFER, WALLERSTEIN, GREENSTEIN and PARKER (HELFER and WALLERSTEIN 1959 *a*, PARKER, GREENSTEIN, HELFER and WALLERSTEIN 1961). They have found that these stars have abundances of Na, Mg, Si, Ca, Sc, Ti, Cr, Mn, Fe and Ni which are slightly increased by factors of about 1.2 when compared with the sun. Barium is apparently overabundant by a factor of about 2. In another star the enrichment of all of the elements mentioned above by a factor of about 2 over the sun has been established. To obtain abundances representative to the galaxy at the present time it is necessary to study the compositions of the luminous and massive O and B stars which have ages $\leqslant 10^8$ years. Such studies based on detailed model atmosphere calculations have been made by ALLER and

JUGAKU (1959). They show that the elements He, C, N, O, F, Ne, Mg, Al, Si, P, S, Cl, and A have the same abundances as those elements in the sun. No determinations are possible for the iron-peak elements and the heavy elements since the atmospheric conditions preclude the appearance of suitable lines in the spectra of these hot stars. A survey of the abundances of a few of the lighter elements in a number of massive stars by HACK (1960) using less precise methods, suggests also that these elements may have solar abundances and that in some cases He, C, N, and O are systematically more abundant than in the sun by a factor of about two. Such abundances are all characteristic of the material in the spiral arms. On the other hand a study by ABT (1960) of two massive supergiant stars which were condensed recently, probably out of gas clouds lying well above the galactic plane, shows that the abundances of some heavy elements beyond the iron-peak (Zr, La, Ce, Eu) which are s-process elements predominantly, are low by factors of the order of 10 relative to the solar composition.

Thus the evidence at present is that there may have been a slight degree of enrichment of the gas in the disk of the galaxy in the last $5\cdot10^9$ years. However, the factor is certainly small. Further, there is a weak indication that at the present epoch the gas above the plane in the lower part of the halo may have a different composition from the gas in the plane as far as some s-process elements are concerned.

3·4. *Stars showing evidence of recent or current nuclear activity.* – We now discuss stars which show evidence of recent or current nuclear activity. Such stars fall into two categories: those in which the conditions are such that it is believed that specific nuclear processes have occurred, or are taking place now in their surfaces, and those in which nuclear activity has taken place in their deep interiors, and the results of this activity are shown in material which has been mixed to the surface.

We consider first the peculiar A-type stars with strong surface magnetic fields, which as a group show the greatest degree of peculiarity, since many elements covering a large range in the periodic table are involved. From the work of BURBIDGE and BURBIDGE (1955 *a* and *b*, 1956 *c*) and HACK (1958, 1961) detailed abundance analyses are now available for five magnetic stars, and the anomalies in characteristic cases are shown in Table II. The effects are very large and can now be considered to be well established. The apparent underabundance of oxygen in a number of these stars has been studied by SARGENT and SEARLE (1962) and it has been found that this element is underabundant with respect to hydrogen by factors of 10 to 100 for the peculiar magnetic stars which show very strong anomalies in the rare earths, Sr, and Cr, by a lesser factor in the stars which show the Si anomaly most strongly, while in the stars where the Mn anomaly appears to be very strong the oxygen

TABLE II. – *Abundances in two magnetic stars.*

Element	Stellar abundance/Solar abundance	
	α^2 *Canum venaticorum*	HD 133029
Mg	0.4	1.4
Al	1.1	2.2
Si	10	25
Ca	0.02	0.05
Sc	0.7	—
Ti	2.6	2.3
V	1.3	3.2
Cr	5.2	10.3
Mn	16	15 (very uncertain)
Fe	2.9	4.1
Ni	3.0	2
Sr	14	11.5
Y	20 (very uncertain)	
Zr	30	40
Ba	$\leqslant 0.9$	—
La	1020 (very uncertain)	200 (very uncertain)
Ce	400	190
Pr	1070	630
Nd	250	200
Sm	410	260
Eu	1910	640
Gd	810	340
Dy	760	460
Pb	1500 (very uncertain)	1500 (very uncertain)

abundance is normal. In the early work by FOWLER, BURBIDGE and BURBIDGE (1955 *b*) a rough estimate of the abundance of Be was made in the magnetic star α^2 *Canum Venaticorum*. A more detailed investigation by BONSACK (1961 *b*) gives a value of H/Be of $6\cdot10^8$, as compared with 10^{10} for the sun. In the early work (BURBIDGE and BURBIDGE 1955 *a*, FOWLER, BURBIDGE and BURBIDGE 1955 *b*, BURBIDGE and BURBIDGE 1958 *a*) a variety of nuclear processes which may explain these anomalies were proposed. These all require for their initiation the acceleration of large fluxes of protons and α-particles in the varying magnetic fields in excited regions in the atmospheres. The new observations have only added to the complexity of the models which are required.

A star of great interest because of its remarkable composition is the B5 star 3 *Centauri A*. First commented on because of the strong phosphorus lines in its spectrum by BIDELMAN (1960), its spectrum has been analysed in detail by SARGENT, JUGAKU and GREENSTEIN (SARGENT and JUGAKU 1961, JUGAKU SARGENT and GREENSTEIN 1961) and the abundances are as follows:

a) About $(84 \pm 10)\,\%$ of the helium in this star is in the form of ^{3}He; *i.e.* $^3\mathrm{He}/^4\mathrm{He} = 5$. The total abundance of helium with respect to normal stars is down to a factor of about 6. An earlier attempt to detect ^{3}He in the magnetic star *21 Aquilae* by BURBIDGE and BURBIDGE (1956 *b*) led to a rather uncertain but positive result and later investigation (STRUVE, WALLERSTEIN and ZEBERGS 1961) has cast doubt on this unless the abundance has varied significantly over a time of about 10 years. However, in *3 Centauri* the ratio is well established and the value given above should be compared with the terrestrial ratio $^3\mathrm{He}/^4\mathrm{He}$ of 10^{-4} to 10^{-5}, and with the upper limit for this ratio in the sun which is about 0.02. To produce this amount of ^{3}He by surface reactions either by building it up through $^1\mathrm{H}(\mathrm{n},\gamma)^2\mathrm{D}(\mathrm{p},\gamma)^3\mathrm{He}$ or by spallation of heavier elements and $^4\mathrm{He}(\mathrm{p},\ \mathrm{d})^3\mathrm{He}$ and $^4\mathrm{He}(\mathrm{p},\ \mathrm{np})\ ^3\mathrm{He}$ demands an immense amount of surface activity and if, for example, spallation were responsible, it appears that all of the surface material would be broken down to the stable isotopes of the lightest elements and an increase in the $^3\mathrm{He}/^4\mathrm{He}$ ratio as distinct from a build-up of ^{3}He would not occur. An upper limit to the D/H ratio in this star is estimated to be 0.01. It is natural to relate this result to that of APPA RAO (1961) who finds that the value of $^3\mathrm{He}/^4\mathrm{He}$ in the primary cosmic rays is 0.41 ± 0.09. Moreover the isotope abundances of material ejected, and presumably built, in solar flares suggests that both ^{2}D and ^{3}He may be produced in appreciable quantities under these conditions. It is of considerable interest that outside the solar system the only method of detecting ^{3}He is by measuring the isotope shift in the stellar spectra and consideration of the uncertainties of this method leads to the conclusion that a universal $^3\mathrm{He}/^4\mathrm{He}$ ratio of $\leqslant 0.05$ to 0.1 would remain (GOLDBERG, MOHLER and MULLER 1958, SCHAEFFER and ZÄHRINGER 1962) undetected. The most efficient method of building ^{3}He is by the incomplete proton-proton chain reactions mentioned earlier, going on under thermal conditions in a stellar interior. It is tempting to suppose that such a mechanism has indeed contributed to the large abundance of ^{3}He in *3 Centauri* (cf. BURBIDGE and BURBIDGE 1962) though if this were the case much ^{3}He must have been produced in a previous generation of stars. Furthermore, production of ^{3}He in this way would probably only be of importance in the contraction phases in massive stars.

b) The abundances of other elements in *3 Centauri* are as follows: C, Ne, Si, A, and Ca have normal abundances. O is underabundant by a factor of 6. N, P, Fe, and Kr are overabundant by factors of 5, 100, 4, and 1300, while the identification of Ga which is not certain as yet leads to an overabundance ratio for this element of 8000. These effects in some of the most unexpected elements are presumably due to some form of surface nuclear activity. However, very few of the observed effects can be accounted for by comparatively low-energy reactions. One which may be of importance is $^{32}S(n,np)^{31}P$ which was proposed by BIDELMAN (1960) to explain the anomaly in the abundance of phosphorus. This requires neutrons of only about 9 MeV bombarding energy. By invoking surface nuclear processes to explain the anomalies it is implied by analogy with the peculiar A stars that electromagnetic acceleration processes arising in strong surface magnetic fields are occurring. However, no strong magnetic field has been detected in this star.

Preliminary work on three other peculiar magnetic stars, *ϰ Cancri*, *HR* 465, and *53 Tauri*, by BIDELMAN (1962) and BIDELMAN and ALLER (1962) has led to the identification of P, Kr, Ga, Xe an possibly Hg. For these elements it is fairly certain that the identification of the lines means that the abundances are anomalously high. If the identification of Hg in *HR* 465 is correct it is possible that the dominant isotope is ^{204}Hg. The analysis of *53 Tauri* (BIDELMAN and ALLER 1962) shows that Ga is overabundant by a factor of at least 100 and the elements Mn, Sr, Y and Zr are found to be overabundant by a factor of about 10.

It is well known that ^{2}D, Li, Be, and B are destroyed in stellar interiors so that within the framework of the theory of nucleosynthesis in stars these elements must be produced in the stellar atmospheres, or in regions where non-thermal conditions give rise to reactions which will produce them. While some production may take place in supernova outbursts in the ejected shells, a probable place of origin is in the early stages of stellar evolution before stars have reached the main sequence. Thus a likely place to look for these elements is in the *T Tauri* stars which lie just to the right of the main sequence and are contracting on to it. At later stages these elements will be mixed into the interior and destroyed. A study of the abundance of lithium in stars of this type has been made by HUNGER (1957) and by BONSACK and GREENSTEIN (BONSACK and GREENSTEIN 1960, BONSACK 1961 *a*). It has been found that in twelve cases the lithium abundance when compared with the metals is increased by factors of about 100 over the value for the sun, and is comparable with the abundance of lithium in the stony meteorites. On the other hand the lithium abundance has been determined by BONSACK (1959) in about 40 stars which have passed through the contraction stage and are now either main sequence stars or giants. A considerable range in the abundance of this element over factors of 10^3 to 10^4 is found when it is compared with vanadium which is not expected to show abundance differences in this range of stars. It is probable that this range is due to *a*) variation in the amounts of the perishable light nuclei produced in the early history of the star, and *b*) the time passed by the star in a phase when it has an outer convective zone through which such elements can be mixed into the interior and destroyed. There is some evidence in this direction since stars which are thought now to have the most extensive convection zones tend to have smaller amounts of lithium.

A theoretical discussion of the production of Li, Be, and B in the early stages of star formation was given in the previous section.

Finally in this section we turn to stars in which nuclear processes in the interior relating to their whole evolution have gone on, or are going on now. Following the evolution of a star of slightly greater than solar mass into the giant branch, helium burning commences followed by carbon burning and oxygen burning, and also the conditions become ripe for (α,n) reactions to give a source of neutrons which build the elements in the neutron-capture chain on the long time scale (*s*-process). If the elements resulting from these nuclear processes are mixed to the surfaces, abundance anomalies will be seen. In the case of more massive stars these evolutionary effects will take place more rapidly after the stars leave the main sequence. A summary of the observational situation up to 1957 was given by B^2FH. We summarize here only the more recent investigations and the general results in this field.

A number of stars which are deficient in hydrogen and contain large abundances of He, C, and other light elements have been studied and a discussion of these was given in earlier work (B^2FH, BURBIDGE and BURBIDGE 1958 *b*). A recent investigation of the giant star *R Coronae Borealis* by SEARLE (1961) shows that the C/H ratio is about 10 as compared with a value of about $9 \cdot 10^{-5}$ in normal stars, while the C/Fe ratio is increased over that in normal stars by a factor of about 25. Indirect arguments suggest that besides carbon the bulk of the atmospheric material is He, O, and Ne. Thus two possible compositions for this star calculated as mass fractions are:

Hydrogen	0.000 5	0.000 5
Helium	0.91	0.65
Carbon	0.067 5	0.066 5
Other elements	0.022 5	0.283 5

On the other hand, those elements heavier than Fe have abundances compared with Fe which are entirely normal. Thus the star has probably either burned up almost all of its hydrogen, or has ejected its hydrogen envelope, and is currently burning helium. Some of the helium which has been processed to carbon has then been mixed to the surface. The C/N ratio is rather high, and the ^{12}C ^{13}C bands are absent, indicating that the ^{12}C has not passed through a hydrogen-burning zone in reaching the surface.

Another star which has the velocity characteristics of the halo population and may have evolved to the horizontal branch of the H-R diagram has been investigated by WALLERSTEIN, STONE, and WILLIAMS (1962). It apparently has a He/H ratio of 10 and carbon is overabundant by a factor of about 20.

However there is some uncertainty in the analysis because of possible peculiarities in the atmosphere.

The so-called carbon stars are giant stars in which the carbon bands are very prominent. Apart from *R Coronae Borealis* little is known about the H/He abundance in these stars. However, the $^{12}C/^{13}C$ ratio was estimated by McKellar (cf. B^2FH) from the carbon bands in some carbon stars to be about 4, and if carbon produced in helium-burning cores has passed slowly through a hydrogen-burning zone then the $^{12}C/^{13}C$ ratio should be 4.6. More recent investigations of the $^{12}C/^{13}C$ ratio by Wyller (1960) and Climenhaga (1960) have suggested that this ratio may have been under-estimated and that $^{12}C/^{13}C$ ratios greater than $5 \div 10$ are found in some stars. This and other less direct arguments suggest that, in some cases, either the hydrogen-burning zone is absent, or that ^{12}C has reached the surface rapidly through a very thin hydrogen-burning zone, as had already been suggested earlier for some star by B^2FH.

Evidence for production of the *s*-process elements in stars has been afforded by the S stars and the Ba stars, and nearly all of this work was described B^2FH. The only star in this group for which an abundance analysis has been made is HD 46407 (Burbidge and Burbidge 1957) where the over-abundances of the *s*-process elements were established. More recently Greenstein (1962) has made a preliminary qualitative study of the brightest of the Ba stars, *ζ Capricorni*, and has concluded that the *s*-process elements are strongly enhanced while the *r*-process elements are not.

Later stages of evolution are seen in stars which are hot subdwarfs and white dwarfs. A discussion of the observations and the information available on compositions which shows that such stars frequently show helium and features attributable to elements built in carbon- and oxygen-burning and the α-process was given by B^2FH. A summary of white dwarf spectra has recently been given by Greenstein (1958). For a survey of the spectra of evolved stars in general we may refer also to a survey by Greenstein (1961). No detailed and reliable abundance analyses for any highly evolved stars comparable to those made for normal stars are yet available. The reason for this is that model-atmosphere calculations for such stars are still lacking, and the comparative methods which can be used with some confidence for stars of more normal compositions are totally unreliable in these cases.

The final topic which bears on evidence for element synthesis is that of the spectra and light curves of supernovae and their places of origin in galaxies. This field is unique in that except for very recent observations and some of the very early work (Baade 1938, Baade and Zwicky 1938, Minkowski 1939) the large part of the observational material remains unpublished and the only recent survey (Zwicky 1958) is incomplete. Much material on the large number of supernovae which have been discovered, especially in the program of ZWICKY and his collaborators, has not been made generally available.

Two questions need to be answered concerning the places of occurrence of supernovae. First it is necessary to determine the distribution of supernovae among the types of galaxy in which they occur, and second, the places of occurrence of supernovae within the galaxies need to be studied. Some preliminary answers to the first question are now available (BERTAUD 1961, VAN DEN BERGH 1959). Comparing 53 supernovae which have exploded in galaxies which have been classified (some data are available for 99 supernovae) with the proportions of galaxies of different types (obtained by classification of 806 bright galaxies) gives the result shown in Table III which has been

TABLE III. – *Distribution of supernovae among galaxies of different types.*

	Relative frequency of galaxies of different type from Shapley-Ames catalogue of bright galaxies (%)	Relative frequency of supernovae among galaxies of different types (%)
Ellipticals	23.8	13.2
Irregulars	3.4	5.7
Spirals	72.8	81.1
	Relative frequency of spirals of different types of that fraction of the total number of spirals which have been classified	Relative frequency of supernova in spirals which have been classified
Sa	16.4	0.0
Sb	33.6	35.1
Sc	50.0	64.9

taken from the paper by BERTAUD (1961). These data suggest that supernovae occur preferentially in galaxies of types Sc and Sb whereas there is a deficit for elliptical galaxies and Sa galaxies. A similar conclusion based on only 18 supernovae was drawn earlier by BAADE (1938). This conclusion is strengthened by the fact that in 6 galaxies (5 Sc and 1 Sb) two or three supernovae have occurred in the last 60 years (a total of 16 supernovae in 6 galaxies) so that if we take them all together the rate of occurrence in those 6 has been one in every 22 years in each galaxy. This is to be compared with the three supernovae which have been detected in our own galaxy in the last 100 years, although two of these occurred in A.D. 1572 and 1604 respectively. The average rate at which supernovae explode in a galaxy of type similar to our own which may be an Sb, has been variously estimated to between one every 50 years and one every 500 years. An early discussion of this problem was given by ZWICKY (1942). This number was originally thought to be of great importance in calculating the rate at which the heavy elements were synthesized if this process were going on at a steady rate and it is still of importance in the supernova-origin theory of radio emission and cosmic rays.

Apart from a preliminary discussion by REAVES (1953) no study of the places of origin of supernovae, which are of importance in determining the population characteristics, has been published.

The criteria obtained from light curves and supernova spectra which are used to classify the supernovae have never been spelled out in detail. Authors have relied on the early brief statement by MINKOWSKI of the classification (MINKOWSKI 1941) and his detailed account of the spectra then available (MINKOWSKI 1939). The ideas at that time were also given by HUBBLE (1941) and the literature has been summarized by GAPOSCHKIN (1957). Modification of the early ideas by BAADE and MINKOWSKI have never been published. Consequently much confusion remains. However, we attempt to specify the criteria as follows:

Type II (Population I) have spectra in which hydrogen, helium, and carbon features can be identified (GREENSTEIN 1962); very high expansion velocities are seen and the light curve shows a steep rate of decline (5^m or more in the first 100 days). The absolute magnitudes at maximum appear in general to be greater for Type II supernovae.

In the case of Type I (Population II) the spectra present a serious problem in that practically no features have been positively identified, and the light curves may show an exponential form with a less steep rate of decline than those of Type II after an initial drop from maximum.

With these criteria 30 supernovae have been classified as one type or the other. Eleven are of Type II, 5 having been found in galaxies of type Sb and 6 in Sc galaxies. The 19 Type I supernovae are distributed among galaxies as follows: 5 have been found in elliptical galaxies, 1 in an Sb, 11 in Sc galaxies and 2 in irregular galaxies.

Until recently, light curves were obtained by photographic methods sometimes without good calibrations, and the colors were not measured in a consistent way. The bulk of these observations remains unpublished. Thus considerable uncertainties exist as to the colors and changes in color as the supernova shell expands, and also as the supernovae get fainter the magnitudes become very uncertain. Some data showing the variety of light curves has been given by ZWICKY (1939). A study of a recent Type II supernova by ARP (1961) in which many photoelectric observations were combined with photographic measures has now been published. This supernova reached an absolute magnitude at maximum in the blue of -19.1.

4. – The role of neutrino emission in stellar evolution.

Over most of its life history a star is a generator of neutrinos. In the hydrogen-burning phase the neutrino-producing reactions are

$$\mathrm{p}+\mathrm{p} \to \mathrm{d}+\mathrm{e}^+ +\nu \quad \text{(mean neutrino energy 0.26 MeV)},$$

$$^{13}\mathrm{N} \to {}^{13}\mathrm{C}+\mathrm{e}^+ +\nu \quad \text{(mean neutrino energy 0.72 MeV)},$$

$$^{15}\mathrm{O} \to {}^{15}\mathrm{N}+\mathrm{e}^+ +\nu \quad \text{(mean neutrino energy 0.98 MeV)}.$$

The energy carried away by neutrinos is only a small part of that carried by radiation; it is 2% of the 26.22 MeV released in the conversion of $4\,{}^1\mathrm{H} \to {}^4\mathrm{He}$ in the form of radiation in the p-p chain, and 6% of the 25.04 MeV in radiation released in the CN-cycle reactions.

The processes which give rise to anti-neutrinos are

$$\mathrm{n} \to \mathrm{p}+\mathrm{e}^- +\bar{\nu}$$

and they occur in the s- and r-process chains when nuclei β-decay to stable isotopes. Here again the energy carried away by anti-neutrinos is very small. Thus neutrinos play an insignificant part in stellar evolution until the high-temperature phases are reached.

GAMOW and SCHOENBERG (1941) were the first to point out that when a star has evolved so that its core is at high temperature and density, neutrino emission by other reactions may become of importance in its further evolution. This is because under these conditions the rates of the neutrino-producing reactions become significant, and the great penetrating power of neutrinos corresponding to a mean free path in matter of about 10^{20} g/cm² means that the energy carried by them will be lost immediately, whereas the diffusion time for radiation is, in comparison, very long.

GAMOW and SCHOENBERG pointed out that reactions involving the absorption and emission of electrons, *i.e.*

$$(1)\qquad \begin{cases} (A, Z)+\mathrm{e}^- \to (A, Z-1)+\nu\,, \\ (A, Z-1) \to (A, Z)+\mathrm{e}^- +\bar{\nu}\,, \end{cases}$$

would occur at different stages on different elements as the temperature increased. This mechanism was termed the Urca process. Since about $\frac{2}{3}$ of the energy in the transition is carried away by the neutrinos they act as a powerful cooling agent if such transitions dominate over other energy-generating processes. It was therefore argued that this mechanism would lead to the rapid collapse of a stellar core, and they proposed this as a basic mechanism for a supernova outburst. However, the work of HOYLE (1946) and B^2FH showed that the implosion due to the iron-helium transition will take place in a time which is given by the free-fall time and is considerably shorter than the collapse time determined by the Urca process according to GAMOW and SCHOENBERG. Thus it has been argued that although the final phases of evolution are speeded up by this process, it is not responsible for the catastrophic implosion.

In the Urca process different nuclei take part at different stages, depending on their β-decay energies. Recent calculations by CHIU 1961 *b*) have led to the conclusion that for elements with $A < 60$ at a temperature of about $6 \cdot 10^8$ degrees, the only elements which may lead to significant energy losses are ^{33}S and ^{35}Cl (^3He, which gives the largest energy loss, will not be present in a star under these conditions). When the temperature has risen to about $1.2 \cdot 10^9$ degrees CHIU estimates that the energy losses amount to 10^4 to 10^6 erg/g s, while in the range $2.4 \cdot 10^9$ to $6 \cdot 10^9$ degrees, *i.e.* in a region where the e-process elements are formed, the dominant element giving losses by the Urca process will be ^{56}Mn. However, these losses are still many orders of magnitude less than the gravitational energy release, so that it is clear that the Urca process is not of significance in these final evolutionary phases.

On the other hand, other neutrino emission processes may be of much greater importance. Following the development of a theory of a universal Fermi interaction, a coupling term of the form $(e\nu)(e\nu)^+$ arises (FEYNMAN and GELL-MANN 1958, SUDARSHAN and MARSHAK 1958) and it was pointed out by PONTECORVO (1959) that the existence of such an interaction makes possible the process of neutrino bremsstrahlung:

$$e^- + (A, Z) \rightarrow e^- + \nu + \bar{\nu} + (A, Z) \,. \tag{2}$$

The first calculation of this process in astrophysical circumstances was made by GANDEL'MAN and PINAEV (1960) who showed that for temperatures less than about $2 \cdot 10^9$ degrees this process dominates over the Urca process. In addition to this it was pointed out by CHIU, MORRISON and STABLER (CHIU and MORRISON 1960, CHIU and STABLER 1961) that the creation of neutrinos by pair annihilation

$$e^- + e^+ \rightarrow \nu + \bar{\nu} \tag{3}$$

by the photo-neutrino process

$$\gamma + e^- \rightarrow e^- + \nu + \bar{\nu} \tag{4}$$

and by the photon-photon processes

$$(5) \qquad \begin{cases} (a) & \gamma+\gamma \rightarrow \nu+\bar{\nu}\,, \\ (b) & \gamma+\gamma \rightarrow \gamma+\nu+\bar{\nu}\,, \end{cases}$$

must also be important. However, GELL-MANN (1961) has shown that in the case of local coupling $\gamma+\gamma \rightarrow \nu+\bar{\nu}$ has a vanishing rate to lowest order in the coupling constant, though the process $\gamma+\gamma \rightarrow \gamma+\nu+\bar{\nu}$ is not forbidden. On the other hand it has been pointed out by MATINYAN and TSILOSANI (1961) that this argument proving the forbiddeness of $\gamma+\gamma \rightarrow \nu+\bar{\nu}$ no longer applies if one of the photons is replaced by a Coulomb field, *i.e.* if we consider the disintegration of a γ quantum into a neutrino-antineutrino pair in a Coulomb field:

$$(6) \qquad \gamma+(A,Z) \rightarrow \nu+\bar{\nu}+(A,Z)$$

and they have calculated the rate for this process. Also an estimate has been made of the rate of $\gamma+\gamma \rightarrow \nu+\bar{\nu}$ if a nonlocality exists.

Calculations of the rates of processes (3), (4), and (5*b*) have been made by CHIU, MORRISON and STABLER (CHIU and MORRISON 1960, CHIU and STABLER 1961, CHIU 1961 *a*) and we show in Table IV the rates for these processes for

TABLE IV. – *Rates of energy loss by neutrino emission* (*) *energy loss in* erg/g s, *at a density of* 10^6 g/cm³.

Temperature (°K)	Process			t_{cool} (s)
	Annihilation (3)	Photoneutrino (4)	Photon-photon (5*b*) (approximate)	
$5 \cdot 10^8$	$10^{0.7}$	10^2	$10^{3.5}$	10^{13}
$1 \cdot 10^9$	$10^{7.4}$	$10^{4.2}$	10^7	10^9
$2 \cdot 10^9$	$10^{11.6}$	$10^{8.6}$	10^{11}	10^5
$2.5 \cdot 10^9$	$10^{12.6}$	$10^{10.4}$	10^{12}	10^4

(*) In the last column is entered the relaxation time in seconds for losing the full thermal energy content of the material by neutrino emission.

a range of temperatures calculated by CHIU and MORRISON. More detailed results for the photo-neutrino and pair annihilation processes are given by CHIU and STABLER for both the degenerate and nondegenerate cases. The photo-neutrino process is greatly reduced when the electrons are degenerate, but process (5*b*), which is only calculated approximately, is independent of density and degeneracy. RITUS (quoted by MATINYAN and TSILOSANI 1961) has obtained a much greater value for the rate of the photo-neutrino loss process than that given in Table IV but no details of his calculations have been given. Extensive calculations by CHIU of the pair-annihilation process have been tabulated in a form which makes them suitable for stellar evolution computations.

It is clear from these results that the cooling rate is greatly speeded up over that obtained when only the Urca process was considered. For the core of a star which has been heated to a temperature of 2.10^9 degrees the relaxation time for cooling is reduced to 10^4 to 10^5 s. Since the star remains in mechanical equilibrium, the neutrino loss will result in shrinking at an increasing rate. When the characteristic time for neutrino-energy loss becomes less than the time for gravitational re-adjustment, which is given by the free-fall time $(G\varrho)^{-\frac{1}{2}}$, the star will cease to be in mechanical equilibrium and will implode. In the earlier sections we have described the final stages of evolution of a massive star and have attributed the implosion and the subsequent explosion to the transition from iron to helium. All of the nucleosynthesis processes remain unchanged when these neutrino emission processes are taken into account, though these determine the rate at which the star evolves. Approximate calculations still show that the iron to helium transition point is reached before the neutrino energy losses lead to catastrophic collapse. Further investigations in which attempts are made to consider this phase of evolution for a very simple stellar model, based on the massive pre-supernova model of HOYLE and FOWLER (1960), have been made by CHIU and FULLER (CHIU 1961 *c*, CHIU and FULLER 1962). They have reached the conclusion that in this case the implosion due to neutrino emission will occur at a temperature lower than that required for the iron-helium phase change, *i.e.* below about $6.9 \cdot 10^9$ degrees. This result is obviously very uncertain since the initial conditions, and particularly the density in this region, can only be estimated very crudely.

MATINYAN and TSILOSANI have compared the rates of processes (2) as calculated by GANDEL'MAN and PINAEV with (6). They conclude that, for example, for a star consisting of ^{24}Mg (6) will dominate for $T \gtrsim 6 \cdot 10^8$ degrees and $\varrho \approx 10^5$ g/cm^3. Thus for stars of very low luminosity which are in an advanced stage of chemical evolution, the energy losses by neutrino processes of this type may determine the evolutionary time-scale.

Apart from their role in stellar evolution some speculations have been made concerning the importance of neutrinos in the overall energy balance of the Universe. It is well known that there is a discrepancy between the mean mass-energy density in the Universe, which for a number of acceptable cosmological models is of the order of 10^{-29} g/cm^3 (6 keV/cm^3), while the mass density of visible matter is only about $3 \cdot 10^{-31}$ g/cm^3. It has been suggested that perhaps neutrinos contribute significantly to the overall energy density (PONTECORVO and SMORODINSKY 1962). Experimental estimates of the neutrino-energy density can be made by assuming that in the experiment of REINES and COWAN (1958) the effect which was observed when the reactor was switched off is due to interstellar and intergalactic neutrinos. This argument shows that experimentally the upper limit to the energy density of anti-neutrinos in the energy range 3 to 10 MeV is about 10^3 MeV/cm^3. Clearly therefore the real upper limit at the present is set by the cosmological estimates of the mean density. It should also be emphasized that modern ideas concerning the condensation of galaxies and stars lead to the conclusion that the discrepancy between the density of visible matter and the cosmological value may well be due to the fact that the bulk of the material remains uncondensed or is present in unobservable forms other than neutrinos. The role of neutrinos in several cosmological models has been discussed by WEINBERG (1962). Finally HARRISON, WAKANO and WHEELER (1958), in considering the fate of large masses which at the end of thermonuclear evolution reach their very lowest energy state, have speculated that the dissolution of nucleons into neutrinos might occur (cf. the following section). Such an effect would violate the law of conservation of nucleon number, but leave unaffected most other conservation laws.

ADAMS, RUDERMAN and WOO (preprint 1962) have recently pointed out that neutrino pair emission can take place through the decay of quantized plasma waves (plasmons) in a relativistic plasma for $T > 10^7$ degrees and $\varrho > 10^5$ g/cm^3. The efficiency of this process is such that energy losses by neutrino emission from stars even at the red-giant and carbon-burning stages is of significance and must be taken into account in constructing evolutionary models. This effect, since it is important at a comparatively early stage of a star's evolution as compared with the neutrino processes previously described, makes it possible to contemplate an astronomical test of the universal theory of Fermi interactions. Thus calculation of theoretical models and comparison with observation of stars at the giant and carbon-burning stages may enable an upper limit to be set to the energy loss by neutrino processes.

5. – Matter at high densities.

5'1. *Properties of matter at high densities.* – In recent years there has been a considerable revival of interest in the properties of matter at high densities. As was originally pointed out by LANDAU (1932) we can consider in general terms the case where the pressure that stabilizes a cold star is due to an ideal Fermi gas with the conditions:

a) the pressure is due either to electrons or nucleons;

b) ideal Fermi particles; no correction for interaction between particles;

c) limiting situation is reached when the A particles in a sphere of radius a are packed so tightly that the inner linear dimensions of the region of confinement of one particle $\sim a/A^{\frac{1}{3}}$ are smaller than the Compton wavelength of the particle, or the energy of the particle is relativistic and the rest energy plus kinetic energy is proportional to momentum and independent of mass;

d) a Newtonian treatment of gravitation is used with no corrections for general relativity.

He concluded that the Fermi energy of the particles will dominate over the gravitational energy when the total number of particles is comparatively small, and an equilibrium will be reached at nonrelativistic energies. For a higher total mass the energy is driven to the relativistic limit.

The first detailed investigations of stellar models in which such conditions apply were made by CHANDRASEKHAR (1938). He considered the equilibrium of stellar matter under gravitational forces at densities sufficiently low so that the electron pressure is all important and the nuclei only contribute inert mass. He considered both relativistic and nonrelativistic electron energies and Newtonian gravitation. He thus derived a critical mass which is given by $(5.73/\mu^2)\,\boldsymbol{M}_\odot$ greater than which there is no equilibrium configuration (μ is the average molecular weight of the stellar material). This is the well-known critical mass for white dwarfs which plays an important role in modern theories of stellar evolution.

On the other hand a second critical point is reached when the density becomes so high that inverse β-decay makes neutrons, the nuclear structure disappears, and the nuclear binding energy is small compared with the Fermi kinetic energy of the neutron gas. Such neutron configurations were first considered by OPPENHEIMER, SERBER and VOLKOFF (OPPENHEIMER and SERBER 1938, OPPENHEIMER and VOLKOFF 1939) who treated an ideal neutron gas in both relativistic and nonrelativistic cases and used a general relativistic formulation of the equations of hydrostatic equilibrium. OPPENHEIMER and VOLKOFF (1939) thus found an observable critical mass of $0.7\,M_\odot$ for a neutronstar. For observable masses greater than this no equilibrium neutron configurations exist.

An attempt to consider the whole range of densities has been made by HARRISON, WAKANO and WHEELER (1958). Starting with the end point of normal thermonuclear evolution (^{56}Fe) they used at higher densities semi-empirical formulae for nuclear binding energies, and at very high densities the approximation of an ideal Fermi gas, nuclear forces being neglected. They also used the general relativistic formulation of the hydrostatic equations. Their results are shown in Fig. 5 and demonstrate rather nicely the existence of the two critical masses of CHANDRASEKHAR and of OPPENHEIMER and VOLKOFF in a single plot of mass against central density.

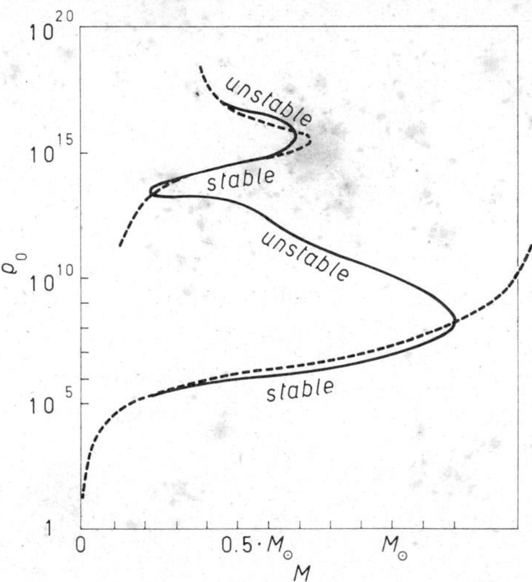

Fig. 5. – Mass of a star made out of cold catalyzed matter in solar units as a function of $\varrho_\odot$, the central density in gm/cm³, as deduced from integrations made by WHEELER and his colleagues. The lower dashed curve shows the connection between mass and central density which was deduced by CHANDRASEKHAR without the use of general relativity, where the pressure derives entirely from an ideal electron gas. The upper dashed curve represents the results of Oppenheimer and Volkoff derived by integrating the general relativity equation of hydrostatic equilibrium for an idealized pure Fermi neutron gas. The two transitions on this curve from state equilibrium to unstable equilibrium mark what might be called two « crushing points ». The lower curve marked « stable » where it runs under the upper curve marked « stable » is metastable with respect to contraction to a high density; however, a potential hill has to be surmounted to pass from the less compact condition of stability to the more compact one. From the more compact configuration it appears conceivable by surmounting another potential hill to pass to a configuration of still lower energy. However, little is understood about this.

While the properties of configurations at very high densities are of considerable interest in their own right, and may be of great cosmological significance (cf. HARRISON, WAKANO and WHEELER 1958) they may be of more direct importance from the point of view of stellar evolution. Thus the early work on neutron stars was carried out at a time when BAADE and ZWICKY (BAADE and ZWICKY 1934 and 1938, ZWICKY 1938 and 1939) had proposed that the energy release in a supernova explosion might result from the formation of a neutron star. Present ideas, described in an earlier section, suggest that the implosion may leave a neutron configuration, though the details are far from clear. It is with this background that CAMERON (1959 *d*) and SALPETER and HAMADA (SALPETER 1960, HAMADA and SALPETER 1961) have re-investigated neutron configurations. Quite a different reason for studying the properties of matter at ultra-high densities has been advanced by AMBARTSUMIAN and SAAKYAN (1960, 1961). This is the idea of AMBARTSUMIAN that the initial form of matter out of which stars and galaxies are formed is not, as is commonly supposed, very low density gas, but is gas at exceedingly high densities.

In the final phases of nuclear evolution the central region of a star is exceedingly hot, but as was discussed in the previous section, the cooling by neutrino processes takes place very rapidly. During the collapse phase it is not clear what the run of temperature will be, but provided that it drops to values such that the thermal energy is very small compared with the nuclear binding energy, and with the Fermi energy of the electrons, then at densities greater than about 10^{12} g/cm^3 one is dealing with an almost pure neutron core. To consider the further evolution of such a configuration it is necessary to determine the equation of state in detail. If the nuclear interactions are neglected then the neutron pressure $p_n \propto \varrho^{\frac{5}{3}}$, since the neutrons constitute a non-relativistic degenerate Fermi gas. However, as the density increases to about 10^{11} g/cm^3, the attractive nucleon-nucleon interaction potential begins to come into play and depresses the pressure below the Fermi gas value. The actual pressure depends very strongly on the details of the nuclear potential. SALPETER has discussed the equation of state for a wide range of densities at zero temperature. CAMERON has used an approximate equation of state based on a simple three-body effective nuclear potential developed by SKYRME. By integrating the general relativistic equations of hydrostatic equilibrium he has constructed neutron-core models. The work of SALPETER indicates that for densities between about $1.2 \cdot 10^{12}$ g/cm^3 and $1.5 \cdot 10^{16}$ g/cm^3 the simple expression used by CAMERON is satisfactory, while outside this range this approximation breaks down. The results of CAMERON show that the maximum observable mass for a neutron star is increased over that originally obtained by OPPENHEIMER and VOLKOFF to about 2 $M_\odot$ (corresponding to a proper mass of about 3 $M_\odot$). For this mass the radius of the star is about 8 km. This maximum mass has a central density rather close to the maximum density for which the equation of state remains valid, so that the numerical value is rather uncertain. However it is probably significantly larger than the CHANDRASEKHAR critical mass.

A neutron star would cool off very rapidly following its formation in a supernova outburst, and if the neutron configuration alone remained it would presumably be undetectable. However, if an envelope outside the neutron core were stable, it is possible that such a star with a neutron core might show some activity which would make it detectable.

At all densities there appears to be some admixture of other nucleons in a neutron configuration, and at densities above about 10^{15} g/cm^3 a gradual transformation to a gas of protons and hyperons begins. AMBARTSUMIAN and SAAKYAN (1960, 1961) have discussed the transformations which take place in the range of density 10^{12} to 10^{18} g/cm³. At densities of 10^{16} to 10^{17} g/cm³ there are appreciable concentrations of Λ^0, Σ^+, Σ^-, Σ^0, Ξ^- and Ξ^0 particles. At densities greater than about $5 \cdot 10^{16}$ g/cm³ the small separations between baryons mean that strong short-range repulsive forces become of importance, and at the highest densities the bulk of the mass may be concentrated in a hyperon core. ZEL'DOVICH (1961) has pointed out that short-range repulsive forces lead to an infinite pressure at finite density which gives a sound velocity $> c$. Consequently the equation of state at very high densities must be modified in order to satisfy the requirements of the theory of relativity. Thus he obtained an equation of state of the form $p \approx n^2$ where n is the baryon density, which applies at very high densities. It is not clear, however, whether such high density configurations play any part in astrophysics.

Recent calculations relating to the CHANDRASEKHAR critical mass have been carried out by SALPETER and HAMADA (SALPETER 1961, SALPETER and HAMADA 1961). They have considered corrections to the equation of state for a degenerate electron gas due to:

a) the classical Coulomb energy of an ion lattice with uniformly distributed electrons;

b) Thomas-Fermi deviations from uniform charge distributions of the electrons;

c) exchange energy and spin-spin interactions between electrons.

The first of these corrections is the largest. At very high densities the correction to the pressure is a multiplying constant factor which is 0.994, 0.986 and 0.960 for $Z = 2$, 6 and 26. Also at very high densities there are restrictions on possible values of A and Z due to inverse β-decays and high density reactions.

Stellar models for such configurations have been calculated using the normal hydrostatic equations for assumed compositions of ^{4}He, ^{12}C, ^{24}Mg, ^{28}Si, ^{32}S and ^{56}Fe. The limitations on composition come because high density reactions, which are discussed in the following sub-section, lead to the conversion of H to ^{4}He at densities above $5 \cdot 10^4$ g/cm³, ^{4}He to ^{12}C at densities above $8 \cdot 10^8$ g/cm³, and ^{12}C to ^{24}Mg at densities above $6 \cdot 10^9$ g/cm³. The models have smaller radii and masses than the simple Chandrasekhar models. The deviations are largest—of the order of 20%—for a central density of 10^5 g/cm³ and a mass of $0.1\,M_\odot$.

For the higher density models the most important effects are due to inverse β-decays. Instead of the Chandrasekhar limit of $1.46\,M_\odot$ for a star composed of ^{4}He, which is reached as the central density tends to infinity, maximum masses are obtained for finite central densities. The lowest possible value of the critical mass is $1.01\,M_{\odot}$ for a star of optimum composition; for a star composed of ^{56}Fe it is $1.11\,M_{\odot}$, and it is largest for a star composed of ^{12}C where it is $1.40\,M_{\odot}$. The values of the central density for these configurations lie in the range 10^9 to 10^{10} g/cm^3. Thus a star with a mass less than $1.01\,M_{\odot}$ cannot contract beyond a critical finite central density of the order of 10^9 g/cm^3 or less. On the other hand a star with a mass greater than $1.40\,M_{\odot}$ must contract to central densities exceeding 10^{10} g/cm^3 as soon as its nuclear fuel is exhausted. These revised models suggest that stars in the range 1.01 to $1.40\,M_{\odot}$ may reach the explosive stage described by HOYLE and FOWLER (1960) rather than the much narrower range between 1.0 and $1.1\,M_{\odot}$ which those authors deduced from Chandresakhar's models.

HAMADA and SALPETER have also constructed some models in which a dense core is surrounded by a hydrogen-rich envelope. In these the radius is considerably increased as compared with the increase in mass.

5·2. *Nuclear reactions at high densities.* – At high densities electron shielding cuts off the nuclear potential barriers quite near to the nuclear surface. Under these circumstances the classical turning points of low-energy ions are very insensitive to the bombarding energy. Nuclear reaction rates then become very insensitive to temperature but very sensitive to density. The possible importance of nuclear reactions which take place under these conditions was pointed out by SCHATZMAN (1948, 1958). Calculations of the reaction rates for helium burning, $^{14}\mathrm{N}(\alpha,\gamma)^{18}\mathrm{F}(\beta^+\nu)^{18}\mathrm{O}$ and $^{16}\mathrm{O}(\alpha,\gamma)^{20}\mathrm{Ne}$, and for the ions ^{12}C, ^{16}O, ^{24}Ne, ^{32}Mg, ^{38}Si, ^{44}S, ^{50}A, and ^{56}Ca reacting with themselves, in regions of high density, have been carried out by CAMERON (1959 *e*). The results are given in the form of ignition diagrams. The conditions of temperature and density which have been chosen range from $5\cdot10^6$ to $1.6\cdot10^8$ degrees and from 10^6 to 10^{10} g/cm^3 for the light elements, and low temperatures and densities in the range 10^8 to 10^{11} g/cm^3, for heavy ions reacting with themselves. CAMERON originally proposed that the onset of reactions of this type might provide a mechanism for nova outbursts. However, the observational evidence now strongly suggests that all novae are members of binary systems, so that whatever the mechanism of a nova outburst may be, it must be intimately related to the binary nature of the system, perhaps through the capture of material from the companion. Consequently it does not seem probable that such reactions are important in explaining nova outbursts. On the other hand reactions of this type may be of significance in the late stages of stellar evolution when high densities $\gtrsim 10^6$ g/cm^3 may be reached in a degenerate zone. Whether such processes will lead to violent outbursts is an open question. Although the nuclear physics of the problem is fairly straightforward, this is another example of a situation in which the astrophysical importance of the process cannot be assessed until theoretical evolutionary tracks are taken much further.

The rates of nuclear reactions in hydrogen at high densities and low temperatures have been calculated by ZEL'DOVICH (1957) using the approach of solid state physics. He points out that at densities of 10^4 to 10^6 g/cm^3 and temperatures below 10^6 degrees hydrogen is a solid, so that the protons are located at the lattice points of a crystal and thermal motions can be neglected so that normal thermonuclear reactions will not occur. The rate of the reaction $^1H(p, e^+\nu)^2D$ is then calculated by considering the tunnel effect in this situation. The height of the Coulomb barrier is unaffected, but the wide low-energy region which is normally most difficult to overcome is got rid of because of the material pressure. Following the first reaction in the proton-proton chain the reaction $^2D(p, \gamma)^3He$ goes very rapidly. This is followed at these densities by $^3He(e^-, \nu)^3T$ and finally $^3T(p, \gamma)^4He$. The rate of energy production at densities of $5 \cdot 10^5$ g/cm^3 and $0.7 \cdot 10^5$ g/cm^3 is found to be 1000 erg/g s and 0.25 erg/g s respectively. ZEL'DOVICH has pointed out that such reactions place a limit on the compression of cold hydrogen. Even at a density of $0.7 \cdot 10^5$ g/cm^3 at zero temperature, a temperature of about $1.5 \cdot 10^6$ degrees will be reached in about $6 \cdot 10^7$ years.

KIRZHNITS (1960) has considered the aggregate state of matter in white dwarfs and has argued that owing to the high density, even at a temperature as high as 10^7 degrees, the Coulomb coupling is so rigid that matter cannot be considered to be in a plasma form, but it must be a condensate which may not be able to evaporate. If it is in this form then the concentration of hydrogen may be several per cent, whereas if it is in the form of plasma, the rate of energy generation calculated for high densities is incompatible with the small luminosity and long time scales for white dwarfs unless the concentration of hydrogen is less than 10^{-5} (WILDHACK 1940).

5·3. *Beta-decay processes under extreme conditions of density and temperature.* – The extreme conditions of temperature and density which may arise in stellar interiors may change the rates for certain β-processes. The following effects have been investigated:

a) Beta-decay from excited states. The possibility that excited states in some nuclei are occupied and that these excited states can β-decay with half-lives which are much different from the terrestrial measured half-lives of the ground states has been discussed by CAMERON (1959 *c*). He pointed out that if the ground state is already rapidly undergoing such transitions then it is unlikely that the total decay rate will be changed significantly. However, if the ground state decays very slowly, some significant accelerations in the rates are possible. CAMERON has considered how this may affect the production of critical elements in the *s*-process chain, which is taken to occur during the carbon-burning process at a temperature of about $7 \cdot 10^8$ degrees. In the case of technetium, which is seen in the S stars, it is argued that at this temperature the lower excited states will be occupied, and the first and third excited states of ^{99}Tc can decay to the ground state of ^{99}Ru by allowed transitions; the calculations give partial half-lives for these transitions of $8 \cdot 10^{-3}$ years. Thus under these conditions ^{99}Tc decays in a day or two, and in this case it is not this isotope of technicium which is seen. CAMERON argues therefore that the technicium which is seen must be ^{97}Tc, which has a half-life of $2.6 \cdot 10^6$ years, so that this is the indicator of the time for s-process material to reach the surface after production, rather than $2.1 \cdot 10^5$ years which is the half-life for ^{99}Tc. This chain of argument leads to further difficulties since ^{97}Tc does not lie in the s-process chain. Thus it is necessary to produce it from decay of an excited state of ^{97}Mo (which lies on the main s-process path). This in turn requires that the material has been subjected to high temperatures for a period of the order of 10^2 years after the s-process has terminated. Such complications can be avoided if the neutron sources appear at lower temperatures. The conditions under which neutrons can be supplied were discussed earlier.

b) Electron capture. The capture of electrons from continuum orbits in stellar interiors has been investigated by DUMEZIL-CURIEN and SCHATZMAN (DUMEZIL-CURIEN and SCHATZMAN 1950 and 1951, DUMEZIL-CURIEN 1951, SCHATZMAN 1953 and 1958). The reactions discussed were:

$$p + e^- \rightarrow n + \nu ,$$

$$^3He + e^- \rightarrow {}^3T + \nu ,$$

$$^7Be + e^- \rightarrow {}^7Li + \nu .$$

The second and third of these are of especial interest since at high densities they occur in the modified proton-proton chain reactions. A new calculation of processes of this type using the universal Fermi interaction has recently been made by BAHCALL (1962). The total rate for allowed electron capture in a star (λ_{star}) as compared with that on the Earth (λ_{earth}) can be written in the form:

$$\frac{\lambda_{star}}{\lambda_{earth}} = N_K + R ,$$

where N_K is the probable number of electrons in the K shell in the atom in a stellar interior, and R is the ratio of the capture rate from a continuum orbit to the capture rate from a bound orbit. The calculations of N_K by COX and EILERS (quoted by BAHCALL 1962) and R by BAHCALL show that in some cases at high temperatures and for fairly large Z, ($R \approx 200\, Z^{-2}$, $Z \geqslant 20$) the stellar electron capture rates can be orders of magnitude different from the terrestrial values. These results may have some minor effects in element synthesis. At present however, calculations of the degree of ionization of heavy elements in stellar interiors are incomplete.

Another effect which may occur in stellar interiors is the possibility that a nucleus may, if it is completely ionized, β-decay by creating an electron in a bound orbit instead of a continuum state (BAHCALL 1961).

c) The effect of atomic binding energies. The total energy for β-decay on Earth is the sum of the nuclear energy release W_0 and the difference between the initial and final atomic binding energies. If W_0 is less than mc^2 the terrestrial β-decay can only occur because the electron binding energy increases when the charge increases from Z to $(Z+1)$. For ^{187}Re, $(W_0 - mc^2)$ is probably less than 5 keV, so that it could not β-decay in a star if it was completely stripped of electrons. A similar situation probably also obtains for ^{241}Pu (BAHCALL 1962). Neither of these effects is expected to be of any appreciable importance in nuclear astrophysics.

d) The exclusion principle. The free electron densities in stellar interiors may sometimes be so large that the number of states available to a decay electron is significantly reduced according to the exclusion principle. The calculations of this effect by BAHCALL (1962) for the interior of a white dwarf, in which the electrons are represented by degenerate Fermi statistics, and at an electron density of $7\cdot10^{29}/\text{cm}^3$ for which the Fermi energy is 200 keV, show that a number of nuclei which normally decay by emitting electrons with kinetic energies $\leqslant 200$ keV will not be able to decay. These nuclei are ^{3}H, ^{32}Si, ^{35}S, ^{63}Ni, ^{66}Ni, ^{79}Se, ^{93}Zr, ^{95}Nb, ^{107}Pd, ^{129}I, ^{151}Sm, ^{171}Tm, ^{187}Re, ^{191}Os, ^{203}Hg, ^{210}Pb, ^{227}Ac, ^{228}Ra, ^{241}Pu, ^{246}Pu, and ^{249}Bk. In real astrophysical circumstances this interesting effect may not be of significance, since in white dwarf configurations under normal circumstances nuclear activity has ceased. It is conceivable, however, that such effects will have to be taken into account in stars in which these nuclei are synthesized either in regions of high density, or under more normal conditions, and then mixed into regions of exceedingly high density. Recently BAHCALL (1961) has pointed out that the effects of the exclusion principle and β-decay from excited states may allow us to determine the temperature at which the *s* process has gone on if isotope abundances of Zr could be determined observationally in S stars.

6. – Age determinations from radio-active nuclei.

6·1. *The age of the elements in the solar system.* – It was first pointed out by RUTHERFORD (1929) that knowledge of the radioactive decay properties of uranium and thorium derived from terrestrial samples could be used to determine the period which has elapsed since the formation processes of these elements ceased. Since this was done before the development of modern ideas of energy production and nucleosynthesis, he believed that such calculations referred to the period since the terrestrial material was removed from the sun, since it was supposed that the uranium was formed in the sun. Recent work has shown that the elements have been formed in the stars over the age of the Galaxy, which is dated from a zero point corresponding to the formation of the first stars, so that these arguments appear now to give us information on this age. Consequently this part of the subject is now commonly referred to as nuclear cosmochronology. An excellent summary of the early ideas, and comparisons of time-scales for the galaxy obtained by different methods at the present time has been given by FOWLER (1962).

The isotopes ^{232}Th, ^{235}U, and ^{238}U have mean lives of 20.1, 1.03, and $6.51 \cdot 10^9$ years respectively. The other isotope which has a comparable mean life is ^{40}K ($1.87 \cdot 10^9$ years). Now ^{232}Th, ^{235}U, and ^{238}U are all synthesized by the *r*-process which took place in a supernova outburst. On the other hand ^{40}K lies in the *s*-process capture chain, and this was synthesized in a star at a different stage of evolution. We devote our attention here to the uranium and thorium. Some discussion of the time scale indicated by ^{40}K was given by B^2FH. In order to derive an age for these isotopes presently existing in the solar system data are required as follows:

1) The relative abundances in the solar system at present.

2) The initial relative abundances with which these isotopes were formed.

3) A detailed understanding of the astrophysical conditions under which they were produced.

It is possible, for any given model, to use these data to derive an age for the elements, and hence a minimum age for our Galaxy, provided that the material has always resided within the Galaxy. The range in the results which have been derived, starting with those given by B^2FH, and followed by the extensions of this work by FOWLER and HOYLE (1961) and FOWLER (1962), the investigations of KOHMAN (1961) and of DICKE (1962) have been due in large part to the different assumptions which have been made under heading 3).

As far as 1) and 2) are concerned B^2FH used the Suess-Urey values $^{235}\mathrm{U}/^{238}\mathrm{U} = 0.0072$ and $^{232}\mathrm{Th}/^{235}\mathrm{U} = 3.5$. Now in order to calculate the production ratios we must consider the details on the r-process. In this process isotopes with A up to about 270 are built. Contributions to the abundance of ^{232}Th come not only from the β-decay of the material produced at $A = 232$, but also from β-decay and subsequent α-decay of material at $A = 236$, 240, 244, 248 (89 % per cent α-decay and 11 % spontaneous fission) and 252 (97 % α-decay and 3 % spontaneous fission). For $A \geqslant 256$ spontaneous fission alone occurs. Thus if all of these nuclei were built in equal abundance, the progenitor abundance factor would be $(4+0.89+0.89\times0.97)=5.75$. For ^{235}U there are 6 contributors $A=235$, 239, 243, 247, 251, 255 and the factor is 6. Contributions to ^{238}U come from $A=238$, 242, 246 and 250 (10% α-decay, 90% spontaneous fission) so that the factor is 3.1. Thus if the abundances were constant the initial values of $^{232}Th/^{238}U = 5.75/3.1 = 1.85$ and $^{235}/U^{238}U = 6/3.1 = 1.94$. However, the detailed calculations of abundances in this region (B^{2}FH, BECKER and FOWLER 1959) show that they are not constant, and if we weight by the appropriate factors the $^{232}Th/^{238}U$ ratio is finally found to be 1.65. In the same way the uniform abundance hypothesis would give $^{235}U/^{238}U = 6/3.1 = 1.94$, but the smaller stability and lower production rate for odd A relative to even A reduces this to 1.65, the exact agreement with $^{232}Th/^{238}U$ being accidental. FOWLER and HOYLE assigned relatively small probable errors to these ratios and adopted $^{232}Th/^{238}U = ^{235}U/^{238}U = 1.65 \pm 0.15$. CAMERON (1962) has used the value of 1.65 for $^{232}Th/^{238}U$, but he has argued that the value of $^{235}U/^{238}U$ should be reduced to about 1.4.

A number of different assumptions can be made about the sequence of astrophysical events which led to the production of uranium and thorium. The simplest is to consider that these elements were formed in a single supernova and that some of the ejected material after mixing with, and being diluted by the interstellar gas, condensed and formed the solar system. This was one of the models originally considered by B^{2}FH. Using the equations for radio-active decay and these initial and present-day abundance ratios the time that has elapsed since this event is then easily calculated and it is found to be $6.6\cdot19^9$ years. Any changes in the value for the initial $^{235}U/^{238}U$ ratio which might be reasonable would only change this value very slightly, in the sense of reducing it if the ratio is reduced below 1.65.

On the other hand as has been described earlier a considerabl body of evidence now exists which suggests that for stars comparable with or less than the age of the sun the compositions are rather similar. While this evidence, strictly speaking, does not apply to the r-process elements, since few abundance determinations are available for these rare elements outside the solar system, it does strongly suggest that by the time that the sun condensed, sufficient element synthesis and mixing had occurred so that the material is made up of the results of many separate synthesis events. If we accept that this also applies to the r-process elements, then a more complex model is required. In such a model it must be supposed that synthesis has gone on continuously since the first stars to condense built elements, and ejected them.

The rate at which this continuous synthesis has gone on can be roughly determined by investigating the abundances of the elements as a function of the time at which they condensed into stars, *i.e.* by relating ages to compositions. The results of these investigations were described earlier. A number of other methods are also of importance. The studies of the rate of star formation in the neighborhood of the sun assuming that it is proportional to the first or second power of the density of the gas out of which the stars form (SCHMIDT 1959, SALPETER 1959, MATHIS 1959) all show that the rate at the present time is very much less than it was in the past. (It must be remembered that other parameters, such as the initial turbulence and temperature of the gas, which may also have a considerable bearing on the rate of star formation have been neglected in these studies.) The mass fraction of the gas in the Galaxy, which was originally unity and is now down to about 0.02 in the form of neutral atomic hydrogen (with an unknown but probably considerable amount of molecular hydrogen) also shows that the rate of star formation must have decreased considerably since the first stars condensed. Also one possible explanation of the He/H ratio being as large as it is, is to suppose that the activity early in the history of the Galaxy was very much greater than it is now.

All of these arguments show that there has been a decrease in the rate of element synthesis, but the details remain rather obscure. One possible assumption, which can be made intuitively, and is also suggested by the work on the rate of star formation, is to suppose that star formation and evolution have decreased exponentially with time (cf. BURBIDGE 1958, SCHMIDT 1961, SALPETER 1959). With this model the heavy element content increases linearly with time and eventually saturates. Another is to suppose that almost the whole of the element synthesis was restricted to a very short period 10^8 to 10^9 years at the very beginning. Such a hypothesis is compatible with present interpretations of the compositions of old stars. It is also not in conflict with the work on the rate of star formation based on the luminosity function of stars in the neighborhood of the sun, because if significant element formation is limited to this short period, it can only have involved very massive stars, so that the mass spectrum of star formation was very different from the form it has now. However, it would be in conflict with the theoretical arguments of HOYLE and FOWLER (1960) that r-process elements are produced in supernovae of Type I which arise when stars with masses of 1.2 to 1.5 $M_\odot$ explode, since the first generation of such stars will take at least $3 \cdot 10^9$ years to evolve to the exploding stage. Clearly for such a model the time scale is almost the same as that derived earlier for synthesis in a single supernova, and as far as the Galaxy is concerned this time scale is unreasonably short. Finally it can be supposed that the elements in the solar system are made up of two components, those which were produced at an early epoch together with an admixture of material which has been made in a supernova outburst just prior to the formation of the solar system. This is the viewpoint currently adopted by CAMERON (1962).

The detailed calculations of FOWLER and HOYLE (1961) and FOWLER (1962) who have assumed an exponential decrease in the rate of star formation give a clear picture of the present situation. They have considered several different values for the time constant in the exponential. Their calculations show that the duration of nucleosynthesis is independent of the assumed rate of decline over rather wide limits. In this work an abundance ratio $^{232}Th/^{238}U = 3.8 \pm 0.3$ has been used, and this represents a slight adjustment from the values originally used by B^2FH. Three separate epochs are involved. They are:

I) The time ($\bar{t}_E$) which elapsed before the first stars synthesized elements and ejected them.

II) The time (Δ) during which element synthesis took place at an exponentially decreasing rate, some of the material being ejected and a fraction retained in the evolved star.

III) The time (t_s) which has elapsed since the material was condensed into the solar system. At this time the material was removed from continuous synthesis and has since been undergoing free decay.

With this division the age of the elements is ($\Delta + t_s$), while the age of the Galaxy is ($\bar{t}_E + \Delta + t_s$). The value for Δ is found to be $(7.7 \pm 2) \cdot 10^9$ years where the probable errors appear to be quite generous. t_s is estimated to be about $4.7 \cdot 10^9$ years. Thus the age of the elements is $(12.4 \pm 2) \cdot 10^9$ years. The results are shown in Fig. 6 which has been taken from the paper by FOWLER (1962). The evolution time $\bar{t}_E$ is still the most uncertain quantity of the three, and with an assumed value of $(3 \text{ to } 4) \cdot 10^9$ years FOWLER and HOYLE originally obtained an age for the Galaxy of $(15^{+5}_{-3}) \cdot 10^9$ years. However, there may be reason to increase this, and the age which seems reasonable at present is $(20 \pm 4) \cdot 10^9$ years.

Clearly the long waiting period $\bar{t}_E$ is of critical importance in considering the age of the Galaxy. Without it the age of the elements is approximately equal to the age of the Galaxy. It can be reduced either by supposing that massive stars give rise to the r-process elements or by other arguments. DICKE (1962) and BRANS and DICKE (1961) have in recent years described cosmological models in which the constant of gravitation is secularly varying, and decreasing as the age increases. Since the luminosity of a star $L \propto G^7$ if Kramers opacity is assumed, this means that in the past all stars were brighter and consequently their evolutionary times were shorter. On this assumption and making certain other changes in the arguments given above —he assumes that massive stars provided heavy elements and halo population stars contribute—DICKE concluded that the age of elements and the Galaxy lies in the range $(7.5 \text{ to } 11.1) \cdot 10^9$ years. This determination is based on the ratio $^{235}U/^{238}U$ alone. More recently DICKE (preprint) has obtained an age of between 7 and $8 \cdot 10^9$ years by using a model in which the heavy elements were produced in massive halo stars. This is essentially a sudden synthesis model.

Another model which can be used in discussing the Th and U ages is obtained by assuming that material can be interchanged between galaxies. Such a model was developed by FOWLER and HOYLE. In this it is assumed that

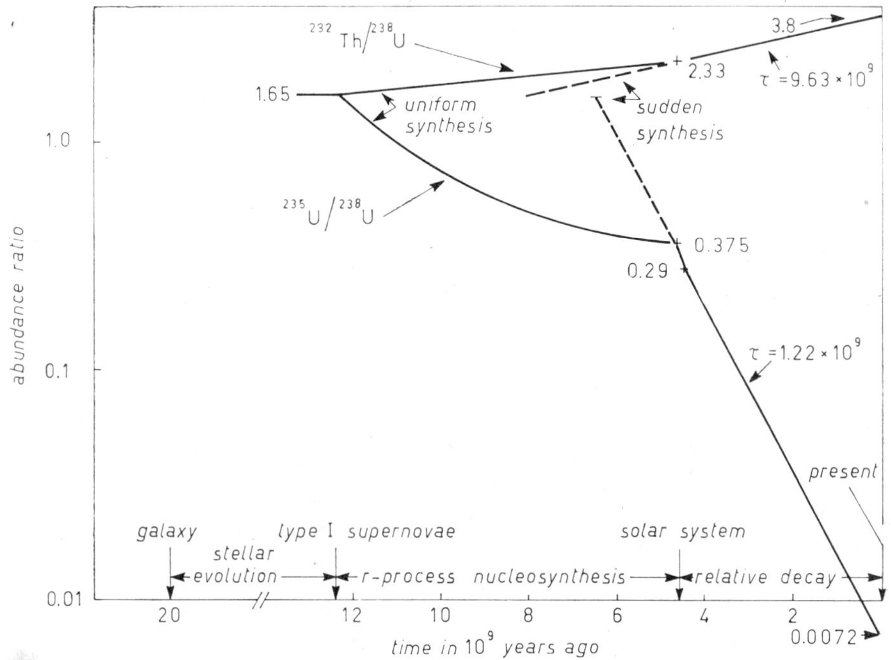

Fig. 6. – The abundance ratios ^{232}Th/^{238}U and ^{235}U/^{238}U as a function of time for solar-system material before and after formation of the system. The curves are adjusted to the present day ratio and to the calculated ratio for production in each r-process event. The calculations indicate continuous synthesis in the Galaxy from $12.4 \cdot 10^9$ to $4.7 \cdot 10^9$ years ago. Type I supernovae are assigned an evolutionary life-time of $7.6 \cdot 10^9$ years leading to an age for the Galaxy of $20 \cdot 10^9$ years. The dashed lines represent the extension of simple decay back to a hypothetical event of sudden synthesis. Concordant dates are not obtained for ^{232}Th/^{238}U and ^{235}U/^{238}U. The most important point is the increase in Galactic age stamming directly from the hypothesis of uniform synthesis.

the r-process material is characteristic of extragalactic material. It is then possible to overcome the difficulty that the abundances of the r-process elements are too high by a factor of about 100 relative to the elements produced in other processes in the models of B^{2}FH and the later modifications, if the energetics of the light curves of supernovae are determined from the californium hypothesis, and if the normal frequencies of supernovae are used. If the r-process elements contaminate the whole of the material in the Universe, since, as we have mentioned earlier in another connection, the cosmological mean density may be about 100 times the density of luminous material in the Universe, this discrepancy is removed. For the steady-state cosmological model it is possible to use this model to calculate the value of the Hubble constant from the uranium and thorium data. The resulting value for H^{-1} (FOWLER and HOYLE 1961) is $(11 \pm 6) \cdot 10^9$ years. The difficulties with this model are severe, however, since it is required that supernovae eject r-process elements into the intergalactic medium with a high efficiency. Although the ejection velocities are high enough to do this if there is little deceleration of the supernova shells, many supernovae are known to have gone off in regions of high gas density where the interaction with the interstellar medium must have stopped ejection from the galaxy.

6·2. *The age of the solar system.* – We review briefly some age determinations obtained from radio-active nuclei in the solar system. PATTERSON and his colleagues (PATTERSON 1956, PATTERSON, TILTON and INGHRAN 1955) determined the isotope ratio of ^{206}Pb, ^{207}Pb, and ^{208}Pb relative to ^{204}Pb in a number of iron and stony meteorites. Measurements have also been made in oceanic lead. The first three lead isotopes contain contributions from the decay of ^{238}U, ^{235}U and ^{232}Th respectively, whereas ^{204}Pb is nonradiogenic. Now if it is assumed that the meteorite parent bodies were formed at the same time, and that they originally contained lead and uranium with the same isotopic composition as the Earth, then various lead isotope ratios may be expected to arise, due to the occurrence of different U/Pb and Th/Pb ratios in different parent bodies. These assumptions, together with the observed isotope ratios in different meteorites and the terrestrial ^{235}U/^{238}U or ^{232}Th/^{238}U ratios enable a time t_m to be determined; this is the time which has elapsed since the meteorite parent bodies formed. PATTERSON showed that the slope of the plot of the ratio ^{207}Pb/^{204}Pb against ^{206}Pb/^{204}Pb for different meteorites, which plot he called an isochron, determined t_m with great accuracy. He found that $t_m = (4.55 \pm 0.07) \cdot 10^9$ years. This determination is of the greatest importance, since it gives an age for the solar system with the assumptions made above, which is far more certain than anything that can be obtained, for example, by applying the theory of stellar evolution to derive an evolutionary age for the sun (such determinations are bedevilled by the uncertainties concerning the initial solar composition etc.). Now from the theory of nucleosynthesis, we know that the synthesis of all of the lead isotopes has occurred in the s-process, and in addition the radiogenic contributions to ^{206}Pb, ^{207}Pb and ^{208}Pb come from the uranium and thorium built in the r-process. Thus the additions to these isotopes began when the first r-process stars ejected material, and continued until the material condensed to form the solar system. These isotope ratios, under the assumptions of continuous production, have been discussed by B^{2}FH and by CLAYTON, FOWLER, HULL and ZIMMERMAN (1961). The age of the solar system of $4.7 \cdot 10^9$ years which was used in the preceding discussion is slightly greater than $4.55 \cdot 10^9$ years because it is necessary to take into account the time which elapsed between the separation of the solar-system material from the surrounding interstellar material, and the formation of the primary planetary bodies.

New developments which enable the chronology of the early history of the solar system to be described in some detail came with the discovery by REYNOLDS (1960 *a*, *b* and *c*) of the anomalous abundance of ^{129}Xe in chondritic meteorites, and the discovery by MURTHY (1960) of the anomalous abundance of ^{107}Ag in iron meteorites. An extensive discussion of meteorite ages has been given by ANDERS (1962). It is thought that ^{129}Xe arises from the decay of ^{129}I with a mean life of $2.5 \cdot 10^7$ years and ^{107}Ag from the decay of ^{107}Pb with a mean life of 10^7 years. In addition it has been proposed by FISH, GOLES and ANDERS (1960) that ^{26}Al with a mean life of $1.0 \cdot 10^6$ years has been largely responsible for the heating of the parent bodies of meteorites. It should also be added that smaller anomalies in abundance of other Xe isotopes have been found. A discussion by KURODA (1960) suggests that these may be due in part to excesses in the atmospheric abundances of Xe due to fission by ^{244}Pu.

The explanation for the ^{129}Xe anomaly assumes that when the I/Xe ratio was finally fixed in the meteorite some ^{129}I remained from a nucleosynthesis process which had taken place not long before. If Δt is the interval between the end of nucleosynthesis and the fixing of the ratio I/Xe then it has been shown that

$$\Delta t = 5.7 \cdot 10^7 \left[\log \left(\frac{^{127}\text{I}}{^{129'}\text{Xe}} \right) + \log \left(\frac{^{129}\text{I}}{^{127}\text{I}} \right)_0 \right] \text{years,}$$

where ^{127}I is a convenient abundance standard which can be determined also in the meteorite, $^{129'}$Xe is the amount of ^{129}Xe attributed to the decay of ^{129}I, and $(^{129}\text{I}/^{127}\text{I})_0$ is the ratio of the abundance of the iodine isotopes at the end of nucleosynthesis. Both isotopes are mainly produced in the *r*-process, *i.e.* in supernovae. REYNOLDS assumed a single sudden nucleosynthesis event and calculated Δt to be $3.5 \cdot 10^8$ years. However, as we have seen previously, a more probable situation is that in which there has been continuous synthesis up to the point at which the material first began to condense. This was taken into account by WASSERBURG, FOWLER and HOYLE (1960) and their result, modified by some changes in the values of $(^{129}\text{I}/^{127}\text{I})_0$ obtained from calculation of *s*- and *r*-processes by FOWLER and his colleagues, is that $\Delta t = 1.5 \cdot 10^8$ years. This then must be the interval which elapsed between the end of the nucleosynthesis which produced those isotopes and the formation of large solar-system bodies. For the case of ^{107}Ag MURTHY found on the basis of a model of continuous nucleosynthesis a value of Δt of $4.5 \cdot 10^7$ years.

In these cases, and also when considering the possible presence of ^{26}Al, we are faced with time scales which appear to be very short as compared with what is usually assumed (cf. B^2FH) for the material to be separated and go through the initial phase of evolution of the solar nebula to the point at which the planetesimals formed. Two different proposals have been made to explain these results. The most detailed are those of FOWLER, GREENSTEIN and HOYLE (1962) who have argued that the production of these isotopes took place, not in the normal galactic synthesis processes, but in that phase in the evolution of the solar system in which there was a considerable amount of nuclear activity due to the production of high-energy protons, which gave rise to spallation reactions, and, secondarily, through such reactions to a flux of neutrons and hence neutron-induced reactions. This phase is thought by them to be responsible also for the synthesis of D, Li, Be, and B. The production of ^{26}Al comes from reactions of the type ^{26}Mg(p, n)^{26}Al, ^{27}Al(p, pn)^{26}Al and ^{28}Si(p, 2pn)^{26}Al. Over a time scale of the order of 10^7 years sufficient ^{26}Al is produced to account for the heating required in the parent bodies of the meteorites. ^{107}Pb is probably produced by neutron capture on ^{106}Pd and ^{108}Pd, and ^{129}I by spallation of the nuclei in the range $A = 130$ to 148.

CAMERON (1962) has taken a very different approach to explain these anomalies. He has used them as evidence that the solar-system material was subjected to a sudden period of nucleosynthesis just prior to the condensation of the solar system. Further he argues that some of the material of the meteorites was first subjected to neutron irradiation in the primative sun, so that, for example, the abundances of the xenon isotopes have been considerably affected by neutron-capture processes, the neutrons coming from the burning of deuterium in the primitive sun; in this case there may be a slight depletion of ^{124}Xe and ^{126}Xe in the sun and a considerable increase in both ^{128}Xe and ^{130}Xe, the former at the expense of ^{127}I and the latter at the expense of ^{129}Xe. Qualitatively such effects may explain the anomalies in these isotopes. Similar arguments are used for the silver isotopes.

In a model of this type it is required that the material out of which the solar system formed was synthesized partly in a continuous synthesis process over the age of the Galaxy, and partly by a sudden synthesis process which took place over a period no longer than $3 \cdot 10^8$ years before the formation of the solar nebula took place. CAMERON apparently implies that in this single process of nucleosynthesis both the *s*- and *r*-processes occurred in the same star. At the end of this nucleosynthesis only a short period $< 10^7$ years must have elapsed before the meteorite parent bodies formed, if ^{26}Al is to be responsible for the thermal heating. With such a picture the extinct radio-activities ^{129}Xe and ^{107}Ag were produced in this last period of activity. Finally, MURTHY and UREY (1962) have also proposed that a supernova which occurred very shortly ($\sim 2 \cdot 10^6$ years) before the formation of solid bodies in the solar system was responsible for the extinct radio-activities.

If such models are used it is hard to see why large-scale abundance anomalies are not seen in the solar system relative to other stars in the sense that the solar system abundances are peculiar. Furthermore, although the time-scale for contraction of the sun may be much shorter than was originally thought, due to convection processes, present estimates still suggest that the mixing of material, followed by the whole condensation process from the first stage of gravitational contraction of the solar nebula to the formation of solid bodies, takes much longer than the times of order 10^6 to 10^7 years which are required for these models.

7. – High-energy nuclear astrophysics.

In a review of nuclear astrophysics there is clearly some overlap with topics which are normally discussed in reviewing cosmic-ray problems. The regions of common interest are in the chemical composition and nuclear interactions of the primary cosmic rays, and in modern ideas concerning the places of origin of cosmic rays. As far as the chemical composition is concerned, the fact that the heavy elements are more abundant relative to hydrogen and helium has now been established and has been described by various authors, while excellent reviews of cosmic-ray origin theories have recently been given by GINZBURG and SYROVATSKY (1961) and MORRISON (1961). It is our intention here to describe only those developments which relate to the role of nuclear-energy generation and release in normal supernovae, and possibly under other conditions and the deductions which can be made from the observations concerning the existence of fluxes of high-energy particles.

The discovery of strong sources of nonthermal radio-emission both inside and outside the Galaxy and the observational confirmation of the prediction by SHKLOVSKY that such sources are radiating through the synchrotron mechanism led to a very considerable advance in our understanding of cosmic-ray origin problems. The application of the synchrotron theory to radio sources for which distances and hence dimensions and flux emitted are known, has led to determination of the minimum total energy in the form of relativistic particles and magnetic energy which must reside in these sources.

From the observations and the synchrotron theory it is only possible to deduce the existence of relativistic electrons and positrons. It has been suggested that such particles can arise alone in β-decay processes following the production of β-unstable nuclei in nucleosynthesis at the time of a supernova outburst (BURBIDGE 1956 *b*) and that they can be accelerated in the supernova shell. However, it appears doubtful whether sufficient particles can be generated in this way, and detailed calculations are required to settle this point. The annihilation of nucleons and anti-nucleons could also give such a flux (BURBIDGE 1956 *b*, BURBIDGE and HOYLE 1960) but the difficulties in such a model will be described later. If electrons are not generated by such mechanisms then it is clear that a flux of protons must also be generated at the same time. Calculations suggest that in general the energy content of the proton flux must then be very much greater than that in the electron flux. While the energy losses of the electrons in the relativistic-energy range are due almost entirely to synchrotron radiation and leakage of the particles from the region of production, the losses of the protons are due to nuclear collisions with gas and dust in the region of confinement and to leakage. In nuclear collisions meson production occurs and the final stable products are neutrinos, γ-rays, electrons, and positrons. The neutrinos and γ-rays will immediately escape, but the electrons and positrons will be trapped in the magnetic field in the source and will radiate by the synchrotron mechanism. This mechanism of secondary production of electrons and positrons was proposed independently by GINZBURG (GINZBURG and SYROVATSKY (1961) and references given there) and by BURBIDGE (1956 *a* and *b*), and it is of importance for all of the non-thermal radio sources in which the density of gas and dust is sufficiently high for nuclear collisions to occur at an appreciable rate. The efficiency of the process is fairly low since only about 5% of the energy in a nucleon-nucleon collision at energies $\sim 10^9$ eV or greater will be transferred to electrons and positrons. A reasonable estimate is that where a proton flux is important in a radio-source its total energy is about 100 times the electron-positron energy.

Where this mechanism applies, the time scale for a source is determined not by the lifetimes of the electrons and positrons as given by their synchrotron losses but by the time scale for the proton flux. Since the densities of gas and dust in the radio-sources are low and the rate of nuclear collisions is correspondingly small, the time scale is determined by the rate of leakage of the proton flux from the vicinity of its place of origin. Thus it appears that the bulk of the protons will always escape and contribute to the general cosmic-ray flux.

The spectra of the radio-sources are all of the form $P(\nu) \propto \nu^{-\chi}$ where χ lies in general between about 0.3 and 1.2, the bulk of the sources having spectral indices between 0.5 and 1.0. The range of frequencies covered is in general 38 MHz and 3200 MHz. In a few cases the spectral indices may change slightly in going from low to high frequencies. The index χ is simply related to the particle spectral index n, where $N(E) \propto E^{-n}$, by the relation $n = 2\chi + 1$.

We now turn to a discussion of the different types of source.

7·1. *Supernova remnants.* – The most important developments have come from the study of the *Crab nebula* which is the remnant of the supernova of A.D. 1054. The object is a strong source of both radio and optical synchrotron radiation, thus indicating that a large flux of high-energy electrons and/or positrons with energies in the range 10^8 to 10^{12} eV is present, while the average magnetic field has been estimated by most authors to lie in the range 10^{-4} to 10^{-3} gauss (cf. WOLTJER 1958). A number of other sources in the Galaxy have been identified by MINKOWSKI (1959) with supernova remnants of different ages ranging from a few hundred to about 50 000 years and HARRIS (1961) has described the observational situation with respect to them. Thus radio-astronomy has shown that there are currently large reservoirs of high-energy particles in such sources.

In the case of the *Crab nebula* a model in which it is supposed that the proton flux provides the particle energy source has been proposed (BAADE (1958). This would overcome the problem associated with the fact that many generations of electrons must have been produced since the supernova exploded. However, there are severe energetic difficulties associated with this model which currently requires a total energy $\sim 10^{50}$ erg in particles and magnetic field, because to give this now the energy conditions in the early stages would be almost impossible to fulfil.

The generation of a large flux of particles and a magnetic field in a supernova outburst has been the subject of much discussion but no satisfactory model has yet been developed. As far as the production of a particle flux is concerned two different mechanisms can be mentioned. The Soviet theoreticians GINZBURG, SHKLOVSKY, PIKELNER and SYROVATSKY (cf. GINZBURG and SYROWATSKY 1961) have argued that the acceleration of a flux of high-energy particles takes place by a statistical mechanism of the type proposed by FERMI, in the expanding shell of a supernova. They have also argued that the acceleration mechanism may preferentially accelerate the heavier ions, thus accounting for the composition of the primary cosmic rays. On the other hand, in the model of COLGATE *et al.* (COLGATE and JOHNSON 1960, COLGATE, GRASBERGER and WHITE 1962), the acceleration takes place in the explosion following the implosion when the outermost part of the star is blown off with relativistic velocity.

There is, as yet, no model which will account for the form and estimated strength of the magnetic field in a supernova remnant as instanced by the field in the *Crab nebula.* It seems clear now that the field cannot simply be the expanded field of the star nor can it originate simply by the action of the outward moving shell on the interstellar magnetic field; there must be amplification of an initial field both by the expanding shell and also by turbulent activity in the central regions subsequent to the explosion.

There are also signs of activity in the *Crab* at the present time. Small bright light ripples were observed by BAADE (BAADE 1958, WOLTJER 1958) to be moving in the *Crab* in the central region with velocities of the order of 0.2 to 0.3 c. As was suggested by SHKLOVSKY these may be manifestations of fast hydromagnetic waves. These features suggest that the central faint star which is thought to be the remnant of the supernova is still ejecting energy. This star is presumably a highly evolved star, possibly with a neutron core, as was originally proposed by BAADE and ZWICKLY (1934 and 1938). If the generation of electrons which radiate visible synchrotron radiation has not gone on continuously through secondary interactions of a primary proton flux, or through continuous acceleration processes, then it is natural to suppose that the central star has produced them. If it has been doing this then the energy sources present in the remnant after a supernova has exploded must be rather large.

7·2. *Radio stars.* – Until recently, the only radio-source emitting either thermal or nonthermal radio emission which had been identified with an optical star was the sun. However, in the last year, SANDAGE and MATTHEWS (1962) have made optical identification of three nonthermal radio-sources with faint stars having apparent magnitude of 16^m to 17^m. These identifications have been made possible because radio positions have been determined for a number of sources with accuracies of $\pm 5''$ in each co-ordinate. The stars are exceedingly puzzling. Only two spectral features, at $\lambda\,3830$ and $\lambda\,4686$, have so far been measured in one of these stars. Very faint nebulosity has also been seen in two cases. The colors of the stars suggest that they may be highly evolved stars. No distance estimates have so far been possible for them.

Whether they are supernova remnants of a different type from those discussed previously, or whether they are simply very much older than objects such as the *Crab*, the fact that they are emitting synchrotron radiation in radio frequencies, and may also be emitting visible synchrotron radiation, means that they still must contain considerable energy sources of nuclear or gravitational origin. Estimates based on the distribution of radio-sources suggest that there cannot be more than one or two hundred such stars in the Galaxy. If as many as this do exist, they may contribute heavily to the output of high-energy protons, electrons, and positrons in the Galaxy.

7·3. *Normal Galaxies.* – The primary cosmic-ray flux in our own Galaxy is probably made up of several components (MORRISON 1961). GINZBURG and SYROVATSKY (1961) have argued that the contributions which come from peculiar stars of various types as manifested by peculiar abundances and other evidence of surface nuclear activity (these were discussed in the section on compositions) must be rather small. They concluded that the supernovae make the major contribution to the primary flux. It also appears that an extragalactic component may make a significant contribution (BURBIDGE 1962). The energy density and energy spectrum of cosmic rays determined from observations at the top of the earth's atmosphere have frequently been reviewed. The synchrotron radiation from the Galaxy and its surrounding halo arises when electrons and positrons radiate in the interstellar magnetic fields. The particles are produced either in the cosmic-ray sources directly, by secondary processes in the sources, or by secondary processes in the interstellar gas. The uncertainties in the strength of the interstellar magnetic field in disk and halo —for the disk values between $3.\cdot 10^{-5}$ and 10^{-6} gauss are under discussion— make it uncertain as to what the total energy in the electron-positron component must be to explain the observed radio emission. However, the bulk of these particles must have energies 10^8 to 10^{10} eV. The spectral index of the galactic radio emission $\chi \approx 0.5$.

Evidence of the presence of fluxes of high-energy particles in external galaxies can be derived only from radio data. The power levels are such that only a few of the comparatively nearby spiral and irregular galaxies have been detected (HANBURY, BROWN and HAZARD 1959 and 1961). At present there is no evidence for radio-emission from normal elliptical or SO galaxies. However, calculations based on the power radiated by normal spiral galaxies suggest that they contain fluxes of high-energy particles which are comparable with those in our own Galaxy.

7·4. *Peculiar radio-Galaxies.* – One of the most significant contributions of radio astronomy to high-energy astrophysics has been the detection of galaxies in which exceedingly large amounts of energy are present in the form of relativistic particles. While the total power output of a normal galaxy such as our own in radio-frequencies is of the order of 10^{38} erg/s, the strong extragalactic sources for which distances have been determined are radiating in radio-frequencies at power levels in the range 10^{40} to $7\cdot 10^{44}$ erg/s. These are to be compared with the average power radiated by normal thermal processes (*i.e.* from stars) in a galaxy which is of the order of 10^{43} to 10^{44} erg/s.

From the synchrotron theory and the observational data it is possible to calculate the total energy which must be present in the form of radiating particles for different assumed values of the magnetic field. The total energy in particles and magnetic field which is required is a minimum when the energy of the particles (E_p) is equal to the magnetic energy (E_m). Calculations of this type for all of the sources for which data are available have been made by BURBIDGE (1959) and MALTBY and MOFFET (1961). The results show that these minimum values range from about $2\cdot 10^{56}$ erg to $2\cdot 10^{60}$ erg and $4\cdot 10^{60}$ erg for the two strongest identified sources, *Cygnus A* and *Hercules A*. For about half of the sources the minimum energies lie in the range 10^{58} to 10^{59} erg. The numbers are calculated on the assumption that the total proton energy is 100 times the total electron-positron energy. Apart from the few small sources which are located in the central regions of galaxies where the total energies lie in the range 10^{54} to 10^{56} erg, the characteristic features of these sources are that their dimensions are very much greater than those of optical galaxies, and that frequently they appear to be double. Typical sizes are about 100 kpc with separations between the two components sometimes of the order of 250 kpc. The largest source is *Centaurus A* which is centered about the optical galaxy NGC 5128 and has a dimension of about 0.7 Mpc.

The optical galaxies about which these sources are located have no obvious common features as far as their structures or spectra are concerned. Following the first detailed optical studies by BAADE and MINKOWSKI, they made the suggestion that *Cygnus A*, NGC 1275 and NGC 5128 were each made up of pairs of galaxies in collision (1954) and the idea became prevalent that a strong radio-source will appear when two galaxies collide. However, the optical interpretations were never clear-cut. Moreover the theoretical arguments which are based on the amounts of energy which can be released in a collision, and also arguments based on the frequency of collisions between galaxies, made this suggestion appear improbable. With the recent identification of a considerable number of elliptical galaxies as radio-sources the objections to this hypothesis become very strong, because, even if a collision between two elliptical galaxies occurs, very little energy will be released since the gas content of such systems is very small. Moreover, a number of the strong sources are associated with single galaxies. Thus we conclude that this hypothesis must be abandoned.

The problem is that of understanding how a very large flux of relativistic particles can be generated in a galaxy so that the electron-positron component can later radiate in a magnetic field in a very large volume most of which lies outside the stellar system. The equipartition values for the mean magnetic field derived from the calculations described previously all lie in the range 10^{-5} to 10^{-4} gauss. These are unreasonable large for field strengths in regions outside visible galaxies and in many cases they are unreasonably large for galactic magnetic fields. Consequently it must be supposed that the average field strengths in these radio-sources are much weaker than this, perhaps $\sim 10^{-6}$ gauss or less (BURBIDGE 1962). This in turn means that the total particle energies which are required to explain the observed fluxes of radio-emission must be increased over the equipartition values by factors of $(B_0/B)^{\frac{3}{2}}$, where B_0 is the equipartition value and B is the field strength assumed reasonable for an intergalactic region. The energy requirement may be lessened to some extent by supposing that the total energy of the proton flux is less than 100 times the energy of the electron-positron flux. In any case, the argument leads to the result that particle fluxes with total energies in the range 10^{56} to 10^{62} erg must be generated in violent events in galaxies.

The only energy sources which are known to exist, or which may be speculated on, are:

I) Annihilation energy which is released when matter and anti-matter collide. Such a process is attractive because it will directly provide radiating particles without a proton flux. However, further acceleration of these particles will be required in order to give the radio-spectra. Such an idea was proposed by Burbidge (1956 *b*) and Burbidge and Hoyle (1960). However, the difficulties associated with postulating the presence of anti-matter are severe. There is no independent evidence for the presence of anti-matter in the universe, and, more seriously, it is not clear how matter and anti-matter originally separated so that condensations of matter of one kind could occur and later interact to produce a radio-source (cf. Burbidge and Hoyle 1960).

II) Some part of the magnetic and rotational energy in a galaxy might be converted to energetic particles. A mechanism of this type has been outlined by Hoyle (1961), who has argued that a galactic flare rather similar in principle to a solar flare might occur. The difficulty with this mechanism is that galaxies which are very massive and which contain large amounts of gas are required, but it appears that there is very little gas present in many of the galaxies which are the nuclei of radio-sources.

III) The release of gravitational energy. If collapse to extremely high-density configurations takes place in a large number of stars or in a single massive configuration, the release of gravitational energy might be large enough to provide the energy required. However, there is at present no understanding of the way in which a large number of stars could go through this collapse phase essentially simultaneously, and the idea that very massive configurations ($\sim 10^6 M_\odot$) could condense, evolve and collapse has never been investigated.

IV) The release of energy in supernova explosions. Although, as has been discussed earlier, the mechanism by which a supernova gives rise to a large flux of high-energy particles is only qualitatively understood, the observations must be accepted, so that is appears plausible to suppose that a large number of supernovae can give rise to the energy required in the extragalactic sources. As was described earlier, in a supernova explosion both nuclear and gravitational energy are released. In order to produce a strong radio-source by this mechanism it is necessary for the rate of supernova outbursts to become very large. This is the viewpoint adopted by Shklovsky (1960). If the frequency of supernovae is to be very high this means that the periods of evolution of the pre-supernova stars must be very short. Thus the galaxy must contain large numbers of massive stars which are continuously being formed. However, the observations show that in elliptical galaxies the stars on the average have small masses, $\sim 1 M_\odot$, and there is no evidence that star formation is going on. A different mechanism has been proposed by Burbidge (1961). Here it is argued that in the central regions of elliptical galaxies the star density is probably very high so that the separations between the stars are exceedingly small. Star densities in the range 10^6 to $10^7/\mathrm{pc}^3$ are assumed. Under these conditions a supernova which goes off in the central region following the normal processes of evolution may trigger a chain reaction of stars which explode over a time scale of the order of a few hundred years. The explosions will cease to propagate when the star density decreases sufficiently. Such a mechanism will explain the gross features of many radio-sources and is not in conflict with the optical observations. The particles which are generated in such a galactic explosion must then escape from the galaxy and radiate in the magnetic fields outside.

As was pointed out earlier, the strength of the magnetic field in the radio-source must be much smaller than the equipartition value, *i.e.* $E_p \gg E_m$. Under these conditions the time scale for a strong radio-source is only of the order of 10^6 years after which the bulk of the particles will have escaped into the general intergalactic medium. For the region centered on the *Virgo* cluster of galaxies with a radius $\sim$15 Mpc which contains our own Galaxy and may form a single physical unit, it is possible to calculate the rate at which intergalactic cosmic rays have been injected by short-lived radio-sources. If this rate remained constant over a time of the order of 10^{10} to $2\cdot10^{10}$ years the contribution to the primary cosmic-ray flux may be comparable to the cosmic-ray energy density in our own Galaxy (BURBIDGE 1962).

It is clear from this discussion that the sources of energy which are required to explain the nonthermal radio sources and which may very well be nuclear in origin, and the fluxes of particles themselves, are of importance in the general field of nuclear astrophysics.

* * *

I am indebted to many friends and colleagues who have provided much material prior to publication. This review was begun at the University of Chicago, continued at the University of California ad completed at the Leiden Observatory. I also wish to thank CAROL ZINK for helping to prepare and for typing the manuscript.

BIBLIOGRAPHY

ABT, H. A.: 1960, *Ap. J.*, **131**, 99.
ADGIE, R. L. and HEY, J. S.: 1957, *Nature*, **179**, 370.
ALLER, L. H. and CHAMBERLAIN, J. W.: 1951, *Ap. J.*, **114**, 52.
ALLER, L. H. and FAULKNER, D. J.: 1962, *Pub. Astr. Soc. Pacific*, **74**, 219.
ALLER, L. H. and GREENSTEIN, J. L.: 1960, *Ap. J. Supp.*, **5**, 139.
ALLER, L. H. and JUGAKU, J.: 1959, *Ap. J. Supp.*, **38**, 109.
ALPHER, R. A. and HERMAN, R. C.: 1950, *Rev. Mod. Phys.*, **22**, 153.

AMBARTSUMIAN, V. A. and SAAKYAN, G. S.: 1960, *A. J. URSS*, **37**, 193; *Soviet Astronomy*, **4**, 187.
AMBARTSUMIAN, V. A. and SAAKYAN, G. S.: 1961, *A. J. URSS*, **37**, 785; *Soviet Astronomy*, **5**, 601.
ANDERS, E.: 1959, *Ap. J.*, **129**, 327.
ANDERS, E.: 1962, *Rev. Mod. Phys.*, in press.
APPA RAO, M. K. V., 1961, *Phys. Rev.*, **123**, 295.
ARP, H. C.: 1959, *A. J.*, **64**, 441.
ARP, H. C.: 1961, *Ap. J.*, **133**, 883.
ARP, H. C.: 1962, *Ap. J.*, **135**, 311.
ATKINSONS, R.: 1931, *Ap. J.*, **73**, 250, 308.
ATKINSONS, R. and HOUTERMANS, F.: 1928, *Z. Physik*, **54**, 656.
BAADE, W.: 1938, *Ap. J.*, **88**, 285.
BAADE, W.: 1958, Private communication.
BAADE, W. and MINKOWSKI, R.: 1954, *Ap. J.*, **119**, 206.
BAADE, W. and ZWICKY, F.: 1934, *Proc. Nat. Acad. Sci.*, **20**, 259.
BAADE, W. and ZWICKY, F.: 1938, *Ap. J.*, **88**, 411.
BAHCALL, J. N.: 1961, *Phys. Rev.*, **124**, 495.
BAHCALL, J. N.: 1962, *Phys. Rev.*, **126**, 1143.
BASCHEK, B.: 1959, *Zs. f. Ap.*, **48**, 95.
BASHKIN, S., KAVANAGH, R. W. and PARKER, R.: 1959, *Phys. Rev. Lett.*, **3**, 518.
BASHKIN, S. and PEASLEE, D. C.: 1961, *Ap. J.*, **134**, 981.
BECKER, R. A. and FOWLER, W. A.: 1959, *Phys. Rev.*, **115**, 1410.
BELL, G. D. and KING, R. B.: 1960, unpublished.
BERGH, S. VAN DEN: 1959, *Ann. d'Ap.*, **22**, 123.
BERTAUD, C.: 1961, *Ann. d'Ap.*, **24**, 516.
BESKOV, F. and TREFFENBERG, L.: 1947, *Arkiv Mat. Astr. Fys.*, 34 A, Nos. 13 and 17.
BETHE, H. A.: 1939, *Phys. Rev.*, **55**, 103, 434.
BETHE, H. A. and CRITCHFIELD, C. L.: 1938, *Phys. Rev.*, **54**, 248.
BIDELMAN, W. P.: 1960, *Pub. Astr. Soc. Pacific*, **72**, 24.
BIDELMAN, W. P.: 1962, *Sky and Telescope*, **23**, 140; *A. J.* **67**, 111.
BIDELMAN, W. P. and ALLER, L. H.: 1962, private communication.
BONDI H., GOLD, T. and HOYLE, F.: 1955, *Observatory*, **75**, 80.
BONSACK, W. K.: 1959, *Ap. J.*, **130**, 843.
BONSACK, W. K.: 1961*a*, *Ap. J.*, **133**, 340.
BONSACK, W. K.: 1961*b*, *Ap. J.*, **133**, 551.
BONSACK, W. K. and GREENSTEIN, J. L.: 1960, *Ap. J.*, **131**, 83.
BOOTH R., BALL, W. P. and MAC GREGOR, M. H.: 1958, *Phys. Rev.*, **112**, 226.
BOSMAN-CRESPIN, D., FOWLER, W. A. and HUMBLET J.: 1954, *Bull. Soc. Roy. Sci. de Liège*, **9**, 327.
BRANS, C. and DICKE, R. H., 1961, *Phys. Rev.*, **124**, 925.
BROWN, R. E.: 1962, *Phys. Rev.*, **125**, 347.
BURBIDGE, E. M. and BURBIDGE, G. R.: 1955*b*, *Ap. J.*, **122**, 396.
BURBIDGE, E. M. and BURBIDGE, G. R.: 1956*a*, *Ap. J.*, **124**, 116.
BURBIDGE, E. M. and BURBIDGE, G. R.: 1956*b*, *Ap. J.*, **124**, 655.
BURBIDGE, E. M. and BURBIDGE, G. R.: 1957, *Ap. J.*, **126**, 357.
BURBIDGE, E. M. and BURBIDGE, G. R.: 1959, *Ap. J.*, **129**, 513.
BURBIDGE, E. M. and BURBIDGE, G. R.: 1961, *Scientific American*, **204**, 111.
BURBIDGE, E. M. and BURBIDGE, G. R.: 1962, *J. Phys. Soc. Japan*, 1 *Supp.*, A **3**, 161.

BURBIDGE, G. R.: 1956*a*, *Phys. Rev.*, **103**, 264.
BURBIDGE, G. R.: 1956*b*, *Ap. J.*, **124**, 416.
BURBIDGE, G. R.: 1958, *Pub. Astr. Soc. Pacific*, **70**, 83.
BURBIDGE, G. R.: 1959, *Paris Symposium on Radio Astronomy*, p. 54, Stanford University Press.
BURBIDGE, G. R.: 1960, *Ap. J.*, **131**, 519.
BURBIDGE, G. R.: 1961, *Nature*, **190**, 1053.
BURBIDGE, G. R.: 1962, *Prog. Theor. Phys.*, **27**, 999.
BURBIDGE, G. R. and BURBIDGE, E. M.: 1955*a*, *Ap. J. Suppl.*, **1**, 431.
BURBIDGE, G. R. and BURBIDGE, E. M.: 1956*c*, *Ap. J.*, **124**, 130.
BURBIDGE, G. R. and BURBIDGE, E. M.: 1958*a*, *Ap. J.*, **127**, 557.
BURBIDGE, G. R. and BURBIDGE, E. M.: 1958*b*, *Handbuch der Physik*, **51**, 134.
BURBIDGE, G. R. and HOYLE, F.: 1960, *Nuovo Cimento*, **4**, 558.
BURBIDGE, E. M., BURBIDGE, G. R., FOWLER, W. A. and HOYLE, F.: 1957, *Rev. Mod. Phys.*, **29**, 547 (B^2FH).
CAMERON, A. G. W.: 1954, *Phys. Rev.*, **93**, 932.
CAMERON, A. G. W.: 1955, *Ap. J.*, **121**, 144.
CAMERON, A. G. W.: 1957, *Atomic Energy of Canada*, Ltd. CRL-41.
CAMERON, A. G. W.: 1958*a*, *Ann. Rev. Nuclear Sci.*, **8**, 299.
CAMERON, A. G. W.: 1958*b*, *Atomic Energy of Canada*, Ltd. CRP-652.
CAMERON, A. G. W.: 1959*a*, *Ap. J.*, **129**, 676.
CAMERON, A. G. W.: 1959*b*, *Ap. J.*, **130**, 429.
CAMERON, A. G. W.: 1959*c*, *Ap. J.*, **130**, 452.
CAMERON, A. G. W.: 1959*d*, *Ap. J.*, **130**, 884.
CAMERON, A. G. W.: 1959*e*, *Ap. J.*, **130**, 916.
CAMERON, A. G. W.: 1960, *Ap. J.*, **131**, 521.
CAMERON, A. G.W.: 1962, *Icarus*, **1**, 13.
CAUGHLAN, G. R. and FOWLER, W. A.: 1962, preprint.
CHANDRASEKHAR, S.: 1938, *Stellar Structure*, University of Chicago Press.
CHANDRASEKHER, S. and HENRICH, L.: 1942, *Ap. J.*, **95**, 288.
CHIU, H. Y.: 1961*a*, *Phys. Rev.*, **123**, 1040
CHIU, H. Y.: 1961*b*, *Annals of Physics*, **15**, 1,
CHIU, H. Y.: 1961*c*, *Annals of Physics*, **16**, 321.
CHIU, H. Y. and FULLER, H. W.: 1962, preprint.
CHIU, H. Y. and MORRISON, P.: 1960, *Phys. Rev. Lett.*, **5**, 573.
CHIU, H. Y. and STABLER, R. C.: 1961, *Phys. Rev.*, **122**, 1317.
CLAYTON, D. D. and FOWLER, A. W.: 1961, *Annals of Physics*, **16**, 51.
CLAYTON, D. D., FOWLER, W. A., HULL, T. E. and ZIMMERMAN, B. A.: 1961, *Annals Physics*, **12**, 331.
CLIMENHAGA, J. L.: 1960, *Pub. Dom. Astr. Obs.*, **11**, No. 16.
COLGATE, S. A., GRASBERGER, W. H. and WHITE, R. A.: 1962, *J. Phys. Soc. Japan, Suppl.*, A **3**, 157.
COLGATE, S. A. and JOHNSON, M. H.: 1960, *Phys. Rev. Lett.*, **5**, 235.
COX, A. N. and BROWNLEE, R.: 1961, *Sky and Telescope*, **21**, 252.
COX, A. N. and EILERS D.: unpublished, quoted by J. N. BAHCALL.
COX, J. and GIULI, R.: 1961, *Ap. J.*, **133**, 755.
COX, J. and SALPETER, E. E.: 1961, *Ap. J.*, **133**, 764.
CRAWFORD, J. A.: 1953, *Pub. Astr. Soc. Pacific*, **65**, 210.
DICKE, R. H.: 1962, *Rev. Mod. Phys.*, **34**, 110 and preprint.
DOSTROVSKY, I., FRAENKEL, Z. and RABINOWITZ, P.: 1960, *Phys. Rev.*, **118**, 791.

DUMEZIL-CURIEN, P.: 1951, *Ann. d'Ap.*, **14**, 40.
DUMEZIL-CURIEN, P. and SCHATZMAN, E.: 1950, *Ann. d'Ap.*, **13**, 80.
DUMEZIL-CURIEN, P. and SCHATZMAN, E.: 1951, *Ann. d'Ap.*, **14**, 46.
FEYMMAN, R. P. and GELL-MANN, M.: 1958, *Phys. Rev.*, **109**, 193.
FISH, R. A., GOLES, G. and ANDERS, E.: 1960, *Ap. J.*, **132**, 243.
FOWLER, W. A.: 1954, *Mem. Soc. Roy. Sci. de Liège*, **14**, 88.
FOWLER, W. A.: 1958*a*, *Ap. J.*, **127**, 551.
FOWLER, W. A.: 1958*b*, *Pontificae Academiae Scientiarum Scripta Varia*, **16**, 269.
FOWLER, W. A.: 1959*a*, *Mem. Soc. Roy. Sci. de Liège*, **16**, 207.
FOWLER, W. A.: 1959*b*, *Modern Physics for the Engineer*, Vol **2**, New York.
FOWLER, W. A.: 1962, *Proc. Rutherford International Jubilee Conference*, p. 640.
FOWLER, W. A., BURBIDGE, G. R. and BURBIDGE, E. M.: 1955*a*, *Ap. J.*, **122**, 271.
FOWLER, W. A., BURBIDGE, G. R. and BURBIDGE, E. M.: 1955*b*, *Ap. J. Supp.*, **2**, 167.
FOWLER, W. A., GREENSTEIN, J. L. and HOYLE, F.: 1962, *Geophysical Journal R.A.S.* **6**, 148.
FOWLER, W. A. and HOYLE, F.: 1961, *Annals of Physics*, **10**, 280.
FRANK-KAMENETSKII, D. A.: 1961, *A. J.URSS*, **38**, 91; *Soviet Astronomy*, **5**, 66.
GAMOW, G.: 1939, *Nature*, **144**, 575, 620.
GAMOW, G.: 1953 *Math-Fys, Medd.*, **27**, No. 10.
GAMOW, G. and SCHOENBERG, M.: 1941, *Phys. Rev.*, **59**, 539.
GANDEL'MAN, G. M. and PINAEV, V. S.: 1959, *Žurn. Èksp. Teor. Fiz.* **37**, 1072; *JETP*, **10**, 764.
GAPOSCHKIN, C. PAYNE: 1957, *Galactic Novae*, Amsterdam.
GELL-MANN, M.: 1961, *Phys. Rev. Lett.*, **6**, 70.
GIBBONS, J. H., MACKLIN, R. L., MILLER, P. T. and NEILER, J. H., 1962, *Phys. Rev.* in press.
GINZBURG, V. L. and SYROVATSKY, S. I.: 1961, *Prog. Theoretical Phys. Suppl.*, No. 20.
GOLD, T.: 1960, *Ap. J.*, **132**, 274.
GOLDBERG, L., MOHLER, O. C. and MULLER, E. A., 1958, *Ap. J.*, **127**, 320.
GOLDBERG, L., MULLER, E. A. and ALLER, L. H., 1960, *Ap. J. Suppl.*, **5**, 1.
GOVE, H. E., LITHERLAND, A. E. and CLARK, M. A.: 1961, *Nature*, **191**, 1381.
GOVE, H. E., LITHERLAND, A. E. and FERGUESON, A. J.: 1961, *Phys. Rev.*, **124**, 1944.
GRASBERGER, W., 1960, UCLR 6196.
GREENSTEIN, J. L., 1958, *Handbuch der Physik*, **50**, 161.
GREENSTEIN, J. L.: 1961, *Stars and Stellar System*, Vol. 6, *Stellar Atmospheres*, Ed. by J. L. GREENSTEIN, pag. 676, Chicago, Ill.
GREENSTEIN, J. L., 1962, private communication.
HACK, M.: 1958, *Mem. Soc. Astr. Italiana*, **29**, 263.
HACK, M.: 1960, *Mem. Soc. Astr. Italiana*, **30**, 000.
HACK, M.: 1961, *Mem. Soc. Astr. Italiana*, **31**, 280.
HAMADA, T. and SALPETER, E. E.: 1961, *Ap. J.*, **134**, 683.
HANBURY BROWN, R. and HAZARD, C.: 1959, *M. N. Roy. Astr. Soc.*, **119**, 297.
HANBURY BROWN, R. and HAZARD, C.: 1961, *M. N. Roy. Astr. Soc.*, **122**, 479.
HARRIS, D.: 1961, *Thesis*, California Institute of Technology.
HARRISON, K., WAKANO, M. and WHEELER, J. A.: 1958, *La Structure de l'Univers* (Editor, R. STOOPS) Bruxelles, Solvay conference Reports.
HASELGROVE, C. B. and HOYLE, F., 1959, *M. N.*, **119**, 112.
HAYAKAWA, S.: 1960, *Pub. Astr. Soc. Japan*, **12**, 115.
HAYASHI, C.: 1961, preprint.
HAYASHI, C. and CAMERON, R. C.: 1962, *Ap. J.*, in press.

Hayashi, C., Jugaku, J. and Nishida, M.: 1959, *Prog. Theoretical Physics*, **22**, 531.
Hayashi, C. and Nishida, M.: 1956, *Prog. Theoretical Physics*, **16**, 613.
Hayashi, C., Nishida, M., Ohyama, N. and Tsuda, H.: 1958, *Prog. Theoretical Physics*, **20**, 101.
Hayashi, C., Nishida, M., Ohyama, N. and Tsuda, H.: 1958, *Prog. Theoretical Physics*, **22**, 101.
Hayashi, C. Nishida, M. and Sugimoto, D.: 1961, *Prog. Theoretical Physics*, **25**, 1053.
Helfer, H. L. and Wallerstein, G.: 1959*a*, *Ap. J.*, **129**, 347.
Helfer, H. L. and Wallerstein, G.: 1959*b*, *Ap. J.*, **129**, 920.
Helfer, H. L., Wallerstein, G. and Greenstein, J. L., 1959, *Ap. J.*, **129**, 700.
Helliwell, T. M., 1961, *Ap. J.*, **133**, 566.
Holmgren, H. D. and Johnson, R. L.: 1959, *Phys. Rev. Lett.*, **2**, 275, 381.
Hoyle, F.: 1946, *M. N. Roy. Astr. Soc.*, **106**, 343.
Hoyle, F.: 1954, *Ap. J. Suppl.*, **1**, 121.
Hoyle, F.: 1959, *M. N.*, **119**, 124.
Hoyle, F.: 1960, *Quart. JRAS*, **1**, 28.
Hoyle, F.: 1961, private communication.
Hoyle, F. and Fowler, W. A.: 1960, *Ap. J.*, **132**, 565.
Hubble, E.: 1941, *Astr. Soc. Pacific.*, **53**, 141.
Hughes, D. J. Spatz, W. B. D. and Goldstein, N.: 1949, *Phys. Rev.*, **78**, 1781.
Hunger, K.: 1957, *A. J.*, **62**, 294.
Ito, K.: 1961, *Prog. Theoretical Physics*, **26**, 990.
Johansson, S. A. E.: 1959, *Nuclear Physics*, **12**, 449.
Johnson, H. M.: 1959, *Pub. Astr. Soc. Pacific*, **71**, 425.
Johnson, W. H. and Bhanot, V. B.: 1957, *Phys. Rev.*, **107**, 1669.
Jugaku, J., Sargent, W. L. W. and Greenstein, J. L.: 1961, *Ap. J.*, **134**, 783.
Kavanagh, R. W.: 1958, *Bull. Amer. Phys. Soc.*, II, **4**, 444.
Kinman, T.: 1959, *M. N. Roy. Astr. Soc.*, **119**, 538.
Kirzhnits, D. A.: 1960, *Žurn. Ėksp. Teor. Fiz.*, **38**, 503.
Klein, O.: 1947, *Arkiv Mat. Astr. Fysik*, **34** A, No. 19.
Kohman, T. P.: 1961, *J. Chemical Education*, **38**, 73.
Kuroda, P. F.: 1960, *Nature*, **187**, 36.
Landau, L.: 1932, *Phys. Zeits. URSS*, **1**, 285.
Limber, D. N.: 1958, *Ap. J.*, **127**, 363, 387.
Macklin, R. L., Inada, T. and Gibbons, J. H.: 1962, *Nature*, in press.
Macklin, R. L., Lazar, N. H. and Lyon, W. S.: 1957, *Phys. Rev.*, **107**, 504.
Maltby, P. and Moffet, A.: 1961, private communication.
Mathis, J.: 1957, *Ap. J.*, **126**, 493.
Mathis, J.: 1959, *Ap. J.*, **129**, 259.
Matinyan, S. G. and Tsilosani, N. N.: 1961, *Žurn. Ėksp. Teor. Fiz.*, **41**, 1681; *JETP*. **14**, 1195.
Miller, J. M. and Hudis, J.: 1959, *Ann. Rev. Nuclear Sci.*, **9**, 159.
Minkowski, R.: 1939, *Ap. J.*, **89**, 156.
Minkowski, R.: 1941, *Pub. Astr. Soc. Pacific*, **53**, 224.
Minkowski, R.: 1959, *Paris Symp. on Radio Astronomy*, p. 315.
Morgan, W. W.: 1959, *A. J.*, **64**, 432.
Morrison, P. M.: 1961, *Handbuch der Physik*, **46**/I, 1.
Murthy, V. R.: 1960, *Phys. Rev. Lett.*, **5**, 539.
Murthy, V. R. and Urey, H. C.: 1962, *Ap. J.*, **135**, 626.
Nishida, M.: 1960, *Prog. Theoretical Physics*, **23**, 896.

NISHIDA, M. and SUGIMOTO, D.: 1962, *Prog. Theoretical Physics*, **27**, 145.
OPPENHEIMER, J. R. and SERBER, R.: 1938, *Phys. Rev.*, **54**, 540.
OPPENHEIMER, J. R. and VOLKOFF, G. M.: 1939, *Phys. Rev.*, **55**, 374.
OSTERBROCK, D. E.: 1953, *Ap. J.*, **118**, 529.
OSTERBROCK, D. E. and ROGERSON, J. B.: 1961, *Pub. Astronom. Soc. Pacific*, **73**, 129.
PARKER, R., GREENSTEIN, J. L., HELFER, H. L. and WALLERSTEIN, G.: 1961, *Ap. J.*, **133**, 101.
PATTERSON, C.: 1956, *Geochimica and Cosmochimica Acta*, **10**, 230.
PATTERSON, C., TILTON, G. and INGHRAN, M.: 1955, *Science*, **121**, 69.
PONTECORVO, B. M.: 1959, *Žurn. Ėksp. Teor. Fiz.*, **36**, 615; *Sov. Phys. JETP*, **9**, 1148.
PONTECORVO, B. M. and SMORODINSKY, J.: 1962, preprint.
REAVES, G.: 1953, *Pub. Astr. Soc. Pacific*, **65**, 243.
REED, G. W., KIGOSHI, K. and TURKEVICH, A.: 1962, *Geochimica and Cosmochimica Acta*, in press.
REEVES, H.: 1959, *Phys. Rev. Lett.*, **2**, 423.
REEVES, H.: 1962, *Ap. J.*, **135**, 779.
REEVES, H. and SALPETER, E. E.: 1959, *Phys. Rev.*, **116**, 1505.
REINES, F. and COWAN, C. L.: 1958, *Proc. Int. Conf. Atomic Energy*, Geneva.
REYNOLDS, J. H.: 1960*a*, *Phys. Rev. Lett.*, **4**, 8, 351.
REYNOLDS, J. H.. 1960*b*, *Z. Naturforschung*, **5** a, 1112.
REYNOLDS, J. H.: 1960*c*, *J. Geophys. Res.*, **65**, 3843.
SAKASHITA, S., ONO, Y. and HAYASHI, C.: 1959, *Progress Theoretical Physics*, **21**, 315
SALPETER, E. E.: 1952, *Ap. J.*, **115**, 326.
SALPETER, E. E.: 1953, *Ann. Rev. Nuclear Sci.*, **2**, 41.
SALPETER, E. E.: 1955, *Phys. Rev.*, **97**, 1237.
SALPETER, E. E.: 1957, *Phys. Rev.*, **107**, 516.
SALPETER, E. E.: 1959, *Ap. J.*, **129**, 608.
SALPETER, E. E.: 1960, *Annals of Physics*, **11**, 393.
SALPETER, E. E.: 1961, *Ap. J.*, **134**, 669.
SANDAGE, A. R. and MATTHEWS, T.: 1962, private communication.
SANDAGE, A. R. and WALLERSTEIN, G.: 1960, *Ap. J.*, **131**, 538.
SARDAGE, A. R.: 1962, *Ap. J.*, **135**, 333, 349.
SARGENT, W. L. W. and JUGAKU, J.: 1961, *Ap. J.*, **134**, 777.
SARGENT, W. L. W. and SEARLE, L.: 1962, *Ap. J.*, in press.
SCHAEFFER, O. A. and ZÄHRINGER, J.: 1962, *Phys. Rev. Lett.*, **8**, 389.
SCHATZMAN, E.: 1948, *Journ. Phys. et Rad.*, **9**, 46.
SCHATZMAN, E.: 1953, *Ann. d'Ap.*, **16**, 162.
SCHATZMAN, E.: 1958, *White Dwarfs*, Amsterdam.
SCHMIDT, M.: 1959, *Ap. J.*, **129**, 243.
SCHMIDT, M.: 1961, *Proc. La Plata Conference on Stellar Evolution* and private communication.
SCHMIDT, R. A., SHARP, R., MOSEN, A. W., DUFFIRLD, R. B. and ZUMWALT, L.: 1962, *Geochimica and Cosmochimica Acta*, in press.
SCHWARZSCHILD, M.: 1958, *The Structure and Evolution of Stars*, Princeton N. J.
SCHWARZSCHILD, M. and HÄRM, R.: 1958, *Ap. J.*, **128**, 348.
SCHWARZSCHILD, M. and HÄRM, R.: 1961, *Ap. J.*, **66**, 45.
SCHWARZSCHILD, M. and HÄRM, R.: 1962, *Ap. J.*, in press.
SCHWARZSCHILD, M. and UPTON, J.: 1962, private communication.
SEARLE, L.: 1961, *Ap. J.*, **133**, 531.
SHIMA, M.: 1962, private communication.

SHKLOVSKY, I. S.: 1960, *Astr. J. URSS*, **37**, 945.
STANLEY, E. J. and PRICE, R.: 1956, *Nature*, **177**, 1221.
STRUVE, O., WALLERSTEIN, G. and ZEBERG, V.: 1961, *Pub. Astr. Soc. Pacific*, **73**, 220.
SUDARSHAN, E. G. and MARSHAK, R. E.: 1958, *Phys. Rev.*, **109**, 1860.
TRAVING, G.: 1962, *Ap. J.*, **135**, 439.
TANNER, N. W.: 1959, *Phys. Rev.*, **114**, 1060.
VON WEIZSÄCKER, C. F.: 1938, *Physik. Z.*, **39**, 633.
WALLERSTEIN, G.: 1962, *Ap. J. Suppl.*, **6**, 407.
WALLERSTEIN, G. and CARLSON, M.: 1960, *Ap. J.*, **132**, 276.
WALLERSTEIN, G., GREENSTEIN, J. L., PARKER, R., HELFER, H. L. and ALLER, L. H.: 1962, private communication.
WALLESRTEIN, G., STONE, Y. H. and WHITNEY, Y. A.: 1962, *Ap. J.*, **135**, 459.
WASSERBURG, G. J., FOWLER, W. A. and HOYLE, F.: 1960, *Phys. Rev. Lett.*, **4**, 112.
WEINBERG, S.: 1962, *Nuovo Cimento*, in press and preprint.
WEINREB, S.: 1962, private communication from D. S. HEESCHEN.
WHEELER, J. A.: 1958, private communication to W. A. FOWLER.
WILSON, O. C.: 1959, *Ap. J.*, **130**, 496.
WILDHACK, W.: 1940, *Phys. Rev.*, **57**, 81.
WYLLER, A. A.: 1960, *Astrophysica Norvegica*, **7**, No. 2.
WOLTJER, L.: 1958, *Bull. Astr. Soc. Netherland*, **14**, 39.
ZEL'DOVICH, I. B.: 1957, *Žurn. Ėksp. Teor. Fiz.*, **33**, 991.
ZEL'DOVICH, I. B.: 1961, *Žurn. Ėksp. Teor. Fiz.*, **41**, 1609; *Sov. Phys. JETP*, **14**, 1143.
ZWICKY, F.: 1938, *Ap. J.*, **88**, 522.
ZWICKY, F.: 1939, *Phys. Rev.*, **55**, 726.
ZWICKY, F.: 1942, *Ap. J.*, **28**, 96.
ZWICKY, F.: 1959, *Handbuch der Physik*, **51**, 766.

Observed Abundance Anomalies Indicating Nucleosynthesis in Individual Stars.

E. M. Burbidge

University of California - San Diego, Cal.

From the discussion of the time scales for the different processes required to account for the production of the isotopes of all elements from hydrogen, we may estimate roughly the evolutionary lifetimes for stars while they are building a given set of elements. We know that these range from 10^{10} years for H-burning down to some minutes or seconds for the *e*-processes. Consequently, we may predict that the best chances of seeing the products of nucleosynthesis in individual stars will occur for H-burning, He-burning and the *s*-processes (excluding the *x*-process, for which it is more difficult to estimate a time-scale). The α-process and the ^{12}C, ^{16}O, etc. reactions and the *e*-process occur so rapidly that there will not be much chance of seeing a star in which these have occurred unless such a star has (perhaps by undergoing an explosion that leaves a core or by steadily ejecting its outer parts) settled down into a white dwarf configuration.

Various types of stars in which the products of H-burning and He-burning may now be seen on the stellar surfaces were discussed in B^2FH and there has been some recent work also.

1. – Stars apparently of spectral type O or B with weak or absent hydrogen lines.

HD 12448 (Popper's star), HD 160641 (Bidelman's star), HD 168476 (Thackeray's star) have been known for many years. They show no evidence for the presence of any hydrogen at all; He and C lines are strong in all three, O lines are absent in HD 168476. Aller found that in HD 160641 the C, N, Ne were all more abundant, relative to O, than in normal B stars while He has apparently completely replaced H.

Such stars are actually rather rare, since the luminous B stars have been sampled to fairly large distances from the Sun.

Hot subdwarf stars studied by GREENSTEIN and MÜNCH also show that H is probably depleted and considerably replaced by He in the atmosphere, while anomalies among the light elements appear; in particular the nitrogen abundance may be high.

2. – A-type stars weak or lacking in hydrogen.

The classical case is *υ Sgr*, of which I reproduce a spectrum by GREENSTEIN (Fig. 1). The lack of a Balmer jump and the absence of H lines, compared to a normal star, are striking. No abundance analysis has ever been

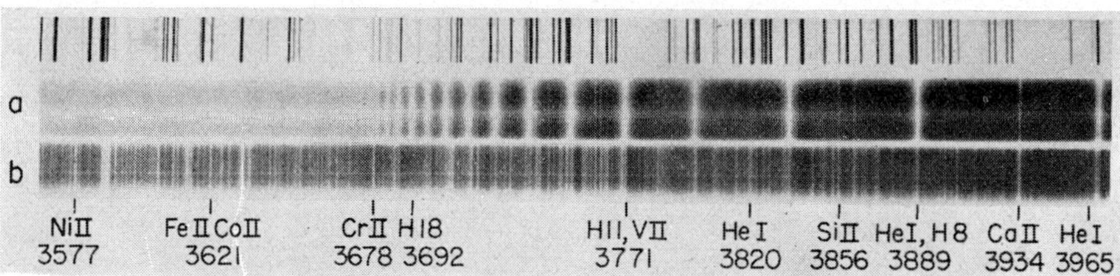

Fig. 1. – *a*) Normal *A*-type star, *η Leonis*, showing strong Balmer lines of hydrogen and a strong Balmer discontinuity at the series limit. *b*) Peculiar star, *υ Sgr*, in which hydrogen has a much smaller abundance than normal. Spectra obtained by J. L. GREENSTEIN at McDonald Observatory.

carried out; such determinations are difficult since the absence of H makes such a radical change in the structure of the atmosphere by reducing the normal opacity sources.

Sex 7 is a high-velocity *A* star recently studied by WALLERSTEIN, STRUVE and WILLIAMS (1962). Comparison with Sirius gives a ratio of metals to hydrogen about the same in both stars. From absolute determinations, He/H in *Sex 7* is about 10 (instead of about 1/10 as normal) and C/H is about 10^{-2} which is about 20 times the normal ratio. Unless peculiar effects have been introduced — *e.g.* because *Sex 7* might be a rapidly rotating star viewed pole-on — then it may be an originally metal-poor star at a late evolutionary stage in which mixing and/or mass loss have brought the He produced in H-burning and C produced in He-burning to the surface. The absence of Ba (giving an upper limit for Ba/Fe which is below normal) is in line with this suggestion.

3. – Wolf-Rayet stars.

As long ago as 1943, 5 years after the discovery of the CN-cycle, Gamow suggested that the carbon and nitrogen sequences in classical WR stars might be due to the action of the CN-cycle. However, many workers (*e.g.* Underhill) have considered that the great spectral differences are not due to abundance differences but are excitation effects. Here we have manifestly stars whose atmospheres are not in local thermodynamic equilibrium and thus Underhill might well be right. However, there are various lines of argument which make it tempting to suggest that we are indeed seeing here the results of interior nuclear reactions.

As Prof. GRATTON pointed out, it is the WN stars which occur mainly in associations and, if they are evolved stars, must therefore have originally been very massive, since the associations are young, so that the stars must have gone through their lifetimes rapidly. Some masses are available — *e.g.* *V 444 Cyg*, a well known binary, in which the WR star is about 10 $M_\odot$, while the O-star is more massive; this suggests that the WR star was *originally* more massive than the O-star and hence evolved more rapidly and mass loss has reduced it to its present value while stripping it of its outer envelope. WR stars are certainly losing mass at present.

Aller has suggested that the abundance ratios He:C:N are 20:1/20:1, while H/He is less than normal. This suggests that C and N have gone through the CN-cycle, which produces more N than C; the nearly equal C to N ratios in average solar neighborhood material suggest that not much cycling through the CN-cycle has occurred.

The WC stars (Aller's suggested abundances are here He:C:O=17:3:1, N is not seen at all) would then be remnants of stars in which He-burning ($3\alpha \rightarrow {}^{12}C$) has occurred in the center and either the outer envelope was lost or mixing occurred which did *not* carry the ^{12}C thus produced through a region where the CN-cycle was operating.

4. – White dwarfs.

Spectra of these, which must contain virtually no internal nuclear fuel, may show only He and no H or other peculiar features indicating evolution even through the α and e-processes. A discussion was given in B^2FH; recent surveys may be found in Greenstein's articles in the *Handbuch der Physik* (1958) and in the Kuiper compendium volume on *Stellar Atmospheres* (1961). However, abundances have not been estimated, as this is very difficult in these stars.

The Early Stages of Stellar Evolution.

E. SCHATZMAN

Institut d'Astrophysique - Paris

I. – Contraction Towards the Main Sequence.

1. – Introduction.

In the following, the basic idea is to take into consideration the law of conservation of angular momentum. It is well known that the law of conservation of angular momentum is one of the fundamental laws of physics; when we consider the problems of stellar evolution, we are immediately facing the problem of angular momentum, as a consequence of the distribution of equatorial velocities of stars in the H-R diagram.

Struve has shown already in 1948-49 (Stellar Evolution) the remarkable distribution of the equatorial velocities along the main sequence. Stars of the upper part of the main sequence have an average equatorial velocity of 200 km s^{-1}; stars of the lower part of the main sequence have an average equatorial velocity which is much smaller, certainly less than 30 km s^{-1}; the transition from the high velocities to the small velocities takes place in a very small range of spectral types.

Let us consider first the stars with high equatorial velocity. We have to face at present two theories of the stellar formation. *a*) The theory which is usually accepted is that stars are formed by contraction of diffuse matter. As we shall see more precisely later, a very small initial angular velocity is sufficient to produce a star of high equatorial velocity. *b*) The theory of AMBARTSUMIAN (1947) of the origin of stars from extremely dense matter. An object of mass M, radius R, cannot store an angular momentum higher than a certain limit. The maximum angular velocity ω_{max} is given by

$$\omega_{max}^2 \simeq \frac{GM}{R^3}.$$

And the maximum angular momentum is given by

$$I_{\max} = G^{\frac{1}{2}} K M^{\frac{3}{2}} R^{\frac{1}{2}} ,$$

where KMR^2 is the moment of inertia of the star ($K = 1/13.5$ for polytrope $n = 3$). If the star had grown from very dense matter, its equatorial velocity would have decreased like $\bar{\varrho}^{-1/3}$, $\bar{\varrho}$ being the mean density, supposing no exchange of momentum with the matter from outside. The maximum equatorial velocity is of the order of

$$V_{\text{eq max}} \simeq 10^{7.615} \varrho^{-1/6} m_*^{1/3} ,$$

where m_* is expressed in solar units.

Assuming an initial density 10^{15}, we find for $m_* \simeq 10$, that an equatorial velocity $V_{\text{eq max}} \simeq 3 \cdot 10^{10}$ cm s^{-1} would drop, for $\bar{\varrho} = 1$, to an equatorial velocity $V_{\text{eq}} \simeq 3 \cdot 10^5$.

Consequently, even when forgetting all the problems of the expansion of very dense matter to ordinary stellar matter, we are met with the difficulty that the angular momentun of a B star is about 300 times larger than the angular momentum of the same amount of very dense matter.

If it were true that stars are formed by expansion of very dense matter, the fast rotating stars could obtain their angular momentum by accretion of nebular matter. However, it has been shown that accretion is not a very efficient process. Consequently, it is difficult to imagine such a succession of very unlikely processes.

One more possibility would be the formation of close binary stars, and the merging of these stars into each other as an effect of gravitational radiation, the importance of which has been recently stressed by KRAFT, MATHEWS and GREENSTEIN (1962). If we could imagine the birth of stars in a cluster as a consequence of the merging of double stars into each other, this would solve the angular momentum problem, but we would still be left with the problem of the increase of size.

From now on, we shall consider only the problem of the contraction.

2. – The contracting star.

2·1. *The model.* – We shall consider, in the following, an object which is already a star. It is well known (SPITZER and MESTEL) (1956) that the contraction of a sphere of gas can be prevented by the presence of the magnetic field, but we shall suppose that the transition from the nebular gas to the star is possible, without considering the difficulties in understanding this intermediate phase.

The problem of the contracting object is not fully solved. On the one hand, we have some idea of how a nonrotating object is contracting; but we know much less about the structure of a contracting and rotating object.

In order to make the physics understandable, we shall use a simplified picture, which is justified only because elaborate calculations have been made. Let us first consider a contracting star, the structure of which is described by radiative transfer. An homology transform will help us to have an idea of how the luminosity changes as a function of the effective temperature.

Assuming inside the star an absorption coefficient

$$\varkappa = \varkappa_0 \varrho^\alpha T^\beta \,, \tag{1}$$

we obtain the relation between the surface quantities, the luminosity L and the effective temperature T_{eff}:

$$L \sim T_{\text{eff}}^{4(\beta+3\alpha)/(3\alpha+\beta-2)} \,. \tag{2}$$

In order to choose properly the exponents α and β, we consider the interpolation values of α and β which can explain the main sequence. As we have

$$L \sim M^{3-\beta-\alpha} R^{\beta+3\alpha} \,,$$

along the main sequence

$$L \sim M^4 \,, \qquad R \sim M^{0.75} \,.$$

For $M \geqslant M_\odot$, we obtain the relation

$$\beta = 5\alpha - 4 \,.$$

As (α, β) have a value which is in between $(0, 0)$ (diffusion by the free electrons) and $(1, -3.5)$, we consider that, depending on the region of ϱ and T, we have

$$\beta = -3.5\alpha \,,$$

which leads to the couple of values

$$\alpha = \frac{4}{8.5} \qquad \beta = -\frac{14}{8.5} \,. \tag{A}$$

With these values, we obtain for L,

$$L \sim T_{\text{eff}}^{8/19} \,,$$

or

$$L \sim T_{\text{eff}}^{0.421} .$$

However, the calculations of COX and BROWNLEE (1961), (Fig. 1), and of CAMERON and EZER (1962) indicate, near the main sequence, a variation of L proportional to T_{eff}^2 (to be compared to the main sequence itself, $L \sim T_{\text{eff}}^{6.4}$).

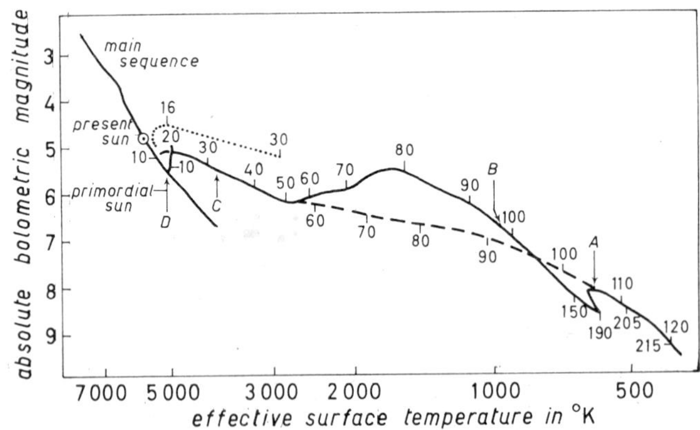

Fig. 1. – Evolutionary tracks for radiative contracting stars, according to COX and BROWNLEE. The tracks are labeled in millions of years before the main sequence is reached. The track of HENYEY, LELEVIER and LEVEE is outdated by improved values of the opacity and energy and energy-generation laws, but the general shape of their curve is still valid. – – – no deuterium; ——— deuterium burning included; . . . track by HENLEY, LELEVIER and LEVEE.

Let us assume that during this contracting phase, when the structure of the star is governed by radiative transport, the luminosity depends on the mass in the same way as on the main sequence. It should be noticed that for the same effective temperature, these stars have a higher luminosity than when nuclear reactions are the only energy source. To represent these properties, we shall take

$$L_* = 10^{0.18} M_*^{2.75} (T_{\text{eff}}/5\,000)^2 ,$$

where L_* and M_* are the luminosity and the mass in solar units. This can be accounted for by taking interpolation values of α and β given by the two relations, one deduced from the properties of the main sequence

$$\beta = 5\alpha - 4 ,$$

the other from the condition $L \sim T_{\text{eff}}^2$:

$$\beta = -3\alpha - 2 ,$$

which leads to

$$(B) \qquad \alpha = 0.25 \,, \qquad \beta = -2.74 \,.$$

The two solutions (A) and (B) can be seen on Fig. 2, which represents the plane (α, β).

However, it is doubtful whether the structure of the contracting star is governed by radiative transfer or not. It is known that on the right of the H-R diagram, stars have a large convective zone. The suggestion has been tested with great care by CAMERON and EZER (1962). Their results legitimate again the use of homology transform, which can lead to interesting results, which are not numerically correct, but give a very interesting physical insight on the structure of these convective stars.

We shall consider a polytrope, in which the relation between pressure and density is given by

$$(3) \qquad P = K\varrho^{\gamma} \,.$$

The structure of the star is described by the differential equation

$$(4) \qquad \frac{1}{\xi^2}\frac{\mathrm{d}}{\mathrm{d}\xi}\left(\xi^2\frac{\mathrm{d}\theta}{\mathrm{d}\xi}\right) = -\theta^n \,,$$

n, λ, θ are defined by the following relations

$$(5) \qquad \begin{cases} n = \dfrac{1}{\gamma - 1}\,, \\ r = \lambda\xi \,, \\ \varrho = \varrho_c \theta^n \,, \\ \lambda^2 = K \dfrac{\gamma}{\gamma - 1}\dfrac{\varrho_c^{\gamma-2}}{4\pi G}\,, \end{cases}$$

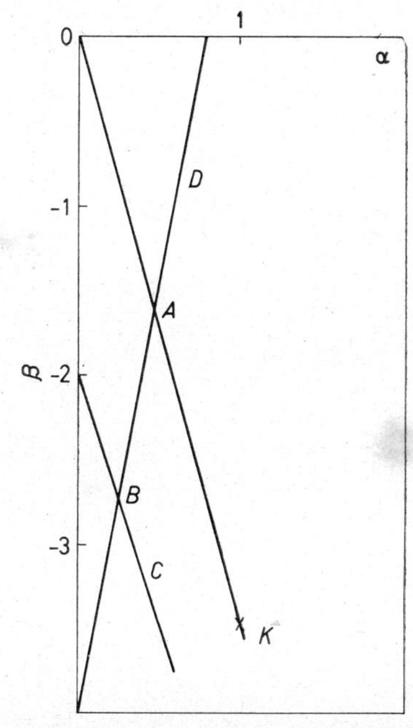

Fig. 2.– The value of the exponents α and β in the expression of the absorption coefficient by a power law. 0 corresponds to scattering and K to the Kramers law of absorption. The line $0K$ corresponds to a possible interpolation between scattering and the Kramers law of absorption. The line D expresses the relation between α and β for the main sequence; the line C expresses the relation between α and β for the contracting star.

r is the distance to the center of the star, ϱ_c, the central density, G the constant of gravitation.

The radius of the star is given by

$$(6) \qquad R = \lambda\xi_S \,,$$

and the mass of the star is given by

$$M = 4\pi\lambda^3 \varrho_c \left(-\xi^2 \frac{d\theta}{d\xi}\right)_s . \tag{7}$$

The polytropic equation is integrated from $\xi = 0$, $\theta = 1$ to the point $\xi = \xi_s$ where θ vanishes, which defines the surface.

We shall assume that the convective star is bounded by a radiative atmosphere. We shall assume that the constant K is given by the values of the temperature and density at the lower boundary of the radiative atmosphere. Let us proceed to the calculations of this constant's value.

Assuming that, in the atmosphere, the absorption coefficient is given by the relation

$$\varkappa = \varkappa_0 \varrho^\alpha T^\beta , \tag{8}$$

in which α and β have values which are different from those in eq. (1), we can estimate the value of the exponents α and β and the value of the coefficient $\varkappa_0$ in the atmosphere.

Let us assume that the main absorbant is the negative hydroge ion. The number of negative hydrogen ions is given with obvious notations by the Saha equation:

$$\frac{[H][e]}{[H^-]} = 4 \frac{(2m_e kT)^{\frac{3}{2}}}{h^3} \exp[-(\chi/kT)] ,$$

χ is the ionization potential of the negative hydrogen ion. The number of electrons is given by the ionization equilibrium of the metals. We shall write:

$$\frac{[M^+][e]}{[M]} = \frac{g^+}{g_0} 2 \cdot \frac{(2\pi m_e kT)^{\frac{3}{2}}}{h^3} \exp[-(\chi_M/kT)] ,$$

where χ_m is the ionisation potential of the metal. The number of negative ions is, therefore, per cubic centimeter:

$$[H^-] = [H][e] \frac{h^3}{4(2\pi m_e kT)^{\frac{3}{2}}} \exp[\chi/kT] .$$

Supposing that the metals are only partially ionized we have:

$$[e] = [M^+] ,$$

and

$$[e] = [M]^{\frac{1}{2}} \left(\frac{2g_+}{g_0}\right)^{\frac{1}{2}} ,$$

therefore

$$[H^-]=[H][M]^{\frac{1}{2}}\left(\frac{2g^+}{g_0}\right)^{\frac{1}{2}}\frac{1}{4}\frac{h^{\frac{3}{2}}}{(2\pi m_e kT)^{\frac{3}{4}}}\exp\left[-\frac{\chi_M-2\chi}{2kT}\right].$$

We can see that the absorption coefficient per gram varies $\varrho^{\frac{1}{2}}$. An interpolation expression for the exponent of the temperature is

$$\beta=-\frac{3}{4}+\frac{\chi_M-2\chi}{2kT}. \tag{9}$$

Taking for the ionization potential of the metals $\chi_m=6$ to 8 eV, for the negative hydrogen ion $\chi=0.75$ eV, and chosing $(5040/T)=1.4$, we find:

$$\beta=7 \quad \text{to} \quad 10\,.$$

We shall take the couple of values

$$\begin{cases}\alpha=\frac{1}{2}\,,\\ \beta=5\,,\end{cases} \tag{10}$$

and use the interpolation formula

$$\varkappa=10^{-16.72}\varrho^{\frac{1}{2}}T^5\,. \tag{11}$$

This is not a very good approximation, as can be judged from the tables of Cox, or from the tables of Vitense. However, it will be a sufficient approximation for our purpose.

With a power law, it is possible to build an atmosphere. Assuming a relation

$$T^4=T_0^4(\tau+q(\tau))\,, \tag{12}$$

between the temperature T at optical depth τ and T_0 at the surface, it is possible to obtain the solution

$$\varrho^{\alpha+1}=\frac{g\mu(\alpha+1)}{\varkappa_0 kT_0^{\beta+1}(\frac{3}{2})^{(\beta+1)/4}(\tau+q)^{(\alpha+1)/4}}\int_0^\tau(\tau+q)^{(\alpha-\beta)/4}\,\mathrm{d}\tau\,, \tag{13}$$

where g is the surface gravity and μ the mean molecular weight. It is easy now to obtain the boundary condition for the atmosphere and the convective zone. We have for the radiative gradient

$$\frac{\mathrm{d}\log T}{\mathrm{d}\log P}=\frac{1}{4}\frac{kT}{g\mu}\frac{1+\mathrm{d}q/\mathrm{d}\tau}{\tau+q(\tau)}\varkappa\varrho\,. \tag{14}$$

Assuming $q=\text{const}$, the condition at the boundary of the convective zone,

$$(15)\qquad \frac{\alpha+1}{\beta-\alpha-4}\left[\left(\frac{\tau_c+q}{q}\right)^{(\beta-\alpha-4)/4}-1\right]=\frac{1}{n+1},$$

defines the optical depth at the top of the convective zone. Assuming that we can derive from ϱ and T at the boundary the value of K, which, in eq. (3), discribes the inernal constitution of the star, we use the relation

$$K=\frac{kT}{\mu}\,\varrho^{(-1/n)},$$

and obtain

$$(18)\qquad K=ET_0^{1+((\beta+1)/n(\alpha+1))}M^{-1/n(\alpha+1)},$$

with the value of the constant E:

$$(17)\qquad E=\left(\frac{k}{\mu}\right)^{1+1/n(\alpha+1)}\left(\frac{3}{2}\right)^{\frac{1}{4}(1+(\beta+1)/n(\alpha+1))}(\tau+q)^{\frac{1}{4}+(\beta-3)/4n(\alpha+1)}\left(\frac{\varkappa_0(n+1)}{4G}\right)^{1/n(\alpha+1)}.$$

The value of $(\tau+q)$ is given by the solution of eq. (15). We consider two cases, $n=2.5$ and $n=\frac{5}{3}$. They correspond to $\gamma=1.4$ and $\gamma=1.6$; the interest of this last value will come out later. The first case corresponds to a fairly cool star, and the second case to a star nearer to the main sequence.

We have the following results:

$$(19)\qquad n=2.5,\quad K=ET_0^{2.6}M^{-4/15}R^{8/15};\qquad \tau=0.725,$$

$$(19b)\qquad n=\tfrac{5}{3},\quad K=ET_0^{3.4}M^{-0.4}R^{0.8};\qquad \tau=1.073,$$

and the corresponding values of E,

$$n=2.5;\qquad \log_{10}E=7.612+1.266\log\left(\frac{1}{\mu}\right),$$

$$n=\frac{5}{3};\qquad \log_{10}E=7.430+1.4\log\left(\frac{1}{\mu}\right).$$

We can eliminate R, K and ϱ_c between the expressions of the radius (8) of the mass (9), of the constant of the polytropic relation (19), and of the luminosity:

$$(20)\qquad L=4\pi R^2\sigma\tfrac{3}{2}T_0^4,$$

(σ is the constant of Stefan.)

It is found that L and T are related by the following relation:

$$L \sim T_0^f, \tag{21}$$

with the following value of f:

$$f = 4 + 2\left(1 + \frac{\beta + 1}{n(\alpha + 1)}\right)\frac{n(\alpha + 1)}{(3 - n)(\alpha + 1) - 2}\,. \tag{22}$$

It is interesting to consider several values of n. We have the following results (TABLE I).

TABLE I.

γ	n	f
–	3	− 6.5
1.4	2.5	−11.6
1.6	–	∞
–	–	70

The slope of the relation changes with n; in other words, when the gas is fully ionized, the star moves upwards from right to left, then it moves vertically ($f = \infty$), and then from left to right. This is in qualitative agreement with the result of CAMERON and EZER (1962), and provides a physical interpretation. We note here the interest of the value $n = \frac{5}{3}$ which corresponds to a vertical displacement of the star in the H-R diagram.

We want now to derive the numerical dependance of L from the numeric coefficients. To obtain this, we collect all the formulae we need:

1) The relation between luminosity, radius and surface temperature

$$L = NR^2 T_0^4 \qquad \text{with} \qquad \log N = -\,2.964\,. \tag{23}$$

2) The expression of K, derived from the surface properties, as given by eq. (19) or (19*b*)

$$K = EM^{-(4/15)}R^{8/15}T_0^{2.6} \qquad \text{with} \qquad \log E = 7.612\,, \tag{24}$$

or

$$K = EM^{-0.4}R^{0.8}T_0^{3.4} \qquad \text{with} \qquad \log E = 7.430\,.$$

3) The expression of the mass, derived from the properties of the polytropes. For $n = 2.5$, we have

$$M = m\varrho_c^{0.1}K^{\frac{3}{2}} \qquad \text{with} \qquad \log m = 11.37\,; \tag{25}$$

another expression of m is

$$m = \left(\frac{\xi_S^2(n+1)}{4\pi G}\right)^{\frac{3}{2}} \frac{4\pi}{3} \frac{\bar{\varrho}}{\varrho_c} . \tag{26}$$

For $n = 2.5$, $\xi_s = 5.35$; $\varrho_c/\bar{\varrho} = 23.4$.

4) The expression of the radius, derived from the properties of the polytropes. We have from $n = 2.5$,

$$R = Q\varrho_c^{-0.3} K^{\frac{1}{2}} \quad \text{with} \quad \log Q = 4.04 . \tag{27}$$

By elimination of R, ϱ_c, K we obtain, when $n = 2.5$

$$L = NE^{-6} m^{-3.6} Q^{-1.2} M^{5.2} T_0^{11,6} . \tag{28}$$

For the case $n = \frac{5}{3}$, we obtain

$$T = M^{0.2355} \left(\frac{3\varrho_c}{4\pi\bar{\varrho}}\right)^{0.118} \left(\frac{3\pi G}{2\xi^2 E}\right)^{0.294} , \tag{29}$$

which gives

$$T = 2660 M_*^{0.2355} , \tag{30}$$

for the above mentioned value of E, $\log E = 7.612$.

For the case $n = 2.5$, and introducing the luminosity, L^* and the mass M_* in solar units, and $(T_{\text{eff}}/5000) = T_*$, we obtain

$$L_* = 10^{2.7} M_*^{5.2} T_*^{-11.6} . \tag{31}$$

A useful expression can be written, using the expression of M and Q:

$$L = NE^{-6} \left(\frac{4\pi G}{(n+1)\xi_S^2}\right)^6 \left(\frac{4\pi}{3} \frac{\bar{\varrho}}{\varrho_c}\right)^{-3.6} M^{5.2} T_0^{-11.6} . \tag{32}$$

However, it should be noticed that the value of the constant E which has been obtained is obviously an overstimate. If we consider the shape of the adiabatics, it can be seen that near the surface, the pressure decreases faster. For the same temperature, the surface pressure, along the adiabatic, is smaller than it should be if γ were a constant all through the star. To obtain the real value of K, we have to take artificially a larger pressure, which corresponds to a smaller value of E. On the other hand, the concentration of matter can be smaller, which would correspond to a larger value of M. Altogether, it is not easy to decide exactly, and we think it is better to adjust more or less the value of L to the one obtained by Cameron and Ezer. We shall take from now on:

$$L_* = 10^{-0.87} M_*^{5,2} T_*^{-11.6} . \tag{33}$$

On the Hertzsprung-Russell diagram, log T_{eff}, log L_*, we plot our homology result, the modified result, and Cameron and Ezer's result (Fig. 3). On another figure, we plot the contracting path for two values of the mass, with the adopted scale (Fig. 4). Several other lines which can be found on Fig. 4 will be explained later. First of all, we plot the contracting path, when the structure of the star is governed by radiation. Adjusting to Cameron and Ezer's results, we take

$$L_* = 10^{0.18} M_*^{2.75} T_*^2 . \tag{34}$$

A star contracts quickly towards the main sequence until the radiative processes govern again the structure of the star.

The use of the homology relations gives the possibility of understanding easily the behaviour of the convective star. When it contracts, the entropy of the gas changes, the constant of the adiabatics, K, decreases like $R^{0.2}$. The convective zone takes place about at a constant optical depth; as the geometrical dimension de-

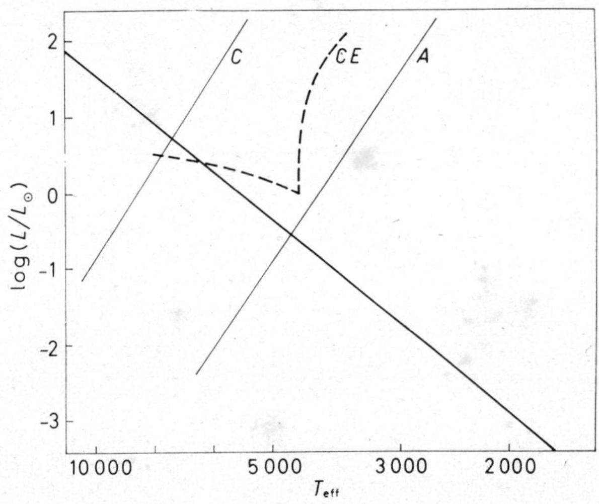

Fig. 3. – Hertzsprung Russel diagram. C is the calculated curve eq. (31), A is adjusted to agree approximately eq. (33) with the curve of CAMERON and EZER, called CE.

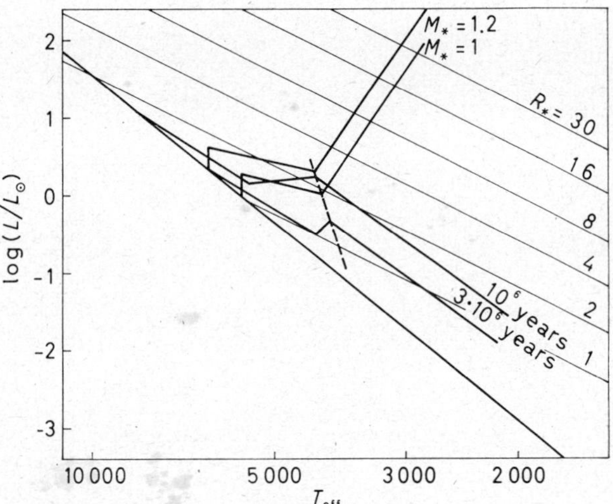

Fig. 4. – H.-R. diagram. The tracks for $M_* = 1.2$ and $M_* = 2$ have been drawn; the lines of constant R, and the curves $t = 10^6$ and $t = 3 \cdot 10^6$ years.

creases, the absorption coefficient must increase. The increase of the absorption coefficient is due essentially to a small increase of the temperature. We therefore have a large decrease of the radius with only a small increase of the temperature. *The peculiar aspect of the evolution track is due to the behaviour of the absorption coefficient as a function of the temperature.*

CAMERON and EZER (1962) have performed the exact calculations of a convective sun. Their exact results allowed us to formulate the description which has been given above.

It is interesting to test whether the star can be convective to the center. If we calculate the actual gradient, we obtain, at the center:

$$\frac{\mathrm{d}\log T}{\mathrm{d}\log P} = \frac{3}{16\pi acG}\left(\frac{\mathfrak{R}}{\mu}\right)^2 \varkappa\varrho T^{-2}\left(\frac{1}{r}\frac{\mathrm{d}r}{\mathrm{d}t}\right). \tag{35}$$

assuming for the absorption coefficient, pure scattering, $\varkappa = 0.4$, a density $\varrho = 1$ for a stellar radius $R = 3R_\odot$, a central temperature $T = 2.10^6$ degrees, we calculate $[(\mathrm{d}\log T)/(\mathrm{d}\log P)] \simeq 2$.

The gradient changes with the radius. The rate of change of the radius is proportional to a certain power of R:

$$\frac{1}{r}\frac{\mathrm{d}r}{\mathrm{d}t} = \frac{R}{GM}\frac{L}{M}. \tag{36}$$

The luminosity changes in the homology transform like $T^{-11.6}$; the radius changes like $T^{-7.8}$. The central temperature changes like R^{-1}, and the radiative gradient like $L \sim T^{-11.6}$, supposing that the absorption coefficient is due to pure scattering.

Above a certain radius, all configurations are fully convective, from the center to the surface.

Cameron and Ezer have noticed that for $R > 20R_\odot$, for the solar mass, the total energy is positive, the absolute value of the gravitational potential energy being less than the sum of the thermal, ionization and dissociation energies of the material. Such models are purely formal and are in the region of gravitational collapse.

2·2 *The time scale of contraction.* – It results from the structure of these convective stars, that the time scale of contraction, from $R = 20R_\odot$ to $R = 2R_\odot$ is much shorter than the classical Kelvin scale.

We have the following relations:

1) Luminosity as a function of R and T:

$$L_* = 10^{-0.234} R^{*2} T^{*4}. \tag{37}$$

2) Luminosity along the contracting track

(38) $$L_* = 10^{-0.87} M_*^{5.2} T_*^{-11.6} .$$

3) Expression of the luminosity as a function of the change in gravitational energy:

$$L = -c \frac{GM^2}{R^2} \frac{dR}{dt} ,$$

or

(39) $$L_* = -c \cdot 10^{15} \frac{M_*^2}{R_*^2} \frac{dR_*}{dt} .$$

the constant c is given by eq. (44).

We have, along the contracting track the following relation between T_* and R_*:

(40) $$T_*^{15.6} = 10^{-0.636} M_*^{5.2} R_*^{-2} ,$$

and we have therefore by elimination of T_* between (38) and (40)

(41) $$L_* = 10^{-0.4} R_*^{1.487} M_*^{1,333} ,$$

from which we deduce the time needed for the contraction:

(42) $$t = c \cdot 10^{15.4} M_{*0}^{0.666} \frac{1}{2.487} \left(\frac{1}{R_{*0}^{2.487}} - \frac{1}{R_{*1}^{2.487}} \right) .$$

We can practically neglect the upper limit. We have therefore, in millions of years

(43) $$\frac{t}{10^6 \text{ years}} = c \cdot 10^{1.6} M_*^{0.666} R_{*0}^{-2.487} ,$$

with

(44) $$c = \frac{3}{5-n} \frac{3\gamma - 4}{3(\gamma - 1)} ;$$

for $\gamma = 1.4$, we have $c = 0.2$, therefore

$$\frac{t}{10^6 \text{ years}} = 10^{0.9} M_*^{0.666} R_{*0}^{-2.487} .$$

The contraction to $R \simeq 3R_\odot$ would take about $4.5 \cdot 10^5$ years (Cameron and Ezer give $5 \cdot 10^5$ years).

Let us consider now the contractions to the minimum radius. The minimum radius is about at the intersections of the convective track and the radiative track; that is to say at the intersections of

$$L_* = 10^{-0.87} M_*^{5.2} T_*^{-11.6} , \tag{38}$$

and

$$L_* = 10^{0.18} M_*^{2.75} T_*^{2} . \tag{34}$$

This leads to the following relations, along the minimum radius

$$\left\{ \begin{aligned} L_* &= 10^{1.36} T_*^{17.28} , \\ R_* &= 10^{0.80} T_*^{6.64} , \\ M_* &= 10^{0.43} T_*^{5.55} , \\ R_* &= 10^{0.29} M_*^{1.195} . \end{aligned} \right. \tag{45}$$

The contraction to the minimum radius takes more and more time, when the mass decreases. We have

$$\frac{t}{10^6 \text{ year}} = c \cdot 10^{0.88} M_*^{-2.31} , \tag{46}$$

for $c = 0.2$

$$\frac{t}{10^6 \text{ year}} = 10^{0.18} M_*^{-2.31} . \tag{46'}$$

It is intersting to consider, in the H-R diagram, the line which is reached in $3 \cdot 10^6$ years.

We have the relation

$$R_{*0}^{2.487} = \frac{c}{(t/10^6 \text{ year})} 10^{1.6} M_*^{0.666} , \tag{47}$$

from which we derive the relation

$$L_* = 10^{1.264} \left(\frac{c}{(t/10^6 \text{ year})} \right)^{0.896} T_*^{5.75} , \tag{48}$$

with $c = 0.2$

$$L_* = 10^{0.636} \left(\frac{t}{10^6 \text{ year}} \right)^{-0.896} T_*^{5.75} . \tag{48'}$$

We find a boundary which is very similar to the region where stars are found in extremely young clusters, above the main sequence.

To be complete, we calculate the age sequence during the radiative phase. In the radiative phase we have

$$L_* = 10^{0.18} M_*^{2.75} T_*^2 , \tag{34}$$

$$L_* = 10^{-0.234} R_*^2 T_*^4 , \tag{37}$$

from which we derive

$$-c' \cdot 10^{15} \frac{M_*^2}{R_*^2} \frac{dR_*}{dt} = 10^{0.59} M_*^{5.5} R_*^{-2} , \tag{49}$$

where c' is given by the relation (44), applied to the star in radiative equilibrium. We deduce the time

$$\Delta t = c' \cdot 10^{13.93} M_*^{-3.5} \Delta \left(\frac{-1}{R_*^3}\right) . \tag{50}$$

The total time in units of 10^6 years is

$$\left(\frac{t}{10^6 \text{ year}}\right) = c \cdot 10^{-.88} M_*^{-2.31} - c' \cdot 10^{0.44} M_*^{-686} + c' \cdot 10^{0.43} \frac{M_*^{-3.5}}{R_{*1}^3} . \tag{51}$$

To reach the main sequence, we must consider the time needed to reach the radius $R_{*1} = 2R_*$ (main sequence) and add the time necessary to reach at constant surface temperature the main sequence.

$$\left(\frac{t}{10^6 \text{ year}}\right) = c \cdot 10^{0.88} M_*^{-2.31} - c' \cdot 10^{0.44} M_*^{-6.86} + c' \cdot 10^{-0.47} M_*^{-5.75} + c' \cdot 10^{0.86} M_*^{-2.75} . \tag{52}$$

The results have been collected in Table II, assuming $c' = 0.5$, which corresponds to a polytrope $n = 3$ and $\gamma = 1.5$.

TABLE II.

Log (M_*)	t (millions of years)
− 0.05	3.71
0	5
0.05	6.71
0.10	8.94

The constant c' has been adjusted to obtain for $M_* = 1$ the same value as CAMERON and EZER. ($5 \cdot 10^6$ years for the sun).

This time scale is much shorter than the old contracting scale of Rayleigh-Jeans. This is due to the two facts: (1) the first part of the contraction takes place during a phase of very high luminosity, (2) the last part of the contraction takes place with luminosity larger than the luminosity of the stars homologous to the sun. This last property is due to the change of internal structure from contractive to nuclear generation of energy.

The concentration, and therefore the temperature are higher in the contracting star. A look at the equation of radiative transport shows that the opacity being smaller in the contracting star, the luminosity is larger.

2.3. *Effect of rotation.* – Let us consider now how the contraction proceeds during mass loss.

We shall suppose, for the time being that the mass loss takes place only through the equator. As the proto-star contracts, its angular velocity increases, until the gravity at the equator is balanced by the centrifugal force. From that time on, the star contracts losing matter through the equator. Let us call KMR^2 the moment of inertia of the star. The conservation of angular momentum gives

$$\mathrm{d}(KMR^2\omega) = \tfrac{2}{3}R^2\omega\,\mathrm{d}M\,. \tag{53}$$

If the evolution takes place, as said above, with equality between the gravity and centrifugal force, we have

$$\frac{GM}{R^2} = \omega^2 R\,, \tag{54}$$

from which we deduce

$$\omega^2 = GMR^{-3}\,, \tag{55}$$

and

$$\frac{3}{2}\frac{\mathrm{d}M}{M} + \frac{1}{2}\frac{\mathrm{d}R}{R} = \frac{1}{K}\frac{\mathrm{d}M}{M}\,. \tag{56}$$

The mass changes with the radius, according to the relation

$$M \sim R^\lambda\,, \tag{57}$$

with

$$\lambda = \left(\frac{2}{K} - 3\right)^{-1}. \tag{58}$$

The values of λ, as a function of the polytropic index are given in Table III.

TABLE III.

n	$1/K$	$1/\lambda$
0	2.3	2
1	3.79	4.58
1.5	5	7
2.5	9	15
3	13.5	24

When the structure of the star is governed by radiative equilibrium, we can assume a polytropic index 3, with a corresponding value $1/\lambda' = 24$; when the structure of the star is governed by convection, the polytropic index is lower. In the H-R diagram, the track is changed. The relation between the luminosity and the temperature is given by eq. (59):

$$\log L = \frac{4(1 - 2.75\lambda)}{2 - 2.75\lambda} \log T \,. \tag{59}$$

For $\lambda = \lambda' = (1/24)$, $L \sim T^{1.875}$ instead of $L \sim T^2$ as in eq. (34).

When the structure of the star is governed, by convective equilibrium, we can assume a polytropic index 2.5, with a corresponding value $\lambda = 1/15$. The star evolves *above* the track without mass loss (Fig. 5) with

$$L \sim T_{\text{eff}}^{-1.80} \,. \tag{60}$$

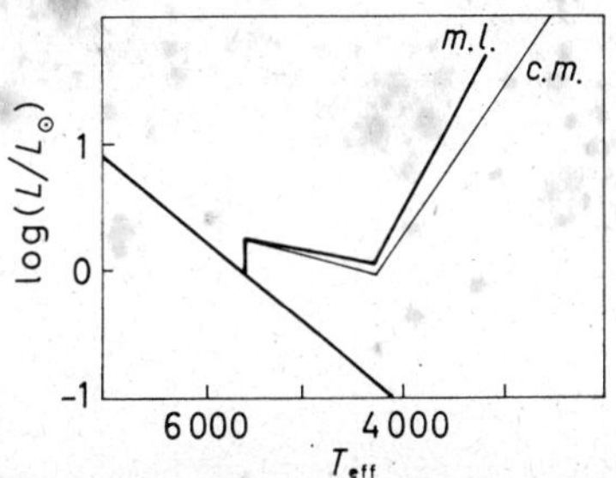

Fig. 5. – H.-R. diagram. The tracks towards the main sequence with mass loss (m.l.) and with constant mass (c.m.). The final mass is $M_{*f} = 1$.

Let us consider now the rate of contractions during mass loss. For a star which contracts very slowly, it is possible to neglect the acceleration of the motions and to apply the *virial theorem*:

$$L = -\frac{\mathrm{d}E}{\mathrm{d}t} \,, \tag{61}$$

E is the total energy:

$$E = -(3\gamma - 4)U \,, \tag{62}$$

and U is the internal energy

$$U = \int_0^M c_V T \, \mathrm{d}M \,. \tag{63}$$

c_V is the specific heat at constant volume.

The variation of internal energy is

$$\delta U = \int_0^M c_V \, \delta T \, \mathrm{d}M + c_V T_0 \, \delta M \,. \tag{64}$$

The surface temperature can be neglected, consequently

$$\delta U = \int_0^M c_V \, \delta T \, \mathrm{d}M \,, \tag{65}$$

and we obtain for the rate of energy production

$$L = cGM \frac{\mathrm{d}}{\mathrm{d}t} \frac{M}{R} \,. \tag{66}$$

For a rotating object,

$$L = GM \frac{\mathrm{d}}{\mathrm{d}t} \left\{ \left(\frac{3\gamma - 4}{3(\gamma - 1)} - \frac{1}{2} K \right) \frac{M}{R} \right\} , \tag{67}$$

KMR^2 being the moment of inertia. The energy of the rotation decreases the energy available for radiation. We can as well take c smaller.

Let us call M_f the final mass of the star, M_m the mass at minimum, R_f the final radius of the star, R_m the radius at minimum.

We calculate the time t,

$$t = c \frac{10^{1.906}(1 - \lambda)}{0.33\lambda + 2.487} M_m^{0.666} R_m^{0.333} R^{-0.333\lambda - 2.487} \,. \tag{68}$$

The radius at minimum is given by

$$R_m = 10^{0.304} M_f^{2.01} \,. \tag{69}$$

As we have

$$M_m = M_f (R_m / R_f)^{\lambda'} \,, \tag{70}$$

with $\lambda' = 1/24$ and $\lambda = 1/15$, we obtain finally the time needed to reach minimum,

$$t_m = c \cdot 10^{0.743} M_f^{0.423} \, 10^6 \text{ years} . \tag{71}$$

For $c = 0.155$, $M = M_\odot$, we find

$$t_m = 0.85 \cdot 10^6 \text{ years} ,$$

instead of $1.5 \cdot 10^6$ years without mass loss. We can conclude that the existence of the mass loss increases appreciably the speed of the evolution.

3. – Atmospheric structure of a rotating star.

Let us define the « surface » of the star by the point where the optical thickness, perpendicularly to the equipotentials is equal to τ_0, and consider the position of the convective zone.

The gravity vanishes at the equator. Taking the Roche equipotential surface as a reference surface, the gravity, as a function of the distance to the Roche equipotential, and as a function of the distance to the equatorial plane, can be calculated.

Calling x the distance to the equator, y the distance to the surface, we have for the potential:

$$\psi = -\frac{1}{2} G \frac{M}{R^3} (2y^2 - 2\sqrt{3}xy) , \tag{72}$$

R is the radius of the equator. The gravity, as a function of x and y is given by the relations:

$$\begin{cases} -\dfrac{\partial \psi}{\partial y} = (4y - 2\sqrt{3}x) \dfrac{1}{2} \dfrac{GM}{R^3} , \\ -\dfrac{\partial \psi}{\partial x} = -2\sqrt{3}y \dfrac{1}{2} \dfrac{GM}{R^3} . \end{cases} \tag{73}$$

If we are near to the surface, $y \ll x$, and as a first approximation, the gravity is

$$g = -\sqrt{3}x \frac{GM}{R^3} . \tag{74}$$

The optical depth at the boundary of the convective layer is given by

$$\frac{\tau + q}{q} = \left(1 + \frac{\beta - \alpha - 4}{n + 1}\right)^{4/(\beta - \alpha - 4)} .$$

With the values used in Section **2**,

$$\beta = 5\,, \quad \alpha = \tfrac{1}{2}; \qquad n = 2.5\,, \quad q = \tfrac{2}{3}\,, \qquad \text{we obtain} \qquad \tau = 1.273\,.$$

The surface of reference is defined by $P = \text{const}$, as we have

$$\nabla P = -\varrho \nabla \psi\,.$$

The change of pressure is zero along an equipotential. Therefore, we study the optical depth of the surface $P = \text{const}$.

We calculate the pressure at optical depth τ:

$$P = \frac{\Re}{\mu}\left(\frac{g\mu(\alpha+1)}{\varkappa_0 R}\right)^{1/(\alpha+1)} T_0^{-(\beta-\alpha)/(\alpha+1)} \left(\frac{3}{2}\right)^{-(\beta+1)/4(\alpha+1)} \cdot \\ \cdot \left[\frac{q^{((\alpha-\beta)/4)+1} - (\tau+q)^{((\alpha-\beta)/4)+1}}{((\alpha-\beta)/4)+1}\right]^{1/(\alpha+1)}. \tag{75}$$

From eq. (75), we derive the relation (76) which gives the optical depth along a surface of constant pressure as a function of gravity g. τ_0 is the optical depth of the isobar at the point of gravity g_0.

$$\left(\frac{q}{\tau+q}\right)^{(\beta-\alpha-4)/4} - 1 = -\frac{g_0}{g}\left[1 - \left(\frac{q}{\tau_0+q}\right)^{(\beta-\alpha-4)/4}\right]. \tag{76}$$

Let us take τ_0 very small. We can develop the right-hand side of (76) to first order in τ_0 and obtain $\tau = \tau_1$:

$$\left(\frac{q}{\tau_1+q}\right)^{\frac{1}{8}} = 1 - \frac{g_0}{g}\frac{3}{16}\tau_0\,, \tag{77}$$

τ_1 becomes infinite for

$$g = \frac{3}{16} g_0 \tau_0\,, \tag{78}$$

as we have

$$g_0 = -\frac{q}{4}\frac{GM}{R^2}\,,$$

we obtain for the distance x to the equator:

$$x = \frac{27R}{64\sqrt{3}}\tau_0 = 0.244 R\tau_0\,. \tag{79}$$

If we take for exemple $\tau_0 = 0.01$ we obtain $x = 0.00244R$. At this point, the optical depth of the « isobar » becomes infinite. We can notice, especially, that the convective zone reaches the isobar near this point.

If we consider the surface as defined by the layer of constant optical thickness, it can be noticed that this surface rises very high above the isobar. If τ_1 is the optical thickness of the isobar, τ_0 the optical thickness of the surface, the position z of the surface is given by

$$z = \int \frac{d\tau}{\varkappa_0 \varrho^{\alpha+1} T^\beta} . \tag{80}$$

Taking into account the result of VAN ZEIPEL (1924), that the flux is proportional to the gravity, and calling H_0 the scale height at the pole, we obtain,

$$z = \left(\frac{3}{2}\right)^{\frac{1}{4}} H_0 \left(\frac{g_0}{g}\right)^{\frac{3}{4}} \left[-\log\left(\frac{g}{g_0} - \frac{3\tau_0}{16}\right) + \frac{3}{16}\tau_0 \left[\left(\frac{g}{g_0} - \frac{3\tau_0}{16}\right)^{-1} - 1\right]\right] . \tag{81}$$

Let us write

$$x = \frac{27}{64\sqrt{3}} R\tau_0(1+\varepsilon) , \tag{82}$$

we obtain

$$y = \left(\frac{3}{2}\right)^{\frac{1}{4}} H_0 \left(\frac{16}{3\tau_0}\right)^{\frac{3}{4}} (1+\varepsilon)^{-\frac{3}{4}} \left[\log\frac{16}{3\tau_0} - \log\varepsilon + \frac{1}{\varepsilon} - \frac{3\tau_0}{16}\right] . \tag{83}$$

We have drawn in Fig. 6 the shape of the optical surface of the star. The limit of the convective zone is at an optical depth τ_c. The geometrical depth y_c of the convective zone is a function of g:

$$y_c = \left(\frac{3}{2}\right)^{\frac{1}{4}} H_0 \left(\frac{g_0}{g}\right)^{\frac{3}{4}} \left[-\log\left(\frac{g}{g_0} - \frac{3\tau_0}{16}\right) - \right.$$
$$\left. - \log\frac{\tau_c}{\tau_0} + \left(1 - \frac{3}{16}\frac{g_0}{g}\tau_0\right)^{-1} - \left(\frac{\tau_c + q}{q}\right)^{\frac{1}{8}}\right] . \tag{84}$$

It can be found that y_c vanishes for

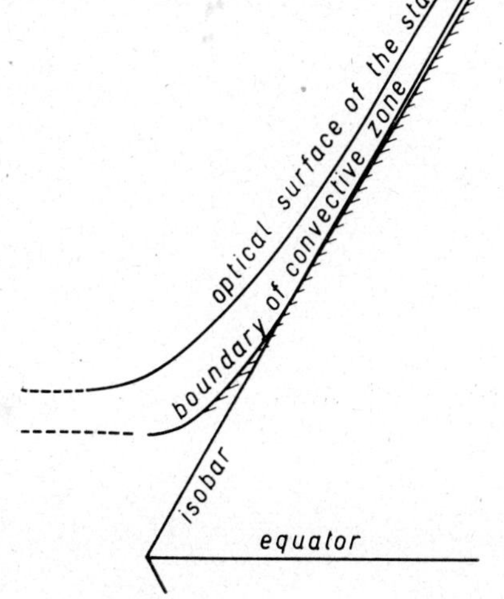

Fig. 6. – The shape of the atmosphere near to the equator of the rotating star. The connection of the atmosphere with the ring is sketched by the dashed line.

$\varepsilon = 4.34$, which corresponds to

$$x_c = 0.013\,R\,. \tag{85}$$

We can say that below $x_c = 0.013R$, the theory of the atmosphere becomes unrealistic. The convective zone really merges with the interstellar matter, and goes into a circumstellar ring (Fig. 6).

4. – The flow in the contracting star.

It is quite obvious that the equatorial loss which we have just described implies a flow of matter from the center to the surface. To simplify the problem of the flow we can consider it equivalent to finding the lines of flow of a source at the center, and a sink at the equator. It should be necessary even with this simplification, to treat the problem of the lines of flow inside a Roche surface. However, as it does not seem from the kinetical point of view, that there is any important difference between the sphere and the Roche surface, and as the properties of the function on the sphere are very well known it has been considered easier to treat the problem of the sphere. When the motion is stationary, we have only to consider the continuity equation

$$\operatorname{div} \varrho v = 0\,. \tag{86}$$

If the viscosity is dominant, v derives from a potential. We have solved the problem for the case $\varrho = \text{const}$. As we can neglect the accelerations, we can use the solution even when $\varrho = \varrho(r)$. We shall see later the difference with the case where v and not ϱv derives from a potential.

We assume that there is a source, of power one, at the center, and a sink, of power 2, along the equator; we add, to this potential, a potential satisfying the equation

$$\nabla^2 U_3 = 0\,, \tag{87}$$

and such that the flux across the surface of the sphere is zero. The potential of the charge $+1$ is

$$U_1 = \frac{1}{r}\,. \tag{88}$$

The potential of the charge -2 is obtained by adding the differential along the equator

$$U_2 = -\frac{2}{\pi}\int\limits_{-\pi/2}^{+\frac{1}{2}} \frac{\mathrm{d}\varphi}{(a^2 + r^2 - 2ar\sin\theta\sin\varphi)^{\frac{1}{2}}}\,. \tag{89}$$

The radial component of the gradient of the potential $U_1 + U_2$ is

$$\frac{\partial}{\partial r}(U_1 + U_2) = -\frac{1}{r^2} + \frac{2}{\pi}\int_{-\pi/2}^{+\pi/2} \frac{(r - a\sin\theta\sin\varphi)\,d\varphi}{(a^2 + r^2 - 2ar\sin\theta\sin\varphi)^{\frac{3}{2}}}. \tag{90}$$

On the sphere of radius a, the radial component is

$$\frac{\partial}{\partial r}(U_1 + U_2) = -\frac{1}{a^2} + \frac{2}{\pi a^2}\frac{1}{2^{\frac{3}{2}}}\int_{-\pi/2}^{+\pi/2} \frac{d\varphi}{(1 - \sin\theta\sin\varphi)^{\frac{1}{2}}}. \tag{91}$$

To obtain the solution U_3, we apply the theory of spherical harmonics. The interior potential is

$$\psi = \sum A_n \left(\frac{r}{a}\right)^n P_n(\cos\theta), \tag{92}$$

where A_n are arbitrary coefficients, and P_n the Legendre polynomial of order n; the radial component of the gradient is

$$\frac{\partial\psi}{\partial r} = \sum \frac{nA_n}{a}\left(\frac{r}{a}\right)^{n-1} P_n(\cos\theta). \tag{93}$$

By identifications with a function $f(\theta)$ on the sphere, we have

$$\left(\frac{\partial\psi}{\partial r}\right)_{r=a} = \sum \frac{nA_n}{a} P_n(\cos\theta) = f(\theta). \tag{94}$$

By the standard method, concerning the orthogonal functions, we obtain

$$2\pi\frac{nA_n}{a}\frac{2}{2n+1} = \int_0^{2\pi} d\varphi' \int_0^{\pi} \sin\theta'\,d\theta'\,P_n(\cos\theta')\,f(\theta'), \tag{95}$$

from which we derive the radial component of the gradient:

$$\frac{\partial\psi}{\partial r} = \frac{a}{r}\sum\left(\frac{r}{a}\right)^n \frac{2n+1}{4\pi}\int_0^{2\pi} d\varphi' \int_0^{\pi} f(\theta')P_n(\cos\theta')P_n(\cos\theta)\sin\theta'\,d\theta'. \tag{96}$$

To find another expression of the gradient, we use the development

$$\frac{1-h^2}{(1-2hz+h^2)^{\frac{3}{2}}} = \sum(2n+1)h^n P_n(z), \tag{97}$$

and the addition formula for spherical functions:

$$P_n(\cos\alpha) = P_n(\cos\theta)P_n(\cos\theta') + \tag{98}$$

$$+ 2\sum_{1}^{n} \frac{\overline{n-m}!}{\overline{n+m}!} P_n^m(\cos\theta)P_n^m(\cos\theta')\cos m(\varphi' - \varphi''),$$

were

$$\cos\alpha = \cos\theta\cos\theta' + \sin\theta\sin\theta'\cos(\varphi' - \varphi''). \tag{99}$$

Considering the fact that the integration of $\cos(\varphi' - \varphi'')$ would give zero, we can write

$$\int_0^{2\pi} d'\varphi \int_0^{\pi} f(\theta')P_n(\cos\theta')P_n(\cos\theta)\sin\theta'\,d\theta' = \int_0^{2\pi} d\varphi' \int_0^{\pi} f(\theta')P_n(\cos\alpha)\sin\theta'\,d\theta', \tag{100}$$

and we derive the general expression of $\partial\psi/\partial r$

$$\frac{\partial\psi}{\partial r} = \frac{a}{r}\,\frac{a(a^2 - r^2)}{4\pi}\int_0^{2\pi} d\varphi' \int_0^{\pi} f(\theta'). \tag{101}$$

We have only to carry in this last expression, for $f(\theta')$, the gradient of the source and the well given by the expression (91), but with the opposite sign.

Let us consider now the situation close to the surface. We have to calculate an integral

$$S = \int_0^{2\pi} d\varphi' \int_0^{\pi} \frac{f(\theta')\sin\theta'\,d\theta'}{(r^2 - 2ar\cos\alpha + a^2)^{\frac{3}{2}}}. \tag{102}$$

Near $r = a$, $\theta' = \theta$, we develop the denominator. We calculate

$$r^2 - 2ar\cos\alpha + a^2 = (\sigma - a)^2 + ar[(\theta' - \theta)^2 + \sin^2\theta\,\varphi'^2], \tag{103}$$

and obtain

$$S = \int_{-\pi}^{+\pi} d\varphi' \int_0^{\pi} \frac{f(\theta)\sin\theta\,d\theta'}{(a-r)^3\left[1 + (ar/(a-r)^2)[(\theta' - \theta)^2 + \sin^2\theta\varphi'^2]^{\frac{3}{2}}\right]}. \tag{104}$$

Putting

$$\begin{cases} \sin\theta\,d\varphi'\,d\theta' = \varrho\,d\varrho\,d\omega\,\dfrac{(a-r)^2}{ar}, \\ \theta' - \theta = \varrho\cos\omega\,\dfrac{a-r}{\sqrt{ar}}, \qquad \sin\theta\varphi' = \varrho\sin\omega\,\dfrac{a-r}{\sqrt{ar}}, \end{cases} \tag{105}$$

we obtain

$$S = \frac{f(\theta)}{(a-r)^3} \iint \frac{(a-r)^2}{ar} \frac{\varrho\, d\varrho\, d\omega}{(1+\varrho^2)^{\frac{3}{2}}}, \tag{106}$$

and finally

$$S = \frac{2\pi f(\theta)}{ar(a-r)}, \tag{107}$$

and consequently

$$\frac{\partial\psi}{\partial r} = \frac{a+r}{2ar^2}\left[1 - \frac{1}{\pi}\int_0^{2\pi} \frac{1}{2^{\frac{3}{2}}} \frac{d\varphi}{(1-\sin\varphi\sin\theta)^{\frac{1}{2}}}\right]. \tag{108}$$

To obtain the flux near the surface, we shall first calculate the potential near the surface, next obtain the angle-dependent term from the Laplacian

$$\frac{1}{r}\frac{\partial U}{\partial\theta} = -\frac{1}{\sin\theta}\int_0^{\theta}\left(\frac{1}{r}\frac{\partial}{\partial r}\left(r^2\frac{\partial V}{\partial r}\right)\right)\sin\theta\, d\theta . \tag{109}$$

We have

$$U = \frac{1}{r} - \frac{1}{a} + \frac{2}{\pi}\left[\frac{1}{a}\int_{-\pi/2}^{+\pi/2}\frac{d\varphi}{\sqrt{2}\,(1-\sin\theta\sin\varphi)^{\frac{1}{2}}} - \right. \tag{110}$$

$$\left. -\int_{-\pi/2}^{+\pi/2}\frac{d\varphi}{(a^2+r^2-2ar\sin\theta\sin\varphi)^{\frac{1}{2}}}\right] + \frac{2}{\pi}\int_a^r \frac{a+r}{2ar^2}\left[1-\frac{2}{\pi}\int_{-\pi/2}^{+\pi/2} 2^{-\frac{3}{2}}\frac{d\varphi}{(1-\sin\theta\sin\varphi)^{\frac{3}{2}}}\right] dr .$$

To the lowest power in $(r-a)$, we have

$$U = \frac{(r-a)^2}{a^3}\left[1 + \frac{1}{2\pi}\int_{-\pi/2}^{+\pi/2}\frac{d\varphi}{\sqrt{2}(1-\sin\theta\sin\varphi)^{\frac{3}{2}}}\right], \tag{111}$$

and we obtain the velocity along the surface

$$v_\theta = \frac{1}{\sin\theta}\int_0^{\theta}\frac{2}{a}\left(1+\frac{1}{2\pi}\int_{-\pi/2}^{+\pi/2}\frac{d\varphi}{\sqrt{2}(1-\sin\theta\sin\varphi)^{\frac{3}{2}}}\right)\sin\theta\, d\theta . \tag{112}$$

This can be expressed with help of the elliptic integrals. Let us consider the variable t:

$$\sin\varphi = -\cos 2t = -1 + 2\sin^2 t . \tag{113}$$

We have to consider the integral Q:

(114) $$Q = \int_0^{2\pi} \frac{d\varphi}{(1 - \sin\theta \sin\varphi)^{\frac{3}{2}}} = \int_0^{\pi} \frac{2\,dt}{(1 + \sin\theta - 2\sin\theta\sin^2 t)^{\frac{3}{2}}} .$$

Put

$$\frac{2\sin\theta}{1 + \sin\theta} = x .$$

We have

(115) $$Q = \frac{2}{(1+\sin\theta)^{\frac{3}{2}}} \int_{-\pi/2}^{+\pi/2} \frac{dt}{(1 - x\sin^2 t)^{\frac{3}{2}}} .$$

We use the classical notations for complete elliptic integrals. Introducing $J(x)$:

$$J(x) = \int_0^{\pi/2} \frac{dt}{(1 - x\sin^2 t)^{\frac{3}{2}}} ,$$

we write:

$$v_\theta = \frac{2-x}{x} \int_0^x \frac{2}{a^2} \left(1 + \frac{1}{\sqrt{2}} \frac{1}{\pi} \frac{J(2\sin\theta/(1+\sin\theta))}{(1+\sin\theta)^{\frac{3}{2}}}\right) \frac{x\,dx}{\sqrt{1-x}\,(2-x)^2} .$$

If we use the complete integrals $K(x)$ and $B(x)$,

$$K = \int_0^{\pi/2} \frac{dt}{(1 - x\sin^2 t)^{\frac{1}{2}}} , \qquad B = \int_0^{\pi/2} \frac{\cos^2 t\,dt}{(1 - x\sin^2 t)^{\frac{1}{2}}} ,$$

we obtain:

$$v_\theta = \frac{2-x}{x} \int_0^x \frac{2}{a^2} \left(1 + \frac{1}{4\pi} (2-x)^{\frac{3}{2}} \left(K + \frac{x}{1-x} B\right)\right) \frac{x\,dx}{\sqrt{1-x}\,(2-x)^2} .$$

Near $x = 0$ and $\sin\theta = 0$, we have for v_θ

$$v_\theta = \frac{1}{a^2} \left(1 + \frac{1}{2^{\frac{3}{2}}}\right) \sin\theta .$$

Near $\theta = \pi/2$, $x = 1$, the velocity becomes infinite, as should be expected. The radial velocity, near the surface vanishes like $r - a$:

(120) $$v_r = \frac{2(a-r)}{a^3} \left[1 + \frac{1}{2\pi} \int_{-\pi/2}^{+\pi/2} \frac{d\varphi}{\sqrt{2}\,(1 - \sin\theta\sin\varphi)^{\frac{3}{2}}}\right] .$$

The velocity v_θ has been tabulated (Table IV). The curve can be found in Fig. 7.

TABLE IV.

θ	$a^2 v_\theta$
0	0
6° 38	0.1507
14° 48	0.3470
20° 52	0.5000
28° 08	0.702
37° 7	0.990
51° 7	1.522
65° 8	2.61

We change now from the velocity v_θ calculated for $\varrho = cte$ to the material velocity u_θ, when ϱ is not a constant. The velocity, near the surface, increases like ϱ^{-1}. The question arises naturally of finding where the differential velocity will vanish. Let us consider how long it takes for the viscosity to act in such a way that the atmosphere moves with a uniform velocity over one scale height.

Between time t_v, length L_v, mean path l and mean velocity ξ, we have the approximate relation

$$L_v^2 t_v^{-1} \simeq \tfrac{1}{3}\xi l \,. \tag{121}$$

The molecular mean free path l is approximately

$$l = 1.6 \cdot 10^{-9} \varrho^{-1} \,. \tag{122}$$

From eq. (121) and (122), we derive, for $\xi = 5 \cdot 10^5 \text{ cm s}^{-1}$, which corresponds to a temperature of a few thousand degrees,

$$L_v^2 t_v^{-1} \simeq 10^{-3.68} \varrho^{-1} \,.$$

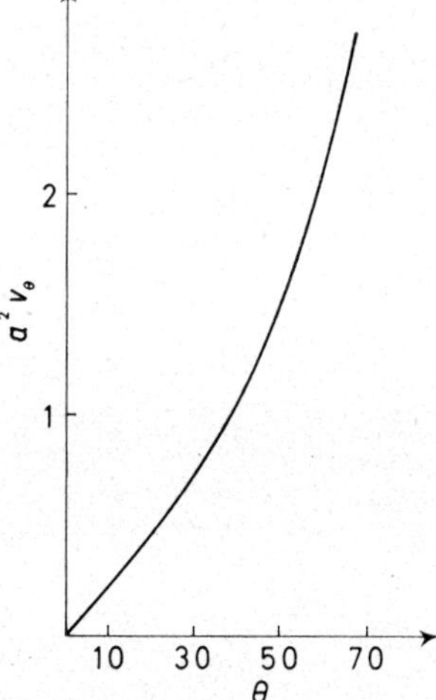

Fig. 7. – The dimensionless quantity $a^2 v_\theta$ as a function of the colatitude θ. The velocity increases towards the equator.

With $\varrho = 10^{-7}$, $L_v \simeq 10^7$ cm, we derive

$$t_v \simeq 10^{10.58} \simeq 10^3 \text{ years} \,.$$

We can see that in 10^6 years, uniform velocity is established to a layer of density 10^{-4}.

Let us now consider the material velocity u_θ. We assume that ϱu_θ has the same properties as v_θ. Assuming that the mass loss dM/dt occurs over the solid angle 4π, near to the center, we obtain

$$u_\theta = v_\theta \frac{1}{4\pi} \frac{dM}{dt} \frac{1}{\varrho}, \tag{123}$$

dM/dt is of the order of $10^{-1} M_\odot\ 10^{-6}\ \text{years}^{-1} \simeq 10^{19}\ \text{g s}^{-1}$.

For $\varrho \simeq 10^{-4}$ (limit of uniform motion), a radius $a = 10^{11}$ cm we obtain

$$u_\theta = 0.80(a^2 v_\theta)\ \text{cm s}^{-1}. \tag{124}$$

The time needed for the atmosphere to be carried from the pole to the equator is of the order of 10^{12} s. To be more precise, we have from 90° to θ a time of transport $t_\theta^{90°}$ given by eq. (125):

$$t_\theta^{90°} = \frac{4\pi \varrho a^3}{(dM/dt)} \int \frac{d\theta}{a^2 v_\theta}. \tag{125}$$

The times are given in Table V for:

$$\varrho = 10^{-4}, \qquad dM/dt = 10^{19}\ \text{g s}^{-1}.$$

TABLE V.

θ	$\int \frac{d\theta}{a^2 v_\theta}$	t (10^6 years)
90	0	0.0000
80	8.25	0.0006
70	34	0.0023
60	76.2	0.0054
50	135.7	0.0095
40	217.5	0.0152
30	330.5	0.0233
20	496.3	0.0350
10	796	0.0560
5	1096	0.0770
2.5	1396	0.098
1.25	1696	0.119

We see that with this rate of mass loss, after 10^5 years, about 10^{-4} of the stellar surface, near the polar caps, has been diluted over the whole surface of the star. In case of production of surface nuclear reactions, it would divide the abundances of peculiar elements by 10000.

II. – The Role of Magnetic Activity During Star Formation.

5. – Introduction.

As it has been mentioned before, contractions of stars towards the main sequence should always lead to fast rotating stars. The problem is to get rid of the excess of angular momentum.

Surface mass loss, as it was suggested by FESSENKOFF (1949) could do it; but the angular momentum per gram at the surface of the star is only 10 times the mean angular momentum per gram for the whole star. The differential equation reads:

$$\frac{d}{dt} KMR^2\omega = \frac{2}{3} R^2\omega \frac{dM}{dt}, \tag{126}$$

with $R \sim M^{\frac{1}{3}}$, we have:

$$\omega \sim M^n, \tag{127}$$

with:

$$n = \left(\frac{2}{3}\frac{1}{K} - \frac{5}{2}\right). \tag{128}$$

For the favorable case $1/K = 13.5$, $n = 6.5$ and we see that a decrease of ω by a factor 100 can be obtained only by a mass decrease by a factor 2.

Three objections can be made to this theory of the loss of angular momentum (1st), such a large mass loss could not escape observations; (2nd) it does not explain the difference between the upper part and the lower part of the main sequence; (3rd) the known mechanisms of mass ejection, like the so-called stellar wind, have a low efficiency.

A possibility, which has been suggested by SPITZER (1956), was the coupling between the star and the surrounding matter, through the magnetic field of the star. The difference between upper and lower main sequence would have resulted from the existence of H II regions around early type stars.

The question of how far the magnetic field carries the matter can be solved only by comparing the order of magnitude of the different times in the equation of motion. Let us compare the kinetic energy, the magnetic energy, and the viscous force. The viscosity is considerably reduced by the magnetic field, so that the boundary is essentially given by the equality of the magnetic energy and the kinetic energy. If we call ϱ_{im} the density of interstellar matter, and assume the magnetic field $B_0(a/r)^3$ of a dipole, we obtain:

$$\varrho_{im} r^2\omega^2 \simeq B_0^2 \left(\frac{r}{a}\right)^{-6}, \tag{129}$$

from which we derive:

$$r^8 \simeq \varrho_{im}^{-1} \omega^{-2} B_0^2 a^6 \,. \tag{130}$$

We calculate the order of magnitude of the rate of loss of angular momentum

$$\frac{\mathrm{d}}{\mathrm{d}t} KMa^2\omega \simeq -\varrho_{im}^{-3/8}\, \omega^{3/4}\, B_0^{5/4}\, a^{15/4} \,,$$

from which we derive the order of magnitude of the time scale,

$$t = K\omega^{1/4} \varrho_{im}^{-3/8} Ma^{-7/4} B^{-5/4} \,. \tag{131}$$

The time scale is shorter, the greater is ϱ_{im}. However, r should not be smaller than a, and we have the maximum value:

$$\varrho_{im} \leqslant B_0^2 \omega^{-2} a^{-2} \,, \tag{132}$$

from which we derive:

$$t \geqslant K\omega a^{-1} M B_0^{-2} \,. \tag{133}$$

For the greatest speed of rotation, $\omega = 10^{-3}$ and $a = 10^{11}$, $B_0 = 100$ G, $m = 2 \cdot 10^{33}$

$$\varrho < 10^{-12} \,, \tag{134}$$

and

$$t \geqslant 2 \cdot 10^{14} \text{ s} \,. \tag{135}$$

However, for such high densities, if they could at all exist around a star, there would not be any difference between the stars, and no H II region could ever appear.

For smaller densities, the time t is much too long; for example, for $\varrho = 10^{-20}$,

$$t \simeq 2 \cdot 10^{17} \text{ s} \,. \tag{136}$$

The idea of associating the loss of angular momentum to the magnetic activity of the star, came out essentially from the fact, that, if in the equation of conservation

$$\frac{\mathrm{d}}{\mathrm{d}t}(KMR^2\omega) = \frac{2}{3} R^2 \omega \,\mathrm{d}M \,, \tag{137}$$

we substitute to R in the right hand side a distance d, much greater, we increase the rate of loss of angular momentum in a very large proportion. The magnetic field can carry jets of matter in such a way that the ejected matter is lost for the star at a large distance. We combine, in such a way, the mass loss and the magnetic effects, so that we increase the efficiency of the mass loss by the action of the magnetic field. Taking example on the Sun, we are led to the idea that jets of matter are related to the electromagnetic activity. Consequently, we have, in order to reach a conclusion, to consider a theory of the magnetic activity, and to study a large variety of associated phenomena.

It turns out that we are led to a solution of the problem of angular momentum, and find, in the meantime, an interpretation of the distribution of peculiar abundance among the spectral types.

The steps of the reasoning are the following:

1) The electromagnetic activity is related to the rotation and to the existence of a convective zone of some importance;

2) Stars with a large convective zone are distributed on the H-R diagram on the right of a line which is almost vertical, and crosses the main sequence in the spectral type F;

3) The matter which is ejected is carried by the magnetic field of the rotating star and lost at a large distance. Stars with a convective zone do have electromagnetic activity and experience mass loss; stars without a convective zone do not have electromagnetic activity, and do not lose matter.

4) Stellar flares, shock waves of various strengths are at the origin of surface nuclear reactions, and give the possibility of explaining the formation of peculiar abundances at the surface of the star.

5) Further evolution restricts the possibility for the stars to keep these peculiar abundances at the surface to a narrow region near to spectral type A, with lesser possibility near to spectral type B2-B3.

To explain all these facts, we have to use results derived from a variety of physical theories, and to apply them to astrophysics. We can either begin with the astrophysics or with the physics. It seems more reasonable to begin with the physics and to get afterwards only to astrophysics.

6. – Theory of the acceleration of charged particles by magneto-hydrodynamic shock waves.

6·1. *Shock waves*. – Classically, a shock wave is defined as a discontinuity in a fluid; all quantities: velocity, density, temperature, pressure, entropy, are discontinuous across the surface of discontinuity, or shock front. When there is a magnetic field present, the magnetic field also is discontinuous.

By considering the equations of conservation of mass, momentum, and energy, and the electromagnetic equations, Maxwell's equations and Ohm's law, it is possible to find the relations between all physical quantities, across the shock front.

The detailed theory of hydromagnetic shock waves is extremely complicated. But we are interested here especially in one of the simplest cases, the case when the magnetic field is perpendicular to the direction of propagation of the wave or *perpendicular shock.*

Let us write the classical equations relating medium 1 ahead of the shock front, to medium 2 behind the shock front. For the non relativistic case, we have

$$\varrho_1 v_1 = \varrho_2 v_2 \qquad \text{or} \qquad [\![\varrho v]\!] = 0\,,$$

$$\left[\!\!\left[p + \varrho v^2 + \frac{B^2}{8\pi} \right]\!\!\right] = 0\,,$$

$$\left[\!\!\left[E\varrho v + \frac{1}{2}\varrho v^3 + pv + v\frac{B^2}{4\pi} \right]\!\!\right] = 0\,,$$

$$\left[\!\!\left[\frac{B}{\varrho} \right]\!\!\right] = 0\,.$$

Subscript 1 refers to the medium ahead of the shock, subscript 2 to the medium behind the shock. v_1, ϱ_1, B_1, p_1, are the velocity, the density, the magnetic field and the pressure. E is the internal energy of the gas. The shock front is supposed here to be at rest with respect to the co-ordinate system.

For the ideal gas

$$E = \frac{p}{(\gamma - 1)}\,, \tag{139}$$

and we have the relations

$$\left(p_2 + \frac{B_2^2}{8\pi}\right)\left[\frac{1}{2\varrho_1} - \frac{1}{2\varrho_2} - \frac{1}{(\gamma-1)\varrho_2}\right] - \left(p_1 + \frac{B_1^2}{8\pi}\right)\left(\frac{1}{2\varrho_2} - \frac{1}{2\varrho_1} - \frac{1}{(\gamma-1)\varrho_1}\right) = \frac{B_1^2}{8\pi}\frac{\gamma-2}{\gamma-1}\frac{\varrho_2 - \varrho_1}{\varrho_1^2}\,, \tag{140}$$

and

$$v_1 - v_2 = \left\{\left[\left(p_2 + \frac{B_2^2}{8\pi}\right) - \left(p_1 + \frac{B_1^2}{8\pi}\right)\right]\left[\frac{1}{\varrho_1} - \frac{1}{\varrho_2}\right]\right\}^{\frac{1}{2}}. \tag{141}$$

Putting, as it is usual

$$\frac{\varrho_2 - \varrho_1}{\varrho_1} = \bar{\eta}\,, \tag{142}$$

we obtain

$$(143)\qquad \frac{1}{\varrho_2}\left(p_2+\frac{B_2^2}{8\pi}\right)\left(\frac{\bar{\eta}}{2}-\frac{1}{\gamma-1}\right)+\frac{1}{\varrho_1}\left(p_1+\frac{B_1^2}{8\pi}\right)\left(\frac{\bar{\eta}}{2(\bar{\eta}+1)}+\frac{1}{\gamma-1}\right)= =\frac{B_1^2}{8\pi}\frac{1}{\varrho_1}\frac{\gamma-2}{\gamma-1}\bar{\eta}\,,$$

from which we derive for $\gamma=\frac{5}{3}$

$$(144)\qquad \left(p_2+\frac{B_2^2}{8\pi}\right)\frac{\bar{\eta}-3}{2}+\left(p_1+\frac{B_1^2}{8\pi}\right)\left(\frac{4\bar{\eta}+3}{2}\right)=-\frac{\bar{\eta}(\bar{\eta}+1)}{2}\frac{B_1^2}{8\pi}\,,$$

Then

$$(145)\qquad p_2+\frac{B_2^2}{8\pi}=\frac{(p_1+(B_1^2/8\pi))(3+4\bar{\eta})+(B_1^2/8\pi)\bar{\eta}(\bar{\eta}+1)}{3-\bar{\eta}}\,,$$

from which we derive the velocity v_1,

$$(146)\qquad v_1^2=\frac{\bar{\eta}+1}{3-\bar{\eta}}\left[5\,\frac{p_1}{\varrho_1}+\frac{(6+\bar{\eta})B_1^2}{8\pi\varrho_1}\right],$$

and the ratio of the temperatures,

$$(147)\qquad \frac{T_2}{T_1}=\frac{(3+4\bar{\eta})+\bar{\eta}^3(B_1^2/8\pi p_1)}{(3-\eta)(\bar{\eta}+1)}\,,$$

and the ratio of the velocity behind the shock front to the sound velocity behind the shock:

$$(148)\qquad \frac{v_2^2}{\gamma\mathfrak{R}T_2}=\frac{1+s_1+\frac{1}{6}\bar{\eta}}{s_1+\frac{4}{3}\bar{\eta}s_1+\frac{5}{18}\bar{\eta}^3}\,,$$

with

$$(149)\qquad s_1=\frac{\gamma p_1}{(B_1^2/4\pi)}\,.$$

6·2. *Structure of the shock front.* – These relations apply only between the physical quantities ahead and behind the shock. The shock itself is naturally not infinitely thin, it has a thickness, governed by the mean free path of the particles.

An interesting case, the importance of which will appear later is the case when the Larmor radius of the electrons is much smaller than the mean free path. This latter case has been studied by GARDNER *et al.* (1958) and by R. LÜST.

To treat this case, one considers, in the first approximation, a plasma which is neutral, and one neglects the collisions.

Let us reproduce here briefly the theory of GARDNER. We assume that the fluid is neutral. The conservation of matter gives

$$\varrho v = m \, , \tag{150}$$

where m is a constant.

The conservation of momentum gives

$$\varrho v^2 + p + \frac{B^2}{8\pi} = P \, , \tag{151}$$

and we replace the energy condition by the adiabatic relation for two degrees of freedom

$$p = \alpha \varrho^2 \, , \tag{152}$$

therefore we have

$$B^2 = 8\pi \left[P - mv - \alpha \frac{m^2}{v^2} \right] . \tag{153}$$

The equation of the motion of the fluid in the directions y perpendicular to the direction of propagation, gives

$$v \frac{\mathrm{d}}{\mathrm{d}x} \varrho_i v_i = n e_i \left(E_y - \frac{vB}{c} \right) , \tag{154}$$

where E_y is a constant. As we have

$$\frac{\mathrm{d}B_z}{\mathrm{d}x} = - \frac{4\pi}{c} \left(n_i v_i e_i + n_e v_e e_e \right) , \tag{155}$$

we obtain finally,

$$\frac{\mathrm{d}^2 B_z}{\mathrm{d}x^2} = 4\pi n_e e^2 \left(\frac{1}{m_+} + \frac{1}{m_-} \right) \left(\frac{E_y}{v} - \frac{B}{c} \right) . \tag{156}$$

As v is a function of B, we can write

$$+ 4\pi n_e e^2 \left(\frac{1}{m_+} + \frac{1}{m_-} \right) \left(\frac{E_y}{v} - \frac{B}{c} \right) = - \frac{\partial \Phi(B)}{\partial B} \, , \tag{157}$$

and we have to solve the equation

$$\frac{d^2B}{dx^2} = -\frac{\partial\varphi}{\partial B}. \tag{158}$$

The function φ can easily be represented as a function of B (Fig. 8). The corresponding oscillations of B can be represented as a function of x (Fig. 9). For the oscillations of small amplitude, the wave length is:

$$\lambda = \frac{c}{\omega_p}, \tag{159}$$

where

$$\omega_p^2 = \frac{4\pi N_e e^2}{m_e},$$

is the plasma frequency.

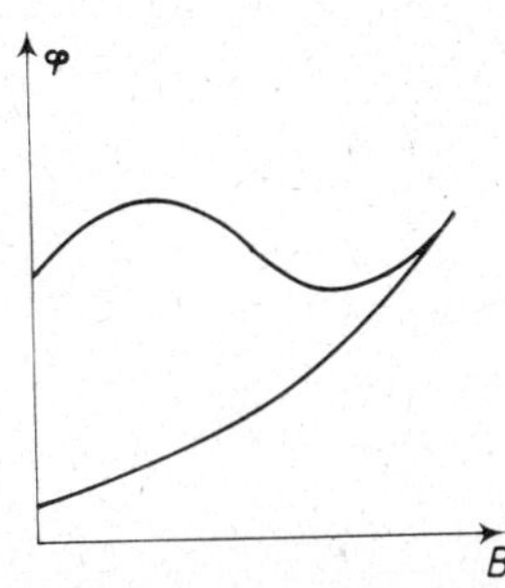

Fig. 8. – The « potential » $\varphi(B)$ as a function of B. The oscillation takes place near the minimum of $\varphi(B)$.

To have such oscillations, two conditions have to be fulfilled:

1) the Larmor radius must be smaller than the characteristic wave length;

2) the Larmor radius must be smaller than the mean free path.

This leads to the following two conditions:

$$\frac{TN_e}{B^2} \ll 10^{16.161} \qquad \frac{N_e^2}{B^2\,T^3} \ll 10^{14.328}. \tag{160}$$

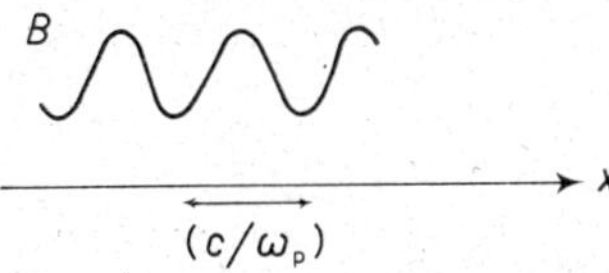

Fig. 9. – The oscillations of B.

They have been represented on Fig. 10 for $B = 100$ G. They limit a region where fluctuations of the magnetic field can be present. The fluctuations of this kind, as it has been shown, can be as large as 3 times the initial magnetic field.

The effect of collisions has not been properly considered in this case. It must

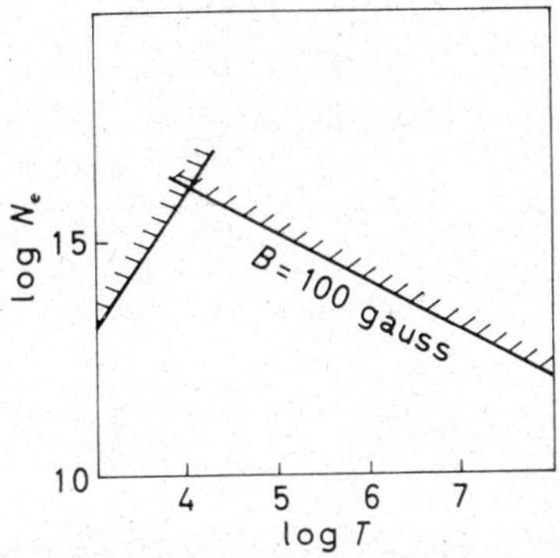

Fig. 10. – The region of existence of the fluctuations of the magnetic field.

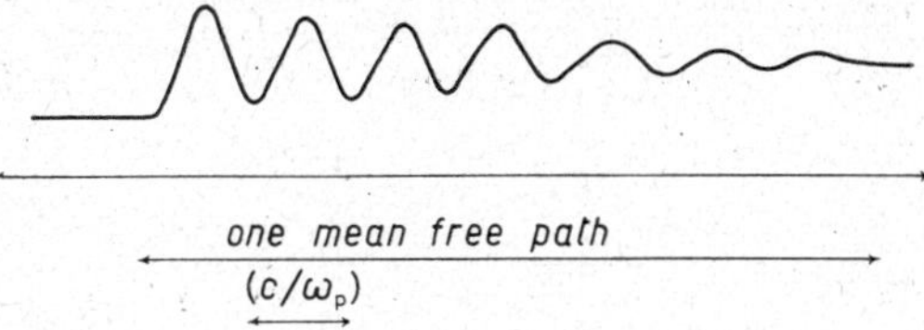

Fig. 11. – Schematic view of the fluctuations of B when the collisions are taken into account.

correspond to a damping of the oscillations of the magnetic field, in such a way that, in a shock, the passage from $v_1\varrho_1 T_1 B_1$ to $v_2\varrho_2 T_2 B_2$ takes a few mean free paths, but with fluctuations of B in this interval (Fig. 11).

6.3. *The effect of the shock on superthermal particles.* – Let us first consider a particle whose Larmor radius is larger than the mean free path, and consider what happens to its velocity v when it crosses the shock front.

To this effect, we shall use two different co-ordinate systems, both systems moving with the matter, with velocity V behind the shock front and $(1+\bar{\eta})V$ ahead of the shock front.

The particle of mass m_A in system 1 has, at the intersection with the shock front, the velocity

$$v_A \sin\psi\,, \qquad v_A \cos\psi\,, \qquad 0\,.$$

(Figure 12). In system (2), the velocity has new components, given by the relativistic transform from system (1) to system (2):

$$\frac{v_A \sin\psi + \bar{\eta}V}{1+\bar{\eta}(Vv_A\sin\psi/c^2)}\,, \qquad \left(1-\frac{\bar{\eta}^2V^2}{c^2}\right)^{\frac{1}{2}}\frac{v_A\cos\psi}{1+\bar{\eta}(Vv_A\sin\psi/c^2)}\,, \qquad 0\,.$$

Let us assume that $\bar{\eta}$ is small, so that the change in ψ is small, neglecting higher order times in $(V/c)^2$, we get

$$\Delta_1\psi = \bar{\eta}\frac{V}{v_A}\cos\psi\,. \tag{161}$$

The corresponding change in velocity can be obtained, and is given by the relation

$$\Delta v = \bar{\eta}V\sin\psi\left(1-\frac{v_A^2}{c^2}\right). \tag{162}$$

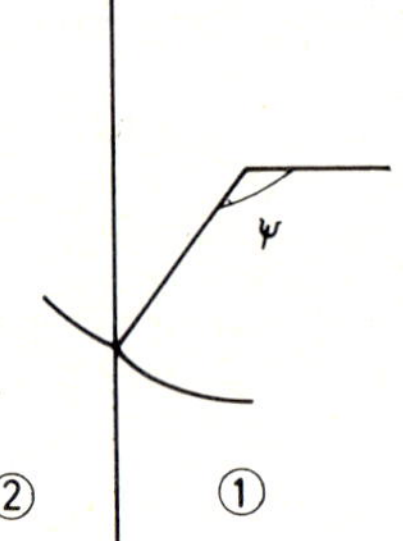

Fig. 12. – Definition of the angle ψ of entry from one region to another.

We see that each time the particle crosses the shock front its velocity increases. The energy change, in one turn, is

$$\Delta E = m_A v_A \bar{\eta} V\left(1-\frac{v_A^2}{c^2}\right)^{-\frac{1}{2}}\sin\psi\,, \tag{163}$$

for a particle of mass m_A.

When considering the total change of angle on one turn, it is found that it is equal, for weak shocks, to

$$\Delta\psi = 2\bar{\eta}\frac{V}{v_A}\cos\psi - \frac{2\pi + 2\bar{\eta}\psi}{\sin\psi}\frac{V}{v_A}. \tag{164}$$

When considering the motion of the shock front, it can be seen that the angle ψ is always decreasing, so that finally, if nothing else happens, the particle is left behind by the shock front. PARKER (1958) has shown that the energy increases in the ratio of the magnetic fields, like in the adiabatic case. This is unfortunate as we can calculate that if a particle could go with a shock front its energy could become very large. Taking $\overline{\sin\psi} = \frac{1}{2}$ we find, if V is the Alfvén velocity

$$\frac{dE}{dx} = \frac{1}{2}\frac{\bar{\eta}}{2\pi}\frac{v_A eB}{c} = \frac{1}{4\pi}\bar{\eta}eB\left(1 - \frac{m_A^2c^4}{E^2}\right), \tag{165}$$

for $\bar{\eta}$ and B constants, we derive

$$E^2 - m_A^2c^4 = \left(\frac{\bar{\eta}eBx}{4\pi}\right)^2. \tag{166}$$

Numerically, this gives for protons,

$$E_{eV} = \left[(9\cdot10^8)^2 + \left(\frac{300}{4\pi}\bar{\eta}Bx\right)^2\right]^{\frac{1}{2}}. \tag{167}$$

As an example, if we take $x = 10^9$ cm, $B = 50$ G, $\bar{\eta} = 10^{-3}$, we obtain

$$E_{eV} = 6\cdot10^8 \text{ eV}.$$

The remarkable acceleration of particles during solar flares, leads to the idea that some mechanism should act, associated with the shock. We can easily see that, to keep the particles going with the shock front, it is only needed to deflect their motion, without energy changes, in such a way that the angle ψ between the particle and the shock front does not decrease. Random fluctuations of the angle ψ could provide the necessary mechanism. Naturally, some of the particles would be deflected under the proper angle, and would keep obtaining energy from the shock, and some other particles would be lost.

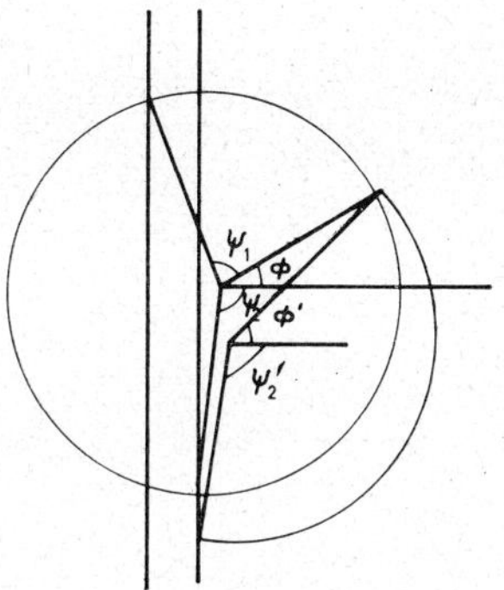

Fig. 13. – Definition of the position angle Φ for a deflection.

Let us consider exactly the geometry of a deflection (Fig. 13). With the notations used for the position angles, we have to consider four cases

$$(168)\qquad \begin{cases} 0 < \Phi < \psi_1 \,, \\ 0 < \Phi < \psi_2 \,, \\ \psi_3 < \Phi < \pi \,, \\ \psi_4 < \Phi < \pi \,. \end{cases}$$

In these four cases, for the deflection $\delta\theta$, we have the relations:

$$(169)\qquad \begin{cases} \text{I} & \begin{cases} -\cos\psi_1 + \cos\Phi - \cos(\Phi + \delta\theta) + \cos\psi_2' = \dfrac{V}{v}[\psi_1 + \delta\theta + \psi_2'] \,, \\ -\cos\psi_1 + \cos\psi_2 = \dfrac{V}{v}[\psi_1 + \psi_2] \,, \end{cases} \\ \text{II} & \begin{cases} -\cos\psi_1 + \cos\Phi - \cos(\Phi - \delta\theta) + \cos\psi_2' = \dfrac{V}{v}[\psi_1 + \delta\theta + \psi_2'] \,, \\ -\cos\psi_1 + \cos\psi_2 = \dfrac{V}{v}(\psi_1 + \psi_2) \,, \end{cases} \\ \text{III} & \begin{cases} -\cos\psi_3 + \cos\Phi - \cos(\psi - \delta\theta) + \cos\psi_4' = \dfrac{V}{v}[2\pi - \psi_3 - \psi_4' + \delta\theta] \,, \\ -\cos\psi_3 + \cos\psi_4 + \cos\psi = \dfrac{V}{v}(2\pi - \psi_3 - \psi_4') \,, \end{cases} \\ \text{IV} & \begin{cases} -\cos\psi_3 + \cos\Phi - \cos(\Phi + \delta\theta) + \cos\psi_4' = \dfrac{V}{v}(2\pi - \psi_3 - \psi_4' + \delta\theta) \,, \\ -\cos\psi_3 + \cos\psi_4' = \dfrac{V}{v}(2\pi - \psi_3 - \psi_4) \,, \end{cases} \end{cases}$$

We derive for

$$\psi' = \psi + \delta\psi \,,$$

with $\delta\theta$ and $\delta\psi$ small,

$$(170)\qquad \begin{cases} \delta\theta \sin\Phi - \sin\psi_2 \,\delta\psi_2 = \dfrac{V}{v}[\delta\theta + \delta\psi_2] \,, \\ -\delta\theta \sin\Phi - \sin\psi_2 \,\delta\psi_2 = \dfrac{V}{v}[\delta\theta + \delta\psi_2] \,, \\ -\delta\theta \sin\Phi - \sin\psi_4 \,\delta\psi_4 = \dfrac{V}{v}[\delta\theta - \delta\psi_4] \,, \\ \delta\theta \sin\Phi - \sin\psi_4 \,\delta\psi_4 = \dfrac{V}{v}[\delta\theta - \delta\psi_4] \,, \end{cases}$$

and then, taking now Φ with its sign,

$$(171)\qquad \begin{cases} \delta\psi_2 = \dfrac{(\sin\Phi - (V/v))\delta\theta}{\sin\psi_2 + (V/v)}, \\[2ex] \delta\psi_4 = \dfrac{(\sin\Phi - (V/v))\delta\theta}{\sin\psi_4 - (V/v)}. \end{cases}$$

If we can suppose, $v \gg V$ we have a simpler algebra. The dispersion of $\delta\psi$ is

$$(172)\qquad [\overline{(\delta\psi)^2}]^{\frac{1}{2}} = \frac{1}{\sqrt{2}}\,\frac{\delta\theta}{\sin\psi}.$$

To keep the particle going with the shock, we should have at each turn its total deflection $\delta\psi$ larger than

$$\delta\psi_{\min} = \frac{2\pi V}{v_{\mathrm{A}}}\,\frac{1}{\sin\psi}.$$

To obtain these random deflections, let us assume the existence of clumps of the magnetic field, of size ΔL and magnetic field $B + \Delta B$. When crossing such a clump, the particle turns of an angle

$$(173)\qquad \theta = \frac{\Delta L\,\omega_{\mathrm{B}}}{v_{\mathrm{A}}} = \frac{m_{\mathrm{A}} c v_{\mathrm{A}}}{e\Delta L}\,(B + \Delta B)(1 - \beta^2)^{\frac{1}{2}}.$$

The deflection $\delta\theta$ is

$$(174)\qquad \delta\theta = \frac{m_{\mathrm{A}} c v_{\mathrm{A}}}{e\,\Delta L\,\Delta B}\,(1 - \beta^2)^{\frac{1}{2}}.$$

To obtain the statistical properties of the deflection, we shall consider the probability that, after N turns, the deflection is greater than the minimum deflections needed. The minimum needed is

$$(175)\qquad \delta\psi_{\min} \simeq \int \frac{2\pi V}{v_{\mathrm{A}}}\,\frac{1}{\sin\psi}\,\mathrm{d}N,$$

and the dispersion after $\mathrm{d}N$ turns is obtained by adding the square of the independent dispersions. On one turn, the particle meets n clumps,

$$(176)\qquad n \simeq 2\pi\,\frac{m_{\mathrm{A}} c v_{\mathrm{A}}}{eB\,\Delta L(1 - \beta^2)^{\frac{1}{2}}},$$

and the dispersion $\Delta\psi_0$ of the statistics is given by

$$\Delta\psi_0^2 = \int \left(\frac{e\,\Delta L\,\Delta B}{m_A c v_A}\right)^2 (1-\beta^2)\,\frac{2\pi m_A c v_A}{eB\,\Delta L(1-\beta^2)^{\frac{1}{2}}}\,\frac{\mathrm{d}N}{2\sin^2\psi}\,. \tag{177}$$

The number of turns is given simply by the time $\mathrm{d}t$, divided by the period $2\pi/\omega_B$

$$\mathrm{d}N = \frac{\omega_B}{2\pi}\,\frac{\mathrm{d}x}{V}\,. \tag{178}$$

We obtain then

$$\delta\psi_{\min} = \int \frac{2\pi + 2\bar{\eta}\psi}{2\pi}\,\frac{\mathrm{d}x}{v_A} = \int \frac{2\pi + 2\bar{\eta}\psi}{2\pi}\,(1-\beta^2)^{\frac{1}{2}}\,\frac{\mathrm{d}x}{v_A}\,, \tag{179}$$

with the value of β derived from the expression (166) of the energy:

$$\beta^2 = \left(\frac{\bar{\eta}eBx\sin\psi}{2\pi}\right)^2 \left[m_A^2 c^4 + \left(\frac{\bar{\eta}eBx\sin\psi}{2\pi}\right)^2\right]^{-1},$$

we obtain

$$\frac{eB}{mc_A}\,\frac{(1-\beta^2)^{\frac{1}{2}}}{c\beta} = \frac{2\pi}{\bar{\eta}\sin\psi}\,\frac{\mathrm{d}x}{x}\,; \qquad \delta\psi_{\min} = \frac{2\pi}{\bar{\eta}\sin\psi}\log x\,. \tag{180}$$

In the same way, we obtain $\Delta\psi_0^2$

$$\Delta\psi_0^2 = \int \frac{eB\,\Delta L}{m_A c}\left(\frac{\Delta B}{B}\right)^2 \frac{eB}{m_A cV}\,\frac{1-\beta^2}{v_A}\,\frac{\mathrm{d}x}{2\sin^2\psi} =$$

$$= \left(\frac{eB}{m_A c}\right)^2 \frac{\Delta L}{Vc}\left(\frac{\Delta B}{B}\right)^2 \frac{1}{2\sin^2\psi}\int \frac{2\pi}{\bar{\eta}eB\sin\psi}\,\frac{\mathrm{d}x}{x}\,(1-\beta^2)^{\frac{1}{2}}\,,$$

$$\Delta\psi_0^2 = \frac{\pi}{\bar{\eta}\sin^3\psi}\,\frac{eB\,\Delta L}{m_A cV}\left(\frac{\Delta B}{B}\right)^2 \frac{1}{2}\log\frac{E - m_A c^2}{E + m_A c^2}\,. \tag{181}$$

The probability of having the deflection $\Delta\psi$ larger than $\delta\psi_{\min}$ is given by

$$p = \mathrm{Erf}\left[\frac{\delta\psi_{\min}}{\Delta\psi_0}\right]. \tag{182}$$

When $\delta\psi_{\min}$ is large, p goes to the limit

$$p = \frac{1}{2\sqrt{\pi}}\,\frac{\Delta\psi_0}{\delta\psi_{\min}}\exp\left[-\left(\frac{\delta\psi_{\min}}{\Delta\psi_0}\right)^2\right], \tag{183}$$

and we have

$$(184)\quad \left(\frac{\delta\psi_{\min}}{\Delta\psi_0}\right)^2 =$$

$$= \frac{(2\pi + 2\bar{\eta}\psi)\sin\psi}{\bar{\eta}(\Delta B/B)^2(\Delta L/a_V)} \frac{\log^2\left((E^2 - m_A^2c^4)/(E^7 - m_A^2c^4)\right)}{\log\left[\left((E - m_Ac^2)/(E_0 - m_Ac^2)\right)\cdot\left((E_0 + m_Ac^2)/(E + m_Ac^2)\right)\right]},$$

where we have written

$$a_V = \frac{m_A cV}{eB}.$$

For low energies, $E = m_Ac^2 + E_{kin}$, and we have

$$(185)\quad \left(\frac{\delta\psi_{\min}}{\Delta\psi_0}\right)^2 = \frac{(2\pi + 2\bar{\eta}\psi)\sin\psi}{\bar{\eta}(\Delta B/B)^2(\Delta L/a_V)}\log\frac{E_{kin}}{E_{kin0}}.$$

We see that the energy spectrum follows essentially a power law,

$$(186)\quad p = \text{const}\cdot\left(\frac{E_{kin}}{E_{kin0}}\right)^{-\nu}.$$

The maximum contribution is given by $\sin\psi = 1$. Therefore, we write the exponent

$$(187)\quad \begin{cases} \nu = \dfrac{2\pi + \bar{\eta}\pi}{\bar{\eta}(\Delta B/B)^2(\Delta L/a_V)}, & \text{for low energies}, \\ \nu = \dfrac{4(2\pi + \bar{\eta}\pi)}{\bar{\eta}(\Delta B/B)^2(\Delta L/a_V)}, & \text{for relativistic energies}. \end{cases}$$

Let us now proceed to an estimate of $\Delta L/a_V$. For the smallest possible clumps, $\Delta L = (c/\omega_p)$, ω_p being the plasma frequency for electrons, and supposing V equal to the Alfvén velocity, we have, for protons $(\Delta L/a_V) = (m_e/m_H)^{\frac{1}{2}}$. This would mean a very high exponent ν, especially if the shock strength $\bar{\eta}$ is small. This would correspond to

$$(188)\quad \Delta L_{\min} \simeq 10^{5.72} N_e^{-\frac{1}{2}},$$

which is very small.

This difficulty can be avoided if we assume that for longer and larger energies, the particles sees longer and larger clumps.

The assumptions concerning the size of the clumps are obviously the weak part of this theory.

However, it should be noticed that another characteristic length is $\Delta L' = c/\Omega_p$, where Ω_p is the plasma frequency of the protons. This length provides a larger value of ν, as it gives $(\Delta L'/a_\nu) \simeq 1$.

6·4. *The injection mechanism.*

6·4.1. Condition of injection. Let us consider now more closely the angle conditions at the shock front. We can see on Fig. 14 the definition of the angles ψ_3 and ψ_4. We have the relations

$$\cos \psi_4 - \cos \psi_3 = \frac{V}{v_A}(2\pi - \psi_3 - \psi_4)\,, \tag{189}$$

ψ_4 is the angle of re-entry in the shock front.

For any value of ψ_3, the minimum value of (V/v) is reached when ψ_4 is minimum. ψ_4 is minimum when the differentials of the two sides of eq. (189) with respect to ψ_4 are equal:

$$\sin \psi_4 = \frac{V}{v}\,. \tag{190}$$

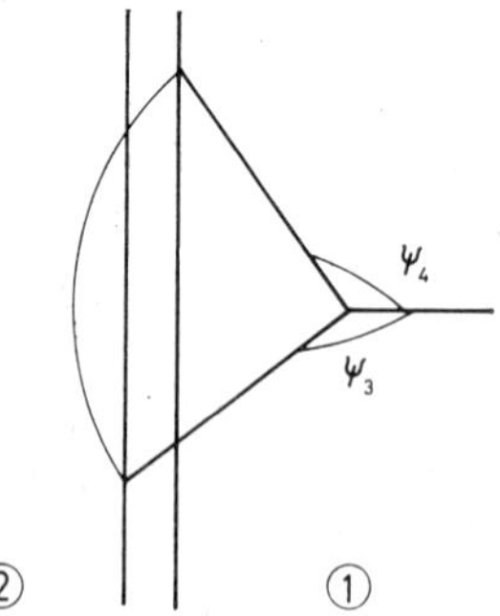

Fig. 14. – Definition of the angles ψ_3 of entry into and ψ_4 of exit from region 2.

We then have a relation between ψ_4 and ψ_3. It can be approximately represented by the equation

$$\psi_4 = \psi_{40} - \tfrac{1}{3}(\pi - \psi_3)\,, \tag{191}$$

with

$$\psi_{40} = 46°\ 20'\,. \tag{192}$$

Let us now consider, at least in order of magnitude, the probability that a particle is accelerated at least once by the shock wave. The particle is accelerated if its velocity v is larger than $(V/\sin \psi_4)$ and this is true for all values of ψ_4 given by the relation (191). To calculate the probability p_s that a particle of mass m_A could take part in the acceleration process we shall suppose a velocity distribution function with two dimensions:

$$f(v)\,\mathrm{d}v = \frac{m_A}{2\pi kT}\, v\,\mathrm{d}v\,\mathrm{d}\psi \exp\left[-\frac{1}{2}\frac{m_A v^2}{kT}\right], \tag{193}$$

where ψ is an angle in the interval 0 to 2π.

If we consider now the definition of the angle ψ_3 which is in the interval $0, \pi$, we see that the identification of ψ and ψ_3 is possible. This means that the particles for which $\pi < \psi < 2\pi$ does not get also into the accelerating process.

For a given angle ψ_3 a particle will be accelerated only if its velocity is large enough, that is to say, if $v > (V/\sin\psi_4)$. We shall have for the probability p_s

$$p_s = \int_0^{\pi} \mathrm{d}\psi_3 \int_{(V/\sin\psi_4)}^{\infty} \frac{m_A}{2\pi kT}\, v\, \mathrm{d}v \exp\left[-\frac{1}{2}\frac{m_A v^2}{kT}\right]. \tag{194}$$

By integrating over v, and taking the approximate expression of ψ_4, we obtain approximately

$$p_s = \frac{3\sqrt{2}}{\pi}\frac{kT}{m_A V^2}\psi_{40}^3 \exp\left[-\frac{1}{2}\frac{m_A V^2}{kT\psi_{40}^2}\right], \tag{195}$$

with $\psi_{40} \simeq 0.8$, we obtain

$$p_s = 0.69\frac{kT}{m_A V^2}\exp\left[-0.77\frac{m_A V^2}{kT}\right]. \tag{196}$$

6.4.2. Gains and losses. The accelerating process can actually take place only if the gains are larger than the losses. GRYZINSKI (1957) has given a very useful expression of the energy losses of a charged particle by collision with the electrons. When the velocity of the charged particle of charge Z_1 is v_A, with v_A much smaller than the thermal velocity of the electrons, the energy loss per centimeter is

$$\frac{\mathrm{d}E}{\mathrm{d}x} = -\frac{8\pi}{3}\frac{e^4}{m_e}\left(\frac{\pi m_e}{8kT}\right)^{\frac{3}{2}} Z_A^2\, v_A N_e. \tag{197}$$

When the velocity v_A of the charged particles is much larger than the thermal velocity of the electrons, the losses are given by

$$\frac{\mathrm{d}E}{\mathrm{d}x} = -\frac{4\pi e^4}{m_e v_A^2} Z_A^2 N_e \log\frac{v_A^2}{\langle v_e^2\rangle}. \tag{198}$$

For the injection problem, v_A is much smaller than the velocity of the electrons, and we shall use the first form of the expression of the energy losses.

As we have seen, the gains, on one turn, are given by

$$\Delta E_{\text{gains}} = \bar{\eta} m_A v_A V(\sin\psi_3 + \sin\psi_4), \tag{199}$$

where $\bar{\eta}$ is the strength of the shock and V the velocity of the shock front.

The losses, on one turn, are obtained by multiplying the losses for one centimeter by the length of one turn.

$$\Delta x = \frac{2\pi m_A c v_A}{Z_A e B}. \tag{200}$$

For the low velocities, the conditions for the gains to be larger than the losses, can be written

$$\bar{\eta} m_A v_A V(\sin\psi_3 + \sin\psi_4) > \frac{16\pi^2}{6}\frac{e^3}{B}\frac{m_A}{m_e} c v_A^2 \left(\frac{\pi m_e}{8kT}\right)^{\frac{3}{2}} N_e Z_A. \tag{201}$$

It should be noticed that this relation can be read as a condition on v_A. The velocity of the charged particle has to be small enough for the losses, proportional to v_A^2, to be smaller than the gains, proportional to v_A. The inequality can be written

$$v_A < \frac{3}{12\pi^2}\frac{m_e}{e^3}\frac{B}{c} V \left(\frac{8kT}{\pi m_e}\right)^{\frac{3}{2}} \frac{\bar{\eta}(\sin\psi_3 + \sin\psi_4)}{N_e Z_A}. \tag{202}$$

We should naturally, for these low velocities, consider the spiral motion of the particle. We shall simply replace $(\sin\psi_3 + \sin\psi_4)$ by the larger quantity $2\sin\psi_3$, and we shall write

$$v_A < \xi \sin\psi_3, \tag{203}$$

putting

$$\xi = \frac{3}{8\pi^2}\frac{m_e}{e^3}\frac{B}{c}\left(\frac{8kT}{\pi m_e}\right)^{\frac{3}{2}} \frac{\bar{\eta}}{Z_A N_e}. \tag{204}$$

This shows that the integral which gives the probability of acceleration has to be modified, to include the existence of an upper limit to v_A

$$p_s = \frac{2}{\pi}\int_0^\pi d\psi_3 \int_{(V/\sin\psi_4)}^{\xi V \sin\psi_3} \frac{m_A}{2\pi kT} v\, dv \exp\left[-\frac{1}{2}\frac{m_A v^2}{kT}\right]. \tag{205}$$

We see that the conditions for any acceleration at all is

$$\xi V \sin\psi_3 > \frac{V}{\sin\psi_4}. \tag{206}$$

Using the expression of ξ the condition can be written

$$\frac{3}{8\pi^2}\frac{m_e}{e^3}\frac{B}{c}\left(\frac{8kT}{\pi m_e}\right)^{\frac{3}{2}}\frac{\bar{\eta}}{N_e Z_A} > \frac{1}{\sin\psi_3 \sin\psi_4}. \tag{207}$$

This is equivalent to a condition on the density, which has to be lower than a certain limit, given by the relation (207). The maximum of $\sin\psi_4 \sin\psi_3$ is reached for $\psi_3 = \pi - 57° \, 11'$ and is:

$$(\sin\psi_3 \sin\psi_4)_{\max} = 0.0795\,, \tag{208}$$

and the following relation is obtained

$$N_e < 10^{-5.298} B T^{\frac{3}{2}} \bar{\eta} Z_A^{-1}\,, \tag{209}$$

or, to compare with eq. (160)

$$\frac{N_e^2}{B^2 T^3} < 10^{10.596} \bar{\eta}^2 Z_A^{-2}\,, \tag{210}$$

and we see that, if this last condition is fulfilled, one of the conditions (160) is also certainly satisfied. This means that if the conditions for injections are fulfilled the smallest possible clumps are also present.

Another condition can also be introduced. In order that the number of accelerated particle is appreciable at all, it is necessary that the exponential term in p_s, $\exp[-\frac{1}{2}(n_A V^2/kT)]$ is not too small. In other words, the velocity of propagation of the shock should not be too large compared to the thermal velocity.

If we put

$$\frac{m_A V^2}{kT} = \alpha\,, \tag{211}$$

the number α has to be smaller than 20 to 30 to obtain an appreciable number of accelerated particles.

Numerically, this gives

$$N_e = \frac{B^2}{T}\frac{m_A}{m_H}\frac{10^{15.66}}{\alpha}. \tag{212}$$

Comparing to eq. (160), we see again that if the conditions for injection are satisfied, the other condition for the presence of small clumps is also satisfied.

Finally, we can give the probability p_s. By developping $\sin\psi_4$ near $\sin\psi_{40}$

and $\sin\psi_3$ near its minimum, given by $\xi\sin\psi_4\sin\psi_3=1$, by replacing $\sin\psi_4$ by ψ_4 and $\sin\psi_3$ by $3(\psi_{40}-\psi_4)$ and finally, by supposing ξ large compared to one, we obtain

$$p_s=\frac{2}{\pi}\frac{kT}{m_A V^2}\,\psi_{40}\left(3\psi_{40}^2-\frac{1}{\xi}\right)\exp\left[-\frac{1}{2}\frac{m_A V^2}{kT\psi_{40}^2}\right]. \tag{213}$$

This represents the properties which we have described above. p_s decreases exponentially very quickly when V increases; p_s is zero for all values of ξ smaller than $\frac{1}{3}(1/\psi_{40}^2)$ that is to say for the larger densities. This last condition is not exactly condition (206), but is very close, so that we shall consider p_s as the expression of the probability of injections.

We also see that it decreases very fast when the mass of the injected particle increases.

6·5. *Acceleration in a density gradient.* – Let us consider a shock propagating across a region where the magnetic field is parallel to the stellar surface. The shock strength changes with height, and we shall consider a simple way of estimating this change.

Let us assume that the magnetic field is constant and parallel to the stellar surface, the velocity of propagation is supposed to be the Alfvén velocity. We write the equation of dissipation in the form:

$$\frac{\mathrm{d}}{\mathrm{d}z}\varrho v^2 V=-D\frac{Vt_0}{\varrho v^3}, \tag{214}$$

where D is a numerical constant and t_0 a characteristic time. The value of D has been estimated by SCHATZMAN (1949*a*) and by OSTERBROK (1961) for magnetohydrodynamic shocks. We obtain the equation for v in the following form:

$$\frac{v'}{v}+\frac{2\pi D\varrho v}{t_0 B^2}=-\frac{1}{4}\frac{\varrho'}{\varrho}. \tag{215}$$

Taking ϱv as a variable, we obtain

$$\frac{(\varrho v)'}{\varrho v}+\frac{2\pi D}{t_0 B^2}\varrho v=\frac{3}{4}\frac{\varrho'}{\varrho}, \tag{216}$$

from which we derive the solution

$$\varrho v=\frac{\varrho^{\frac{3}{4}}}{\int^{z}((2\pi D)/(t_0 B^2))\,\varrho^{\frac{3}{4}}\,\mathrm{d}z}, \tag{217}$$

and we obtain:

$$\bar{\eta}=\frac{v}{V}=\frac{(4\pi)^{\frac{1}{2}}\varrho^{\frac{1}{4}}}{B\int^{z}(2\pi K/t_0B^2)\,\varrho^{\frac{3}{4}}\,\mathrm{d}z}, \tag{218}$$

where the origin of the co-ordinates has been taken arbitrary. From now on, we shall write

$$\bar{\eta}=\eta_0\left(\frac{\varrho}{\varrho_0}\right)^{\frac{1}{4}}. \tag{219}$$

If we carry this expression of the shock strength in the eq. (165) for the energy increase, we obtain

$$(E^2-m_A^2c^4)^{\frac{1}{2}}=\int_{z_1}^{z}\frac{\bar{\eta}eB\,\mathrm{d}z}{4\pi}. \tag{220}$$

Assuming a scale-height H for ϱ, we obtain

$$(E^2-m_A^2c^4)^{\frac{1}{2}}=\bar{\eta}_0\frac{eBH}{\pi}\left(\exp\left[-\frac{z_1}{4H}\right]-\exp\left[-\frac{z}{4H}\right]\right). \tag{221}$$

It will be noticed that the maximum energy which can be obtained by a particle, is finite. Comparing with the expression (166) of the energy, we see that it corresponds to a travelling distance, at constant shock strength over four scale-heights, $4H$.

For example, with $\eta_0=10^{-4}$, $B=10^2$, $H=300$ km we obtain, for protons, an energy $E_{\max}\simeq 6\cdot 10^5$ eV.

6·6. *Energy spectrum.* – To calculate the energy spectrum in a density gradient, we simply carry the expression of the changing shock strength in the expressions which have been used for calculating the energy spectrum.

Let us consider the lower part of the energy spectrum, with nonrelativistic energies. As we have

$$E=m_Ac^2+E_{\text{kin}},$$

we obtain

$$E_{\text{kin}}=\frac{1}{2m_Ac^2}\left(\frac{\bar{\eta}_0eBH}{\pi}\right)^2\exp\left[-\frac{z_1}{2H}\right]\left(1-\exp\left[\frac{-z}{4H^2}\right]\right)^2= \tag{222}$$

$$=E_{\max}\left(1-\exp\left[-\frac{z}{4H}\right]\right)^2.$$

We obtain, in the same way

$$\Delta\psi_0^2 = \frac{\pi}{\sin^3\psi}\frac{eB\Delta L}{m_A cV}\left(\frac{\Delta B}{B}\right)^2\int\frac{1}{\eta}\frac{dz}{z}, \tag{223}$$

and we see that, as long as $E_{kin} \ll E_{max}$, there is no essential change with the results obtained for the case of *no* density gradient.

Acceleration takes place from z_0 to z; but at altitude z, we have particles for which the acceleration has taken place at any altitude, from z_1 to z, with $z_0 < z_1 < z$.

The probability that a particle of energy E will be seen at altitude z is the product of the probability of injection at altitude z_1, by the probability of reaching altitude z.

We have to consider the probability that the particles are left with anergy E by the shock. This is equal to the differential dp of the probability p given by eq. (182):

$$dp = \frac{1}{\sqrt{\pi}}\frac{d\,\delta\psi^{min}}{\Delta\psi_0}\exp\left[-\left(\frac{\delta\psi_{min}}{\Delta\psi_0}\right)^2\right], \tag{224}$$

or, assuming $(\delta\psi_{min}/\Delta\psi_0)^2 = \nu\log(E/E_1)$,

$$dp = \left(\frac{\nu}{4\pi}\right)^{\frac{1}{2}}\frac{dE}{E_1}\log^{-\frac{1}{2}}\frac{E}{E_1}\left(\frac{E}{E_1}\right)^{-(\nu+1)}. \tag{225}$$

Particles are accelerated from level z_1 to level z. In the interval dz_1 we have a certain fraction p_s (eq. (196)) of the number of the particles which is accelerated. Then, a fraction dp is left in the interval $(dE/dz)dz$ of energy. We find therefore the result:

$$dN = N_0\exp[-(z_1/H)]\,dz_1\cdot 0.69\,\frac{kT}{m_A V^2}\exp\left[-\frac{0.77 m_A V^2}{kT}\right]\cdot \left(\frac{\nu}{4\pi}\right)^{\frac{1}{2}}\frac{dE}{E}\log^{-\frac{1}{2}}\frac{E}{E_1}\left(\frac{E}{E_1}\right)^{-\nu}, \tag{226}$$

where

$$E_1 = 0.38\,m_A V^2 = 0.38\,\frac{m_A B^2}{4\pi\varrho},$$

is the energy of injection. We take $\varrho = \varrho_0\exp[-(z_1/H)]$.

We make the following assumptions:

If N_0 is the density where the injection mechanism starts, f the frequency of the shocks, we get the approximate expression for the number of particles of energy E, produced at altitude z, per cubic centimeter, per second, assuming $\varrho T = c^{te}$ in the chromospheric region:

$$(227)\qquad \frac{\mathrm{d}N}{\mathrm{d}t} = fN_0 \cdot 0.69 \frac{4\pi k T_0 \varrho_0}{m_A B^2} \exp\left[-\frac{0.77 m_A B^2}{4\pi\varrho_0 k T_0}\right] \cdot$$

$$\cdot \mathrm{xep}\left[(\nu-1)\left(\frac{z}{H} - 4\left(\frac{E}{E_{\max}}\right)^{\frac{1}{2}}\right)\right] \cdot \left(\frac{E\cdot 4\pi\varrho_0}{0.38 m_A B^2}\right)^{-\nu} \cdot \left(\frac{\nu}{4\pi}\right)^{\frac{1}{2}} \frac{\mathrm{d}E}{E} \log^{-\frac{1}{2}} \frac{E}{E_1} \mathrm{d}z\,.$$

There is a cut-off of the spectrum; the energy is smaller than $E_{\max}(z/4H)^2$ at altitude z.

The total number of particles of energy E produced per second, per cm² is obtained by integrating from $z = 4H(E/E_{\max})^{\frac{1}{2}}$ and gives:

$$(228)\qquad \int \frac{\mathrm{d}N}{\mathrm{d}t}\mathrm{d}z = \int fN_0 \frac{0.69H}{\nu-1} \frac{4\pi k\, T_0\, \varrho_0}{m_A B^2} \exp\left[-\frac{0.77 m_A B^2}{4\pi k\, T_0 \varrho_0}\right] \cdot$$

$$\cdot \left(\frac{E\cdot 4\pi\varrho_0}{0.38 m_A B^2}\right)^{-\nu} \left(\frac{\nu}{4\pi}\right)^{\frac{1}{2}} \log^{-\frac{1}{2}} \frac{E}{E_1} \frac{\mathrm{d}E}{E} \mathrm{d}z\,.$$

Let us consider finally the number of particles of energy larger than $E_{\min}$. We obtain their number by integrating first over z, from $4H$ to infinity, and then from $E_{\min}$ to infinity. We obtain

$$(229)\qquad N = fN_0 H \frac{0.69}{\nu(\nu-1)} \frac{4\pi k T_0 \varrho_0}{m_A B^2} \exp\left[-\frac{0.77 m_A B^2}{4\pi k T_0 \varrho_0}\right] \cdot$$

$$\cdot \left(\frac{E\cdot 4\pi\varrho_0}{0.38 m_A B^2}\right)^{-\nu} \left(\frac{\nu}{4\pi}\right)^{\frac{1}{2}} \log^{-\frac{1}{2}} \left(\frac{E}{E_1}\right).$$

In the following, eq. (229) will be fundamental.

A discussion of the numerical values leads to the following orders of magnitude: with

$$\nu = 2\,, \qquad \frac{0.38 m_A B^2}{4\pi\varrho_0} = 500\ \mathrm{eV}\,,$$

$$\frac{4\pi k T_0 \varrho_0}{0.77 m_A B^2} = 10^{-1}\,, \qquad fN_0 = 10^{11}\,, \qquad H \simeq 10^{7.5}\,,$$

we get for $E \geqslant 5$ MeV, [protons](cm^{-2} s^{-1}) $\simeq 10^{5.1}$.

The case of strong shocks, which is much more favorable has been considered by SCHATZMAN (1963).

7. – Surface nuclear reactions.

The problem of the s-process, at the surface of stars (BURBIDGE, BURBIDGE FOWLER and HOYLE (1957)) implies the formation of neutrons. An excess of heavy elements by a factor 100 can be obtained if each nucleus in the iron group captures 20 neutrons. From the numbers given for the cross-sections by FOWLER, BURBIDGE and BURBIDGE (1955), we find that hydrogen captures 5000 times more neutrons that the metals; to have 20 neutrons captured per metallic nucleus means that 10^5 neutrons have been produced. With a ratio of $2.2 \cdot 10^{-5}$ of the metals to the hydrogen, we find a production of 2 neutrons per hydrogen atom.

The thickness of the column in which the spectral lines are seen is of the order of 10^{22} hydrogen atoms. During the life of the star, about 10^{22} neutrons cm^{-2} at least must have been produced, which represent, in three million years, about 10^8 neutrons s^{-1} cm^{-2}.

BURBIDGE and BURBIDGE (1955) have suggested that these neutrons are produced by (p-n) reactions, especially by the reaction

$$^{14}N(p, n)^{14}O(\beta, \nu)^{14}N\,, \tag{230}$$

which has a threshold of 6 MeV.

We can consider that about any proton of energy larger than 6 MeV will react to produce a neutron. Neutrons produced by reaction (230) are thermalized by collisions with protons, after encountering $1.5 \cdot 10^{24}$ hydrogen nuclei. We must therefore consider that half the neutrons are lost, and that about 10^{25} protons ($E > 6$ MeV) have been formed for each square centimeter of the star.

Therefore, we must consider a process which can produce 10^{11} protons $cm^{-2}\,s^{-1}$ of energy larger than 6 MeV over the whole surface of the star for $3 \cdot 10^6$ years.

Let us consider now the production of deuterons.

The neutrons produced in the chromosphere are sent back towards the photosphere, where they react with protons. We then have approximately a rate of production of deuterons of 10^{10} deuterons $cm^{-2}\,s^{-1}$. As deuterons are very slowly destroyed, we can assume that their number increases linearly with time:

$$[D] \simeq 10^{10}\, t\,.$$

^{3}He will be produced in the lower chromosphere, where the energy spectrum is essentially made of low energy particles. The accelerated protons will react with the deuterons to give ^{3}He. The cross-section is, in the center of gravity system (FOWLER, 1960)

(231) $$\sigma = SE^{-1}\exp\left[-31.28\,\frac{Z_1 A_0 Z^{\frac{1}{2}}}{E^{\frac{1}{2}}}\right] = \frac{S}{E}\exp\left[-\frac{K}{E^{\frac{1}{2}}}\right].$$

With E in keV

$$S = 8\cdot10^{-5}\ \text{barn}\cdot\text{keV}\,,$$

with an energy spectrum

(232) $$\mathrm{d}N = \alpha\,\frac{\mathrm{d}E}{E^{\nu}} = (\nu-1)E_0\,\frac{\mathrm{d}E}{E^{\nu}}\,,$$

we can calculate the total number of ^{3}He produced par second per nucleus of deuterium.

(233) $$n = 2\,\frac{S}{K^{2\nu}}\,(\nu-1)\overline{(2\nu-1)}\,!\,E_0^{\nu-1}N_0\,.$$

Let us consider the number of protons of energy larger than 6 MeV as a reference number:

$$N_0 = 10^{11}\ \text{cm}^{-2}\,\text{s}^{-1}\,.$$

We can write

(234) $$\frac{\mathrm{d}[^3\mathrm{He}]}{\mathrm{d}t} = 10^{-11}[\mathrm{D}]\,.$$

Taking into account the depletion of $[D]$ by production of ^{3}He, we obtain

(235) $$[D] = 10^{21}\big[1-\exp[10^{-11}t]\big]\,,$$

and derive the number of nuclei ^{3}He formed:

(236) $$[^3\mathrm{He}] = 10^{10}t - 10^{21}\big[1-\exp[-10^{-11}t]\big]\,.$$

For $t = 10^{14}$ s, this gives

$$[^3\mathrm{He}] = 10^{24}\,.$$

Therefore, in $3\cdot10^6$ years, we have a production of 10^{24} ^{3}He in a column of $1.5\cdot10^{24}$ hydrogen atoms.

This of the right order of magnitude for explaining the presence of ^{3}He in 3 *Cen* A, found by SARGENT and JUGAKU (1961).

Formation of Li, as observed in stars by BONSACK and GREENSTEIN (1960) seems to be due to fast protons. They attribute the formation of Li to spallation of ^{20}Ne by protons of energy larger than 600 MeV. With the energy spectrum in $E^{-2.5}$ we have used, we have about 10^6 protons $cm^{-2}\,s^{-1}$ of energy larger than 600 MeV.

This would produce, with a cross-section $10^{-27}\,cm^{-2}$ and $2\cdot 10^{20}$ Neon atoms cm^{-2}, about

$$[\mathrm{Li}] \simeq 0.2\,t\ \mathrm{cm}^{-2}\,.$$

With $t = 10^{14}$ s, this gives $2\cdot 10^{13}$ for [Li] instead of $10^{15.5}$ observed.

We could also consider the reaction of ^{3}He and ^{4}He with a cross-section

$$\sigma = \frac{1.5\ \mathrm{keV\ barn}}{E_{\mathrm{keV}}} \exp\left[-\frac{99.5}{E^{\frac{1}{2}}}\right]. \tag{237}$$

A proton of energy E, through an elastic collision with an α at rest, produces an α of energy $E/4$, with a probability of about 1/40 (by taking a ratio of the abundancy of 10^{-1}). Therefore, the fast protons, before being thermalized, produce fast α. We can estimate for an energy spectrum $E^{-2.5}$, their number to about $10^{15}\,cm^{-2}\,s^{-1}$; this corresponds to about $10^{12}\,\alpha\,cm^{-2}\,s^{-1}$.

If we take into account the depletion of ^{7}Li by the ^{7}Li(p, α)^{4}He reaction we obtain

$$\frac{\mathrm{d}[^7\mathrm{Li}]}{\mathrm{d}t} = 10^{-10}\,t - 10^{-12}[^7\mathrm{Li}]\,,$$

which gives for $t = 10^{14}$ s,

$$[\mathrm{Li}] \simeq 10^{16}\ \mathrm{cm}^{-2}\,\mathrm{s}^{-1}\,,$$

which is about the observed number.

If we consider the production of fast protons, and the particles accelerated by elastic collisions with the protons before their thermalization, we find that we are able to interpret the various peculiarities in the abundance of the elements. However, as mentioned by Mrs. BURBIDGE (1962), the depletion of ^{16}O is remarkable and needs an explanation. The ^{16}O(^{16}O, p)^{31}P reaction seems to be very unlikely, as the number of fast ^{16}O is very small. A reaction like the (p, 2n) reaction, leading to ^{15}N might be more important. Further work is needed, with the new model for the energy spectrum of accelerated particles.

8. – The explanation of the peculiar stars in H-R diagram.

8'1. *The loss of angular momentum through electromagnetic coupling.* – As already suggested in Section **5** we suppose that when ejection of matter occurs at the surface of a star which has a magnetic field, the ejected matter is forced to turn with the star up to a critical distance beyond which it is no longer dragged by the magnetic field. Since that distance is much larger than the radius of the star, the loss of angular momentum per unit mass is large; a small mass loss can produce a very large effect. We follow here closely the paper of the author (SCHATZMAN, 1962).

We could describe more precisely the mechanism in the following way: during flares, jets of matter are produced, which, after having locally disturbed the magnetic field, follow the magnetic lines of force up to a point where the magnetic stresses are no longer sufficiently large to force matter to follow the lines of force. At that point, it can be said that matter leaves the star and carries away some angular momentum.

In order to estimate the rate of loss of angular momentum, we proceed in the following way: if B is the magnetic field at a distance r from the star, we suppose that the field drags the matter as long as the tension $(B^2/4\pi R_c)$ of the lines of force, of radius of curvature R_c, can resist the pressure gradient due to the Coriolis force acting on the jet of ionized gas which has the density and the velocity V at the distance r.

$$2\varrho V\omega = \frac{B^2}{4\pi R_c}. \tag{238}$$

For a gas moving along the lines of force at a constant velocity, the shape of force of the magnetic field can be obtained, at least roughly, from the equations of motion. The condition of escape is achieved when the radius of curvature is of the order of the distance of the star. We have taken then

$$2\varrho V\omega \simeq \frac{B^2}{4\pi r}. \tag{239}$$

We have to consider the change of density and magnetic field with distance. In a spherical expansion ϱ varies like r^{-2}; in a cylindrical expansion it varies like r^{-1}. If n is either 2 or 1, we can write

$$\frac{\varrho}{\varrho_0} = \left(\frac{r_0}{r}\right)^n, \tag{240}$$

where r_0 is of the order of the radius of the star.

Similarly, the magnetic field varies like

$$\frac{B}{B_0} = \left(\frac{a}{r}\right)^p. \tag{241}$$

For an ordinary dipole, $p = 3$, and $B_0 a^3$ is the magnetic moment of the dipole of length a. We could also imagine two parallel lines of magnetic polarity north and south. For such a distribution of magnetic charges, we can represent the magnetic field with eq. (104) but then, $p=2$ and we have:

$$\frac{B}{B_0} = \left(\frac{a}{r}\right)^2 .$$

We derive from (239), (240) and (241) the following expression for the distance r where the angular momentum is lost:

$$r^{2p+1-n} = \frac{B_0^2 a^{2p}}{8\pi r_0^n V\omega} . \tag{242}$$

For a magnetic dipole, and spherical expansion, $p=3$, $n=2$, and we have

$$r \simeq a \left(\frac{B_0^2/8\pi}{\varrho_0 V(r_0^2/a)\omega}\right)^{\frac{1}{5}} . \tag{243}$$

Let us now consider the equation of conservation of angular momentum. We suppose that the loss of momentum occurs at a distance r larger than the radius R of the star, and then the equation of conservation of angular momentum (1) becomes:

$$\frac{\mathrm{d}}{\mathrm{d}t}(KMR^2\omega) = \frac{r^2}{R^2} R^2\omega \frac{\mathrm{d}M}{\mathrm{d}t} . \tag{244}$$

If we insert the expression (243) into eq. (244) we obtain the differential equation

$$\frac{1}{M}\frac{\mathrm{d}M}{\mathrm{d}t} + \frac{2}{R}\frac{\mathrm{d}R}{\mathrm{d}t} + \frac{1}{\omega}\frac{\mathrm{d}\omega}{\mathrm{d}t} = \frac{a^2}{KR^2}\left(\frac{B_0^2/8\pi}{\varrho_0 V(r_0^2/a)}\right)^{\frac{2}{5}} \omega^{-\frac{2}{5}} \frac{1}{M}\frac{\mathrm{d}M}{\mathrm{d}t} . \tag{245}$$

If a, R, B_0, ϱ_0, V and r_0 are constants eq. (245) can be integrated to give

$$\left(\frac{\omega}{\omega_0}\right)^{\frac{2}{5}} = 1 + \frac{2}{5}\frac{a^2}{KR^2}\left(\frac{B_0^2/8\pi}{\varrho_0 V(r_0^2/a)\omega_0}\right)^{\frac{2}{5}} \frac{\Delta M}{M} , \tag{246}$$

if we suppose that M changes very little. Let us take for ω, the largest possible value, corresponding to the equality between the gravitational and the centrifugal force at the equator:

$$\omega_1^2 = \omega_{\max}^2 \simeq \frac{GM}{R^3} ,$$

the numerical value of $\omega_{\max}$ is $10^{-3.45}$ for the solar mass and radius.

For ϱ_0 and V we take characteristic values of a jet of ionized matter as suggested by solar flares:

$$\varrho_0 = 2 \cdot 10^{-21}\ \text{g cm}^{-3}; \qquad V = 10^8\ \text{cm s}^{-1}.$$

The magnetic parameters are those of a giant system of bipolar magnetic spots:

$$B_0 = 10^4\ \text{G}; \qquad a/R = 0.1.$$

Finally, with the coefficient

$$K = 1/13.5,$$

of the moment of inertia we obtain

$$\left(\frac{\omega}{\omega_0}\right)^{\frac{2}{5}} = 1 - 1.08 \cdot 10^3 \frac{|\Delta M|}{M}, \tag{247}$$

where $|\Delta M|$ is the mass loss.

We thus see that a very small mass loss can lead to an enormous loss of angular momentum, and that it would be even possible that the star be brought to rest. However, we shall see later that it would take an infinite length of time.

8·2. *The convective zone.* — The physical process we have been considering does not lead in itself to an explanation of the difference between early and late-type stars. However, the process implies both an ejection of matter and a magnetic field. We shall suppose that the ejection of matter is due to the surface activity, similar to the activity at the surface of the sun.

Surface activity is related to a cycle of magnetic activity in the sun. Ejection of matter is due to flares, which occur in the regions of complex magnetic spots, as has been proved by the observations of SEVERNY (1958), HOWARD (1959), LEIGHTON (1959). The neutral-line theory of solar flares (SWEET, 1958; DUNGEY, 1953) does not seem to be confirmed by the observations. If we accept Parker's picture (1955) of solar activity, we are bound to establish a correlation between the presence of the surface convective zone (hydrogen convective zone) and the surface activity.

We shall therefore assume that whenever there exists a convection zone, surface activity is a normal feature of the star, buth that when the convective zone is absent or small, the surface activity is also small or absent even though a strong magnetic field might be present.

As is well known, G, K and M stars have a convective zone of considerable thickness (hydrogen convective zone) and A and B stars have a thin convective zone or none, except that probably the early B stars have a helium convective zone.

The transition from a thick convective zone to a thin convective zone is quite sharp. If we consider the two extreme cases studied by Ch. Pecker-Wimel (1953) and E. Böhm-Vitense (1951, 1953), we can conclude that the actual case is intermediary. The transition probably takes place in the F type stars.

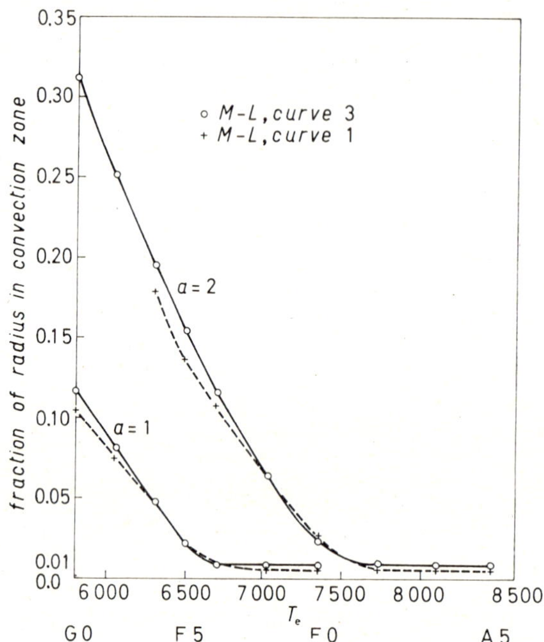

Fig. 15. – Depth of outer convection zone in main sequence stars. (Courtesy N. Baker and R. Kippenhahn).

This has been confirmed by Baker and Kippenhahn (1962). The thickness of the convective zone has been calculated from main sequence stars, for a mixing length equal to 1 or 2 scale heights of the pressure. It is remarkable to notice (Fig. 15) how quickly the convective zone deepens, near to spectral type F2.

This can be extrapolated to other stars.

For stars of a given chemical composition, the line which divides the Hertzsprung-Russell diagram into two regions, region C where stars have a convective zone, and D where stars have a very shallow convective zone or none, is almost parallel to the vertical axis (Fig. 16).

Starting in the F types on the main sequence it reaches the G types in the supergiant range.

It is supposed that when a contracting star goes from region C to region D (Fig. 16), it stops losing angular mo-

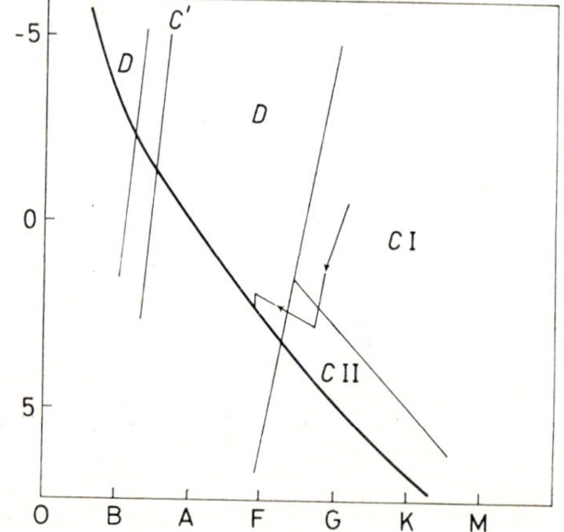

Fig. 16. – Regions in the H-R diagram. Regions C and C' correspond to stars with a convection zone, D without a convective zone. In region C I, stars present electromagnetic and equatorial mass loss; in region C II, only electromagnetic mass loss.

mentum and therefore can reach the main sequence with a high equatorial velocity. The helium convective zone (region C′, Fig. 16) can slow down the stars which happen to reach the main sequence in its range. However, due to its small extent, region C′, has not the power to stop completely the stars which cross it. Therefore, we can only expect a small effect.

We therefore have found what seems to be the basic explanation of the distribution of equatorial velocities along the main sequence: the connection between the convective zone, the magnetic activity (surface activity) and the loss of the angular momentum.

8·3. *Energy balance.* – The energy needed for the exchange of angular momentum between the star and the ejected material is borrowed from the magnetic field. The total energy of a system of magnetic spots is of the order of a, where a is the distance from one spot to the other, and B_0 the magnetic field in the spots. This has to be compared to the energy loss during flares:

$$\frac{1}{2}V^2\,\delta M < \frac{B_0^2 a^3}{8\pi}\,.$$

With the values we have used earlier ($V=10^8$, $a/R=0.1$, $B_0=10^4$) we find that a group of magnetic spots has stored enough energy to allow a maximum mass loss $\delta M < 10^{20.9}$ g. Let us consider a total mass loss of 10^{30} g. This represents a series of 10^9 groups of flares, each group being produced in the same magnetic region. If that succession is to occur in 10^7 years, it requires 10^2 groups of flares a year. Such a frequency is not improbable, from what we know, for example in *UV Ceti* and *T Tauri* stars. If a very small part of the magnetic energy is used to produce the ejection of matter, we can consider the possibility that the number of the magnetic spots is larger, or that flares will start from many different places at the surface of the star. Even an efficiency of 10^{-2} would be sufficient to provide the necessary mass loss.

If we consider now the total magnetic energy at the surface of a magnetic star, which according to the value of a magnetic field is of the order of 10^{40} erg, we see that the energy stored in the magnetic field is quite sufficient to provide the energy necessary for the required mass loss.

8·4. *Magnetic activity.* – If we use Parker's picture (1955) we notice that the poloidal field is built by turbulence and the toroidal field is built by differential rotation. In a three-dimensional co-ordinate system, we suppose that the toroidal component is along the y-axis. The vector potential of the poloidal field has only one component along the y-axis, and the components of the vector potential are $0, A, 0$. Parker writes for the rate of production of the vector potential the expression:

$$\frac{dA}{dt} = \Gamma B + \frac{1}{\mu\sigma} \nabla^2 A , \tag{248}$$

where B is the toroidal component and σ the conductivity. The coefficient has the dimensions of a velocity. We can introduce the coefficient Γ most simply by supposing that the kinetic energy of matter in any vortex is of the order of magnitude of the work done by the Coriolis force acting on the same matter. We therefore have

$$\tfrac{1}{2}\Gamma^2 = \tfrac{1}{2} C_0^2 \omega v_t l_t , \tag{249}$$

where v_t is the velocity of turbulent motion, l_t a characteristic length of the turbulence, and ω the angular velocity of the rotating star. C_0 is an adjustable constant which is not given by the phenomenological theory. We thus write

$$\Gamma = C_0(\omega v_t l_t)^{\frac{1}{2}} . \tag{250}$$

Parker suggests that activity is related to the formation of a migratory dynamo wave. Its wave number k is

$$k = \frac{2\omega'^2}{\Gamma\alpha} , \tag{251}$$

where ω' is the pulsation of the migratory wave, and $\alpha = |\boldsymbol{\nabla} \boldsymbol{V}|$ is a measure of the gradient of the velocity in the differential rotation.

If V is of the order of the equatorial velocity ωR, the gradient of V is proportional to ω:

$$\alpha = C_1 \omega , \tag{252}$$

where C_1 is a constant which is not given by the phenomenological theory.

As the migratory dynamo-wave increases its amplitude with time, it is quite obvious that its wave length should not be much smaller than the radius of the star, otherwise its amplitude would increase by too large a factor during propagation. We make the following choice:

$$k \simeq \frac{2\pi}{(\pi/2)R} = \frac{4}{R} , \tag{253}$$

and derive the relation

$$\omega' = C_1^{\frac{1}{2}} \left(\frac{2\omega\Gamma}{R}\right)^{\frac{1}{2}} . \tag{254}$$

In the case of the sun, with a turbulent velocity in the convective zone of the order of 0.1 km s^{-1}, $v_t = 10^4$ a « mixing length », $l_t = 10^8$ and the angular velocity of rotation of the Sun $\omega = 10^{-5,58}$ we obtain for the angular frequency of the migratory dynamo-wave of period 11.2 years,

$$\omega'_\odot = 10^{-1.17} C_0^{\frac{1}{2}} C_0^{\frac{1}{2}} \omega = 10^{-2.15}\,\omega . \tag{255}$$

$C_1 C_0$ is consequently of the order of 10^{-2}, which seems a reasonable order of magnitude. We shall therefore write

$$\omega' = (2C_1 C_0)^{\frac{1}{2}} (v_t l_t)^{\frac{1}{4}} \omega^{\frac{3}{4}} R^{-\frac{1}{2}} . \tag{256}$$

If the phenomenological relation is correct, we should expect a slow variation of the period of magnetic activity with time. However, the ratio T'/T varies about as $T^{-\frac{1}{4}}$ and is larger when T is smaller. A period of rotation of 1 day would thus correspond to a period of activity of 340 days. It has to be emphasized that in magnetic A stars, where there is no convective zone (or only a small one), we should not expect such a recurrent activity to be present. Such a period of magnetic activity should be expected only in the regions of the H-R diagram where convection is present (C and possibly in region C') (Fig. 16).

The question of the magnitude of the magnetic field is not settled by this theory, since it is a linear theory of the formation of the poloidal and of the toroidal field. We can suppose that the magnetic field is constant. However, we might eventually suppose that it varies like R^{-2} as the star contracts. We shall see below that this is likely to give a reasonable picture.

8'5. *Rate of mass loss*. – At first, we consider the rate of mass loss without contraction. We suppose that we have a period of activity both with a production of N active centers per cycle, each of these centers ejecting matter of the velocity V. An active center may eject ΔM in such a way that $\frac{1}{2}\,\Delta M V^2$ is the fraction $1/n$ of the magnetic energy of the active center

$$\frac{1}{2}\frac{\Delta M}{\Delta t} = -\frac{N}{n}\frac{a^3}{6}\frac{B_0^2}{V^2}\frac{1}{P_{\text{act}}}, \tag{257}$$

where P_{act} is the period of activity. Using relation (119) giving $\omega' = 2\pi/P_{\text{act}}$ we can write

$$\frac{\mathrm{d}M}{\mathrm{d}t} = -\frac{a^3}{3}\frac{B_0^2}{V^2}\frac{N}{n}\frac{1}{2\pi}\left(\frac{2C_0 C_1 \omega^{\frac{3}{2}} (v_t l_t)^{\frac{1}{2}}}{R}\right)^{\frac{1}{2}} . \tag{258}$$

If we use the equation of conservation of angular momentum in the form (108) which gives the following relation between mass and angular velocity:

$$\frac{dM}{dt} = M\frac{KR^2}{a^2}\left(\frac{\varrho_0 V(r_0^2/a)}{B_0^2/8\pi}\right)^{\frac{2}{5}}\frac{d\omega/dt^7}{\omega^{3/5}}, \tag{259}$$

we obtain the relation:

$$\frac{d\omega}{\omega^{27/20}} = -\frac{a^3}{3}\frac{B_0^2}{V^2}\frac{N}{nM}\frac{a^2}{KR^2}\left(\frac{B_0^2/8\pi}{\varrho_0 V(r_0^2/a)}\right)^{2/5}\cdot\frac{1}{2\pi}\left(\frac{2C_0C_1(v\, l_t)^{1/2}}{R}\right)^{1/2}dt. \tag{260}$$

Let us consider the case where a star has reached the main sequence with the angular velocity ω_0. If we suppose that the lengths R_0 and r_0 are both of the order of magnitude of R, the radius of the star, we have

$$\frac{20}{7}\left(\frac{1}{\omega^{7/20}}-\frac{1}{\omega_0^{7/20}}\right) = \frac{4a^{28/5}}{3R^{33/10}MV^{12/5}}\frac{N}{n}\left(\frac{B_0^2}{8\pi}\right)^{7/5}(2C_0C_1)^{1/2}\frac{(v_t l_t)^{1/4}}{K\varrho_0^{2/5}}(t-t_0). \tag{261}$$

If we put

$$t_0 = \frac{K\varrho_0^{2/5}}{(v\, l\,)^{1/4}}(2C_0C_1)^{-1/2}\left(\frac{B_0^2}{8\pi}\right)^{-1/2}\frac{n}{N}\,MV^{12/5}\,\frac{R^{3.3}}{a^{5.4}}\,\omega_0^{-7/20}\,\frac{15}{7}, \tag{262}$$

we can write

$$\frac{\omega}{\omega_0} = \left(1+\frac{t-t_0}{t_0}\right)^{-20/7}, \tag{263}$$

ω is divided by two when $t-t_i = 0.273\ t_0$. If we use the same values as before, we find:

$$0.273\ t_0 = 1.15\cdot 10^9\,\frac{n}{N}\ \text{years}.$$

The value of n is not known; it can be as large as hundred; N is not known, but can be estimated, in case of the Sun, to be of the order of a few thousand. We obtain then a characteristic time of the order of 10^8 years. The time cannot be very much larger nor very much smaller. Its order of magnitude shows the great importance of the electromagnetic mass loss.

If we want to apply this theory to the Sun, we have to take more realistic values of B_0 and a/R. If instead of 10^4 G we take $2\cdot 10^3$ G, we obtain a time scale of $10^{11}\,(n/N)$ years. The angular velocity is divided by a factor 200 in 5.38 t_0. This is of the order of $2.53\cdot 10^{12}\,(n/N)$ years. For $(n/N)\sim 10^{-2}$ we have $2.5\cdot 10^{10}$ years. As we can see, the parameters enter in eq. (262) with large exponents (for example $a^{5.4}$), so that many adjustments are possible to make the time scale fit with the age of the solar system. Therefore, we cannot be too cautious about the application of formula (262) to other stars. Furthermore, the physical quantities can vary with time.

8·6. *Loss of angular momentum during contraction.* – Let us now follow a star during the contraction phase. We start from eq. (244), which we write again:

$$\frac{\mathrm{d}}{\mathrm{d}t}(KMR^2\omega) = a^2\left(\frac{B_0^2/8\pi}{\varrho_0(VR^2/a)\omega}\right)^{2/5}\omega\frac{\mathrm{d}M}{\mathrm{d}t}, \tag{264}$$

and we insert in (264) the expression (258) for the rate of mass loss:

$$\frac{\mathrm{d}}{\mathrm{d}t}(KMR^2\omega)_{\mathrm{magn}} = -a^2\left(\frac{B_0^2/8\pi}{\varrho_0(VR^2/a)\omega}\right)^{2/5}\omega\frac{4a^3}{3}\frac{B_0^2}{8\pi}\frac{1}{V^2}\frac{N}{n}\left(\frac{2C_0C_1\omega^{3/2}(v_tl_t)^{1/2}}{R}\right). \tag{265}$$

Let us consider more specifically for t_0 the value t_m corresponding to the star of mass M, radius R_0 on the main sequence, and magnetic field B_0, where B_0 varies as R^{-2}. (The latter corresponds to conservation of magnetic flux in the star during contraction.) We obtain:

$$\frac{1}{KMR^2\omega}\frac{\mathrm{d}}{\mathrm{d}t}(KMR^2\omega)_{\mathrm{magn}} = -\frac{20}{7}\frac{1}{t_m}\left(\frac{\omega}{\omega_0}\right)^{7/20}\left(\frac{R_0}{R}\right)^{7/2}. \tag{266}$$

The total loss of the angular momentum can be written

$$\left(1-\frac{1}{K}\right)\frac{1}{M}\frac{\mathrm{d}M}{\mathrm{d}t}+\frac{2}{R}\frac{\mathrm{d}R}{\mathrm{d}t}+\frac{1}{\omega}\frac{\mathrm{d}\omega}{\mathrm{d}t} = -\frac{20}{7}\frac{1}{t_m}\left(\frac{\omega}{\omega_0}\right)^{7/20}\left(\frac{R_0}{R}\right)^{7/2}. \tag{267}$$

If we take into account the equations governing gravitational contraction, we have from the equality of gravitational and centrifugal force at the equator,

$$\frac{GM}{R^2} = C_0\omega^2R. \tag{268}$$

Given the relation between luminosity and gravitational contraction,

$$L = -c\frac{GM^2}{R^2}\frac{\mathrm{d}R}{\mathrm{d}t}, \tag{269}$$

(c is the same as in eq. (39) amd (44)) and the mass luminosity relation,

$$L \sim M^iR^{-j}, \tag{270}$$

we can eliminate $\mathrm{d}R/\mathrm{d}t$ and $\mathrm{d}\omega/\mathrm{d}t$ between these equations. We finally obtain:

$$\left(\frac{3}{2}-\frac{1}{K}\right)\frac{1}{M}\frac{\mathrm{d}M}{\mathrm{d}t} = -\frac{20}{7}\frac{1}{t_m}\left(\frac{\omega}{\omega_0}\right)^{7/20}\left(\frac{R_0}{R}\right)^{7/2}+\frac{1}{2t_e}\left(\frac{M}{M_0}\right)^{i-2}\left(\frac{R}{R_0}\right)^{1-j}, \tag{271}$$

where

$$t_0 = \frac{cGM_0^2}{L_0 R_0}, \tag{272}$$

is the time scale of contraction for a star of mass $M_0 R_0$ on the main sequence. Since the mass is changing very little during the final contracting phase, we can write

$$\left(\frac{3}{2} - \frac{1}{K}\right)\frac{1}{M}\frac{\mathrm{d}M}{\mathrm{d}t} = \frac{-20}{7}\frac{1}{t_m}\left(\frac{R_0}{R}\right)^{161/40} + \frac{1}{2t_0}\left(\frac{R}{R_0}\right)^{1-j}, \tag{273}$$

$(\mathrm{d}M/\mathrm{d}t)$ has to be negative. We see that when the star contracts, losing mass from the equator and by electromagnetic coupling, the rate of loss by electromagnetic coupling increases steadily, up to the point where it surpasses the rate of loss through the equator. Then, the loss of mass through the equator stops. On the left hand side of eq. (47) the term $(1/K)$ which came only from the equatorial loss disappears expressing the fact that loss of mass continues only through surface activity. The radius for which the change from one region to the other occurs is given by

$$\frac{R}{R_0} = \left(\frac{40}{7}\frac{t_0}{t_m}\right)^{1/((201/40)-j)}. \tag{274}$$

The value of j is not important, and we may as well write

$$\frac{R}{R_0} = \left(\frac{40}{7}\frac{t_0}{t_m}\right)^{1/5}. \tag{275}$$

With a ratio $[(40/7)(t_0/t_m)] \simeq 10^2$ as can be inferred from the values we have used, we find

$$\frac{R}{R_0} \simeq 2.5\,. \tag{276}$$

Naturally, the numerical value in (276) has no precision, and if the magnetic time scale t_m were larger than t_c (which could be the case, if the magnetic fields were very weak), the loss of matter through the equator would last until the star reaches the main sequence. However, it seems likely that region C (Fig. 16) has to be divided into two by a line running parallel to the main sequence.

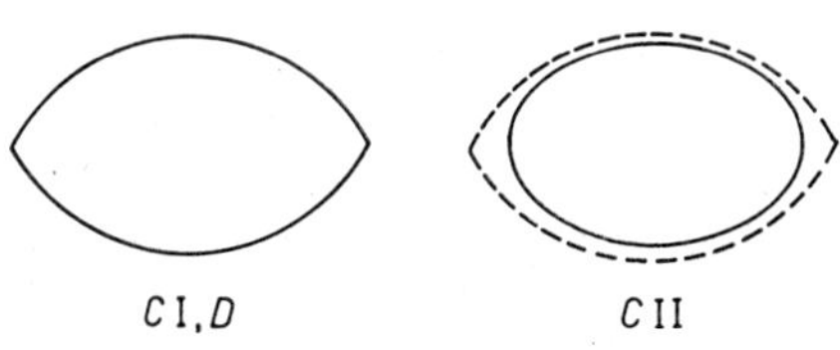

Fig. 17. – Shapes of contracting stars in different parts of the H-R diagram.

Above that line, stars do lose mass through the equator (region C I); below the line, stars undergo only loss of mass through surface activity (region C II). We could in the same way expect a division of region C' into two regions C' I and C' II. We have sketched in Fig. 17, the shape of stars in these regions.

8·7. *Nuclear reactions at the surface of stars and early stages of evolution.* – We have shown, in Sections **6** and **7** that nuclear reactions, due to magnetic activity, occur at the surface of the stars. Therefore, it is essentially only during the crossing of region C that the star can accumulate abnormal amounts of the elements on its surface. The high values of the abundances of certain normally rare elements determined spectroscopically in certain stars has lead to their classification as peculiar. We shall speak of these stars as having P-characteristics.

When a star contracts and loses matter through the equatorial plane, lines of flow carry away part of the surface material. If some were left, it would be in the vicinity of the pole.

If the final contracting phase without activity lasts sufficiently long, we may expect that most of the P-characteristics have been diluted beyond recognition. We can thus explain that in the upper part of the main sequence, the P-characteristics do not appear and we can predict some P-characteristics for stars which have just crossed region C' before reaching the main sequence.

In region C, where the convective zone is present, the higher abundance of the P-elements results from an equilibrium between the formation of the elements at the surface and their dilution inside the convective zone. During the final stage of the contraction, when the star rotates more and more slowly, the magnetic activity decreases and the rate of formation of the P-elements vanishes; the P-characteristics are diluted by mixing with the material of the convective zone, and finally the essential normal abundance of the elements is again found at the surface.

In between the two regions, the type A stars meet the most favorable conditions for the P-characteristics. On both sides of A stars, the conditions are less favorable; B stars have lost matter from the surface, F and G stars have diluted their P-elements in the convective zone.

A rough calculation can be made, leading to a semi-quantitative prediction. We shall suppose that the rate of surface nuclear reactions is proportional to $(\mathrm{d}M/\mathrm{d}t)_{\mathrm{mag}}$; we can suppose that the loss of P-elements through the equator is proportional to their atom number N and to the equatorial mass loss $(\mathrm{d}M/\mathrm{d}t)_{\mathrm{equ}}$; we shall also suppose that the rate of loss by dilution in the convective zone is proportional to the number of atoms N. We can write:

$$\frac{\mathrm{d}N}{\mathrm{d}t} = -\alpha\left(\frac{\mathrm{d}M}{\mathrm{d}t}\right)_{\mathrm{magn}} + \beta N\left(\frac{\mathrm{d}M}{\mathrm{d}t}\right)_{\mathrm{equ}} - \mu N\,, \tag{277}$$

in region C I we can use the stationary solution

$$N_0 = \frac{-\alpha(\mathrm{d}M/\mathrm{d}t)_{\text{magn}}}{\mu - \beta(\mathrm{d}M/\mathrm{d}t)_{\text{equ}}}\,; \tag{278}$$

in region C II we have essentially the time-dependent solution

$$N = N_0 \exp[-\gamma t] - \alpha \exp[-\gamma t]\int_0^t \exp[\gamma t]\left(\frac{\mathrm{d}M}{\mathrm{d}t}\right)_{\text{magn}} \mathrm{d}t\,. \tag{279}$$

In the case where region C II has merged into the main sequence (small magnetic field), we can use the relation (263) which gives the angular velocity as a function of time. We can thus write

$$\left(\frac{\mathrm{d}M}{\mathrm{d}t}\right)_{\text{magn}} = -\frac{M}{t_2}\left(1+\frac{t}{t_0}\right)^{-15/7}, \tag{280}$$

where t_2 is the characteristic time which can be derived from eq. (258).

If we combine such an expression with time-dependent solution (279) we note that the second term

$$J_2 = \alpha \exp[-\gamma t]\int_0^t \exp[\gamma t]\frac{M}{t_2}\left(1+\frac{t}{t_0}\right)^{-15/7} \mathrm{d}t\,, \tag{281}$$

has a maximum.

We can expect that stars will present the P-characteristics with the greatest intensity a little before reaching the main sequence, and as the P-elements are lost in region D, we can expect that stars which just cross the top of region C II will present a maximum amount of P-elements.

As a function of spectral type, we should expect the P-characteristics to vary as sketched in Fig. 18.

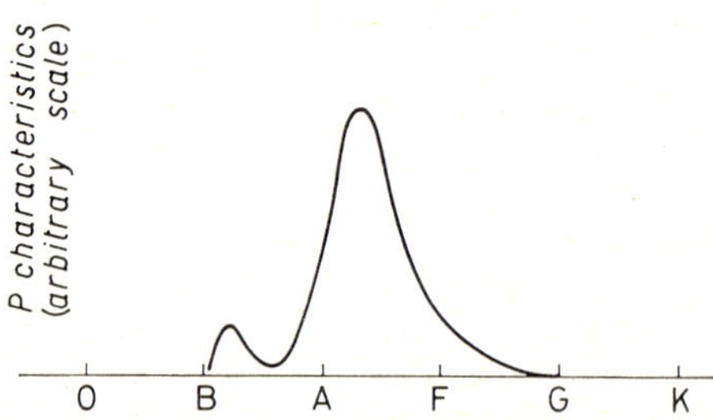

Fig. 18. – The expected distribution of P-characteristics as a function of spectral type.

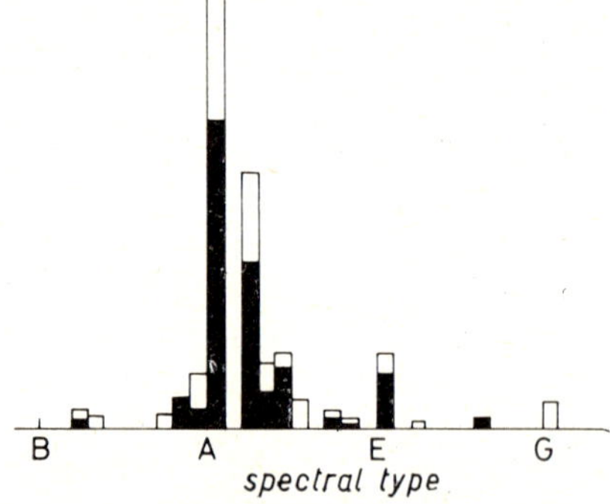

Fig. 19. – Babcock's observations of the frequency of stars with P-characteristics. Solid bars are magnetic; open bars are probably magnetic.

We conclude that the probability of finding a peculiar star is larger for the A0 spectral type. Note the great analogy of the histogram (Fig. 19) of Babcock's magnetic and probably magnetic stars (1958) with the qualitative curve sketched on Fig. 18.

If we assume that there is a statistical distribution of initial magnetic fields, we must conclude that there should also be a statistical distribution of P-characteristics. The greater the P-characteristics, the higher the magnetic field. We can conclude that these stars have a larger magnetic activity than the others, and this activity might have continued in region D for a while. Therefore the A-peculiar stars should rotate more slowly than the others (electromagnetic loss of angular momentum), in agreement with the observed effect.

REFERENCES

AMBARZOUMIAN, V.: 1937, *Izv. Acad. Sciences of Armenia*, Erevan; translated into German in « *Abhandlungen aus der sovjetischen Astronomie* », Folge I, Leipzig (1951).

BABCOCK, H. W.: 1958, *Ap. J. Suppl.*, n. 30, 141.

BAKER, N.: 1962 this Volume, p. 369.

BONSACK, W. K. and GREENSTEIN, J. L.: 1960, *Ap. J.*, **131**, 83.

BURBIDGE, E. M.: (1962), this Volume, p. 174.

BURBIDGE, G. and BURBIDGE, E. M.: (1955), *Ap. J. Suppl.*, **1**, 431.

BURBIDGE, G., BURBIDGE, E. M. and FOWLER, W.: (1955), *Ap. J.*, **122**, 271.

BURBIDGE, G., BURBIDGE, E. M., FOWLER, W. and HOYLE, F.: (1957), *Rev. Mod. Phys.*, **29**, 547.

CAMERON, R. C. and EZER: (1962), Preprint.

COX A. N.: (1962), Private communication.

COX, A. N. and BROWNLEE, R. R.: (1961), *Sky and Telescope*, 31.

DUNGEY, J. W.: (1953), *Phil. Mag.*, **44**, 725.

FESSENKOFF, V. G.: (1949), *Astr. J. U.S.S.R.*, **26**, 62.

GARDNER, C. S., GOERTZEL, H. GRAD, H., MORAWETZ, C. S., ROX, M. H. and RUBIN, H.: 1958, *Nations unies*, vol. **12**, 239.

GRYZINSKI, M.: (1957), *Phys. Rev.*, **107**, 1471.

HAYASHI, C.: (1961), *Publ. Ast. Soc. Japan*, **13**, 450.

HOYLE, F.: (1960), *Quarterly Journal of the R.A.S.*, **28**, 1.

HOWARD, R.: (1959), *Ap. J.*, **130**, 193.

KIPPENHAHN, R.: (1962), this Volume, p. 330.

KRAFT, R. P., MATHEWS, J. and GREENSTEIN, J. L.: (1962), *Ap. J.*, **136**, 312.

LEIGHTON, R. B.: (1959), *Ap. J.*, **130**, 336.

LÜST, R. and SCHLÜTER, A.: (1955), *Zs. f. Ap.*, **38**, 190.

MESTEL, L.: (1959), *M. N.*, **119**, 249.

MESTEL, L. and SPITZER, L., Jr.: (1956), *M. N.*, **116**, 503.

MORAWETS, C. S.: (1961), *Phys. of Fluids*, **4**, 988.

OKE, J. B. and GREENSTEIN, J. L.: (1954), *Ap. J.*, **120**, 384.

OSTERBROCK, D. E.: (1961), *Ap. J.*, **134**, 347.
PECKER-WIMEL, CH.: (1953), *Ann. d'Ap.*, **16**, 321.
PARKER, E. N.: (1955), *Ap. J.*, **121**, 491.
PARKER, E. N.: (1958), *Phys. Rev.*, **109**, 1328.
SARGENT, W. C. W. and JUN JUGAKU: (1961), *Ap. J.*, **134**, 777.
SCHATZMAN, E.: (1949), *Ann. d'Ap.*, **12**, 203.
SCHATZMAN, E.: (1954*a*), *Ann. d'Ap.*, **17**, 300.
SCHATZMAN, E.: (1954*b*), *Ann. d'Ap.*, **17**, 283.
SCHATZMAN, E.: (1959), *I.A.U. Symposium* n. 10 (ed. GREENSTEIN, Paris), 129.
SCHATZMAN, E.: (1962), *Ann. d'Ap.*, **25**, 18.
SCHATZMAN, E.: (1963), *Ann. d'Ap.*, (in the press).
SEVERNY, A. B.: (1958), *News of the Crimean Observatory*, **20**, 22.
SPITZER, L. Jr.: (1949), *Harv. Centennial Symposia* (Cambridge, Mass.), 87.
SPITZER, L. Jr.: (1956), private communication.
STRUVE, O.: (1950), *Stellar evolution* (Princeton).
SWEET, P. A.: (1958), *I.A.U. Symposium* n. 6, 123.
VITENSE, E.: (1951*a*), *Zs. f. Ap.*, **28**, 81.
VITENSE, E.: (1951*b*), *Zs. f. Ap.*, **29**, 73.
VITENSE, E.: (1953), *Zs. f. Ap.*, **32**, 135.

Associations and Very Young Clusters.

L. GRATTON

Laboratorio di Astrofisica e Laboratorio Gas Ionizzati - Frascati

1. – O- and T-associations.

It was known since a long time that very luminous O and early B stars show a tendency to occur only in determined regions of the sky. The same was shown by JOY (1945, 1949) to be true also for the so-called T Tauri stars These are stars whose spectra contain a large quantity of emission lines superimposed over an absorption spectrum of type G and K. Most, if not all of them show irregular variations of brightness (variables of the RW Aurigae class) and are connected with dark and bright nebulosities.

But the term «Stellar association» to indicate the physical relationship corresponding to these facts, was invented by AMBARTSUMIAN (1947, 1949) and has become popular when he showed the great importance of Stellar associations for the evolution of stars. The term «Stellar aggregate» is also used sometimes with more or less the same meaning.

O-associations have been defined by Ambartsumian as loose groups of stars of spectral types O and early B from a dozen to some hundreds in number having diameters ranging between 30 and 200 ps. Stars of different types which are also members of the associations, and (usually) dark or bright nebulosities exist in the same regions of space; often one or more galactic clusters form the nuclei of an O-association.

O-associations show a high concentration towards the galactic plane and therefore are typical objects of Population I or Disk Population; indeed it was shown by MORGAN, WHITFORD and CODE (1953) that they can be used to trace the spiral structure of the Galaxy in the neighbourhood of the Sun.

Although the density of early-type stars is much higher in Stellar associations than in neighbouring regions, the mean density of stars of all spectra types in smaller.

Ambartsumian defined also T-associations as groups of T Tauri stars connected with diffuse nebulosities (dark or bright). Many O-associations contain also a number of T Tauri stars and are, accordingly, also T-associations; indeed (KHOLOPOV, 1959*b*) « T-associations have been discovered up to the present time, in all open clustering of hot stars found within a sphere of 500 pc radius around the Sun. They have also been detected in all more remote O-associations in which search has been conducted ».

Apparently there is no main difference between O-associations and T-associations, with exception of the fact that some T-associations do not contain hot stars.

2. – Space distribution of stellar associations.

The first list of O-associations was published by B. E. MARKARIAN (1952) shortly followed by Morgan, Whitford and Code's Catalogue (1953) of 27 associations. The *Catalogue of Star Clusters and Associations* by ALTER, RUPRECHT and VANYSEK (1958) contains 46 entries and a more recent list by K. H SCHMIDT (1958) based on Hiltner's photometric observations of O and B stars recognizes 62 different groupings.

The recognition of associations as separate entities is not easy since often two or more associations project themselves on the same region of the sky; some of them are also close to each other in space, so that it is some times a question of judgement whether they are to be considered separately or not Unfortunately there is a lack of uniformity in the designation of stellar associations which adds to the confusion. For instance the association called Monoceros I by Morgan and associates is called consistently Monoceros II by the Russian investigators and viceversa. We shall employ here the names given by the American astronomers when they exist.

In Fig. 1 is shown the distribution of 54 O-associations along the galactic equator; these are all those listed by ALTER *et al.* with a few additions. Each association is marked by a circle whose radius is roughly indicative of its apparent dimensions.

Since T Tauri stars are much fainter than O and B stars, T-associations are more difficult to discover and can be followed only to smaller distances The most complete list of T-associations has been published by P. N. KHOLOPOV (1959*a*); it contains 29 « real » and 12 « possible » T-associations.

In Fig. 1 all individual variables of the RW Aurigue class contained in the *General Catalogue* by KUKARKIN and PARENAGO (1958) have been plotted as dots, with the exception of the two large concentrations in Orion and Monoceros containing 266 and 55 variables, respectively, which are indicated by double circles. All these stars have been divided somewhat arbitrarily in 18 groups; a 19-th group of 3 variables in Phoenix contains the only known RW Aur variables with galactic latitude outside the limits of the figure.

For comparison the numbers of the 41 T-associations of Kholopov's list are marked at the corrisponding positions. One can thus get an idea of the arbitrariness of the groupings.

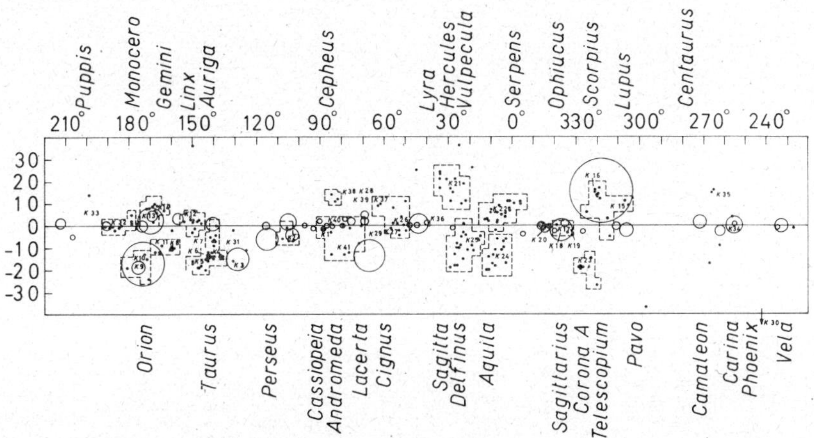

Fig. 1. – Distribution of known O and T-associations along the galactic equator. Circles are O-associations (diameters approximately equal to size). Dots mark the variables of the RW Aur class in Kukarkin and Parenago General Catalogue; numbers preceded by K indicates Kholopov's T-associations.

Distances of O-associations are mostly derived from photometric data. When accurately determined photoelectric magnitudes on the U-B-V system are known for a sufficient number of members of an association, its distance r can be derived as follows.

Denote by small letters the observed and by capital letters the absolute magnitude of a star in a given system; thus for instance,

$$v = V - 5 - 5\log r + A_v(r)\,, \tag{1}$$

r being the distance and $A_v(r)$ the extinction due to interstellar matter in the V-system.

By difference one gets

$$\left\{\begin{aligned} b - v &= B - V + A_b(r) - A_v(r)\,,\\ u - b &= U - B + A_u(r) - A_b(r)\,. \end{aligned}\right. \tag{2}$$

Now consider a group of stars, such as the members of an association, all at the same distance; then $A_b(r)$, $A_v(r)$ and $A_u(r)$ are the same for all the stars of the group. By plotting the observed $b - v$ and $u - b$ and by comparing with a standard $(B-V)$, $(U-B)$ curve (obtained from nearby stars) it is therefore possible to obtain the difference $A_b - A_v$ for the group.

But the ratio between the total absorption in the V-system, $A_v(r)$ and the colour excess $A_b(r) - A_v(r)$ due to interstellar reddening was found to be a constant

$$R = \frac{A_v(r)}{A_b(r) - A_v(r)} = 3.0 \pm 0.2 \,. \tag{3}$$

(HILTNER and JOHNSON, 1956).

The total absorption $A_v(r)$ can thus be found and the observed magnitude and colours of the stars of the group can be corrected for the effect of interstellar absorption.

Finally by comparing the corrected magnitude-colour diagram with a standard absolute magnitude-colour diagram for a similar group of stars, a reliable distance can be determined for the group. If the spectra are known it is, of course, better to compare the corrected apparent magnitude with the absolute magnitude corresponding to the spectral type of each star.

The distances obtained by this method, which is in principle due to W. BECKER (1951), are quite trustworthy, when good photometric data are available. The distances of some O-associations are given in Table I.

TABLE I. – *Distances of some* O-*associations.*
(H. L. JOHNSON and others, 1961)

Associations	$d(pc)$	Note
Cep III	730	
Per I	2250	h and χ Per
α Per	170	
Per II	350	
Ori I	420	
Mon II	750	NGC 2264
Aur I	1260	NGC 1960
Mon I	1660	NGC 2244
CMa II	1500	NGC 2362
Sgr I	1400	NGC 6530
Sct I	1740	NGC 6705
Lac I	580	

The positions in the galactic plane of 45 O-associations have been plotted in Fig. 2. The Sun is at the center; filled circles are well determined data (uncertainty not larger than, say, 15%), open circles are less reliable (uncertainties, in some cases as much as 50%), half filled circles are believed to be of intermediate accuracy. In a few instances the individual distances of clusters belonging to associations as given by H. L. JOHNSON *et al.* (1961) have been plotted, instead than an average distance of the whole association. Taken as a whole, Fig. 2 offers a fair representation of our present knowledge upon the space-distribution of O-associations. The dotted lines show the general position of the three main arms of the spiral structure of the Galaxy in the neighbourhood of the Sun as they are suggested by the observations of stellar associations. Later we will come again to this problem.

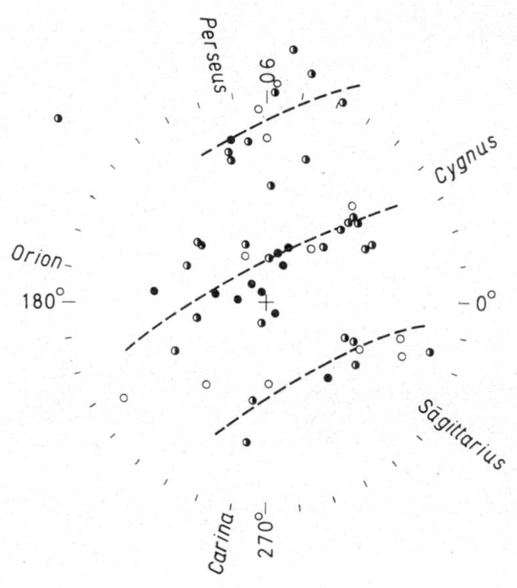

Fig. 2. – Distribution of O-associations on the galactic plane. Filled circles are first class data, half filled circles 2-nd class data and open circles less reliable data. The interrupted lines indicate roughly the spiral arms near the Sun.

3. – Instability of stellar associations.

The most important fact concerning stellar associations is that they are short-lived unstable groups of stars. Several kinds of instabilities come into consideration in connection with associations.

a) *The virial theorem.* Consider a group of stars dynamically interacting; from the *virial theorem*

$$\frac{\mathrm{d}^2 I}{\mathrm{d}t^2} = 2T - \Omega \,, \tag{4}$$

(where I is the moment of inertia, T the total kinetic energy, relative to the center of gravity, and Ω the potential energy due to the mutual attraction of the stars of the group) it follows on quite general grounds that the total energy

$$E = T + \Omega \tag{5}$$

must be negative in order that the group be stable.

The total energy has been evaluated for some associations (KOPYLOV 1953; AMBARTSUMIAN 1954) and it was found to be positive.

When the total energy is positive we may estimate the time of dissolution of a stellar association by simply assuming that all the stars are moving with uniform velocity. Then the association will expand, its radius growing with a velocity egual to the relative velocity of the stars.

Assuming a mean velocity of 5 km/s, it is seen that 10^6 y after the beginning of the expansion the radius will be of 5 pc, after 10^7 y it will have grown to 50 pc and after $3 \cdot 10^7$ y to 150 pc. At this time it is doubtful whether the group of stars will be recognised and certainly after 10^8 y it will have disappeared in the general field of the Galaxy.

b) *Galactic rotation.* If the total energy is only slightly positive or negative, other causes of instability may be important. The disruptive force due to differential galactic rotation was first considered in the case of a cluster by BOK (1934). The theory can be found in text books on stellar dynamics (see, for instance, CHANDRASEKHAR 1942; LINDBLAD 1959).

In his studies of the association in Scorpio-Centaurus, BLAAUW (1946, 1952*b*) assumes more simply that all the stars of the association started at a given moment from an origin O with the same velocity, but in different directions and that their mutual attraction can be neglected relative to the general galactic field. If we suppose that O moves with the circular velocity corresponding to the rotation of the galaxy, it can be easily found that the stars will lie at all times upon ellipses with the center at O. The direction of the major axis of the ellipse corresponding to time t is given by

$$\text{tg}\,\chi = -\frac{1-\cos\varkappa t}{At - ((A-2B)/\varkappa)\sin\varkappa t}, \tag{b}$$

where χ is the angle which the major axis forms with the direction of the galactic rotation, A and B ore Oort's constants, and

$$\varkappa = 2\sqrt{B(B-A)}\,. \tag{7}$$

It is thus seen that χ depends only on t and not on the assumed initial velocity of the stars.

The observed shape of the Scorpio-Centaurus cluster suggests an expansion age of $7.2 \cdot 10^8$ y; the center of the association was at that time at a distance of 2200 pc from the present position of the Sun in the direction of galactic longitude 26° and has moved since then, being at present at about 200 pc in the direction of galactic longitude 318° (Scorpio-Centaurus).

c) Effect of interstellar clouds. The effect of close passages of interstellar clouds upon a group of stars was considered by SPITZER (1958) who found that even a relatively dense galactic cluster would be dissolved by such passages in the rather short time of $2 \cdot 10^8$ y. Although this time is longer than that corresponding to other instabilities, it might have some importance for the denser nuclei of associations.

d) Star evaporation. The evaporation of stars from dense clusters has been considered by AMBARTSUMIAN (1938) and by SPITZER (1940); a full discussion is given in Chandrasekhar's book (1942).

Briefly this effect can be described as follows. Stellar encounters between members of a cluster tend to establish a Maxwellian distribution among their velocities. But the tail of a Maxwellian distribution will correspond to stars possessing velocities greater than the velocity of escape; these stars will therefore leave the cluster very rapidly. The tail of the velocity distribution is thus continuously destroyed and reformed causing a slow desintegration of the cluster. Clearly the time of disintegration is closely related to the *time of relaxation* of the cluster.

The rate of loss by evaporation is certainly negligible for stars of large masses; it might be of importance for stars of very low masses.

4. – Observed expansion of stellar association.

The most convincing direct evidence of the instability of O-associations has been given by Blaauw's discovery (1952*a*) of the general expansion of the association around ζ *Per* (Perseus II).

Figure 3 adapted from Blaauw's paper gives a picture of the individual proper motions of the early-type stars near ζ *Per*. From a discussion of these motions Blaauw derived a velocity of expansion of $0.00268''/\text{y} = 12$ km/s; a revised value by DELHAYE and BLAAUW (1953) gives $0.00236''/\text{y}$. By tracing back the motions of individual stars it is found that a maximum star density was reached about $1.5 \cdot 10^6$ years ago. At that time all the stars were compressed in a fairly small volume of space.

Expansion ages were derived in a similar way for various associations The results are collected in Table II.

Although the expansion of associations as been doubted in some cases (sec for instance, PETRIE 1958; WOLLEY and EGGEN 1958) the present evidence is that O-associations are short-lived formations whose ages cannot be larger than, say, 10^7 years. This is the reason which lead Ambartsumian to affirm that *stars are formed in groups in stellar associations, which are stellar aggregates not older than* 10^7 years.

This conclusion seems to have met general acceptance, at least with respect to the origin of Population I stars.

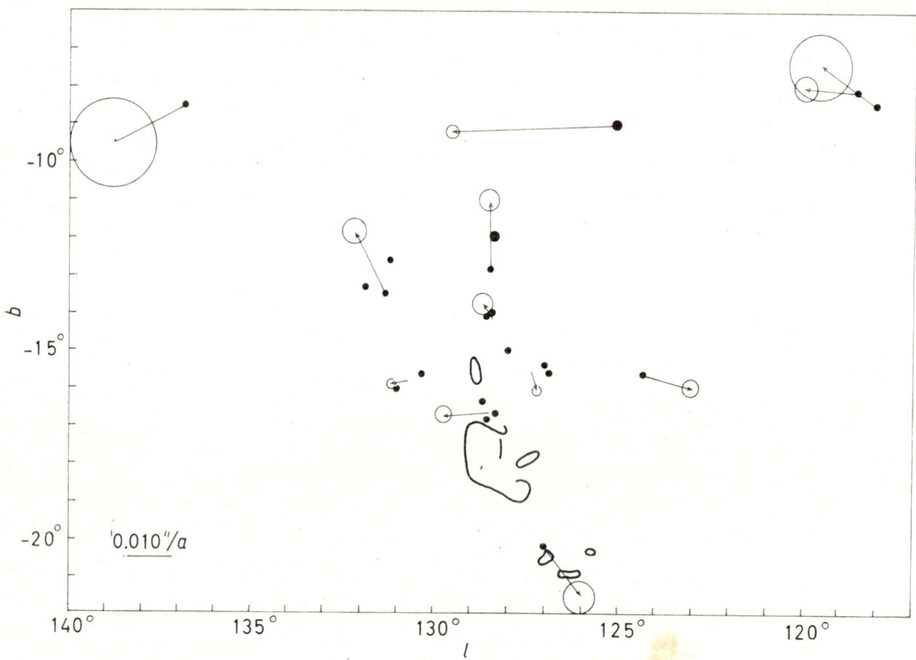

Fig. 3. – Motions of O and early B-stars near ζ *Persei* showing the expansion of the association (adapted from BLAAUW). Arrows indicate the annual proper motions; the circles at the respective points have a radius equal to the errors of the proper motions. Some nebulosities are indicated.

TABLE II. – *Observed expansion of O-associations.*

Association	Exp. ″/y	Vel km/s	Age 10^6 y	Authority
α Persei			8±(3)	DIECKVOSS 1953
Perseus II	0.00236	12	1.5	DELHAYE and BLAAUW 1953
Orion II (centr)	0.0030	7.4	0.3	STRAND 1958
Orion II (ext)			5	
Scorpius II		0.7	72 (1)	BLAAUW 1952*b*
Cepheus I	0.00078	8	4.5 (2)	MARKARIAN 1953
Lacerta	0.00086	8	4.2	BLAAUW and MORGAN 1953

(1) From galactic rotation.

(2) With a somewhat more elaborated method SHTEINS and ABELE (1958) derive an age of $2.8 \cdot 10^6$ y.

Concerning the results of this Section, we may make still the following remarks:

a) No direct data are available for T-associations, so that the conclusion upon their instabilities rests mainly upon the fact that they are found in O-associations. However KHOLOPOV (1959*b*) has remarked that the star density of T-associations not connected with nebulosities is lower and the average diameter larger, than those of T-associations imbedded in nebulosities. This suggests a gradual expansion of T-associations accompanied by the disintegration of the associated nebulosities, which become thus indistinguishable on the stellar background of the Milky Way.

b) As far as the present evidence goes, the members of a stellar association are not born all at the same time, but the process of star formation continues near the association center during a time of the same order of magnitude as the life of the association.

The best instance is Orion I whose expansion age derived from the motion of the stars close to the center of the nebula is of only 300 000 years; but the three stars *AE Aur*, *μ Col* and 53 *Ari*, although several degrees away, are believed by MORGAN and BLAAUW (1954) to be members of the association which moved from the center some $5 \cdot 10^6$ years ago. According to MENON (1958) the whole nebula is expanding as if it were formed some $2.4 \cdot 10^6$ years ago.

c) It has been noted by Ambartsumian that multiple stars of the so called « trapezium » type (after the well known multiple star at the center of the Orion nebula) are to be found only in associations. As these systems are certainly unstable, they constitutes a further argument in favour of the short-livedness of stellar associations (see also, SHARPLESS, 1954*b*). Star chains are also frequent in O and T-associations.

5. – Number and rate of formation of stellar associations.

Obviously, if stellar associations are short-lived, their observed number is related to their rate of formation.

The total number of O-associations nearer than 3 kpc is about 50, but almost half of them are nearer than 1.5 kpc. Assuming 25 associations inside 1.5 kpc and taking into account the fact that the system of O-associations is highly flattened to the galactic plane, the total number of O-associations in the Galaxy is between 10^3 and 10^4.

With a mean life of 10^7 y we get a rate of formation of one Association every 10^4 or 10^3 years.

According to KHOLOPOV (1959*b*) the total number of T-associations in the Galaxy is also 10^4; with a mean life of 10^6 y, we obtain a rate of formation of one T-association every 100 years.

We may, then, inquire whether it is possible that *all* the stars of Population I in our Galaxy have been originated in stellar associations.

Now, as the oldest Population I clusters have an age of, say, 10^{10} years it can be assumed that the process of formation of Population I stars has been going on in our Galaxy since 10^{10} years, during which time, at the present rate, some 10^6 or 10^7 O- and 10^8 T-associations may have been originated.

In a single O-association a number from 10 to 100 stars of early type (and thus of large masses) is known to exist; but the number of stars of smaller mass is certainly much larger. For instance according to MENON (1958) the total mass (diffuse matter and stars) inside the large ring of the Orion region is of the order of 10^5 solar masses (sec also SAVEDOFF 1956 and Section **8**). Thus a number of 10^3 to 10^4 stars is certainly a conservative estimate for an average O-association. In the same way we may consider a hundred stars as a reasonable guess for an average T-associations. We get, by this way, 10^{10} or 10^{11} as the total number of Population I stars which may have been formed in stellar associations. One might then infer that *all* stars do actually originate in stellar associations (BLAAUW, 1958). This conclusion. however, is not accepted by several authors (ROBERTS, 1957; G. BURBIDGE, 1960), mainly on the ground that it is not safe to extrapolate to the past the present rate of formation of stellar associations. But there is no doubt that the number of stars which have been originated in stellar associations is quite significant relative to the total number of stars in the Galaxy.

6. – The stellar content of associations.

Observations of individual members of stellar associations afford an excellent apportunity of studying the properties of stars in the very early stages of their evolution. I do not intend to enter here into too many details; the H-R diagrams of very young clusters and their luminosity functions will be discussed in the next section and T Tauri stars will be treated by Rosino in a special Seminar.

O and early B stars have been known to be very young objects at least since Bethe's discovery of the nuclear reactions responsible for the energy-production in stars. Their presence in stellar associations is thus in a sense another proof that stellar associations are young aggregates of stars.

Among other objects, the following can be mentioned.

53 Ari is a β CMa star connected with Orion I (MÜNCH and FLATHER, 1957); another β CMa star, *θ Oph*, is a member of the Scorpio-Centaurus group.

We may, then, infer that β CMa stars, are not uncommon among the members of stellar associations; apparently this rises some theoretical difficulties concerning the ages of these objects (see, for instance, BLAAUW, 1958, the discussion following the paper). It has also been suggested that the Cepheid *WZ Sgr* be a member of Sagittarius III, but the evidence is rather poor (FERNIE, 1962).

Be stars have been investigated recently by E. MENDOZA (1958) who found that those of low luminosity are probably not connected with associations. But P Cygni stars are very common in stellar associations (AMBARTSUMIAN, 1949; KOURGANOFF, 1957); for instance *P Cygni* itself is prominent in Cygni II (MORGAN, WHITFORD and CODE, 1953) and various P Cygni stars are present in NGC 6231, the nucleus of Scorpio I. In the large association of early type stars in Carina there are several P Cygni stars, like *AG Car*; *η Car*, which I will discuss in a Seminar, is in a sense a P Cygni star. Another famous star, *S Dor*, is a member of an association in the Large Magellanic Cloud.

Wolf-Rayet stars have been found very numerous in stellar associations (AMBARTSUMIAN, 1949; ROBERTS, 1957; HERBIG and MENDOZA, 1960). FEAST (1960) has recently investigated the great nebula NGC 2070 in the Large Magellanic Cloud; a large O-association is imbedded in this nebula and among its brightest members there is a large proportion of Wolf-Rayet stars all of which belong to the nitrogen sequence (WN); this seems to indicate that Wolf-Rayet stars belonging to stellar associations are brighter than normal Wolf-Rayet stars. The Carina association contains also several Wolf-Rayet stars.

ÖPIK (1953) has suggested that there might be some connection between stellar associations and Supernovae. There is not the slightest observational evidence to support this suggestion as far as Supernovae of type I are concerned. But Supernovae of type II are considered to be Population I objects and the possibility is not quite ruled out that *η Carinae* should be considered as a type II Supernova.

It is well known that some late-type supergiants exist in the association Perseus I (BIDELMAN, 1943; SHARPLESS, 1958); others have been found in other associations (BLANCO and GRANT, 1959; T. SCHMIDT-KALER, 1961) It has been suggested that these are stars which have not yet reached the main sequence stage and are still in a phase of gravitational contraction.

The fact that all the above-listed objects are young has certainly important evolutionary implications, although these are not all clear at present.

In this connection we may mention also observations of axial rotation among stars of stellar associations. Quite recently MCNAMARA and LARSSON (1962) investigated the rotational velocities of 50 B0-B3 stars in the Orion association. The average rotational velocity, assuming a random orientation of the axes of rotation, was found to be of 135 km/s, which is considerably lower than that corresponding to field stars (186 km/s). On the contrary from 6 B0-B3 stars in the association Perseus II, observed by TREANOR (1960) one gets an average value of $v \sin i$ of about 143 km/s, which corresponds, for a random orientation of the axes, to a rotational velocity of 182 km/s in apparently good agreement with the field stars.

Since, however, there seems to be strong systematic differences between Treanor's and McNamara and Larsson's measurements, the conclusion that, on the average, the axial rotation of stars in associations is somewhat slower than that of field stars of the same spectral type may be accepted, although with some caution.

7. – Hertzsprung-Russell diagrams and luminosity functions.

Many investigators have obtained colour-magnitude diagrams for O-associations or, rather, for the very young clusters which form their nuclei; probably the most important results are those for Orion I (PARENAGO, 1954; SHARPLESS, 1954; JOHNSON, 1957; KOPYLOV, 1958, see Fig. 4), for NGC 2264 in Monoceros II (WALKER, 1956), NGC 6530 in Sagittarius I (WALKER, 1957)

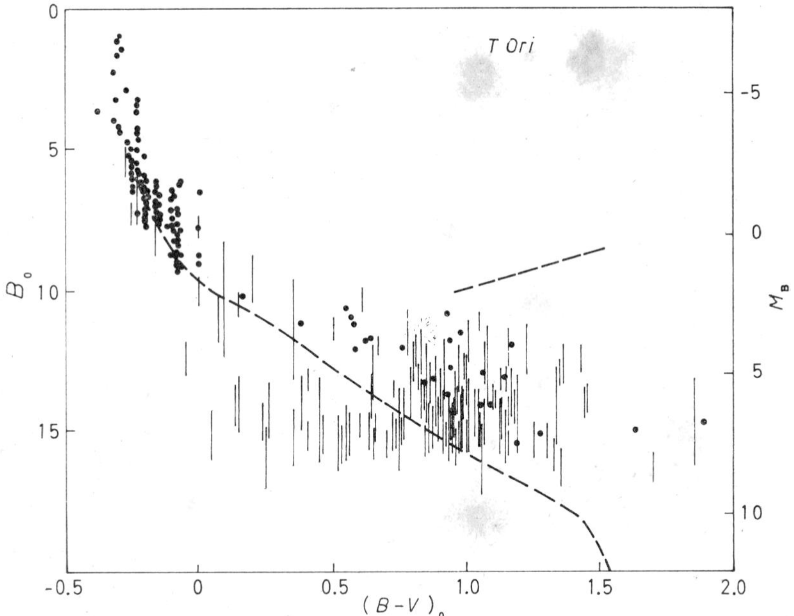

Fig. 4. – Colour-magnitude diagram for the Orion association, according to Parenago (from KOPYLOV, 1958). The vertical lines correspond to variables of the RW Aur class. The zero-age main sequence and the giant branch are shown by dotted lines.

IC 5146 (WALKER, 1959) and NGC 6611 in Serpens I (WALKER, 1961*b*). Fig. 5 gives Walker's diagram for NGC 2264.

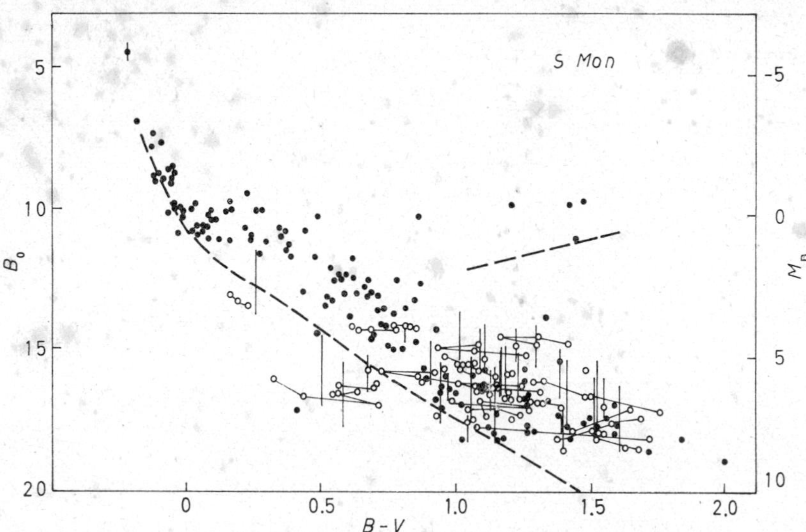

Fig. 5. – Colour-magnitude diagram for NGC 2264 (WALKER, 1956, from KOPYLOV, 1958). Dots are stars of constant brightness, open circles are nebular variables, the vertical bars give the limits of variation of brightness. The zero-age main sequence and the giant branch are shown by dotted lines.

In all cases it was found that the brightest stars fitted the so-called « zero-age » main sequence, determined by JOHNSON and HILTNER (1956) over a range in spectral type from O to A0 approximately. For later types the majority of the cluster stars lie well above the main sequence. Many of these are variables of the T Tauri class or connected with it.

Some suspicions have been recently risen by UNDERHILL (1960), PIK SIN THE and BLANCO (1960) and PIK SIN THE (1960) concerning the membership of the stars above the main sequence to the associations. It was suggested by these authors that these stars might be highly reddened background stars of early type. In a short note WALKER (1961*a*) answers this criticism and although the matter is not definitely settled, it appears that at least part of the stars (and of course the variables) are genuine members of the clusters.

The stars lying above the main sequence are generally believed to be in the contracting stage preceding the main sequence stage, but there seems to be some difficulties connected with the time-scale of the contraction which comes out too long if compared with the age of the association.

From the observed shapes of the main sequences and the theory of gravitational contraction one can compute a « theoretical age » of the associations.

As a rule the ages computed in this way are in satisfactory agreement with those obtained from the internal motions; but, according to SANDAGE (1958*b*) it is difficult to explain theoretically the fact that many cluster stars of faint absolute magnitude are very close to the zero-age main sequence, since they ought to reach the observed values of colour and magnitude only after 10^8 years after the beginning of their contraction.

Recently it was shown by VARSAWSKY (1960) that at least part of this discrepancy is due to an observational effect connected with the spectral energy distribution of the contracting stars. The question was discussed also by HERBIG (1962) who believes that the explanation has to be sought in the fact that the stars in an association do not form all at the same time. As the question will be discussed by others in this Course, I will not pursue further the matter.

The ages obtained from the shape of the main sequence can also be compared with those derived from an age effect upon the hydrogen-line intensities, discovered by STRÖMGREN (1958). This method has been applied by CRAWFORD (1958) and by BAPPU, CHANDRA, SANWAL and SINVHAL (1962).

Strömgren's age effect consists in that, if one plots the H-line intensities with the (U-B) colour indices for the stars of a cluster, there is a trend on the part of the correlation curves to shift towards weaker absorption-line intensities with increasing age. Table III shows this age effect on some clusters, according to Crawford's data.

TABLE III. – *Age effect from H-line intensities.*

Object	Order of H intensity	Spectrum of the earliest main sequence star	Age (10^6 y)
Perseus I	1	O7	1
Scorpio-Cent.	2	B0	4
Orion (*)	2	B0	4
α Persei	3	B3	10
Pleiades	4	B6	80

(*) The very young stars imbedded in the nebula are excluded.

Closely related with the colour magnitude diagram is the luminosity function, *i.e.*, the function which gives the number of stars in any given interval of absolute magnitude.

The importance of the luminosity function for clusters and its meaning for the problem of stellar evolution has been pointed by SALPETER (1955) and has been discussed by several authors (SANDAGE, 1957, 1958*a*; M. SCHMIDT, 1959

VAN DEN BERGH and SHER, 1960). In the case of an association one may expect that the observed luminosity function corresponds to that according to which the stars are formed.

Recently the luminosity function for various clusters has been discussed by VAN DEN BERGH (1961) who finds a complete disagreement between the luminosity function for cluster stars fainter than the absolute magnitude $M_v = +2$, and that of unevolved stars near the Sun derived by SCHMIDT (1959). He concludes that the observed differences can be explained in two ways:

1) the luminosity function of star formation in associations contains more faint stars than the corresponding luminosity function for clusters, or

2) the luminosity function of star formation contains now fewer faint stars than it did originally.

For associations the most important data are those for NGC 2264 (WALKER, 1956) in Monoceros II; this can be compared with Salpeter's initial luminosity function. It is found that the two luminosity functions agree fairly well for stars brighter than $M_v = +3$; for fainter stars there is an excess of cluster stars, which is, of course, to be expected from the Hertzsprung-Russell diagram, since it corresponds to the stars found to lie above the main sequence. A good agreement between Salpeter's initial luminosity function and those observed in some O-associations was found also for the brightest stars by BLAAUW (1958).

8 – Diffuse matter connected with associations.

Almost all stellar associations are known to be connected with large diffuse nebulosities; the Orion I and Sagittarius I associations and the great Carina nebula are among the most beautiful examples. On the other side the region of the double cluster in Perseus shows very little if any evidence of nebulosity (see Fig. 6, 7, 8).

It has been demonstrated in several instances that the hottest stars of the association are responsible for the brightness of the nebula in which they are imbedded. In some cases it was also possible to show that the nebulosity expands with the association.

For the great nebula in Orion, T. K. MENON (1958) found that the motion has a rather irregular pattern; the central part is practically at rest (relative to the circular motion in the Galaxy), but the outer parts expand with a velocity of about 10 km/s. The corresponding expansion age is $2.4 \cdot 10^6$ years, *i.e.*, of the same order as that found for the association Orion I. The results of WILSON MÜNCH, FLATHER and COFFEEN (1959, see also MÜNCH, 1958) are in substantial agreement with Menon's work. These authors found large-scale motions which cause splitting of the nebular lines as large as 25 km/s, superimposed upon a strong turbulence, which varies between 9 and 14 km/s.

MENON (1962) has also investigated the « Rosette » nebula, NGC 2237-2246 which involves the association Monoceros I (NGC 2244). The results are similar to those obtained for the Orion nebula, but a comparison of the density

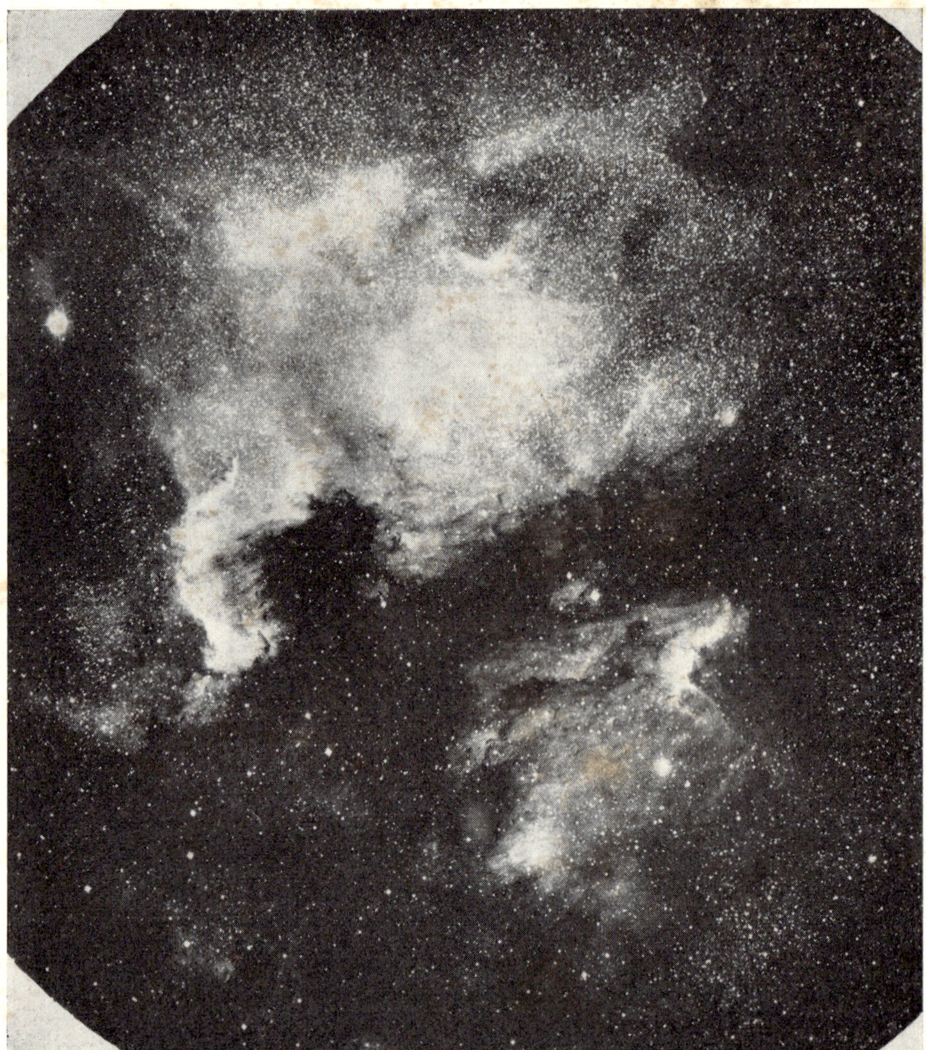

Fig. 6. – The North America nebula (NGC 7000). This very well known Nebula is connected with the system of diffuse nebulosities in Cygnus; several O-associations are known to exist in this region, that associated with NGC 7000 is probably Cygnus II. Herbig found several nebular variables in the region (Asiago obs.).

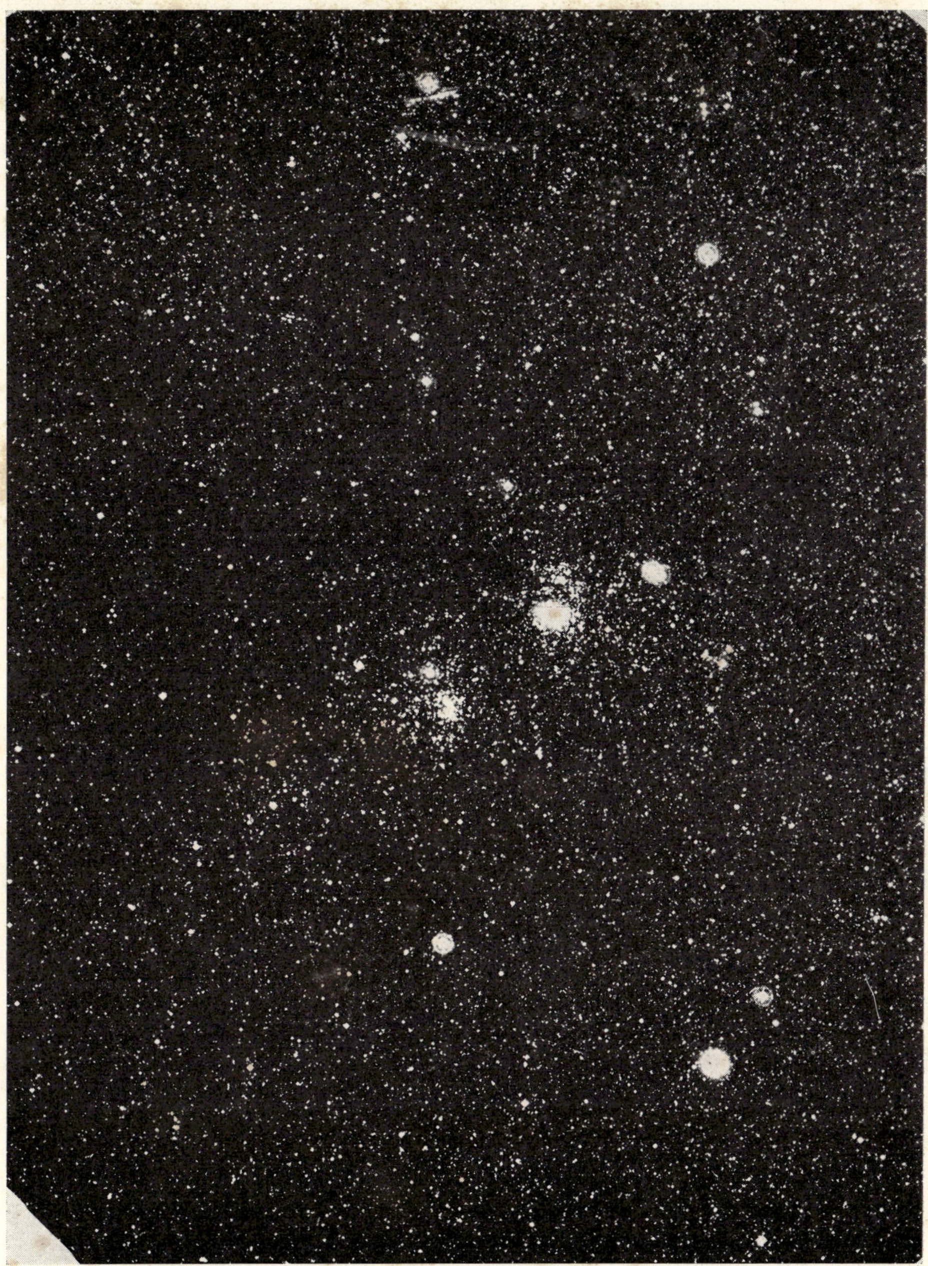

Fig. 7. – The double cluster h and χ *Persei*. This is a beatiful example of an O-association practically devoid of nebulosity. It belongs to the outer spiral arm; its distance is about 2.4 kpc (Asiago Obs.).

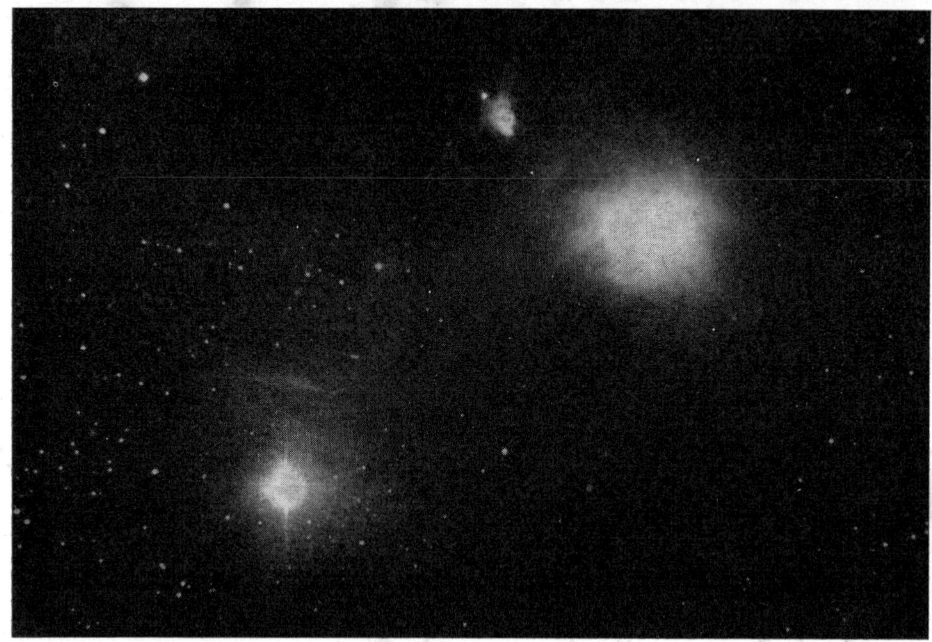

Fig. 8. – The field near γ *CrA*. A fine example of a T-association. The star at the vertex of the small fan-shaped nebula is *R CrA*, while *T CrA* is above it, just at its limit. *S CrA* is not visible, probably because at minimum. The bright irregular nebula (clearly an illuminated part of a larger dark nebula) involves two more *RW Aur* stars *T CrA* and *VV CrA*. The bright star is γ *CrA* a well-known nearby binary. (Cordoba Observatory).

distributions, according to Menon, suggests that there might be certain differences in the ways in which the clusters imbedded in the two nebulae are formed. The ratio between the total mass and the stellar mass is also different in the two associations, being of the order of unity for the Rosette and 10 times larger for the Orion nebula. In both cases the total mass (stars and diffuse matter) was found of the order of 10^5 solar masses. However IKHSANOV (1961) obtains rather smaller masses (10^3 solar masses) for nebulae connected with O-associations.

It is well known that the spiral arms of our Galaxy in the neighbourhood of the Sun suggested by the space distribution of O-associations agree fairly well with the general run of spiral arms found from the radio-observations of interstellar hydrogen. However BECKER (1961) and others (BEER, 1961) found that when the photometric distances of clusters and associations are used, plotting their positions on the same diagram on which the radiastronomically observed HI regions have been plotted, the clusters seem to lie in the spaces between the spiral arms as defined by the HI regions.

On the other side, DIETER (1960) has shown that for 31 out of 40 O-associations investigated by him there is a coincidence with the positions and velocities of the H-regions; only the distances disagree. It seems therefore very probable that the discrepancy is due entirely to the method of determining the distances of the radioastronomically observed H-regions, which is based upon a simple model of circular motion of the hydrogen clouds.

As a matter of fact, from the motion of 23 O-associations, PISMIS (1960) found a general outward motion relative to the center of the Galaxy of the spiral arms near the Sun; although very small (4 km/s) this motion seems well above the observational errors. It may, thus, be concluded that actually the O-associations coincide with regions of high condensation of the interstellar hydrogen.

REFERENCES

ALTER, G., RUPRECHT, J. and VANYSEK, V.: 1958, *Catalogue of Star Clusters and Associations*, Praga Ac. Sc.
AMBARTSUMIAN, V. A.: 1938, *Ann. Lening. St. Univ.*, n. 22, 19.
AMBARTSUMIAN, V. A.: 1947, *Stellar Evol. and Astroph.*, in *Arm. Ak. Dokl.*
AMBARTSUMIAN, V. A. : 1949, *R. A. J.*, **26**, 3.
AMBARTSUMIAN, V. A.: 1954, 5me *Coll. Liège*, 293.
BAPPU, M. K., CHANDRA, S., SANWAL, N. B. and SINVHAL, S. D.: 1962, *M. N.*, **123**, 521.
BECKER, W.: 1951, *Z. f. Ap.*, **29**, 66.
BECKER, W.: 1961, *Z. f. Ap.*, **51**, 151.
BEER, A.: 1961, *M. N.*, **123**, 191.
BIDELMAN, W. P.: 1943, *Ap. J.*, **99**, 61.
BLAAUW, A.: 1946, *Gron. P.*, n. 52.
BLAAUW, A.: 1952*a*, *B. A. N.*, **11**, 405.
BLAAUW, A.: 1952*b*, *B. A. N.*, **11**, 414.
BLAAUW, A.: 1958, *Stellar Populations* (Vatican Obs.), 105.
BLAAUW, A. and MORGAN, W. W.: 1953, *Ap. J.*, **117**, 256.
BLAAUW, A. and MORGAN, W. W.: 1954, *Ap. J.*, **119**, 625.
BLANCO, V. M. and GRANT, G.: 1959, *P. A. S. P.*, **71**, 194.
BOK, B. J.: 1934, *Harv. C.*, n. 384.
BURBIDGE, G. R.: 1960, *The Formation of Stars*, Berlin.
CHANDRASEKHAR, S.: 1942, *Stellar Dynamics*, Chicago.
CRAWFORD, D. L.: 1958, *Ap. J.*, **128**, 185.
DIECKVOS, W.: 1953, *Naturwiss.*, **40**, 505.
DELHAYE, J. and BLAAUW, A.: 1953, *B. A. N.*, **12**, 72.
DIETER, N. H.: 1960, *Ap. J.*, **132**, 49.
FEAST, M. W.: 1960, *M. N.*, **122**, 1.
FERNIE, J. D.: 1962, *Ap. J.*, **135**, 298.
HERBIG, G. H. and MENDOZA, E. E.: 1960, *Ton. B.*, **19**, 21.

HERBIG, G. H.: 1962, *Ap. J.*, **135**, 736.
HILTNER, W. A. and JOHNSON, H. L.: 1956, *Ap. J.*, **124**, 367.
IKHSANOV, R. N.: 1961, *R. A. J.*, **37**, 613 (Am. Ed.).
JOHNSON, H. L., 1957, *Ap. J.*, **126**, 134.
JOHNSON, H. L. and HILTNER, W. A.: 1956, *Ap. J.*, **123**, 267.
JOHNSON, H. L., HOAG, A. A., IRIARTE, B., MITCHELL, R. I. and HALLAM, K. L.: 1961, *Low. B.*, **5**, 133 (n. 113).
JOY, A. H.: 1945, *Ap. J.*, **102**, 168.
JOY, A. H.: 1949, *Ap. J.*, **110**, 424.
KHOLOPOV, P. N.: 1959*a*, *R. A. J.*, **36**, 291 (Am. Ed.).
KHOLOPOV, P. N.: 1959*b*, *R. A. J.*, **36**, 425 (Am. Ed.).
KOPYLOV, I. M.: 1953, *USSR Ak. Nauk Dokl.*, **90**, 975.
KOPYLOV, I. M.: 1958: *I. A. U. Symp.*, n. 10, 41.
KOURGANOFF, V.: 1952, *Astr. N. L.*, n. 64.
LINDBLAD, B.: 1959, *Enc. of Physics*, **53**, 21.
MARKARJAN, B. E.: 1952, *Arm. Ak. Dokl.*, **15**, 11.
MARKARJAN, B. E.: 1953, *Bjur. Sob.*, n. 11, 3.
McNAMARA, D. and LARSSON, H. J.: 1962, *Ap. J.*, **135**, 748.
MENDOZA, E. E.: 1958, *Ap. J.*, **128**, 207.
MENON, T. K.: 1958, *Ap. J.*, **127**, 28.
MENON, T. K.: 1962, *Ap. J.*, **135**, 394.
MORGAN, W. W., WHITFORD, A. E. and CODE, A. D.: 1953, *Ap. J.*, **118**, 318.
MÜNCH, G.: 1958, *3-rd Symp. on Gas Dynamics*, in *Rev. Mod. Phys.*, **30**, 1035.
MÜNCH, G. and FLATHER, E.: 1957, *P. A. S. P.*, **69**, 142.
OPIK, E.: 1953, *Ir. A. J.*, **2**, 219.
PARENAGO, P. P.: 1954, *Stern. Trudy*, 25.
PETRIE, R. M.: 1958, *M. N.*, **118**, 80.
PIK SIN THE: 1960, *Ap. J.*, **132**, 40.
PIK SIN THE and BLANCO, V. M.: 1960, *A. J.*, **65**, 57.
PISMIS, P.: 1960, *Ton. B.*, **19**, 3.
ROBERTS, M. S.: 1957, 8me *Coll. Liège*, 76.
SALPETER, E. E.: 1955, *Ap. J.*, **121**, 161.
SANDAGE, A.: 1957, *Ap. J.*, **125**, 422.
SANDAGE, A.: 1958*a*, *Stellar Populations* (Vatican Obs.), 75.
SANDAGE, A.: 1958*b*, *Stellar Populations* (Vatican Obs.), 129.
SAVEDOFF, M. P.: 1956, *Ap. J.*, **124**, 533.
SCHMIDT, K. H.: 1959, *A. N.*, **284**, 76.
SCHMIDT, M.: 1959, *Ap. J.*, **129**, 243.
SCHMIDT-KALER, T.: 1961, *Z. f. Ap.*, **53**, 28.
SHARPLESS, S.: 1954*a*, *Ap. J.*, **119**, 200.
SHARPLESS, S.: 1954*b*, *Ap. J.*, **119**, 334.
SHARPLESS, S.: 1958, *P. A. S. P.*, **70**, 392.
SHTEINS, K. A. and ABELE, M. K.: 1958, *R. A. J.*, **35**, 70 (Am. Ed.).
SPITZER, L.: 1940, *M. N.*, **100**, 396.
SPITZER, L.: 1958, *Ap. J.*, **127**, 17.
STRAND, K. A.: 1958, *Ap. J.*, **128**, 14.
STRÖMGREN, B.: 1958, *Obs.*, **78**, 137.
TREANOR, P. J., S. J.: 1960, *M. N.*, **121**, 503.
UNDERHILL, A. B.: 1960, *Ap. J.*, **131**, 524.
VAN DEN BERGH, S.: 1961, *Ap. J.*, **134**, 553.

van den Bergh, S. and Sher, D.: 1960, *Tor. P.*, **2**, 201.
Varsawsky, C. M.: 1960, *Ap. J.*, **132**, 354.
Walker, M. F.: 1956, *Ap. J. Suppl.*, **2**, 365.
Walker, M. F.: 1957, *Ap. J.*, **125**, 636.
Walker, M. F.: 1959, *Ap. J.*, **130**, 57.
Walker, M. F.: 1961*a*, *Ap. J.*, **133**, 108.
Walker, M. F.: 1961*b*, *Ap. J.*, **133**, 438.
Wilson, O. C., Münch, G., Flather, E. M. and Coffeen, M. F.: 1959, *Ap. J. Suppl.*, **4**, 199.
v. d. R. Woolley, R. and Eggen, O. J.: 1958, *Obs.*, **78**, 149.

Nebular Variables and Related Objects.

L. Rosino
Osservatorio Astrofisico dell'Università - Padova

Introduction.

The object of this colloquium is the study of a class of irregular variables, normally associated with diffuse nebulae, whose importance from the point of view of stellar evolution has been stressed in recent years. These variables have been called Orion variables, *T Tauri* stars, *RW Aurigae* variables and so on, each name being related to some of their characteristics. For instance, the name « *Orion variables* » comes from the fact that hundreds of variables of this type are found in the Orion Nebula, while the name « *T Tauri* » is connected with the emission features of their spectra.

Henceforward I will call *nebular variables* such stars that satisfy the following two conditions:

1) They are found in stellar associations of clusters, inside or near diffuse nebulosities.

2) They have irregular variations of brightness. The light fluctuations are slow or rapid, continuous or interrupted by periods of constant luminosity.

No restrictions are made with regard to spectra and luminosities. The presence of emission lines is not considered a necessary condition. However, it is a fact that stars with emission lines and dwarf spectra later than *F*, constitute a high percentage (at least 40%) in associations of nebular variables. These stars are more properly called *T Tauri stars* or *T Tauri-like stars* (Herbig, 1962*a*). They form a subtype in the general class of nebular variables, which also includes the *nonemission variables* and the *flare stars.*

1. – Frequency and distribution of nebular variables.

The systematic survey of nebular variables was started recently. Before 1940, with the exception of the Orion Nebula, nebular variables were found more or less by chance. The search is now conducted in two different ways.

a) By intercomparison of photographs centered in suspected regions. The nebular variables are identified by their light variations.

b) By examining objective prism or slitless spectra of stars in obscured regions. In this case only *T Tauri* stars can be identified by their emission lines, but since they are normally associated with nonemission nebular variables their presence is an indicator of the whole association.

In general, when diffuse nebulae with dark matter are thoroughly examined with the proper instruments and techniques, nebular variables are always found. This is the first certain result of the investigations carried out in the last few years. The presence of dust with the gas seems to be a necessary condition.

By looking in the Palomar Atlas it is possible to find at least 60 ÷ 70 diffuse nebulae which are likely to contain nebular variables and the total number

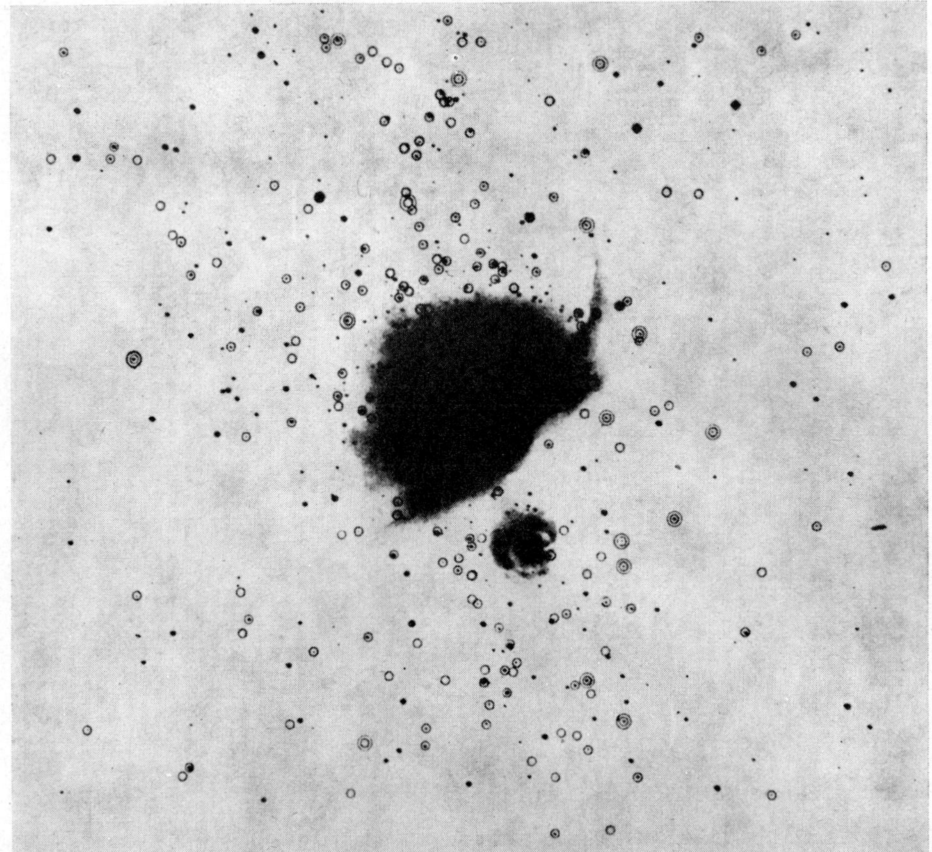

Fig. 1. – Variables around the Orion Nebula. Single circles indicate normal nebular variables, double circles flare stars.

increases to $80 \div 90$ if fields south of $-30°$ are also taken into consideration. Of these regions less than 40 have been investigated. Observations of the other fields are very desirable.

The number of nebular variables found in different nebulosities varies largely from case to case. All other circumstances being equal, this number depends on the following factors:

a) Total number of stars involved in the nebulosity.

b) Distance of the nebula.

c) Absorption inside the nebulosity.

d) Brightness of the nebula.

The influence of factor *a*) is clear. Where the number of stars associated is small, we can not expect to find many variables. Factor *b*) is related to the instrumental possibility of observing stars of a given magnitude, considering

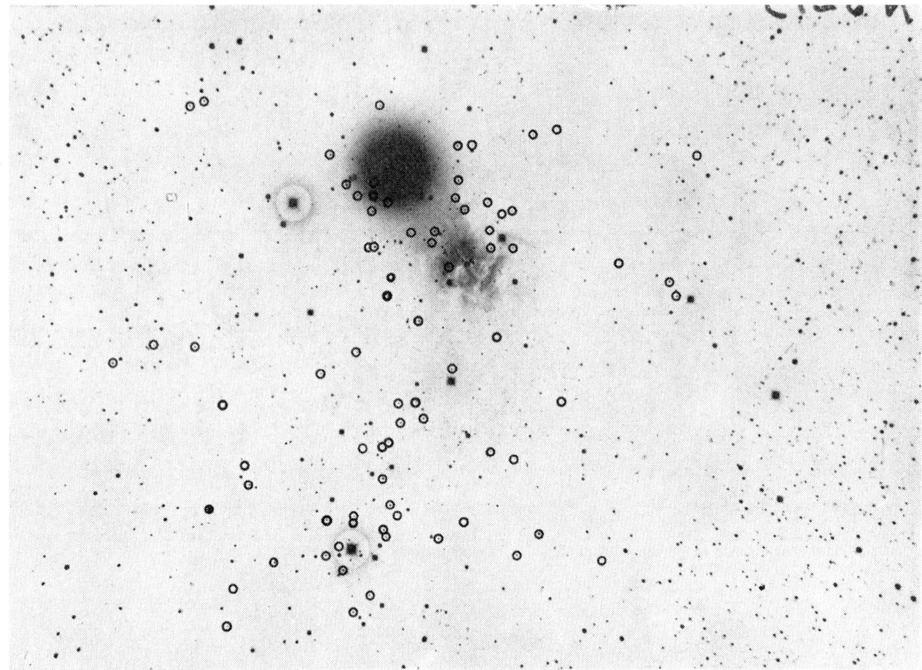

Fig. 2. – Distribution of nebular variables in NGC 2264. The bright star is *S Mon.*

that many nebular variables are objects of low luminosity. Factor *c*) reduces still more the apparent magnitudes of stars buried in dark nebulosities. It is a fact that variable stars in nebulae are found more frequently near the exciting or illuminating stars of higher luminosity, where the dust has been wiped out by radiation, or in the fringes of nebulae where the absorption is falling off rapidly. The extensive use of infrared techniques may reduce, but not eliminate the absorption effects. Finally the surface brightness of nebulae, acting on the contrast, influences the discovery of variable stars. This effect is also reduced by using infrared material or by working in wavelengths where there are no strong emission lines.

When these factors are taken into account the different richness and distribution of known nebular variables can be, in part at least, explained. Figures 1 and 2 show the clustering of variable stars (circles) around the Orion Nebula

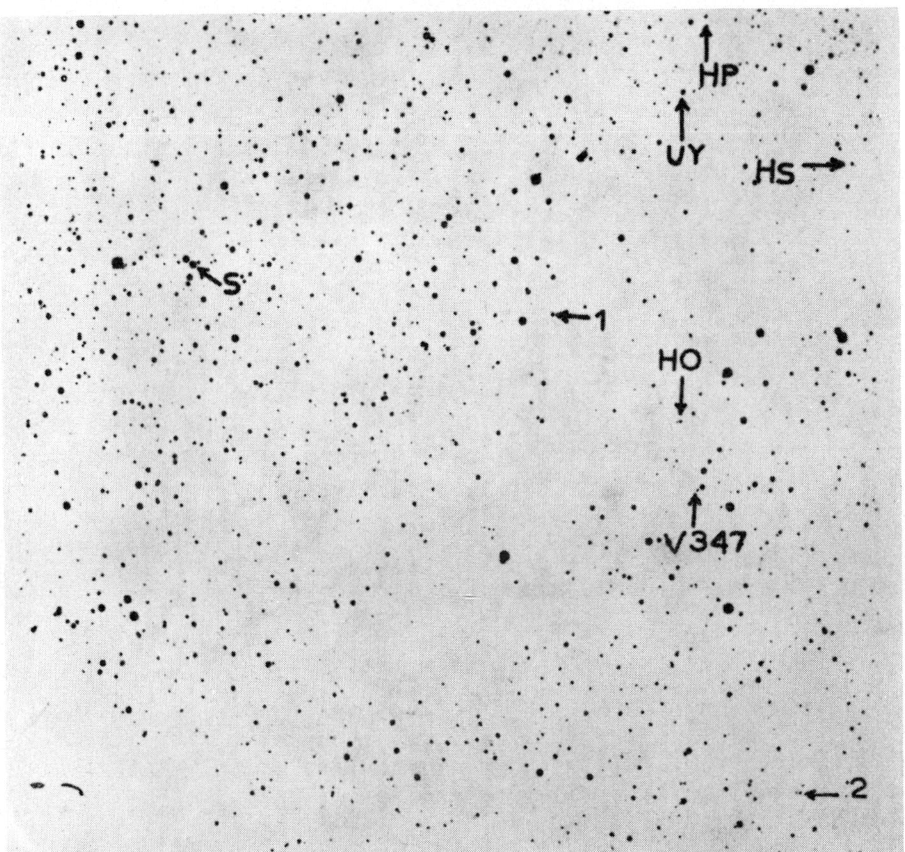

Fig. 3. – Nebular variables at the borders of a dark lane in Orion.

and *S Mon* (NGC 2264). Figure 3 shows how the nebular variable stars are distributed at the borders of dark lanes (Orion). Other regions in which remarkable relations are found between the distribution of dark matter and position of variable stars are NGC 1999, NGC 7023, IC 434, Messier 8, Messier 16, IC 5146 and several others.

2. – Light cur.es.

Although the light curves of nebular variables have certain characteristics in common, it is very difficult to classify them. Apparently the brightness fluctuations of these stars are extremely irregular. Not only is it impossible

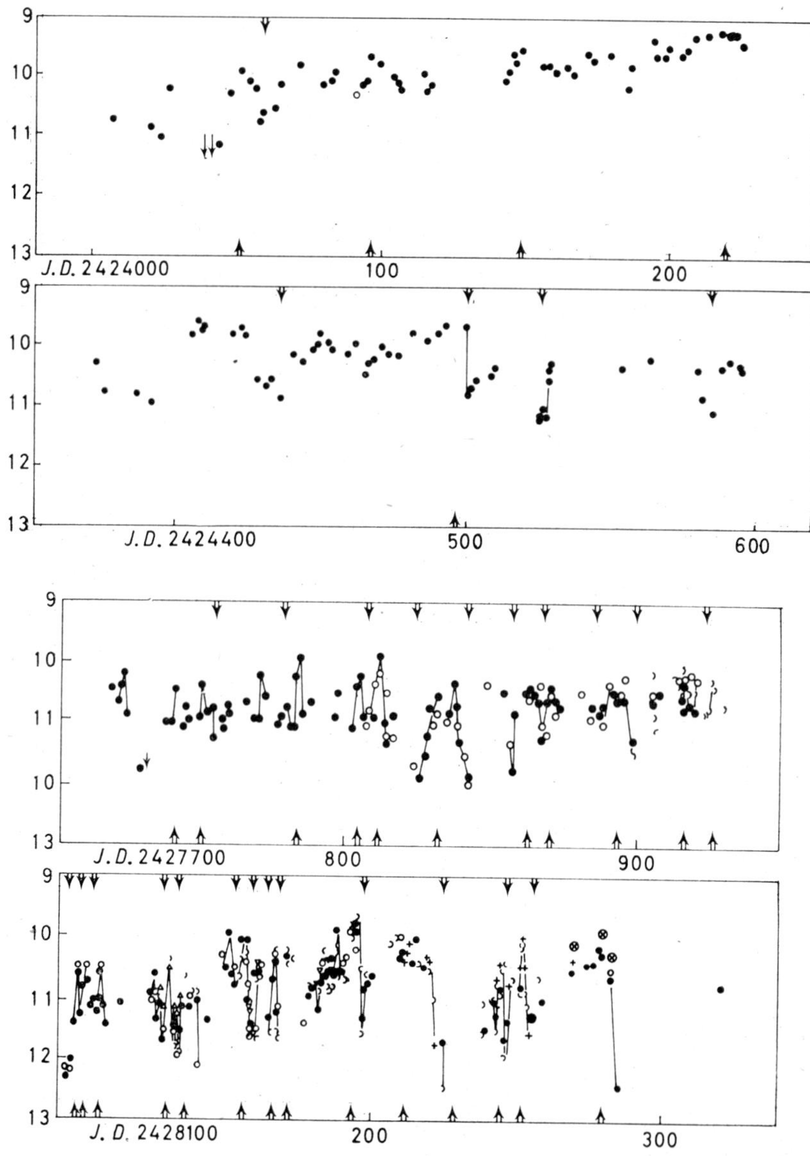

Fig. 4. – Light curve of *RW Aur* in two different epochs (KOLOPOW, 1955).

to predict the light variations for even a short time ahead, but it also frequently happens that a star may suddenly change its «normal» behaviour. A star, for instance, which has for long time had the characteristics of slow irregular variations, may suddenly exhibit very rapid fluctuations of brightness. Remarkable examples are presented by *RW Aurigae* (KOLOPOW (1955), Figure 4) and by *T Orionis* (LACCHINI (1956), Figure 5). The light curves show better than any description what we may expect to happen with nebular variables.

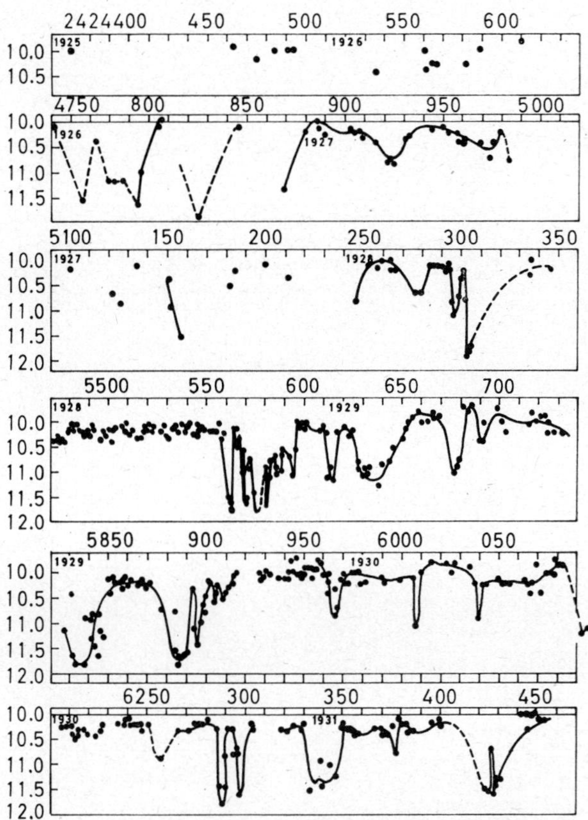

Fig. 5. – Light curve of *T Ori* (LACCHINI, 1956).

An analysis of the photometric material presently available suggests the following classification of the nebular variables from their light curves, due to Parenago:

1) Brightness more frequently near maximum, with irregular deep minima.

2) Irregular variations around an intermediate magnitude.

3) Brightness near minimum with sudden sharp maxima.

4) Erratic fluctuations, without preference for any level, in the range of variation.

A complication in the classification arises from the fact that it is not unusual for a nebular variable to pass from one class to another. Statistical analysis of the light curves of nebular variables have been tried, but without significant results. Apparently there is no relation between the form of the light curve and the absolute magnitude. The presence or absence of emission lines in the spectrum does not seem to influence the characteristics of the light curves. However, recently GÖTZ (1961) has found a relation between the intensity of H_α and the rapidity of the light fluctuations, in the sense that H_α is fainter when rapid variations occur.

The comparison of the light curve of nebular variables in different associations also fails to yield results of any significance. Variable stars in the Orion association show more or less the same irregular fluctuations as the variables in other nebulae. However, there are some slight indications of changes from one nebula to another. For example the variables in NGC 2264 seem to be, in general, more unquiet than those of Orion (ROSINO *et al.*, 1957), while the nebular variables of NGC 7023 present slow variations with superposed rapid changes of brightness of minor amplitude (ROSINO and ROMANO, 1962).

No significant differences have been found in the mean amplitude of nebular variables in Orion ($A = 1^m60$), NGC 2264 ($A = 1^m50$), Taurus ($A = 1^m40$) and NGC 7023 ($A = 1^m70$). The largest amplitudes do not exceed normally 4^m.

A considerable number of the nebular variables are *flare stars*, which belong, in the great majority, to class III. However, their light curves and general characteristics are so peculiar that we shall later devote a special paragraph to them.

3. – Spectra.

The first detailed studies of the spectra of nebular variables in diffuse nebulae are due to A. JOY (1945) who in a classical paper introduced the *T Tauri* type. Later on, extensive spectroscopic researches on emission line nebular variables were made by HARO and ass. (1953, 1954*a*, 1954*b*, 1955) and particularly by Herbig. In a coarse classification the nebular variables can be divided, from the spectroscopic point of view, in three groups:

3·1. *Early type stars with emission lines* (*T Ori*, *V* 380 *Ori*, *R CrA*, *RR Tau*, *etc.*). – The stars of this class have been studied by HERBIG in an important paper (1960). Twenty-six stars, all involved in nebulosities, were examined. These stars are normally the brightest of the nebular association.

They excite or illuminate the diffuse matter around them. The spectra can be divided into the following subclasses:

a) B- or early A-type with hydrogen emission lines, superposed shell spectrum. An example is given by the central star of NGC 7023, HD 200775.

b) Same characteristics, but stronger emission lines, sometimes H and K in emission, *P Cygni* structure (*Z CMa*, BD+61° 154, HD 250550).

c) Stars with composite spectrum. Strong continuum and broad absorption lines of early type, hydrogen in emission, strong emission lines of FeII, TiII, CaII, ScII like advanced *T Tauri* stars.

3·2. *T Tauri and T Tauri-like stars.* – Herbig has recently given (1962*a*) some spectroscopic criteria for the classification of *T Tauri* stars and related objects:

a) The hydrogen lines and *H* and *K* lines of CaII are in emission. Also usually present are the emission lines of FeII, TiII, FeI, CaI and low excitation metallic atoms and the forbidden lines of SII 4068, 4076 and probably 6717 and 6731.

b) Presence of the fluorescent lines 4063 and 4132 of FeI, which are not found in the spectra of other stars.

c) Abnormal strength of Li 6707 in absorption.

Herbig makes a distinction between the *T Tauri* stars which obey the criteria *a*), *b*), *c*) and the *T Tauri-like* stars, whose spectra have conspicuous emission at H_α and *H* and *K* only.

The *T Tauri* stars have an absorption spectrum from F to M. Sometimes, however, the spectrum is veiled by a strong continuum which obliterates the absorption lines. This continuous veiling is particularly strong when the emission lines are intense. When present the absorption lines are shallow and diffuse. They are broader than usually in stars of the same spectral type.

The spectra of many *T Tauri* stars have an excess of ultraviolet radiation, which contributes considerably to the total radiation of these stars. Very little is known on the origin of this ultraviolet continuum. Theories have of course been proposed (free-free transitions, blurring of high order lines of the Balmer series, presence of photospheric « active » areas). The question is still open. However, there is a tendency to admit that outbursts, more or less intense, may be at the origin of this nonthermal violet and ultraviolet radiation.

T Tauri stars and related objects have another peculiarity: an abundance of Lithium 50 to 400 times higher than in the Sun and comparable with the abundance found in stone meteorites. This may be due to the fact that nebular variables are still in the contracting stage of their evolution and that nuclear reactions have not yet started.

3·3. *Nonemission variables.* – Very little is known on the spectra of these stars which represent 60 % or even more of the total number of nebular variables. The spectral types are G, K, M. It is doubtful whether emission lines are permanently absent. Some variables (like *EZ Ori*) that normally do not present emission lines show at times conspicuous emission. It is rather difficult to obtain good spectra of these stars because of their intrinsic faintness. However, it would be highly desirable to have a better knowledge of their spectral characteristics, particularly for the presence of Li 6707 and the comparison with the spectra of normal stars.

4. – Flare stars.

A class of extremely high interest among the nebular variables is that of the « flare stars » or « flash stars » or « rapid variables ». We shall adopt the first name. The flare stars stay normally at minimum, showing small fluctuations of brightness. Their absolute magnitude is low, +8 or less, with a large dispersion between +5 to +11. When the outburst occurs there is a sudden and extremely rapid increase of luminosity and the variable reaches a sharp maximum from which it then returns, more slowly, to minimum. The time to go from minimum to maximum and then again to minimum varies from some minutes to a few hours. The amplitude ranges from $0^m 5$ to $3^m 4$ or even more. We shall not include in the class, for the moment such variables which have small outbursts with an amplitude less than $0^m 5$.

The spectral types of most flare stars range from dK to late dM, with or without hydrogen and CaII emission lines in the quiescent state. When emission lines are present they are comparatively weak. During an outburst the emission lines are strengthened and a strong continuous emission is noticed in the violet and ultraviolet.

Flare stars in *T* associations raise some problems which we shall discuss here briefly:

a) *Relative frequency.* Flare stars have been found in all nebulous regions rich in variables which have so far been systematically surveyed. We quote here: *Taurus region* (RA from 4 h 10 min to 4 h 45 min; D. from +22° to +30°): nebular variables 65, flare stars 12, ratio 18 %. *NGC* 2264 *in Monoceros*: nebular variables 116, flare stars 11, ratio 10 %. *Orion region* (3°, 5 around the *Trapezium*): nebular variables 414, flare stars 40, ratio 10 %. These figures, however, represent a lower limit and the ratio of flare stars to nebular variables will certainly increase in the near future with more systematic searchings. We know, for instance, that HARO, CHAVIRA and MENDOZA (1960) have found 55 flare stars, not yet published, in Orion. If there are no coincidences with the 40 already known, the ratio will reach 20 % instead of 10 %. The conclusion is that the flare phenomenon must be taken seriously into account in any theory concerning the nebular variables.

b) *Frequency of the outbursts in flare stars.* This problem is rather difficult, because two consecutive flares have never been observed in the same star, although several variables have shown more than one. An estimate of the frequency can be gained by the average time between the observation of the first and second flare in a certain star. Taking a mean for all flare variables we conclude that the flare frequency per variable is relatively low, not more than one flare in any 2÷3 days.

c) *Duration of the outbursts.* There seems to be some differences in the mean duration of the flares in different associations. For instance in Taurus and in NGC 2264 (spectral type of the flare stars at minimum M2-M5) we found a duration of 30÷40 minutes. In Orion (K5-K6) more than two hours. HARO and CHAVIRA (1955) have tried to correlate the total time of an outburst with the spectral type at minimum, concluding that there is a definite relation between spectrum and rate of variation, in which also the *UV Ceti* variables are included. According to this relation the rapidity of an outburst increases with increasing spectral type.

5. – Peculiar objects in diffuse nebulae.

Peculiar objects whose nature is somewhat difficult to define are found near or inside diffuse nebulae where associations of variable stars exist. These objects can be classified in the following categories:

a) *Infrared stars.* Are visible in the deep red or infrared. Strong selective absorption or low temperature may be the reason for the extremely high infrared color index.

b) *Infrared stars embedded in minute nebulosities.* (Examples: Haro objects 7*a* and 8*a*, near the Orion nebula). The difference from the preceding objects is that in the infrared a very small nebulosity becomes visible around the stars. Some of these stars embedded in minute nebulosities are variable. Sometimes the infrared stars appear simply nebulous, just slightly different from nearby stars (*OW Ori*).

c) *Small reflection nebulosities without stars.* (Examples: Haro objects 5*a*, 6*a*). They have a continuous spectrum, strong in infrared. No stars are visible inside.

d) *Herbig-Haro objects.* Small nebulous objects whose spectrum shows the following emission features: strong H_α and fairly strong hydrogen lines: [OI] 6300, 6363; [OII] 3727-3729; strong [SII] 6717-6731 and 4068-4076. Accompanying this nebular spectrum is a weak continuum of advanced spectral type, with CaII and FeII in emission, veiled by the nebular spectrum. Probably related with the Herbig-Haro objects are some roundish minute nebulosities with H_α and [SII] 6717-6731 in emission. The most famous Herbig-Haro objects are No. 1, 2 and 3 near NGC 1999. No. 2 represents an unique case: an assembly of semistellar nebulous knots, clearly connected by filaments of gas. In 1952 two of these knots were found to have considerably increased in luminosity. The general opinion was that we were observing the formation of stars from protostars. But it was shown later by HARO and MINKOWSKI (1960) that none of the condensations contains stars brighter than $V = 21$. The Herbig-Haro object no. 1 is also made of small stellar-like condensations without visible star inside. The masses of these condensations are $10^4 \div 10^5$ times smaller than the mass of the Sun, the electron density being $1.3 \cdot 10^4$ to $2.7 \cdot 10^4$ cm^{-3}.

Other peculiar objects related to the nebular variables are the so-called variable nebulae, the most famous of which is NGC 2261, a comet shaped nebula containing in the head the variable *R Monocerotis*. Five other nebulae of this kind have been indicated by HERBIG (1962*b*) in a recent paper. We mention here NGC 1554-5, associated with *T Tauri* and NGC 6729 with *R CrA* These nebulae evidently are illuminated by the variable stars associated with them. However, the spectrum and the color do not match those of the illuminating stars and the variations in brightness do not follow those of the responsible variables. The physical phenomena in the variable nebulae are certainly much more complicated than pure illumination.

6. – H-R diagram.

It is well known that the degree of evolution of a system of stars is well represented by its H-R diagram, which correlates the absolute magnitudes with the spectral types and colors of the stars. At the beginning of their nuclear evolution the stars are distributed over a narrow line which is called *zero-age main sequence*.

Now, before any diagram of stars involved in nebulosities was drawn, it was apparent that something was wrong. Double *T Tauri* stars, studied by JOY (1945), showed a difference ΔS in the spectral class not in agreement with that calculated from the difference Δm of the magnitudes, on the assumption that both components were lying on the zero-age main sequence. However, it was only in 1950, when PARENAGO (1950, 1954) first published his color-

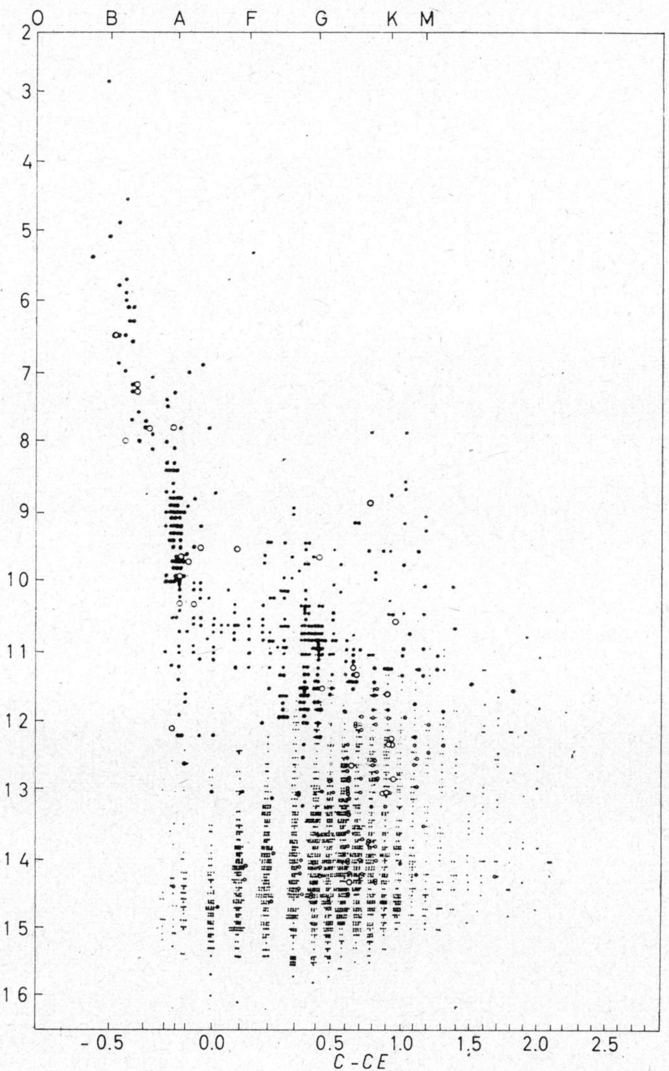

Fig. 6. – H-R diagram of stars in the Orion association (PARENAGO, 1954).

magnitude diagram of the Orion nebula stars, that the high dispersion of the faintest members of the association around the zero-age main sequence became evident (Figure 6). Even by taking account of all the disturbing factors (spurious members, color excess, variability, duplicity, etc.) the diagram did not change its essential form. Some years later, M. WALKER (1956) obtained the H-R diagram of NGC 2264 (Figure 7) and again the deviation of the faintest stars to the right of the zero-age main sequence was evident. The same effect was found by WALKER (1957, 1959, 1961) in NGC 6530, IC 5146 and NGC 6611.

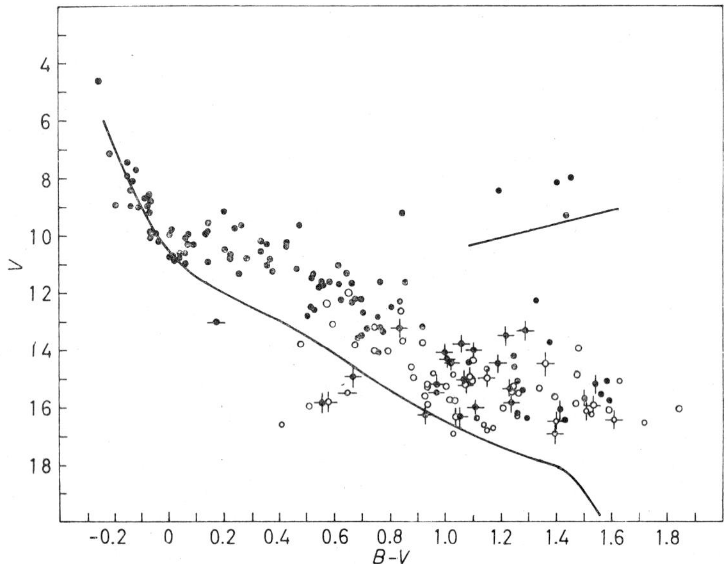

Fig. 7. – Color–magnitude diagram of NGC 2264 (WALKER, 1956).

These diagrams show a normal main sequence from O5 to A0, which abruptly ends near A0. Then the representative points deviate to the right side, running more or less parallel to the zero-age main sequence with a tendency to diverge at the bottom.

The most obvious interpretation of these diagrams is that we have to deal with stars still under gravitational contraction. We are entering now in the last part of this colloquium: the interpretation of the facts hitherto exposed.

7. – Interpretation and conclusions.

Phenomena of star evolution are so complicated that it would be arduous to explain in a simple scheme all of the facts. This is particularly true for the nebular variables. We have a great deal of observational data, but sometimes incomplete and in apparent conflict.

A theory of nebular variables should explain the following facts:

a) That these stars are found in associations in or near nebulosities rich in dark matter.

b) Peculiar structure of the H-R diagram.

c) Presence of emission and nonemission variables in the association.

d) Peculiarities of the spectra and colors.

e) Irregular light variability and its character.

f) Presence of a high proportion of flare stars.

g) Existence of the Herbig-Haro and other peculiar objects.

Rejecting the old hypothesis that nebular associations were simply field stars trapped by nebulae in their random motions, which can hardly be maintained in the light of the present knowledge, the best assumption that immediately explains points *a*) and *b*) is that of the gravitational contraction. The stars of nebular associations are very young stars still inside diffuse clouds of dust and gas where they were formed ($10^6 \div 10^7$) years ago. These stars are now moving towards the main sequence, deriving energy by gravitational contraction. Theoretical researches give the paths of evolution of contracting stars and the contraction times from a starting configuration to the zero-age main sequence. These times vary from $4 \cdot 10^5$ years for a star of terminal type A1V (mass 2.29) to $7 \cdot 10^8$ years for a star of mass 0.65 which terminates at M1. The general lines of the H-R diagrams, with most of the points to the right of the zero-age main sequence, are therefore explained by assuming that the association is not older than $4 \cdot 10^5$ years, which is generally accepted. As Herbig has recently pointed out, it is more difficult to explain why the points are not progressively diverging from the conjunction point with the main sequence, as they should be if the rates of contraction were to get slower for the faintest stars. HERBIG (1962*a*) therefore assumes that the stars of an association may be formed in different epochs and that the process of formation of new stars stops when OB stars are sufficiently near the main sequence to dissipate or evaporate the dusts. The hypothesis of the simultaneous formation does not seem to correspond with the observed facts.

The contraction hypothesis explains also points *c*) and *d*) if we consider that for the conservation of angular momentum, the rotation velocity of newly formed stars increases during contraction. A high rotational velocity is indicated by the shallow and broad absorption lines. Diffuse matter ejected by the star in the first stages of its contraction or residua of the material from which the star was formed may explain the shell spectrum and the forbidden emission lines. In the path of evolution the Herbig-Haro objects should constitute the first step, followed by the variables still involved in minute nebulosities, and then from the *T Tauri* stars and the nonemission variables.

Contraction, however, does not account for the facts connected with variability (light curves, flares, etc.) and for the presence of emission lines of high excitation and ultraviolet excess. To gives reasons for these facts, several hypotheses have been advanced. It was thought that variability might be the effect of occulting dark clouds passing in front of stars (deep minima in the light curves) and interacting with them. The effect of infalling material and the evaporation of dust in the chromosphere would account for the emission lines and outbursts. These « external » causes of variability, however, have lost value in comparison with internal actions. Several years ago AMBARTZUMIAN (1954, 1957) proposed a mechanism that, although obscure in some fundamental points, explains most of the facts of nebular variables and related objects. Ambartzumian's idea was that young stars contain in the deep interior some quantities of a not better specified « prestellar » material. When this material is bronght from the internal to the superficial parts of the star there is a sudden liberation of « nonthermic » energy, which may be directly emitted or transformed into thermal energy. Supposing the convective motions are intense and the prestellar material abundant, it may happen that energy of this type is liberated near the photospheric layers but below them. In this case the energy is converted in thermal radiation and we simply see the erratic changes of luminosity which characterize the light curves of nebular variables. If the explosive material reaches the atmosphere then we have a sudden flash of light (flare), while part of the energy is spent to ionize the external nebular layers from where the emission lines are emitted. We must further suppose that, with increasing evolution, the quantity of pre-stellar material rising to the surface of the star is less and less. At the end we will have only variations of small amplitude interrupted by sudden bursts when pre-stellar material rises from the photosphere and explodes over it. In this picture Herbig-Haro objects are prestellar objects which may give origin, later, to diffuse nebulae and stars.

The weak point of the Ambartzumian's hypothesis is the uncertainty on the real nature of the « pre-stellar » material. Any theory that uses only known laws of physics which can account for sudden and local liberation of nonthermal energy near the photosphere, evidently will give a good explanation of the facts. Magnetic fields, nuclear reactions at the surface and so on have been advanced to substitute the rather indeterminate mechanism of Ambartzumian.

Finally I should like to propose some observations which would be of some value in the study of the nebular variables and related objects:

a) Determination of the position in the H-R diagram of the following types of nebular variables: *T Tauri* with UV emission, *T Tauri*, nonemission variables, flare stars.

b) Study of the small fluctuations of nebular variables. It may be possible that these stars show rapid fluctuations of small amplitude like those observed in *SS Cygni* variables at minimum.

c) Study of the light curves in several colors.

d) Searching of nebular variable stars in new fields.

e) Extensive search of flare stars.

f) Study of the luminosity function of nebular variables.

g) Searching of new Herbig-Haro objects in different stages of evolution and minute emission nebulae.

Some of these programs are presently in progress at Asiago.

REFERENCES

AMBARTZUMIAN, V. A.: 1954, *Comm. Burakan Obs.*, **13**.

AMBARTZUMIAN, V. A.: 1957, in *Non-Stable Stars* (G. H. HERBIG, Ed.), p. 177.

GÖTZ, W.: 1961, *Veröff. Sonneberg*, **5**, 2.

HARO, G.: 1953, *Ap. J.*, **117**, 73.

HARO, G. and TERRAZZAS, L. R.: 1954, *Bol. Tonant.*, **10**, 3.

HARO, G.: 1954, *Bol. Tonant.*, **11**, 11.

HARO, G. and CHAVIRA, E.: 1955, *Bol. Tonant.*, **12**, 3.

HARO, G. and HERBIG, G. H.: 1955, *Bol. Tonant.*, **12**, 33.

HARO, G., CHAVIRA, E. and MENDOZA, E.: 1960, *A. J.*, **65**, 9.

HARO, G. and MINKOWSKI, R.: 1960, *A. J.*, **65**, 490.

HERBIG, G. H.: 1957, in « *Non-Stable Stars* », p. 3.

HERBIG, G. H.: 1960, *Ap. J. Suppl. Series*, **4**, 337.

HERBIG, G. H.: 1962*a*, in *Advances in Astronomy and Astrophysics*, Vol. I (Z. KOPAL, Ed.), pp. 47-103.

HERBIG, G. H.: 1962*b*, *Ap. J.*, **133**, 338.

JOY, A. H.: 1945, *Ap. J.*, **102**, 168.

KOLOPOW, P. N.: 1955, *Per. Zvezdy*, **10**, 390.

LACCHINI, G. B.: 1956, *Mem. Soc. Astr. It.*, **27**, 383.

PARENAGO, P. P.: 1950, *Per. Zvezdy*, **7**, 169.

PARENAGO, P. P.: 1954, *Publ. Sternberg Astr. Inst.*, **25**.

ROSINO, L., GRUBISSICH, C. and MAFFEI, P.: 1957, *Co. Asiago*, n. 82.

ROSINO, L. and ROMANO, G.: 1962, *Co. Asiago*, n. 127.

WALKER, M. F.: 1956, *Ap. J. Suppl. Series.*, **2**, 365.

WALKER, M. F.: 1957, *Ap. J.*, **125**, 636.

WALKER, M. F.: 1959, *Ap. J.*, **130**, 57.

WALKER, M. F.: 1961, *Ap. J.*, **133**, 438.

Notes on Cosmic Radiation.

M. M. SHAPIRO (*)

Nucleonics Division, U. S. Naval Research Laboratory - Washington, D. C.
Weizmann Institute of Science - Rehovoth

Introduction.

The Galaxy may be considered as a quasi-stationary reservoir of magnetically confined cosmic rays randomized in direction by turbulent clouds of plasma. Information about the galactic radiation comes not only from cosmic-ray experiments *per se*, but also from radioastronomical observations. The former are carried out mainly with large balloons near the top of the atmosphere; however, satellites and space probes have begun to play an important role. In addition, much effort is being devoted to studies of extensive air showers deep in the atmosphere. As for the radioastronomical evidence about cosmic rays in remote region of the galaxy, this will be treated in Section 3.

In this Section we shall sketch some salient features of the cosmic radiation incident upon the Earth, *e.g.*, its composition, energy and isotropy. Then we shall discuss implications of the Li-Be-B anomaly and the helium isotope abundances.

1. – Some salient features of the cosmic radiation.

1·1. *Composition.* – Of the various primary components, only the nuclear component is well established. Table I gives the relative abundances of the atoms at thermal energies and at cosmic-ray energies. It can be seen that the ratio of cosmic-ray hydrogen to helium is similar to that for the thermal abundances of these elements in the Sun and nearby stars. All the cosmic-ray nuclei heavier than helium comprise less than 2 per cent of the total, but their relative abundance is nevertheless notably higher than that of the same elements in the general (*i.e.*, thermal) distribution. Table II lists abundances for the

(*) Guggenheim Fellow and visiting professor of physics at the Weizmann Institute in 1962-63.

TABLE I. – *Relative abundances of the atoms at thermal energies and cosmic-ray energies.*

	« General » (*i.e.*, thermal) abundance ([a]) %	Cosmic-ray abundance: At comparable energies nucleon %	At comparable magnetic rigidities ([c]) %	Absol. flux in peter ([b])
Hydrogen	86.6	94	86	600 $\pm$30
Helium	13.3	5.5	13	89 $\pm$ 3
Elements with $Z \geqslant 3$	0.14	0.6	1.4	10.1$\pm$ 0.4

(*a*) According to Cameron's revision of the Table of abundances by SUESS and UREY.
(*b*) Exceeding 4.5 GV, *i.e.*, at geomagnetic latitude 41° N (Texas).
(*c*) peters = particles/m² s sr.

groups of nuclei above helium. The « light » group—Li, Be and B—is enormously overabundant relative to hydrogen; the implications of this anomaly are discussed in Section 2. The medium, heavy, and « very heavy » components are more moderately, but progressively overabundant. A possible explanation for this will be discussed in Section 3.

TABLE II. – *Abundances of the elements relative to* 10^5 *hydrogen atoms.*

	General abundance ([a])	Cosmic-ray abundance ([b])	Ratio $\frac{\text{cosmic-ray}}{\text{thermal}}$ ([a])
Hydrogen	10^5	10^5	1
L-group (Li, Be, B)	$5\cdot10^{-4}$	110	$2\cdot10^5$
M-group (C, N, O, F)	150	400	3
H-group ($Z \geqslant 10$)	15	150	10
« *V.H.* » ($Z \geqslant 20$) ([c])	0.7	40	6)

(*a*) Cameron's revision of the SUESS and UREY data is employed here for the general (*i.e.*, « thermal ») abundances.
(*b*) The cosmic-ray abundances refer to the same *energy* threshold (kinetic energy $E>1.5$ GeV/nucleon).
(*c*) The very heavy (« *V. H.* ») component is a sub-group of the heavy (*H*) component.

A further breakdown of the relative abundances for the *individual* light and medium elements is provided in Table III, which shows the results of the NRL group in Washington, and those of other workers. A noteworthy feature of the cosmic-ray distribution is that carbon is more abundant than oxygen, whereas the reverse is true of the thermal distribution of these elements.

Studies of the isotopic abundances in the cosmic radiation have just begun, and early results for the helium isotopes are discussed in Section 2.

TABLE III. – *Relative abundances of the elements* ($Z \geqslant 3$) *in the primary cosmic radiation* (*).

Relative to all nuclei ($Z \geqslant 3$) (%)		Z	Relative to carbon (%)	
NRL	Others		NRL	Others
5.3	5.2	3	17.6	21.4
2.3	4.3	4	7.6	17.1
7.4	11.9	5	24.6	47.4
30.1	25.1	6	100.0	100.0
9.7	14.9	7	32.2	59.3
19.4	14.5	8	64.4	57.8
2.4	4.0	9	8.0	15.9
15.0	21.4	L	49.8	85.2
61.6	58.5	M	205.0	233.0
23.4	20.1	H	77.7	80.0

(*) After O'DELL, SHAPIRO and STILLER.

Our knowledge of primaries other than nuclei is rudimentary. There is evidence for the arrival of primary electrons in numbers small (about 3 per cent) compared to those of protons (EARL, and MEYER and VOGT). An MIT satellite experiment has detected primary gamma rays with energies above 50 MeV (CLARK and KRAUSHAAR). The collaborative BASJE air shower array on Mt. Chacaltaya in Bolivia has apparently detected primary gamma rays with energies above 10^{14} eV. In addition, experiments are currently designed or under way to detect primary neutrinos and positrons, as well as fast solar neutrons.

1.2. *Particle flux and density* in the vicinity of the Earth, outside the magnetosphere.

Let W = total energy per nucleon

= (kinetic energy E + rest mass energy)/nucleon .

Furthermore, let us adopt an arbitrary threshold energy $W_0 = 1.7$ GeV (corresponding to a mean energy $\overline{W}$ of 5 GeV). Then the following are, respectively, the integral directional intensity $J_{>W_0}$, and the particle density $n_{>W_0}$, at energies exceeding W_0:

$$J_{>W_0} = 0.28 \text{ particle/cm}^2 \text{ s sr}, \quad \text{and} \quad n_{>W_0} = 1.2 \cdot 10^{-10} \text{ cm}^3 .$$

These numbers may be considered as lower limits to the total directional flux and particle density at times of low solar activity in the 11-year cycle.

1.3. *Energy spectra of primary cosmic-ray nuclei* (see Fig. 1). – At kinetic energies of the order of 10^8 eV the spectrum has a peak whose position and magnitude vary with solar activity.

Over the wide energy interval

$$2.5 \cdot 10^9 < W < 10^{14} \text{ eV/nucleon} ,$$

the directional intensity of particles with energy exceeding W is expressible as a simple power of the total energy per nucleon:

$$J_{>W} \propto W^{-1.5} .$$

Within experimental error, the same exponent, -1.5, fits the spectra of hydrogen, helium, and the heavier nuclei.

Above 10^{14} eV, it was formerly thought that the spectrum falls off more sharply (approximately as $W^{-2.2}$), but some recent data give an exponent of -1.8 ± 0.3 even above 10^{15} eV. This suggests that a single power spectrum may be valid up to very high energies.

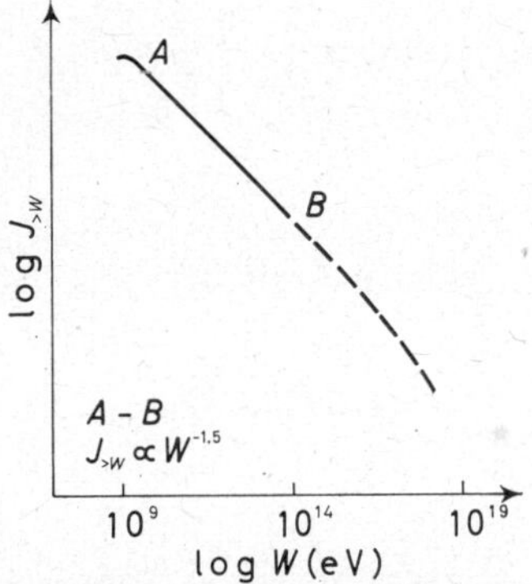

Fig. 1. – Energy spectrum of primary cosmic-ray nuclei. W = total energy per nucleon.

In the energy interval from 10^{12} eV up to nearly 10^{15} eV, information can be obtained from the nuclear interactions (« jets ») observed in huge, balloon-borne stacks of nuclear emulsion. Extensive data on these jets have recently begun to emerge from the International Co-operative Emulsion Flights (ICEF project) organized by the University of Chicago. At energies above 10^{15} eV, extensive air showers detected at mountain altitudes, at sea level, and underground, reveal the incidence of ultra-high-energy particles. Versatile arrays of detectors, including scintillation counters, Čerenkov counters, neutron piles, and multiple plate cloud chambers, as well as ordinary Geiger counters are used. The largest single shower thus far reported (by the MIT group) was estimated to have been initiated by a particle of $6 \cdot 10^{19}$ eV.

As mentioned above, the *mean* energy per nucleon for the arbitrary threshold W_0 taken above is

$$\overline{W} = 5 \cdot 10^9 \text{ eV/nucleon.}$$

The flux of nucleons $F_{>W_0}$ is $0.36/\text{cm}^2$ s sr, so the energy flux φ is given by

$$\varphi = F_{>W_0} \overline{W} = 0.36 \cdot 5 \cdot 10^9 = 1.8 \cdot 10^9 \text{ eV/cm}^2 \text{ s sr} .$$

The *energy density*, for particles moving with the velocity of light c is given by

$$\frac{4\pi}{c}\,\varphi = 0.75\ \text{eV/cm}^3\ .$$

This is a lower limit, and its magnitude is roughly the same as the energy density of starlight.

1'4. *Magnetic confinement.* – At energies exceeding several GeV/nucleon, the energy E of an ion moving in a magnetic field, is related to the charge, the magnetic field, and the radius of the helix it describes, by

$$E_n = 300ZHr = AE\ ,$$

if E, the kinetic energy per nucleon, is expressed in eV, H in gauss, and r in centimeters. A is the mass number.

It is instructive to tabulate r for several high energies of protons and iron-like nuclei. As a representative value of the magnetic field strength in the Galaxy, we have taken $H = 3\cdot10^{-6}$ G. In Table IV the radii are given in light years for ions of energy 10^{12}, 10^{16} and 10^{20} eV. The last column of this

TABLE IV. – *Radii of helices described by ultra-high-energy nuclei of* $Z=1$ *and* $Z=25$ *in a magnetic field of* $3\cdot10^{-6}$ G.

$E_n = AE$	10^{12} eV	10^{16} eV	10^{20} eV
$Z=1$	10^{-3} light years (*)	10 light years	10^5 light years
$Z=25$	$4\cdot10^{-5}$	0.4	4000

(*) 1 light year $=0.95\cdot10^{18}$ cm.

table shows that, for the assumed value of H, protons of 10^{20} eV would not be confined in the Galaxy. Even iron nuclei having the same energy E_n would have a good chance to escape before being destroyed by collision, for in traversing 10^6 light years, they would be penetrating less than 1 gm/cm² of interstellar material. We have tacitly assumed that the galactic halo, as well as the disk, serves as a trapping region. If this were not the case, then particles having energies lower than 10^{20} eV would also escape. Some evidence has been found (LINSLEY and SCARSI) that the highest-energy cosmic-rays are protons. This suggests that they are extra-galactic in origin.

1'5. *Isotropy.* – Above 10^{10} eV the incidence of cosmic rays upon the Earth's magnetosphere is nearly isotropic. This implies strong stirring by magnetic fields in the Galaxy. One might expect some anisotropy at the very highest energies, and this question has been explored with arrays of extensive air shower detectors that reveal directions of arrival. At the Kyoto Conference on Cosmic Rays in 1961, evidence for some anisotropy was presented by the groups from Tokyo, Cornell, and MIT. After acquiring further data, however, the MIT group no longer found a directional effect (Latin-American Symposium on Cosmic Radiation, La Paz, Bolivia, 1962).

2. – Implications of the Li-Be-B anomaly and the helium isotope abundances.

2·1. *The relative abundances of primary lithium, beryllium and boron.* – The universal abundance of the three light elements Li, Be and B is known to be exceedingly low—several atoms per 10^9 hydrogen atoms. In the decade of the 1950's there was a prolonged controversy over the questions: *a*) does the primary cosmic radiation contain an appreciable fraction of these light elements? and *b*) if so, what is the magnitude of this fraction? These questions have an important bearing on the « age » of the radiation and on its propagation through interstellar space, since the Li, Be and B could arise from the collision and breakup of heavier cosmic-ray nuclei. Then the relative intensity of such secondary light nuclei at the top of the Earth's atmosphere would depend upon the average number of collisions suffered by the primordial nuclei, and hence upon the average amount of material they had traversed. Thus, reasonable estimates of the mean path length of cosmic-ray nuclei in the Galaxy, from their sources to the Earth, can be deduced from the abundance of the light element group.

In balloon flight experiments using mainly nuclear emulsion as detectors' various investigators have reported conflicting intensities for the Li-Be-B group, ranging all the way from zero to fluxes approaching those of the heavier atoms. A major difficulty has been that the residual air above the balloon generates secondary light nuclei from the collision breakup of heavier ones. So these secondaries must be separated from the true primaries. The identification of a fast nucleus—*i.e.*, the correct determination of its charge—has also posed experimental difficulties, and the low absolute flux of nuclei heavier than helium has placed severe limitations on the statistical weight of many of the experiments.

At the U.S. Naval Research Laboratory (NRL), O'DELL, SHAPIRO and STILLER were fortunate in getting an emulsion stack exposure at an atmospheric depth of only 2.7 gm/cm². This made the process of extrapolation to the top of the atmosphere much less uncertain than heretofore. Moreover, they gathered an adequate statistical sample, nearly 1000 tracks of cosmic-ray nuclei heavier than helium. Finally, they designed the experiment to meet the problem of charge resolution by employing several independent methods of ionization measurement. Their stack included emulsions of different sensitivities, and a single cosmic-ray nucleus usually passed through several different types of emulsion, leaving tracks of different densities in each. It was thus possible to compare the apparent charge of a cosmic-ray nucleus, deduced from the track it left in one emulsion, with that in another.

The resulting resolution between individual elements in the light and medium groups was good, and, in particular, clean separation was achieved between the crucial elements boron and carbon, segregating the light group of nuclei from all the heavier ones. The relative abundances observed in the emulsion were corrected for the secondaries produced during the ascent of the balloon, and for those produced at altitude in the overlying material, and especially in the residual air. By using appropriate diffusion equations, the ratio of the light elements Li, Be and B to all heavier ones at the top of the atmosphere was found to be 0.18 ± 0.04. From this it was concluded that the primary cosmic-ray abundance of these light elements relative to hydrogen at kinetic energies above 1.5 GeV/nucleon, is approximately $2 \cdot 10^5$ times as great as their relative universal abundance.

Other results on the chemical composition of the primary radiation are shown in Table III.

Next we examine what can be learned from the abundances of the light elements about the average amount of interstallar material traversed by the heavier cosmic-ray nuclei. This material consists mainly of hydrogen, at a density that is not well known. Within the galactic disk, the average density is assumed to be of the order of 1 atom/cm^3, and in the galactic halo it is lower by perhaps two orders of magnitude. To determine what happens when the heavy nuclei collide with hydrogen, one needs the partial cross-sections for the different types of breakup of the various nuclei at cosmic-ray energies. Some experimental information is available, mainly for the inverse process— the collision of fast protons with other target nuclei, but there is a dearth even of such data.

Recently, DANIEL and collaborators at the Tata Institute have calculated the partial cross-sections for the production of many nuclides by breakup in hydrogen. They found good agreement between the calculated values and those experimentally available for a limited number of isotopes. The Bombay group then used the NRL experimental results on the relative abundances of light and heavier nuclei at the top of the atmosphere to deduce the amount of interstellar hydrogen traversed, and they thus obtained a mean path length of (2.5 ± 0.5) g/cm^2.

2·2. *Isotopic composition of cosmic-ray helium.* – The general, *i.e.*, thermal abundance of ^{3}He appears to be very low compared with that of ^{4}He on the

Earth and in the Sun's envelope. In so far it is known—and information is admittedly meager—the same is true of most stars. There is, in fact, only one well-authenticated case of a star having a high abundance of ^{3}He, *i.e.*, *3 Centauri A*, in which the isotopic shifts in wavelength of the helium-I lines suggest that between three quarters and 95 per cent of all the helium in the stellar envelope consiste of ^{3}He. This very exceptional star has been discussed in the lecture by Dr. E. M. BURBIDGE.

It is of great interest to ascertain the isotopic composition of primary cosmic-ray helium. If an appreciable flux of ^{3}He is present, and if it arises mainly as a fragmentation product, then this provides an independent method for estimating the mean path length transversed in space by the primordial radiation. Although ^{3}He could arise as a breakup product of elements heavier than helium, the main contribution would come from the stripping of ^{4}He to give mass-3 particles. We must take into account not only the ^{3}He generated directly in this way, but also the tritium so produced; for the latter decays into ^{3}He with a lifetime of about 12 years, and it would therefore be completely converted to ^{3}He during the long transit time.

The first experiments aimed at measuring the relative abundances of the helium isotopes in the primary radiation were carried out by M. V. K. APPA RAO At energies of about $(200 \div 300)$ MeV/nucleon, he measured the multiple Coulomb scattering along helium tracks as a function of range, and reported values ranging from 0.31 to 0.41 for the ratio $^3\text{He}/(^3\text{He}+^4\text{He})$. These values represent rather large fluxes of ^{3}He, and if the latter is produced entirely by spallation of ^{4}He or heavier nuclei, then the experimental result would imply passage through $(12 \div 14)$ g/cm^2 of interstellar material.

Last year the NRL group (HILDEBRAND, O'DELL, SHAPIRO, SILVERBERG and STILLER) undertook a balloon-borne experiment designed to resolve ^{3}He from ^{4}He by two independent methods. A 2.8-liter stack of Ilford K.2 stripped emulsion was flown on April 21, 1961, at geomagnetic latitude 55° N for 11.2 hours at a mean atmospheric depth of 3.8 g/cm^2. The relative insensitivity of the K.2 emulsion made it possible to deduce the rate of ionization loss from grain counting. Mass identification was accomplished by measuring ionization *vs.* range *and* multiple scattering *vs.* range (constant sagitta technique). By correlating two independent methods, it may be expected that the uncertainty in the isotopic analysis of helium can be reduced. Preliminarly data based on 83 tracks suggest that the ratio $^3\text{He}/(^3\text{He}+^4\text{He})$ is of the order of 0.1, and is unlikely to exceed 0.2.

Let us assume that ^{3}He is present in the primary cosmic radiation reaching the Earth, and that it arises from ^{4}He stripping by interstellar hydrogen. Then we can estimate the thickness of hydrogen transversed with the help of cross-section data from experiments on a (nearly) inverse reaction. INNES bombarded helium with 300 MeV neutrons, and obtained a total cross-section of 10^{-25} cm², corresponding to an interaction mean free path of 16.7 g/cm². In 42 per cent of the collisions, a mass-3 particle was produced. Thus the mean free path for conversion of mass-4 to mass-3 is about 40 g/cm². Using this value, and introducing a correction for the production of ^{3}He from heavier elements, we find that, starting with primordial cosmic rays devoid of ^{3}He, a fraction of about 10 per cent ^{3}He is produced after passage through about 4 g/cm² of hydrogen. Calculations by Hayakawa and by Appa Rao yield a similar result, that a relative abundance of 0.1 implies passage through about 3 g/cm² of hydrogen. On the other hand, the ratio of 0.3 to 0.4 obtained by Appa Rao implies passage through some 13 g/cm².

Now, this long path differs markedly from the path deduced from the NRL result on the elements Li, Be and B. Those results were, however, obtained for nuclei with energies above 1.5 GeV/nucleon. Hence a large ^{3}He flux might be reconciliable with the observed abundance of the light elements if, for some reason, slow cosmic rays travel over a longer path length from their sources to the earth than higher-energy cosmic rays. This explanation is apparently contradicted by recent results of the Rochester group on deuterium at energies comparable to those in the helium investigations. They placed a very low upper limit on the abundance of this nuclide in the cosmic rays. Yet, if we are to interpret a high abundance of slow ^{3}He in terms of a long path length then we should expect to find a considerable flux of deuterium as well. Thus, if Appa Rao's results should be confirmed, then it may be necessary to conclude that most of the ^{3}He is not secondary in origin, but that it originates as such in cosmic-ray sources. It is plain that much additional experimental work on isotopic composition will be required to resolve the present conflicts.

3. – Origin of cosmic rays in supernovae and other sources.

3·1. *Introduction.* – Any theory of cosmic ray origin must account for certain conspicuous features of the radiation. Some of these have been outlined in the preceding Sections, *e.g.*, the composition, energy spectra, and isotropy. The energy density of the radiation in space was shown to exceed $1.2\cdot10^{-12}$ erg/cm³. The heavier cosmic-ray nuclei ($Z\geqslant6$) are progressively enriched relative to hydrogen as compared with the general thermal abundances of the elements in the Sun and nearby stars. Above 10^{10} eV the cosmic-ray intensity is nearly constant in time, and there are georadiochemical indications that this has been true on the geologic time-scale as well.

Thorough stirring and effective storage are provided by the magnetic fields of the Galaxy, at least for particles up to 10^{16} eV. Hence, even sporadic sources that operate with sufficient frequency can provide a quasi-stationary cosmic-ray flux. The gross isotropy ($>10^{10}$ eV) is also well accounted for

— except perhaps at the highest energies — by diffusion in galactic space, since the large volume of the galactic halo is available, as well as that of the disk. However, the time-averaged strength of the sources must suffice to provide enough particles, and the observed energy density.

If a theory can satisfy all of these requirements, it must be taken seriously. If, in addition, it encompasses within its framework reasonable explanations of other phenomena, *e.g.*, the nonthermal radio emission from the galactic disk and halo, and from certain discrete radio socurces, this enhances the power of the theory.

In the last Section we pointed out that the Galaxy can be viewed as a reservoir of magnetically-confined cosmic rays randomized in direction by turbulent clouds of plasma. In the course of their long, winding paths, some of the high-energy particles are lost by nuclear collision, while others escape altogether from the Galaxy.

During the past decade this view has been bolstered by the findings of radioastronomy. In addition to the discrete sources of radio emission there also exists a general galactic radio emission—the « synchrotron radiation » emitted by relativistic electrons spiralling in a magnetic field. At high particle energies in strong magnetic fields, the radiation consists of visible light. At lower electron energies in the same magnetic field, or at high energies in weak magnetic fields, it will be emitted at radiofrequencies. Today, with radio-astronomical evidence pointing to the ubiquitous distribution of relativistic electrons in the Galaxy, it is not difficult to envisage the storage of high-energy *nuclear* particles as well. The question remains: how did these particles originate?

3·2. *Theories on the origin of cosmic radiation.* – Various theories as to the mode of origin and site of origin have been proposed. ALFVÉN believes that the acceleration of ions is a normal property of matter in space, that it is due to variations in the magnetic fields, and is produced by « magnetic pumping ». He considers that acceleration occurs not only in interstellar space, but also in the Earth's magnetosphere, in the interplanetary medium, and in extra-galactic space as well.

Formerly, ALFVÉN concentrated on a local theory — *i.e.*, a solar system theory — of cosmic-ray origin, in the belief that one should thoroughly explore possible mechanisms of acceleration in nearby regions of space where information is most accessible. If this reasonable approach be adopted, then, once a solar origin appears untenable, the same philosophy would prescribe that a thorough exploration of possible *galactic* sources precede attempts to develop a theory of extragalactic origin. Of course, inadequacies in galactic sources may compel us to search for extragalactic ones as well.

Returning to the Sun as a possible source, we know that the sun does produce

« cosmic rays » of modest energies, *e.g.*, 10^8 eV. However, the time-averaged emission of high-energy solar particles contributes only a small part of the cosmic radiation incident upon the Earth. Besides, there is no significant difference between daytime and nocturnal intensities; and when, in its 11-year cycle, the sun is most active, the cosmic-ray flux near the magnetic poles is actually *lower* (by $\sim \frac{1}{4}$) than at minimum solar activity. In addition, it has recently been shown in rocket experiments that the relative abundances of fast nuclei sampled from solar flares differ markedly from the chemical composition of the cosmic radiation. Hence, the Sun cannot be an important source of the bulk of cosmic radiation observed at the Earth.

More active stars than the Sun have been proposed as possible cosmic-ray sources, *e.g.*, flare stars, magnetic stars, *T Tauri* stars. A promising set of candidates has just been found by T. A. MATTHEWS *et al.* — three strong radio emitters of synchrotron radiation — 3C48, 3C196 and 3C286 (*). Each has been optically identified with an object in the same location to within < 10 s of arc, a precision achieved by using a pair of 90 ft. antennas as an interferometer (**).

3.3. *Supernovae as possible sources of cosmic rays.* – The likeliest sources within the Galaxy appear to be supernovae. This theory of cosmic-ray origin is due mainly to V. L. GINZBURG and I. S. SHKLOVSKY, with further contributions by PIKELNER, HAYAKAWA, SYROVATSKY, BURBIDGE, and others.

Altogether, some 120 supernova outbursts have been observed in other galaxies. There are good reasons, however, to believe that a fair number of these stellar explosions have occurred in our Galaxy as well. Supernovae have been classified into Types I and II; we shall be concerned mainly with Type I, which have light curves showing an exponential decay with a period of ~ 55 days.

The best example of an optical object reliably identified as the remnant or a supernova is the Crab Nebula, studied by OORT and his collaborators for many years. Among the most revealing photographs of the Crab Nebula are those made by Baade with the 200 inch Palomar telescope and various light filters. The Crab Nebula is an irregular object of quasi-elliptical shape, some

(*) The Author is grateful to Dr. SANDAGE for private communication of this discovery.

(**) *Note added in proof.* - At first these st rs were thought to be real radio *stars*; however, more recent investigations of radio sources 3C273 and 3C48 by HAZARD, MACKEY, SHIMMINS, SCHMIDT, OKE, GREENSTEIN and MATTHEWS, suggest that they are redshifted galaxies having some 10^{60} erg stored in particles and magnetic field. FOWLER and HOYLE have suggested that the energy source is the gravitational energy released in the collapse of massive stars, $\sim 10^7$ solar masses See *Nature*, **197**, 1037 (1963), and *New Scientist*, no. 332, 681 (1963).

1000 parsecs away. Its « major axis » is about 0.9 parsecs long, its « minor axis », ~ 0.6. In the light of the H_α line, the nebula shows an elaborate filamentary structure. When it is photographed with a filter that masks the emission lines and admits the continuous spectrum, a diffuse, amorphous mass shows up. The peculiar distribution of intensity in this continuous spectrum differs from that of a blackbody, and from the spectra of other gaseous nebulae. Clues to a more satisfactory explanation of this light came in later years from the radio emission of the nebula.

Bright « ripples » have been observed to move out from the central zone of the nebula at prodigious speeds (~ 0.1 of the velocity of light), « almost as though the nebula were breathing ». It has been suggested that these are due to hydromagnetic waves.

The Crab Nebula has an unusual radio spectrum: the intensity remains almost constant over a range of wave-lenghts from 25 cm to 750 cm. SHKLOVSKY showed that this radio emission is not thermal radiation, but the synchrotron radiation (magnetic bremsstrahlung) of relativistic electrons ($\sim (10^8 \div 10^9)$ eV) trapped in the magnetic field of the expanding nebulosity. He also showed that the optical continuum emitted from the amorphous mass of the nebula is the (higher frequency) synchrotron radiation of electrons having energies much higher ($(10^{11} \div 10^{12})$ eV) than those emitting the radio waves. Thus, two widely disparate regions of the electromagnetic spectrum from the Crab Nebula were boldly connected.

GINZBURG pointed out that if the continuum from the Crab Nebula consists of synchrotron radiation, it should be polarized. This prediction was confirmed by means of photoelectric as well as photographic techniques.

Since the light of the continuum proved to be polarized, it seemed reasonable to search for polarization of the radio emission as well. At a wave length of 31 mm, a 7 per cent polarization was found.

The discovery that the continuous light and the radio waves from the Crab are polarized provides evidence that both radiations consist of magnetic bremsstrahlung generated by energetic electrons moving helically in magnetic fields. The processes that accelerate the electrons may be presumed to operate on positive ions as well, thus producing cosmic-ray nuclei. The supernova theory of origin supposes that cosmic rays are accelerated by turbulent plasma clouds in the tenuous remnants of the stellar explosion. After escape, they diffuse into the disk and halo, where they are stirred and stored by magnetic fields that roam the Galaxy.

The nuclear processes that give rise to supernova explosions have been described in the lecture by Prof. BURBIDGE. You will recall expecially the « *r*-process » of rapid neutron capture that leads to the formation of nuclides as heavy as ^{254}Cf. This nuclide is unstable against spontaneous fission, with a half-life of 55 days. After shorter-lived activities have died out, the decay of ^{254}Cf dominates, and its 55-day « signature » can be read in the exponential part of the luminosity-time curve. The explosion itself requires only a few seconds, and the total energy liberated has been variously estimated as $6 \cdot 10^{50}$ ergs (FOWLER and HOYLE, assuming a pre-supernova mass of 1.3 solar masses), or 10^{52} ergs (HAYAKAWA, 10 solar masses). The envelope of the star is ejected with a velocity of > 1000 km/s, and it forms the tenuous gaseous shell in which relativistic electrons are somehow accelerated, and emit synchrotron radiation. The total energy released *after* the onset of the exponential decrease is about 10^{47} ergs.

In addition to the large quantities of heavy elements produced in a supernova explosion, the theory of nucleogenesis (BURBIDGE *et al.*) requires that elements such as carbon and oxygen also be enriched relative to hydrogen. This « over-abundance » is likewise true of the cosmic-ray composition.

3·4. *Frequency of supernovae.* – In our own Galaxy, history records six supernova flareups in the last 1800 years, according to SHKLOVSKY. Four of these have occurred in the past millenium, and of these the last three are celebrated as the Crab Nebula (1054), Tycho's star (1572), and Kepler's (1604). The other three outbursts were observed in the years 185, 369 and 1006. The position of the third supernova (1006) is poorly defined, but the other five events have been closely connected with known radio-sources, *e.g.*, *Taurus A* (the Crab Nebula), and *Cassiopeia A* (the event of 369 A.D., a less certain identification). These stellar explosions all occurred in the galactic disk and in the general neighborhood of the Sun, within a radius of some 10 000 light years. Hence, the *apparent* frequency — once in three centuries — of these outbursts in our galaxy, should be corrected for the large volume in the disk in which outbursts have gone undetected, or unrecorded. This leads to a frequency one order of magnitude higher. Confirmation that supernovae do indeed occur as often as every few decades in some stellar systems comes from observations of 5 galaxies in which a total of 15 outbursts have been seen in 60 years — an average of one per galaxy per 20 years. As a working estimate we shall adopt, for our own galaxy, a frequency of once per century, *i.e.* $3 \cdot 10^{-10}\ s^{-1}$.

3·5. *The supernova theory of the origin of cosmic rays.* – Among the characteristics of supernovae which suggest that their remnants constitute a powerful source of cosmic rays, are the following: an enormous energy release, a copious supply of relativistic electrons, large-scale magnetic fields, extended masses of plasma moving at high velocities, indications of the propagation of hydromagnetic waves (the light « ripples »), and relatively high concentrations of heavier elements.

According to the supernova theory of cosmic-ray origin, the relativistic nuclei and electrons that fill the galactic volume originate in the tenuous remnants of supernova (and perhaps nova) explosions. The particles are somehow injected, and then accelerated by the action of magnetic fields in the supernova shell. They are trapped in the nebular envelope for thousands of years, after which the expanding shell is damped by and dissipated into the surrounding interstellar medium. Then they leak out, first into the galactic disk, and then into the large volume of the galactic halo. The paths of the particles are tortuous, and tend to become isotropic. Meanwhile, the electrons lose energy by magnetic deceleration, the nuclei lose energy through nuclear collisions, and both steadily drop out of the cosmic-ray reservoir in the galaxy. Both must be replenished if this « reservoir » is to be maintained, and the theory asserts that a fresh supply of relativistic nuclei is provided by new supernova explosions.

To ascertain the adequacy of the rate of supply, we can estimate the rate of particle loss and compare it with the output of supernovae. This estimate is very rough, since many astrophysical quantities are known only crudely, *e.g.* the densities of atoms and the magnetic field strengths in various parts of the disk and halo.

Assume that the average density of cosmic rays in the galaxy has been constant in recent geologic time. From observations of the flux of synchrotron radiation at the Earth, and estimates of the magnetic intensity in supernovae (10^{-4} to 10^{-3} G), it is possible to compute the total supply of electrons with energy exceeding 10^8 eV from an « average » supernova as 10^{51}. We shall tentatively assume that the number of relativistic nuclei produced is comparable to the number of electrons. With a frequency of one galactic supernova outburst per century, *i.e.*, of $3\cdot10^{-10}$/s, the overall cosmic-ray source strength of supernovae in the Galaxy is $3\cdot10^{41}$ nuclei/s. Adding a comparable contribution from ordinary novae the number would be $\sim 6\cdot10^{41}$ per s.

Next we estimate the rate of disappearance of relativistic protons from the galactic volume, assuming that it is determined by their rate of collision. The number degraded to subrelativistic energies per cm^3 per s is of the order of $n\nu\sigma c$, where

n = mean density of galactic cosmic-ray nuclei, taken as $1.2\cdot10^{-10}$/cm^3, its value near the Earth;

ν = atomic density averaged over the Galaxy, 0.01 cm^3 (the particles spend most of their time in the halo);

σ = effective cross-section for « destructive collision » of a proton $= 2.8\cdot10^{-26}$ cm^2 taking into account that more than one collision is required for a relativistic proton to lose most of its energy;

c = velocity of the particles, taken equal to that of light.

Hence, the rate of loss is 10^{-27} nuclei/cm$^3 \cdot$s, and in the whole galactic volume of 10^{68} cm^3, the rate is 10^{41}/s. This number may be compared with the average rate of production deduced above.

Suppose that the loss of particles is governed not by collision but by the rate of *escape* — *i.e.*, suppose that the Galaxy is, magnetically, a relatively open system. Then the rate of loss may be several times 10^{41}/s. Even so, the source strength appears adequate to account for the observed cosmic-ray flux.

What of the rate of *energy* production versus energy loss? If supernovae were the only source, then the cosmic-ray energy output required from an average supernova outburst (assumed to occur about once per century) is approximately $(10^{48} \div 10^{49})$ erg depending upon whether the Galaxy is a closed or open system. When these numbers are compared with the *total* energy output ($6 \cdot 10^{50}$ to 10^{52} erg), it appears that only a modest part of the liberated energy suffices to provide the observed cosmic-ray energy flux in the Galaxy.

3.6. *Acceleration of cosmic ray particles.* — By what means are the particles accelerated? GINZBURG favors a statistical mechanism of the Fermi type, in which ions gain energy, on the average, by encounters with irregularities in moving magnetic fields. A charged particle entering the region of a moving magnetic field is subject, in general, to a change in speed as well as direction. Because head-on collisions occur more frequently than overtaking collisions, more particles, on the average, gain energy than lose it.

Fermi showed that this process increases the energy of a relativistic ion at a rate proportional to its energy. This leads to an exponential increase of particle-energy with time, and it also leads to an energy spectrum of the type actually observed. However, if the acceleration takes place, as FERMI postulated, in interstellar space, it is relatively slow. If, on the other hand, it operates within the turbulent confines of a supernova envelope, the collisions are much more frequent, and the process more efficient.

There were other difficulties in Fermi's early theory, which led him to propose the « contracting magnetic bottle » mechanism. If the magnetic lines of force along a segment of spiral arm of the Galaxy converge at the ends, they form an ion trap. If, moreover, the plasma clouds at the two ends of the tube approach each other, the ions are accelerated upon reflection. A similar scheme might help to accelerate particles in the gaseous ejecta of a supernova, with mutually approaching plasma clouds serving as the jaws of a trap.

Other schemes of acceleration have been discussed, *e.g.*, collisions between charged particles and hydromagnetic shock waves. We have alluded briefly to the fast-moving « ripples », or wisps of light that are observed to travel outward several times a year inside the Crab Nebula. They have been imputed to hydromagnetic shocks, and the energy in a single ripple has been estimated as 10^{43} erg.

An intriguing scheme of *hydrodynamic* acceleration is that of COLGATE and JOHNSON, who suggest that cosmic rays are the « blown-off » surface layers of an exploding supernova. The energy liberated in the core of the star generates a powerful shock-wave that speeds up in the decreasing density of the outer layers. The shock wave becomes relativistic well inside the mantle of the star, and the outermost layers are ejected with velocities close to that of light .

Whatever the means of acceleration, it is clear that a copious supply of high-energy electrons exists in objects like the Crab Nebula, and it is certainly plausible that positive ions are likewise accelerated there.

Among the limitations of the supernova theory are the following: It seems doubtful that it can account for particles at energies above 10^{16} eV. Moreover, even at lower energies, the source strength *may* be inadequate if the galactic boundary is a very poor reflector of cosmic rays. ALFVÉN maintains that the ions from a supernova could not reach the Earth in sufficient numbers, *i.e.*, he is not convinced that an adequate transport mechanism has been shown to exist.

Many other ideas on cosmic-ray origin have been advanced (see, for example, P. MORRISON'S article in *Handbuch der Physik*). However, few authors deny that supernova remnants are at least one class of significant sources. Very recently, BURBIDGE has considered the hypothesis that a large part of the cosmic radiation reaching us may originate in the supercluster of galaxies within distances of 10^8 light years.

BIBLIOGRAPHY

A very useful source describing much recent research on cosmic rays and their origin are the Proceedings of the International Conference on Cosmic Radiation (IUPAP) in Kyoto (Sept., 1961). *Journ. Phys. Soc. Japan*, vol. **17**, Supplements A-I, A-II and A-III (1962). References in the present text not otherwise listed in this bibliography will be found in the Proceedings.

V. L. GINZBURG and S. I. SYROVATSKY: *Prog. Theor. Phys.* (*Japan*), Supplement no. 20 (1961).

I. S. SHKLOVSKY: *Cosmic Radio Waves* (Moscow, 1956); Harvard Univ. Press, rev. ENGLISH ed. (Cambridge, Mass., 1960).

S. HAYAKAWA, K. ITO, Y. TERASHIMA: *Prog. Theoret. Phys.* (*Kyoto*), *Suppl.*, **6**, 1 (1958).

E. M. BURBIDGE, G. R. BURBIDGE, W. A. FOWLER and F. HOYLE: *Rev. Mod. Phys.*, **29**, 547 (1957).

A. G. W. CAMERON: *Ann. Rev. Nucl. Sci.*, **8**, 299 (1958).

P. MORRISON: *Handb. d. Phys.*, vol. **46/1**, (Springer, Berlin, 1961), p. 1.

M. M. SHAPIRO: *Science*, **135**, 175 (1962).

G. R. BURBIDGE: *The Origin of Cosmic Rays*, submitted to *Prog. Theor. Phys.* (*Kyoto*) (1962). (The author is grateful to Professor BURBIDGE for the opportunity of reading this paper prior to publication.)

N. U. MAYALL: *Science*, **137**, n. 3523, 13 July (1962).

On Li, Be *and* B *in the primary cosmic radiation*:

F. W. O'DELL, M. M. SHAPIRO and B. STILLER: *International Conference on Cosmic Radiation* (*IUPAP*), *Kyoto* (Sept., 1961), in *Journ. Phys. Soc. Japan*, **17**, *Suppl.*, A-III, 23 (1962).

G. D. BADHWAR, R. R. DANIEL and B. VIJAYALAKSHMI: (private communication from Dr. DANIEL).

On the helium isotopes:

M. V. K. APPA RAO: *Phys. Rev.*, **123**, 295 (1961). Also, Univ. of Rochester Report no. NYO-10029 (Nov., 1961); submitted to *Journ. Geoph. Res.*

B. SHAPIRO, B. HILDEBRAND, F. W. O'DELL, R. SILBERBERG and B. STILLER: *Proc. of Internat. Space Symposium* (*COSPAR*), Washington, D. C. (May, 1962) (in press): Transformations of Complex Cosmic-Ray Nuclei in Galatic Space.

W. H. INNES: *Thesis*, UCRL Report 8040, University of California (1957).

On the Primary Cosmic Radiation:

C. J. WADDINGTON: *Prog. Nucl. Phys.*, **8**, 1 (1960).

M. M. SHAPIRO: *Trans. N. Y. Acad. Sci.*, **20**, 697 (1958).

Handb. d. Phys. (Encyclopedia of Physics), *Cosmic Rays* 1, vol. **46/1**, ed S. FLÜGGE (Berlin, 1961).

On the general, i.e., thermal abundances of the elements:

H. E. SUESS and H. C. UREY: *Handb. d. Phys.*, vol. **51** (Berlin, 1958), p. 269; *Rev. Mod. Phys.*, **28**, 53 (1956).

A. G. W. CAMERON: *Astrophysical Journ.*, **129**, 676 (1959).

L. H. ALLER: *The Abundance of the Elements* (New York, 1961).

The Problem of Eta Carinae.

L. GRATTON

Laboratorio di Astrofisica e Laboratorio Gas Ionizzati - Frascati (Roma)

1. – Introduction.

The importance of the extraordinary southern variable, *η Carinae* for the study of the initial stages of stellar evolution was stressed by me in several occasions (GRATTON, 1955*a*, 1955*b*, 1958*a*, 1958*b*) and, indipendently, by THACKERAY (1956). The fact that according to all evidence *η Car* is a member of a stellar association in Carina was taken as indication that the star is a very young object; indeed, considering its uniqueness among all stars it was thought not impossible that the phenomena shown by it might be those accompanying the formation of a (massive) star or, perhaps, of a group of stars.

It is not the purpose of this Seminar to discuss the evolutionary implications of this hypothesis; I wish merely to demonstrate its plausibility. Indeed it was shown by SANDAGE (1958) on quite general grounds that the probability that we shall ever witness the actual birth of a star is exceedingly small.

However in the case of *η Car* Sandage's argument is much less unfavourable than in the case it was applied by him. First of all, as it will be shown later, *η Car* is a very bright object. If a similar object existed inside a distance of, say, 4.5 kpc it seems difficult that it might have escaped attention; the corresponding volume is $\frac{1}{9}$ the volume of the whole Galaxy, if we consider the flattening of the galactic system.

Secondly the events we are witnessing lasted apparently several centuries; 300 years is a conservative estimate. If only one of them would happen in the Galaxy each 100 years there is still a reasonable chance that we may observe it. But we know that in the Galaxy one O-association is formed each 10^4 or 10^3 years (see my lecture on Stellar Association in this Course). That means that if in each association from 10 to 100 « *η Carinae* events » take place, the probability that we are able to observe one of them is pretty good. As in an average O-association 100 O-B stars and probably 100 times as many less massive stars are formed according to current views, it suffices that one « η *Carinae* event » be associated to the formation of 2 or 3 massive or of some hundred smaller stars.

2. – Light variation.

The story of the light variations of η *Carinae* (called also η *Argus* in the old astronomical literature) is well known. Until 1836 it is summarized in Table I, according to all existing records.

TABLE I. – *Old observations of η Carinae.*

Year	Mag.	Source and authority
1500		Old catalogues do not record the whole region, because too far south.
1600		Houterman observes stars of 4^m in the region, but not η.
1603	4	Bayer's Atlas; source and epoch of observation unknown.
1677	4	Halley's Catalogue of Stars observed at S. Helena.
1685-9	2÷4	Noël (Winnecke, AN 1224).
1752	2	Lacaille, Coelum Stellatum Australis.
1811-5	4	Burchell (Herschel, Res. of Cape Obs.).
1822	6?	Scully (INNES: *Cp. Ann.* **9**, 77 *B*).
1822-3	2	Fallows? (INNES: *Cp. Ann.*, **9**, 77 *B*).
1822-6	2	Brisbane (Star Catalogue).
1825-6		Missing in Burchell's observations at the Cape.
1826-7	2	Rumker (Star Catalogue).
1827	1.0	Burchell (Cp Observ., $\eta = \alpha$ *Cru*).
1828	2	Burchell (Cp Observ.).
1829-33	2	Johnson (Star Catalogue).
1831-2	2	Taylor (Star Catalogue).
1834	1÷2	Herschel (Cp Observ., occasionally $\eta = \alpha$ *PsA*).

Although all these observations are very rough estimates, it can be safely inferred that during the 17th and the 18th centuries and the beginning of the 19th, the star was slowly brightening, with perhaps some occasional outbursts, like those observed in the middle of the 18th century and in 1827.

From 1836 to 1902 the observations were all discussed by INNES (1903) and yielded the light curve shown in Fig. 1. One must note the bright maxima of 1838 and (especially) that of 1843; at the latter epoch the star was noted brighter than *Canopus* and inferior only to *Sirius*. To this stage of great activity, there followed a continuous decrease in brightness from 1856 to 1870; then came a slower decrease and a period of almost constancy at magnitude 7.6. In 1889 η brightened a little, but then it faded slowly again and from 1897 until 1938 it was practically constant at about 8.5. However in 1938 a brightening of about one magnitude was observed (O'CONNELL, 1956) and its present magnitude is about 7.0. It was remarked by THACKERAY (1953) that this brightening is not due to the star itself, which is still at magnitude 8.0, more or less, but to the nebular halo which surrounds it and which was discovered by VAN DEN BOS (1938) and later, independently, by GAVIOLA (1946, 1950) and THACKERAY (1949, 1950).

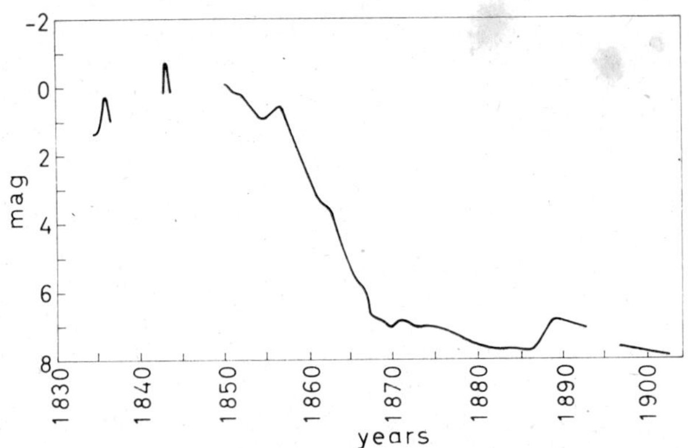

Fig. 1. – Light curve of η *Car* (adapted from INNES, 1903).

This behaviour is very different from that of any other star in our Galaxy and might be sufficient for considering η *Car* as an unique object. It has been suggested that perhaps *S Dor* in the Large Magellanic Cloud and the peculiar variables discovered by HUBBLE and SANDAGE in M31 and M33 (HUBBLE and SANDAGE, 1953) might be objects of the same kind, although the absolute magnitude of η *Car* at maximum was certainly brighter.

3. – The spectrum and its variations.

As is well known the present spectrum of η *Car* (Fig. 2*a*, *b*, *c*) exhibits one of the most extraordinary display of emission lines among stellar spectra.

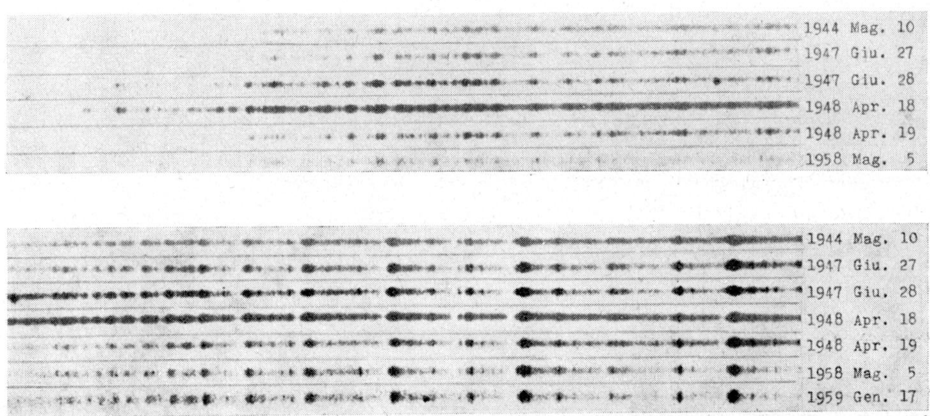

Fig. 2*a*. – The spectrum of η *Car*; upper row: (3200÷3700) Å, lower row: (3650—4150) Å (Cordoba Observatory).

The emission lines are superimposed upon a continuum, which appears on longer exposures, and in which a few absorption lines are visible, displaced to the violet by an amount corresponding to about 475 km/s, relative to the emission.

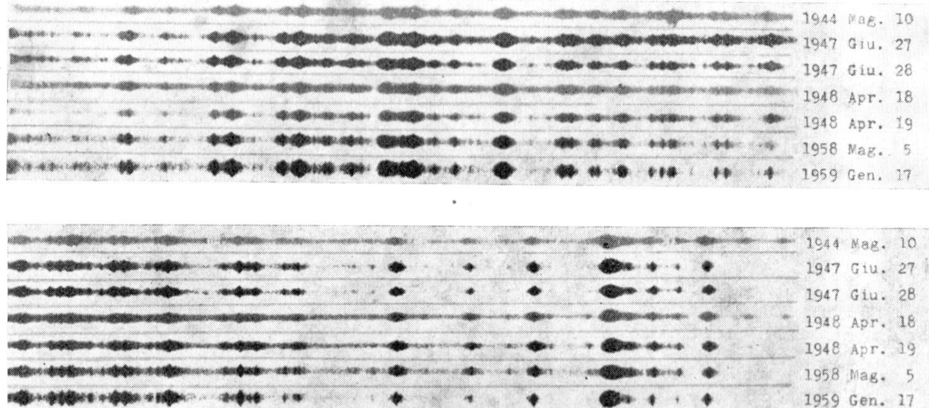

Fig. 2*b*. – The spectrum of η *Car*; upper row: (4100 ÷ 4600) Å, lower row: (4500 — 5000) Å (Cordoba Observatory).

Detailed description of the spectrum have been published by GAVIOLA (1953) and by THACKERAY (1953). Probably the most characteristic features are the numerous and very strong forbidden lines due to quadrupole transitions of

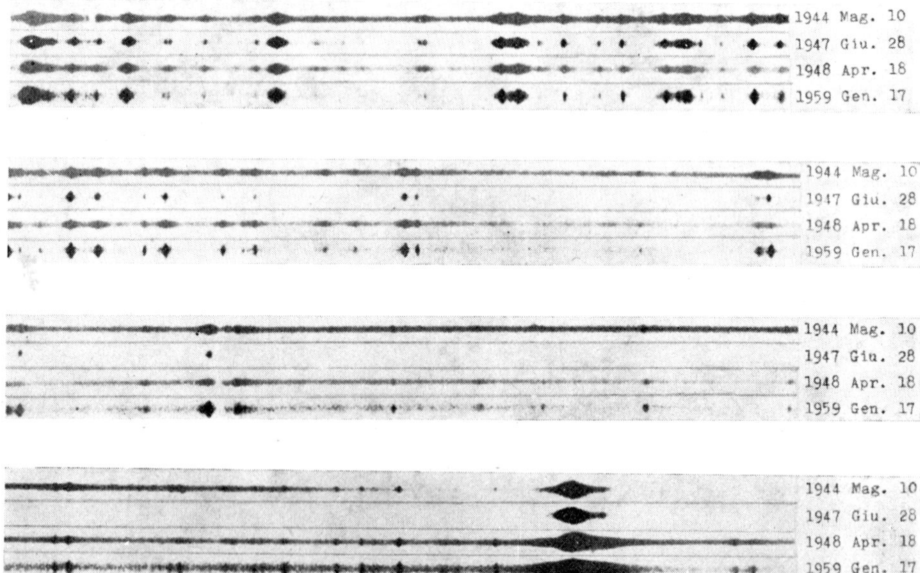

Fig. 2*c*. – The spectrum of η *Car*; visual region (Cordoba Observatory).

FeII; the [FeII] lines were first identified by MERRILL (1929) and are typical of the so-called « η *Carinae* stage » of slow Novae. Hydrogen lines are strong in emission and possess a complicated structure discussed by GRATTON (1958); as they are flanked to the violet by the corresponding displaced absorption lines, the profile assumes the typical « *P Cygni* » aspect.

Several strong lines especially in the near infrared are still unidentified (THACKERAY, 1961*b*).

The general excitation level is not high; it may be estimated to correspond to an excitation temperature around 8000 °K. However a few higher excitation lines due to [NII], [NeIII], [FeIII], ... and several He-lines are quite strong. It is very probable that these lines and, partially, those of hydrogen originate in a different shell from that in which the other lines are formed. The main reason for that is that since the beginning of modern spectrographic observations no changes of the spectrum have been noted; but during a few months in 1948 all the lines of higher excitation and those of He faded away, without any perceptible change of the other lines (and of the whole star). The change was only transitory, as in 1949 the spectrum was normal again. GAVIOLA notes that this change was accompanied by a strengthening of the continuum and by some changes of the structure of the H-lines. According to BIDELMAN (private communication) the old plates of the Chile expedition of the Lick Observatory show very faint or no He-lines at all.

It is very unfortunate that no observations exist of the spectrum during the spectacular variations of brightness of the first half of the last century. The first record of the spectrum is a short notice by LE SUEUR (1869), who observed it visually; the star was at that time approaching its minimum and Le Sueur's description makes it very probable that the spectrum was not very different from the present one.

But in 1893, during the small maximum of brightness, the spectrum was obtained with an objective prism at the Harvard Southern Station. A good analysis of these plates has been published by WHITNEY (1952). Through the kindness of Dr. Bidelman I could obtain a copy of this spectrum which is reproduced in Fig. 3. As can be seen, this is a rich absorption spectrum

Fig. 3. – The spectrum of η *Car* in 1889 (Harvard Observatory).

roughly corresponding to that of an F5 supergiant, with strong *P Cygni* features. According to Whitney, the radial velocity of the absorption, relative to the emission lines, is of − 180 km/s, or − 200 km/s if the FeI lines are excluded.

These changes in the whole are reminiscent of those shown by a slow Nova, but we shall see later that η *Car* cannot be considered a Nova or a Supernova (of Class I), but it is a quite different object.

4. – The expanding nebula around η *Carinae*.

As it was mentioned in Sect. 2, η *Car* is sorrounded by a bright halo, expecially strong in red (hydrogen) light, which extends a few seconds from the central star. It was shown by GAVIOLA (1946) that this halo is expanding at the rate of some hundredths of a second per year.

A detailed discussion of the motion was published by A. RINGUELET (1958), who proved conclusively that the trajectories of the different condensations or «nebulae» diverge from a point which closely coincides with the stellar nucleus (η *Carinae*); moreover from the velocities one finds that the epochs at which the nubeculae were ejected from the nucleus are coincident with those of maximum brightness of the star, within the observational errors. Table II, from Miss Ringuelet's paper, illustrates the situation; see also Fig. 4.

TABLE II. – *Observed motions in the halo around η Car.*

Gaviola's nub.	N° obs.	Exp. V ″/y	Exp. V km/s	t_0	Epoch of max. bright.
C	12	0.029	210	1883 ± 5	
B	11	0.042	300	1895 ± 2	1889
g	2	0.072	510	1898	
h	5	0.046	330	1839 ± 18	1843
d	5	0.051	360	1847 ± 16	1843

The linear velocities were computed assuming a distance of 1500 pc (see later, this Section).

RINGUELET suggests also that the condensations lying at present some 7″ or 8″ from the nucleus might have been ejected during a maximum which occurred in the middle of the 18th century.

From the observed expansion and the radial velocities of the absorption lines the distance of the star can be obtained. Clearly, if r is the distance in parsec, V the expansion velocity in km/s, μ the transverse motion in ″/y and θ the angle of the motion of a condensation relative to the line of sight, then

$$r = 0.21 \frac{V \sin \theta}{\mu} . \tag{1}$$

Now, apart from the factor sin θ, in the present instance there is a considerable doubt concerning which are the right values to take for V, for a given condensations. It may be argued, for instance, that if we consider the motions of nubeculae A and B, the right value to assume for V would be 180 km/s (or 200 km/s), the expansion velocity of the shell ejected in 1889, as measured by Whitney. One obtains on this assumption $r/\sin\theta = 1300$, or 900 pc, respectively.

These values are doubtless of the right order, but we will see presently that they are certainly too small. On the other side, using the velocity of the shell we observe now, 475 km/s, we get $r/\sin\theta = 3440$ and 2370 pc, for the the two nebulae, respectively, which seems definitely too large, even assuming a small, but not too unprobable value of $\sin\theta$ (that is a motion nearly along the line of sight).

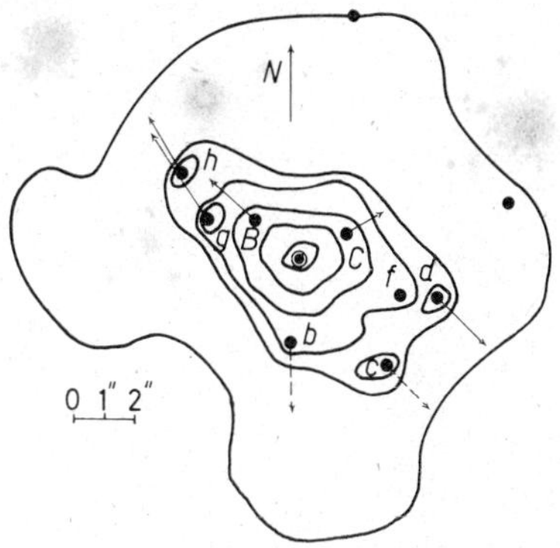

Fig. 4. – Motions in the small nebula around η *Car* (adapted from RINGUELET, 1958).

In 1956 THACKERAY reported the discovery that various portions of the halo showed a remarkable polarization; this was confirmed later by WESSELINK (1958). The polarization is that which one would expect for light from the nucleus, scattered by the halo.

Various authors (GAVIOLA, 1953; THACKERAY, 1953; GRATTON, 1958) tried to get the spectra of the different parts of the expanding nebula. These observations are very difficult, because the light coming from the nucleus especially in bad seeing is too strong at distances of only a few seconds.

The best results are those due to THACKERAY (1961*a*) who observed in excellent conditions. Thackeray's observations refer mainly to nubecula *h* (but see later). He found a very complex spectrum with wide emission lines due to H, HeI, FeII and somewhat sharper forbidden lines of [FeII]; the permitted lines are shifted to the red about +850 km/s, whereas the forbidden lines have a much smaller shift (+220 km/s). Moreover only the permitted lines are polarized.

Thackeray gives a very interesting model to explain these results; according to him the permitted lines are exclusively due to light from the nucleus scattered by free electrons in the halo. The observed radial velocity of these lines is, then, made up by the expansion velocity V, plus the true radial velocity, $V\cos\theta$. On the other side, the unpolarized forbidden lines will show only the true radial velocity of the nubecula; thus $V\cos\theta = 220$ km/s and $V = 850 - 220 = 630$ km/s, $\cos\theta = 220/630 = 0.35$. The corresponding transverse velocity $V\sin\theta$ is 590 km/s with an error (estimated by THACKERAY) of ± 100. Compared with the observed motion of nubecula *h*, this transverse velocity gives the unprobably high value of 2.7 kpc for the distance.

However a much more plausible value can be obtained by noting that at the epoch of Thackeray's observations, nubecula *g*, which left the nucleus in 1889 and moves in the same position angle as *h*, but with a much larger velocity, must have overtaken h; it is thus fairly probable that Thackeray observed actually not *h* but *g*, which is considerably brighter. Adopting Ringuelet's value for the motion of *g*, one gets, on this assumption for the distance of η *Carinae*

$$r = 1700 \pm 300 \text{ parsec},$$

which is very reasonable. Later we shall adopt

$$r = 1500 \text{ parsec},$$

as the most probable value of the distance of η *Carinae.*

5. – The O-association Carina II.

The region of η *Carinae* ranks among the most remarkable in the whole sky for the great number of O and early B-type stars. Several authors have commented upon the extraordinary aspect of this region (Fig. 5), with the intricated structure of NGC 3372, the great " key-hole nebula " extending over it (HERSCHEL, 1843; BOK, 1932). We call Carina II the O-association in the η *Car* region; it is one of the richest O-associations so far discovered. MORGAN and MÜNCH (1953) have given a list of almost 200 bright blue stars in a few square degrees.

The connection of the early type stars with the nebula NGC 3372 has been studied by BOK (1932) and there is no doubt that they are actually imbedded in the nebula and are responsible for its brightness. Clearly the dark lanes are due to obscuring matter formed by nonexcited portions of the nebula, which project upon the bright portions; the whole structure must, thus, have a considerable depth.

In discussing the distance of the association, I derived, from different arguments, a distance modulus of 12.3^{M} (GRATTON, 1955, 1958*b*); at that time I adopted an interstellar absorption of 1.5^{M}, which is definitely too high. Taking 0.7 as the most probable value (BOK, 1953), one gets for the true modulus 11.6^{M}, which corresponds to a distance of a little more than 2000 pc. It is possible however that this is an overestimate, since BOK and VAN WIJK (1952) obtain an average distance of 1500 pc for the B stars in Carina.

As a matter of fact it was shown by HOFFLEIT (1956) that the early-type stars in Carina extend over a considerable depth. She found evidence for two condensations of stars at 1.2 kpc and 2.0 kpc. respectively; the two small clusters NGC 3293 and IC 2944 are at still greater distances 2.1, and 2.6 kpc,

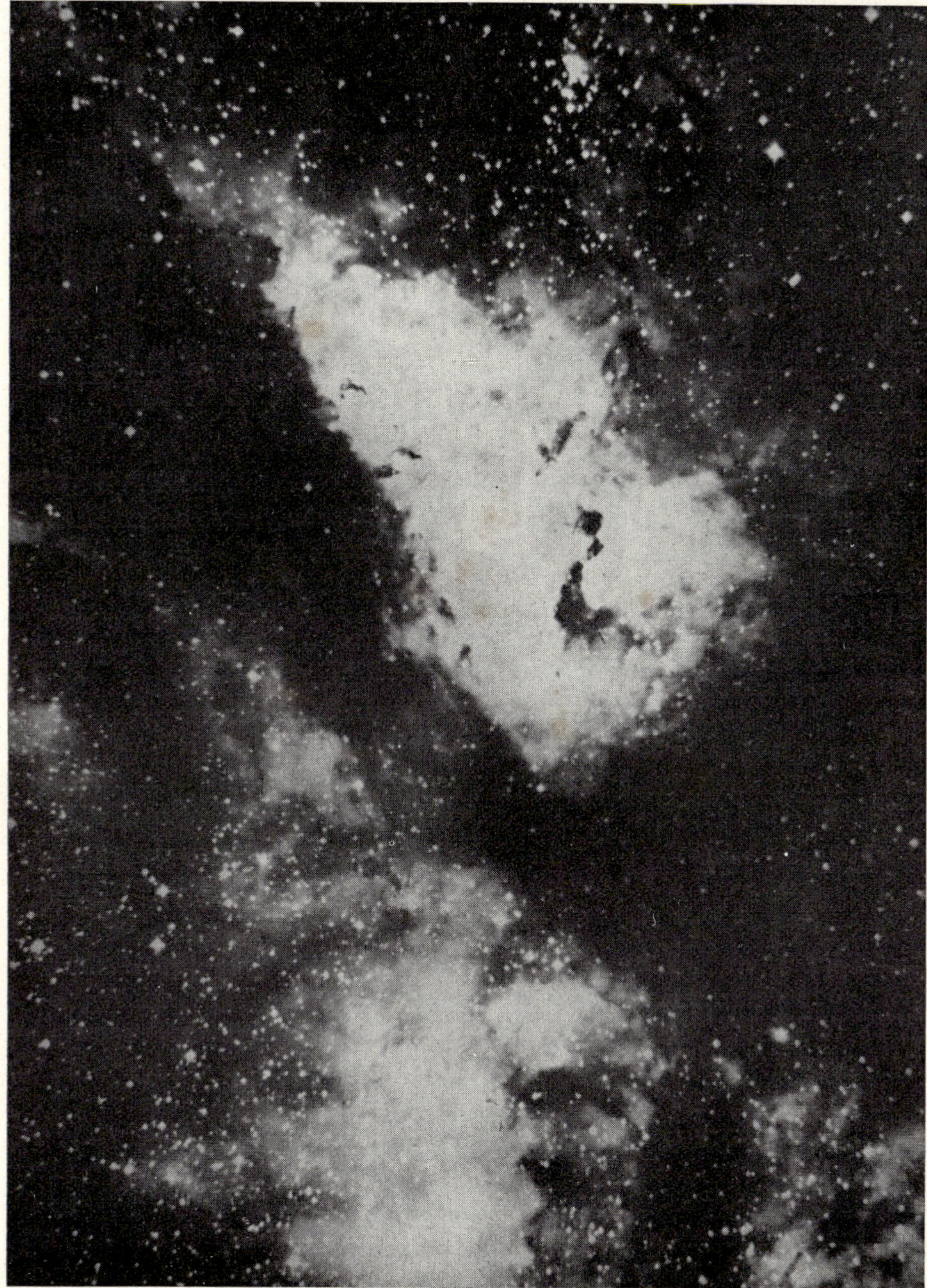

Fig. 5. – The bright nebula NGC 3372 (Cordoba Observatory).

respectively. This is not however my opinion, at least concerning NGC 3293. A third cluster (anonymous) lies at 1.7 kpc. Recently BEER (1961) confirmed the depth of the distribution of early-type stars in Carina; from his results one can estimate two condensations at 1.7 kpc and 2.5 kpc, respectively. It can be remarked that BOK (1959) does not agree with the current view of the existence of an inner spiral arm of the Galaxy running from a distance of, say, roughly 2 000 pc in Sagittarius to 1 500 pc in Carina, but favours the idea of a transverse arm directed more or less away from the Sun in Carina.

At Cordoba I obtained the spectra of several stars from the Morgan and Münch list. Among them there are 3 Wolf-Rayet stars (2 WN and one WC) and two A-supergiants; the rest are main sequence stars (luminosity class V) from O9 to B2. From the strength of the interstellar H- and K-lines I estimate that all these stars are more or less at the same distance, with the possible exception of the WC star and a probable foreground B2V star. The distance modulus of the remaining stars (eleven in number) is 11.4 ± 0.5 (estimated p.e.). Adopting an absorption of 0.7^{M} we get a distance of 1.4 ± 0.3 kpc.

It seems thus that these stars belong to Hoffleit's nearer condensation; this is not unreasonable, since they are all among the (apparently) brightest.

I consider, however, this distance somewhat too small because, due to selection effects, the absolute magnitudes of the stars should be brighter than that of the average main sequence stars; besides, the absolute magnitude I adopted for the 2 Wolf-Rayet stars (-5.5^{M}), although brighter than that commonly accepted, may still be too faint, since FEAST (1960) found several Wolf-Rayet stars among the brightest members of the association around 30 *Dor* in the Large Magellanic Cloud.

For this reason, I would adopt

$$r = 1\,500 \text{ parsec}$$

as the most probable value of the distance of the association Carina II. Incidentally, this value is very close to the mean of Hoffleit's and Beer's value for the distance of the nearer concentration of O stars. There is the possibility that another O-association exists in the same direction, but about 1 kpc farther. According to current views, 1.5 kpc is also the distance of the spiral arm in Carina.

An interesting member of the association Carina II is *AG Carinae;* this is a typical P Cygni star with irregular « novalike » outbursts. Its spectrum varies between B0p and A0p in an irregular way; the velocity of the expanding shell in which the P Cygni-lines are formed is also variable. Another P Cygni star very close to *AG* is *GG Carinae*; this is however an eclipsing binary and a foreground object.

Several finite variables of unknown type are present in this region; but as yet it has not been possible to ascertain whether some of them belong to the T Tauri class.

6. – The membership of η *Car* to the association Carina II.

The fact that η *Car* is situated in the central part of the nebula NGC 3372 and the agreement of the order of magnitude of the distances make it very probable that η *Car* itself is a member of the association Carina II. As, however, some doubt may subsist concerning the distance of the star and of the association, it is important to look for some independent evidence.

Doubtless, the strongest evidence is afforded by the interstellar lines. A simple inspection of the spectra of η *Car* and of several members of the association shows that the strength of the interstellar H- and K-lines is the same; a nearer object, like *GG Carinae*, can be picked out at once, while the probability that η *Car* be situated behind the dense region of the nebula NGC 3372 is very small. We may therefore conclude that it is almost certain that η *Car* is situated inside the association Carina II and is, thus, one of its members.

Another piece of evidence may be found in the connection of η *Car* with NGC 3372. Bok (1932) found that apparently no connection exists between η *Car* and the nebula. But this is not at all surprising, since the temperature of the star, as estimated from the excitation of the emission spectrum, would not be high enough to excite the nebula. However, during the first half of the last century the star was several thousand times brighter and there is also the possibility that it was hotter than now. It seems difficult than this change would not reflect itself upon the nebula if the two are related.

Now, there are of course no photographic documents showing the η *Carinae* region as it was more than a century ago; but fortunately around 1840, one of the keenest visual observers of all times, Sir J. Herschel, spent almost three years observing the nebula and the whole region with the utmost care (Herschel, 1843). I have discussed the visual observations by Herschel and others a few years ago (unpublished) and I draw the conclusion that there is a great probability that the portion of the nebula close to η was much brighter in 1840 than it is now. I have no time to go into details; Fig. 6 and the legend which accompanies it may be inspected for further notice.

Of course, it is well known that old visual observations of nebulae are very unreliable (see, for instance, Holden, 1878); but I feel that in this instance, Herschel's observations are so accurate and his statements so explicit that it is very difficult to admit a large error in his exceptionally beautiful work, probably the best of this kind that has been done before the photographic era.

If real, the changes must have taken place around 1860, when η was rapidly fading from the 1-st to the 7-th magnitude (POWELL, 1864).

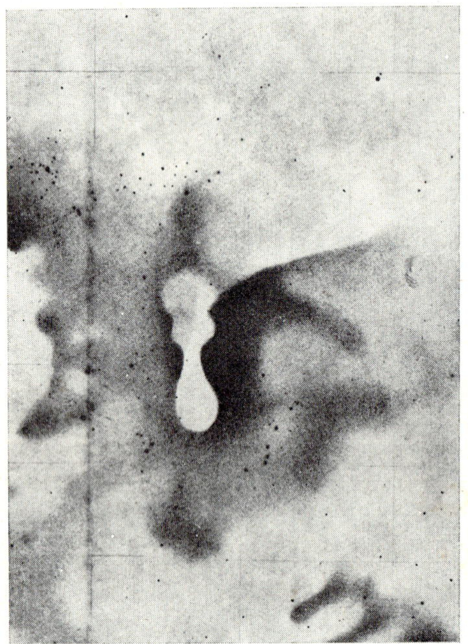

Fig. 6*a*. – Drawing of the central part of NGC 3372 by J. Herschel.

Fig. 6*b*. – Photograph of the central part of NGC 3372 (Cordoba Observatory).

The comparison of the photograph and of Herschel's drawing strongly suggests that some changes have occurred in that part of the nebula which is close to η. Herschel states very cleary that the stars 589, 607, 616, 670 and 664 lie almost exactly at the edge of the dark feature in the form of a « lemniscata » just in the middle of the nebula; he adds that the relative position of the stars and the sharp edge of the nebula might show in future possible changes of the shape of the nebula. Between 1860 and 1870 there was a very hot discussion among astronomers concerning the fact wether the « lemniscata » was closed (as Herschel saw it in 1938) or open (as it is to-day).

Taken as a whole, the present evidence is strongly in favour of the membership of η *Car* to the association Carina II. For this reason I will adopt for the star the same distance that was found for the association, that is

$$r = 1\,500 \text{ parsec},$$

and an apparent visual modulus

$$m - M = 11.6 .$$

7. – Physical properties of η Carinae.

We are now in condition to draw a few conclusions concerning the physical properties of this very peculiar object.

At maximum brightness (in 1843) η was of apparent magnitude $m = -1.0$, or absolute magnitude $M = -12.6$. This is well above the commonly adopted value for Novae (– 7.3, CECCHINI and GRATTON, 1941), but below that of Supernovae. Even now η is an intrinsically bright object ($M_v = -4.6$).

η *Carinae* is certainly a very complicated object. Apart from the expanding halo, which was discussed in Section **4**, it consists of a stellar nucleus of absolute magnitude $M = -3.6$, more or less, and probably not very high surface temperature (of the order of 8000 °K). The nucleus is subject to irregular outbursts corresponding to great increases of its luminosity; during the past outbursts the matter which forms the present halo has been ejected. It may be of some interest to note that although the outbursts of 1843 and 1856 corresponded to a much greater increase of the brightness of the nucleus than that of 1889, yet the velocity of the matter ejected was of the same order (300 km/s).

The nucleus itself is surrounded by an expanding shell which is responsible for the P Cygni lines. The velocity of expansion of the shell is of 475 km/s and thus possibly larger than that of the halo. Its density is certainly very low as is shown by the rich forbidden spectrum of FeII. It has been reported by Aller that an investigation of this spectrum is in progress; I am also studying a large collection of spectra obtained a Bosque Alegre from 1943 to 1958. It may then be hoped that shortly we will know more about the density, temperature and mass of the shell and also about the temperature of the nucleus. Incidentally, I may mention that the spectral energy distribution in the continuum suggests a higher temperature of the nucleus than that adopted.

These are more or less ascertained facts. Let us now speculate a little about other properties.

During the last 250 years, at least, η was very bright, occasionally one of the brightest stars of the sky. The total energy emitted under the form of light might be obtained from the light curve if we knew the bolometric correction. This is unknown, but, if when the star was bright the spectrum was not too different from that of the 1889 maximum, it must not have been too large. Let us assume it to be zero. Then, when the star was, respectively, of magnitude (apparent) 4, 2 and 0, its luminosity was $L = 8 \cdot 10^4 \odot$, $5 \cdot 10^5 \odot$ and $3 \cdot 10^6 \odot$. It is clear that the shorter time of high brightness is then compensated by the greater emissivity; taking, therefore, an average value of $5 \cdot 10^5 \odot$, we get $1.5 \cdot 10^{49}$ ergs, for the total energy emitted during 250 years.

I consider this as a lower estimate and would favour a much higher value on account of the bolometric correction and of the possibility of a more prolonged period of brightness. Furthermore, we ought to take into account the kinetic energy of the ejected matter, which is probably of the same order as the energy emitted in the form of light.

It is then clear that the energy liberated by η *Car* during its period of brightness is more or less the same as that emitted by the Sun during all its life.

Now, the present radius of the nucleus of η *Car* can be estimated from its temperature and luminosity. Again, neglecting the bolometric correction, one finds.

$$R_{\text{nuc}} = 25 R_{\odot} = 2 \cdot 10^{12}\ \text{cm}\ .$$

The gravitational energy liberated by a mass $\mathcal{M}$ contracting from infinity to a radius R is

$$E = \alpha G \frac{\mathcal{M}^2}{R}\ ,$$

where G is the gravitational constant and α is of the order of unity. The mass which has to contract to a sphere of 25 solar radii to liberate $1.5 \cdot 10^{49}$ erg is, thus, $\mathcal{M} = 2 \cdot 10^{34}$ g, or 10 solar masses. If the structure were homologous to the Sun, the time of contraction would be less than 200 years at a luminosity of $5 \cdot 10^{5} \odot$.

The agreement is « too good to be true »; it is evident that things are not quite so simple and we have to learn much more about this wonderful object, η *Car*, before going to conclusions. But I think that this at least may be admitted: the assumption that what was formerly called an « η *Carinae* event » represents the very early stages of a massive star is not conflicting with our current ideas and with observations.

REFERENCES

BEER, A.: 1961, *M. N.*, **123**, 191.
BOK, B. J.: 1932, *Harv. Rep.*, 77.
BOK, B. J.: 1959, *Obs.*, **79**, 61.
BOK, B. J. and VAN WIJK: 1952, *A. J.*, **57**, 213.
CECCHINI, G. and GRATTON, L.: 1941, *Atti Ac. d'Italia, Classe Sc. Fis.*, **13**, 1.
FEAST, M. W.: 1960, *M. N.*, **122**, 1.
GAVIOLA, E.: 1946, *Revista Astronomica*, **18**, 252.

Gaviola, E.: 1950, *Ap. J.*, **111**, 408.
Gaviola, E.: 1953, *Ap. J.*, **118**, 234.
Gratton, L.: 1955*a*, *Revista de la Un. Mat. Argentina*, **17**, 79.
Gratton, L.: 1955*b*, *3-rd Symposium of the I.A.U.* (Non-Stable Stars), 87.
Gratton, L.: 1958, 8me *Coll. d'Astroph.*, Liège, 213.
Gratton, L. and Ringuelet, A.: 1958, *Per. Sv.*, **11**, 35 (Paper presented as the dedication of the Pulkova Observatory).
Herschel, J.: 1843, *Results of Observations at the Cape.*
Hoffleit, D.: 1956, *Ap. J.*, **124**, 61.
Holden, E. S.: 1878, *Washington Observations for 1878.*
Innes, R. T. A.: 1903, *Cp. A.*, **9**, 75 B.
Le Sueur, A.: 1869, *Proc. R. S.*, **18**, 245.
Mervill, P.: 1928, *Ap. J.*, **67**, 391.
Münch, L. and Morgan, W. W.: 1953, *Bul. Obs. Ton.*, **8**, 19.
O'Connell, D. J. K., S. J.: 1956, *Vistas in Astronomy*, **2**, 113.
Powell, E. B.: 1864, *M. N.*, **24**, 171.
Ringuelet, A.: 1958, *Z. f. Ap.*, **46**, 276.
Sandage, A.: 1958, *Stellar Populations*, 79.
Thackeray, A. D.: 1949, *Obs.*, **69**, 31.
Thackeray, A. D.: 1950, *M. N.*, **110**, 524.
Thackeray, A. D.: 1953*a*, *M. N.*, **113**, 211.
Thackeray, A. D.: 1953*b*, *M. N.*, **113**, 237.
Thackeray, A. D.: 1956*a*, *Obs.*, **76**, 103.
Thackeray, A. D.: 1956*b*, *Obs.*, **76**, 154.
Thackeray, A. D.: 1961*a*, *Obs.*, **81**, 99.
Thackeray, A. D.: 1961*b*, *Obs.*, **81**, 102.
van den Bos, W.: 1938, *Un. Obs. Circ.*, n. 100, 522.
Wesselink, A.: *I. A. U. Trans.*, vol. X, 533.
Whitney, C. A.: 1952, *H. B.*, **921**, 8.

Relation between the Evolution of Stars and Galaxies.

E. M. Burbidge

University of California - San Diego, Cal.

1. – Introduction.

Since the study of stellar evolution is now based on a good fundation of observation and theory, it is natural to ask how galaxies evolve. This they must do, because of the evolution of the individual stars of which they are composed, and because of the gradual depletion of the residual gas in a galaxy after its first generation of stars has formed. The time for gravitational contraction is relatively so short for a massive star that the first stars to form and evolve must be the massive ones. Then, as a galaxy ages, the low-mass stars have time to complete their gravitational contraction, while the stellar remnants of the evolved high-mass stars become white dwarfs. The final stage, given a long enough time, would be a galaxy composed wholly of white dwarfs and very low-mass red dwarfs on the main sequence.

2. – Classification of Galaxies.

Galaxies are classified according to their structures in the well-known Hubble sequence, recently revised and extended by Sandage (1961):

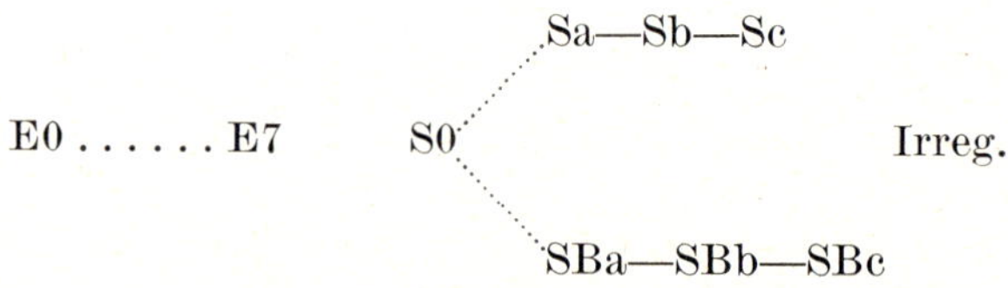

The prominence of the nuclear region of spiral galaxies in comparison with the importance of the spiral arms increases on going from Sc through Sb to Sa type. Roughly, as was suggested several years ago by Shapley (1950), this sequence might represent a possible evolutionary sequence, in the sense: irregular → spiral → S0 → elliptical, because the irregulars in general have more gas and high-luminosity stars as shown by the strength of their H II regions, while this proportion decreases on going through Sc → Sb → Sa types until one comes to the ellipticals, with very little gas and no dust. Let us now consider this change in component population with galactic form.

3. – Stellar population in relation to evolution.

This dates from the classical paper by BAADE (1944) describing the resolution of the elliptical galaxies in the local group into stars whose brightest

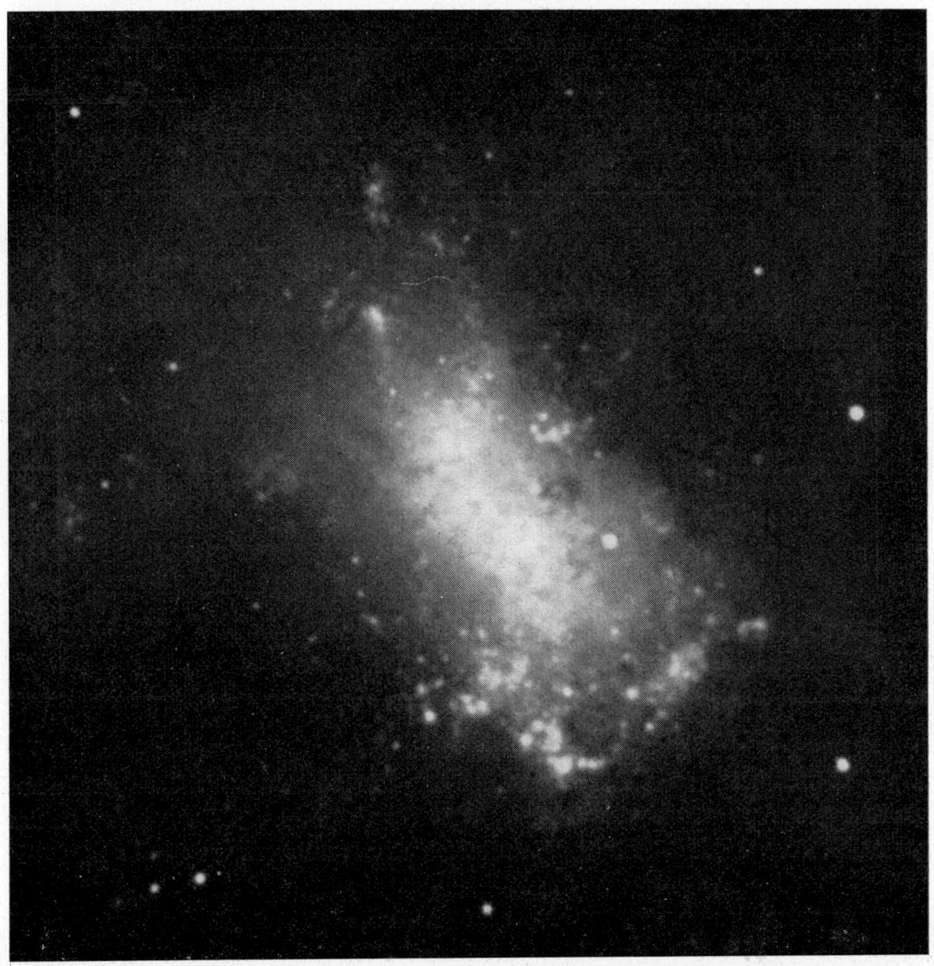

Fig. 1.

members were red giants which he considered to be similar to those in globular clusters and which were all then discussed as « Population II » stars, as distinct from the type of stars found in the solar neighborhood in our Galaxy, whose

Fig. 2.

brightest members are O-B stars and are like the stars found in the spiral arms of other galaxies; these were then called « Population I ». The work on stellar evolution then showed that the globular cluster stars were very old while O-B high-luminosity stars were young.

The nuclear bulge of M31 also contained what were called Population II stars. Wherever this type of stars were found, there was no sign of the existence of any dust and possibly very little gas; this suggested that such regions contained no material out of which new stars could form. The discovery that there were very old clusters of stars which belonged to the disk of our Galaxy, and which were not the same as the globular cluster stars, led however, to the

Fig. 3.

expansion of Baade's concept of the populations into a range of 5: Extreme Pop I; Intermediate Pop I; Disk; Intermediate Pop II; and extreme or halo Pop II. (See *Vatican Symposium on Stellar Populations*, 1957).

The nuclear region of M31 has a stellar population, according to the spectral classification work of MORGAN and MAYALL (1957), similar to the disk population of our Galaxy, *i.e.* similar to the kind of stars in the old galactic clusters M67 and NGC 188. These stars, then, are old but have a heavy-element–to–hydrogen abundance ratio not too different from that of the Sun. The stellar populations in the brighter elliptical galaxies appear to be similar.

Fig. 4.

The fact that a sequence not inconsistent with an age sequence can be set out in the sense irregular $\rightarrow$ spiral $\rightarrow$ S0 $\rightarrow$ E does not mean, however, that such a path need be taken by any particular galaxy. In present-day attempts to make a theory of galactic evolution embracing the observed forms of galaxies, we have to remember that for a long time the observed H-R diagram of field stars was the graveyard of theories of stellar evolution. It was not until the importance of separating the parameters of mass, age, and chemical composition was realized, and a theory developed for the structure of stars with a chemical inhomogenity, that advances were made.

In the evolution of galaxies, it may be expected that the mass of a protogalaxy may be as important in determining the properties of the galaxy as is the case in stars. In the case of a galaxy, initial parameters which may be of very great importance in determining its form and evolutionary history might be its initial angular momentum per unit mass, the initial magnetic field, and perhaps the initial amount of turbulence. Finally, the question must be considered, whether the time scale for galactic evolution might be too long, and the galaxies formed at too nearly the same epoch in the past, for evolution to be deduced as it can in stars by observing groups of different ages.

Let us now look at some different types of galaxies, shown in Figures 1-4:

NGC 4449 – Irregular. Much ionized gas, with high-luminosity aggregates of stars to excite it.

NGC 4736 – Sb with fair-sized nuclear bulge.

NGC 7814 – Edge-on spiral with great flattening and lesser nuclear bulge. Very flat dust layer. Such an object must have large ratio of angular momentum to momentum in the Z-direction.

NGC 5866 – S0 with dust, edge-on but much less flattened than NGC 7814.

It is unlikely that so flattened an object as NGC 7814 could ever evolve into an elliptical; it would have to lose angular momentum to do so.

4. – Estimation of stellar populations in various galaxies.

How can stellar populations in galaxies be estimated? First, by measurement of colors, preferably as a function of position in a galaxy. The integrated color of an assembly of stars containing a main sequence up to $B-V \sim 0.4$ together with red giants is obviously much redder than an assembly containing main-sequence stars up to spectral type A or up to type O and B. The integrated colors of ellipticals, also S0 and those spirals where most of the light comes from the nuclear bulge, do indeed have values of ~ 0.8 to 1.0 (DE VAUCOULEURS, 1961). Integrated spectra will also give the population type; MORGAN and MAYALL (1957) were able to detect the effect of main-sequence stars around early G-type in the ultraviolet spectral region of the nucleus of M31, while the spectra in the blue-green region matches well the spectrum or normal solar-neighborhood-type giants of types G8 III-K3 III. In this connection, the recent work of SPINRAD (1962) is of interest. He believes that the yellow-red spectral regions of such galaxies can be used to indicate the proportion of lower-main-sequence stars present. The strength of the Na I D-lines depends on luminosity in late-type stars; SPINRAD believes that galaxies with strong D-absorption lines have relatively more low-mass dwarfs making up their population. The D-line strength is certainly variable from galaxy to galaxy; the effect of interstellar absorption within the galaxy itself is, however, still uncertain. (N.B.: We have found some indications of a correlation between D-line strength and dust in galaxies.)

An easily measurable line ratio in the red spectral region has been found to be an indicator of stellar population. This is the emission line ratio H_α/[N II]λ6583 (BURBIDGE and BURBIDGE, 1962). This ratio changes from a value near 3 in normal spiral arm regions (it may exceed 3 in particularly bright high-excitation H II regions) to around unity in the nuclear bulges of some galaxies, and even to values much less than unity in the nuclei of those galaxies with large nuclear bulges which have relatively little gas in them (*e.g.* NGC 3623 (Sa), 5005 (Sb), 4736 (Sb)). Other galaxies, however, (irregulars, some Sc and even some Sb galaxies) may have the ratio characteristic of spiral arms right in their nuclei, showing that there is plenty of gas and young high-luminosity stars in the very nuclei. Unless one wishes to interpret the change in spectral-line ratio as an abundance effect in the gas (that in the nuclei being only what has been ejected from evolved stars, and poorer in hydrogen), then this is likely to be an excitation effect, and could be explained by considering the radiation of the gas in spiral arm regions to be excited by the radiation field of hot stars, while in the nuclear region the radiation field comes mainly from K giants, and is therefore relatively unimportant, and the excitation is done by electron collisions in gas which must get its energy from kinetic energy. In M51 the H_α line disappears right in the nucleus while [N II] is quite strong and broad, due to the Doppler broadening presumably from a considerable velocity dispersion in the gas.

An interesting galaxy in this and other respects is the irregular M82. The H_α/[N II] ratio is found to vary from one region to another. Also, the spectrum in part of M82 near the center (which is riddled with dust lanes) indicates fairly high excitation from the emission-line features, while a region on the other side of the principal central dark lane, also containing much dust, changes abrupty to an absorption-line spectrum of early *A* type, suggesting the absence of the most massive O and B stars needed to create extensive H II regions. This needs further study.

Finally, there is another parameter which can be used as an indicator of the population of the faint main-sequence stars in a galaxy. This is the mass-to-light ratio, which can be given (in solar units) if the total mass and luminosity of a galaxy are known. For the Sun itself the ratio is by definition unity. Because of the dependence of luminosity on a power greater than unity of the mass of a star ($L \sim M^{3 \text{ or } 4}$), an assembly of stars containing very many O and B star will have a much lower mass-to-light ratio, and one with large numbers of main-sequence stars no brighter than a limit somewhat below the Sun will have a larger mass-to-light ratio. The presence of much dust or uncondensed neutral or molecular hydrogen will also, of course, reduce M/L in a galaxy.

To sum up, colors, spectra, and the mass-to-light ratio can be used to estimate stellar population in galaxies where the individual stars cannot be resolved. The presence of dust and gas goes with a population containing young stars and this is found in the spiral arm regions of galaxies and in the nuclei of some galaxies. It occurs most extensively in irregular, next in Sc, Sb and finally Sa galaxies. An old population is found in regions devoid of dust and with little gas; this occurs in the nuclei and nuclear bulges of some galaxies of all types, but more predominantly in those of Sa or Sb type, and also occurs throughout E and S0 galaxies.

5. – Masses of Galaxies.

We come now to the parameter which is likely to be of great importance in considerations of galactic evolution — the mass. The most direct way of determining the mass of a galaxy is by observing the rotation curve. The mass distribution is then obtained essentially by solving an integral equation, which can be done in various ways (see SCHWARZSCHILD, 1954; BURBIDGE, BURBIDGE and PRENDERGAST, 1960*a*). From the mass distribution the total mass is obtained. Secondly for a spherical elliptical galaxy (E0) which has little or no angular momentum, it can be assumed that the virial theorem will give for the stars in the center a relation between kinetic and potential energy:

$$2\,\text{K.E.} + \text{P.E.} = 0\,,$$

since the galaxy can be assumed to be in a steady state. Thus an equation giving the mass can be obtained from the observed line-of-sight velocity dispersion in the center of such an elliptical galaxy (assuming symmetry of velocities in 3 dimensions) and the observed light distribution (assuming a constant M/L ratio through the galaxy).

Thirdly the masses of pair of galaxies can be obtained statistically by measuring the differences in line-of-sight velocity between members of the pairs, and assuming random distribution of orbital planes and lines of nodes (*e.g.*, PAGE, 1961, 1962).

Results of determinations by various authors are given in the tables. References can be found in the volume on the I.A.U. Symposium held in Santa Barbara in August, 1961. Also, references will be found in the proceedings of the conference on Groups and Clusters of Galaxies (also Santa Barbara, August 1961), published in *Astronomical Journal*, vol. 66, no. 10 (see table on p. 619), and in the Transactions of the I.A.U. meeting at BERKELEY, 1961, Report of Commission 28.

TABLE I. – *Spirals and irregular galaxies: masses and mass-to-light ratios on distance scale of* $H = 75$ km s^{-1} Mpc^{-1}.

Galaxy	Type	Mass (unit: 10^{10} $M_{\odot}$)	M/L (solar units)
NGC 3556	Irr-Sc	1.4	1.4
L.M.C.	Irr (Sm)	1.3	4
NGC 55	Irr	4	6
M 82	Irr	1.5	7.4
NGC 2146	Scp	1.8	3
M 33	Sc	1.8	11
M 101	Sc	1.4	13
NGC 157	Sc	6.0	1.9
NGC 5248	Sc	3 ÷ 7	2 ÷ 4
NGC 5055	Sbc	5.5	2.8
NGC 2903	Sb	4.9	4
NGC 3646	Sbcp	20 ÷ 30	2.5 ÷ 4
NGC 253	Sc	30	10
NGC 5005	Sb	9	6
M 81	Sb	15	20
M 31	Sb	34	23 ÷ 10
NGC 3623	Sab	20 ÷ 30	10 ÷ 20
S + Irr	Average from pairs	2.0	1.0

The tables show that the masses increase systematically on going from irregular and Sc galaxies, through Sb, Sa and S0, finally being very large for the E0 galaxy M87.

TABLE II. – *Elliptical and* S0 *galaxies: masses and mass-to-light ratios on distance scale of* $H = 75$ km s^{-1} Mpc^{-1}.

Galaxy	Type	Mass (unit: 10^{10} $M_{\odot}$)	M/L (solar units)
Sculptor dwarf	dE	0.0002	2 ÷ 4
Fornax dwarf	dE	0.002	2 ÷ 4
M 32	E2	0.18	13
NGC 4111	S0	1.2	13
NGC 3379	E0	10	12
NGC 3115	S0-E7	11	19
M 87	E0	100	60
E + S0	average from pairs	80	70

The M/L ratios increase in a similar way. The most recent statistical results of PAGE (1962) give a value for M/L for irregular + spiral galaxies that is now in excellent agreement with the results from rotation curves for individual galaxies. The E and S0 galaxies are considered together in Page's statistical results, but it can be seen from Table II that the only two S0 galaxies for which we have individual determinations have masses very similar to the large Sb and Sa galaxies in Table I.

6. – Evolutionary significance of results on M and M/L.

The steady progression in the M/L ratio from irregular and Sc galaxies, through the giant Sb galaxies, Sa and S0, to the giant E0 object (M87) indicates a steady change in stellar population towards a greater proportion of low-luminosity dwarfs (either main-sequence or white dwarfs); which contribute most of the mass but little of the light; the light comes from the evolving red giants mainly. The masses indicate that an irregular or Sc galaxy could only evolve to an elliptical if its mass could increase during its life history. Even more striking, we have seen in Section **3** that the amount of flattening

Fig. 5.

of a galaxy towards the equatorial plane for many spirals suggests a great difference in the angular momentum per unit mass for different forms of galaxy, and makes it unlikely that a flattened spiral could ever evolve into an elliptical.

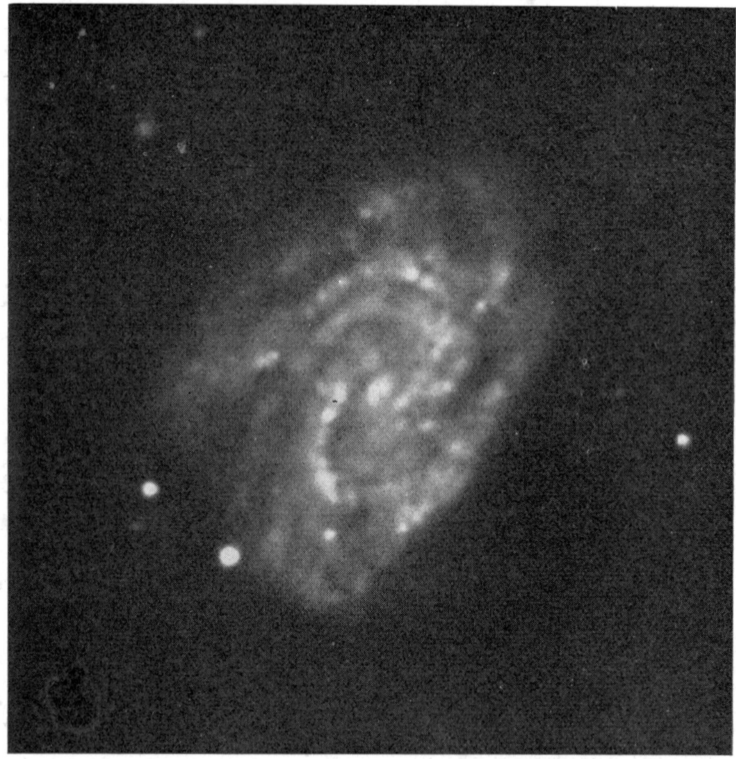

Fig. 6.

However, there are no arguments that can be raised against the possibility of an evolutionary path from irregular → Sc → Sb → Sa → S0 since the flattening of many S0 galaxies is as great as that of spirals. The presence of appreciable rotation in an irregular galaxy must eventually lead to an approach to circular symmetry.

What is the significance of a large difference in mass and M/L ratio between the ellipticals and spirals? Since average diameters of the two kinds of galaxy do not differ, there must be a big difference in density. The lower angular momentum per unit mass of the ellipticals and their higher density suggest that they come from protogalaxies which collapsed more rapidly. Since the rate of star formation must depend on density of the material out of which stars form, and even perhaps on the square of the density, the initial rate of star formation may have been much more rapid in those galaxies we now see as ellipticals, and the available gas and dust have been long ago all used up so that no further new stars could be formed.

Finally, the results already available show that a different mass distribution in spiral galaxies with a similar Hubble type and very similar total masses and M/L ratios, can go with a different appearance of the galaxy. NGC 5055 and NGC 157 (in Table I) show this very well. NGC 5055 has a mass distribution (from the rotation curve) which if plotted as (log density) versus distance from the center, gives a nearly linear plot.

The drop in density from the central region to the boundary of the main luminous region is thus very steep, and goes by a factor $\sim 10^3$. On the other hand, the plot of *density* (not log density) against distance in NGC 157 is itself nearly linear, and the corresponding drop in density from the central region to the boundary is only by a factor of ~ 10.

The photographs of these two galaxies in Fig. 5 and 6 show that the spiral arms in NGC 5055 are thin, rather regular, much branched, while those in NGC 157 are broad and not much wound up (BURBIDGE, BURBIDGE and PRENDERGAST, 1960*a*, 1961*a*).

7. – Barred spirals.

We have said nothing yet about the branching of the spiral sequence into the normal and the barred spirals. A program to determine motions in barred spirals and to derive their masses has led us into more difficulties than were anticipated, mainly because the lack of circular symmetry in the barred spirals makes it hard to determine the orientation of the galaxy in space, and also because there appear to be motions other than circular motions present. The best observed case is NGC 7479 (BURBIDGE, BURBIDGE and PRENDERGAST, 1960*b*). There is an indication that the barred spirals have more angular momentum per unit mass than the normal spirals.

Figure 7 shows NGC 7479, which is very large; the length of the bar is 18 500 pc. The rotation curve along the bar is linear. The barred spiral NGC 1365 is also a very large object, with a total radius of more than 26 000 pc. Observations have not yet been made to determine the motions in its outer part, but the inner nuclear region is in very fast rotation (BURBIDGE, BURBIDGE and PRENDERGAST, 1962) (this nuclear region contains much gas and dust and Pop I Stars).

A study of the streaming motions in the barred spirals has been made by Prendergast and is being prepared for publication.

Fig. 7.

8. – Spiral structure.

This leads us to the consideration of spiral structure itself. It has long been known that the existence of well-marked spiral arms with, on the average, only $1\frac{1}{2}$ turns is a problem in galaxies whose ages may be $\sim 10^{10}$ years, since differential rotation should wind up the arms completely (1 rotation period is 10^8 years). (This problem is quite different from that of the stability of a spiral arm considered by CHANDRASEKHAR and FERMI, in which an isolated arm not in differential rotation was studied.) Appeals to some kind of lag velocity between the gas making up the spiral arm and the stars giving the main mass distribution of the galaxy will not work in the case of NGC 5055, for example, in which the rotation curve was derived from a study of the ionized gas itself, and gave directly the time for winding up the observed gaseous distribution by one turn as $\sim 10^8$ years (BURBIDGE and PRENDERGAST, 1960).

The problem of the maintenance of spiral structure suggests a possible connection with noncircular motions in the nuclear regions of galaxies, which do exist according to recent work.

9. – Peculiar galaxies.

In the final section, I would like to mention some galaxies which either appear not to be in dynamical equilibrium or which give evidence from their forms for being young on their own evolutionary time scale. First, we have NGC 3646 (Figure 8); this is a very large spiral galaxy that has an outer gaseous

Fig. 8.

structure in which the motions cannot be reconciled with Keplerian motion. The dimensions of this outer structure are very large (the diameter of the bright luminous part is 41 000 pc). Perhaps we have evidence here for an interaction between a galaxy and the intergalactic medium (BURBIDGE, BURBIDGE and PRENDERGAST, 1961*b*). NGC 2444-5 is one of the interacting galaxies

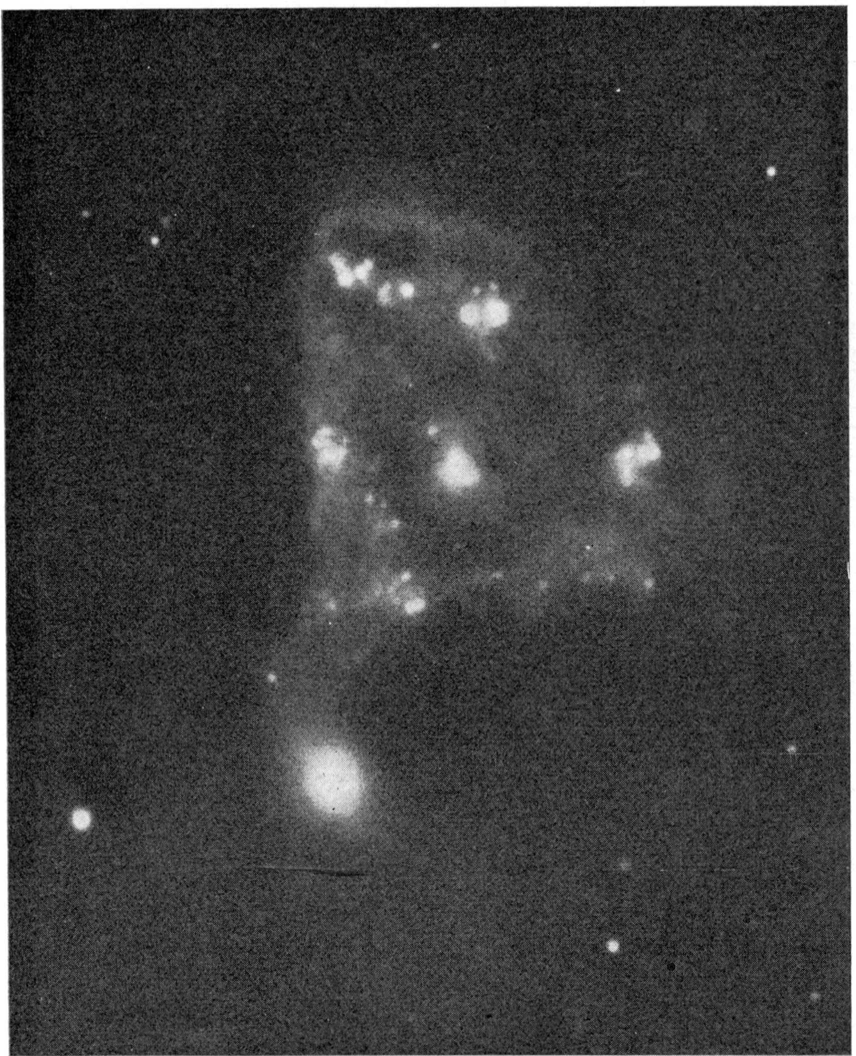

Fig. 9.

catalogued by VORONTSOV-VELYAMINOV (1960), and is shown in Fig. 9. It consists of an apparently normal elliptical galaxy in association with giant complex of H II regions grouped around a knot which may be the nucleus (BURBIDGE and BURBIDGE, 1959*a*). This looks like the formation of a galaxy in the presence of an already existing one.

NGC 4676 (BURBIDGE and BURBIDGE, 1961) is a pair of galaxies, apparently an outlying member of the Coma Cluster, with long projecting tails; it is shown in Fig. 10. The total length of the object with the biggest tail is more

Fig. 10.

than 50 000 pc. Both objects are in fast rotation and contain much ionized gas. It is hard to imagine that such a configuration will last for a long time-scale.

There are in the Vorontsov-Velyaminov Atlas a number of completely unsymmetrical objects which are yet in a state of rapid rotation (an example is NGC 4194, in Fig. 11). These all contain much ionized gas. The presence of considerable rotation in such irregular objects should eventually lead to an approach to circular symmetry; consequently these objects appear to be young. These will be discussed by HOYLE, BURBIDGE and BURBIDGE in a forthcoming publication. One such fast-rotating irregular gaseous object lies in the Hercules Cluster, a cluster which, from study of its kinetic and potential energies and application of the virial theorem, appears either to have a rather short time-scale $\left((1 \div 2)\cdot 10^9 \text{ years}\right)$ or to contain much uncondensed material (BURBIDGE and BURBIDGE, 1959*b*). A colleague of Professor HOYLE has recently studied the shape of a configuration of galaxies arranged in the sky like the

Hercules Cluster and with their random motions. The unsymmetrical shape of the cluster will change to a more symmetrical one, and the time scale for

Fig. 11.

loss of members is of the same size as that estimated by us for passage of a typical member across the dimension of the cluster.

Thus it looks as though there exist within about 100 Mpc of our Galaxy, a certain number of objects that look like newly formed galaxies, or like galaxies in which contraction and star formation on a rapid scale has only recently been taking place.

REFERENCES

BAADE, W.: 1944, *Ap. J.*, **100**, 137.

BURBIDGE, E. M. and BURBIDGE, G. R.: 1959*a*, *Ap. J.*, **130**, 12.
BURBIDGE, E. M. and BURBIDGE, G. R.: 1959*b*, *Ap. J.*, **130**, 629.
BURBIDGE, E. M. and BURBIDGE, G. R.: 1961, *Ap. J.*, **133**, 726.
BURBIDGE, E. M. and BURBIDGE, G. R.: 1962, *Ap. J.*, **135**, 694.
BURBIDGE, E. M., BURBIDGE, G. R. and PRENDERGAST, K. H.: 1960*a*, *Ap. J.*, **131**, 282.
BURBIDGE, E. M., BURBIDGE, G. R. and PRENDERGAST, K. H.: 1960*b*, *Ap. J.*, **132**, 654.
BURBIDGE, E. M., BURBIDGE, G. R. and PRENDERGAST, K. H.: 1961*a*, *Ap. J.*, **134**, 874.

Burbidge, E. M., Burbidge, G. R. and Prendergast, K. H.: 1961*b*, *Ap. J.*, **134**, 237.

Burbidge, E. M., Burbidge, G. R. and Prendergast, K. H.: 1962, *Ap. J.*, **136**, 119.

Burbidge, G. R. and Prendergast, K. H.: 1960, *Ap. J.*, **131**, 243.

Morgan, W. W. and Mayall, N. U.: 1957, *P.A.S.P.*, **69**, 291.

Page T.: *Proc. Fourth Berkeley Symposium on Mathematical Statistics and Probability*, **111**, 277.

Page, T.: 1962, *Ap. J.*, **136**, 685.

Sandage, A. R.: 1961, *Hubble Atlas of Galaxies* (Washington, D.C.: Carnegie Institution of Washington).

Schwarzschild, M.: 1954, *A. J.*, **59**, 273.

Shapley, H.: 1950, *Pub. Univ. Obs. Carnegie Michigan*, **10**, 79.

Spinrad, H.: 1962, *Ap. J.*, **135**, 715.

de Vaucouleurs, G.: 1961, *Ap. J. Suppl.*, **5**, 233 (n. 48).

Vorontsov-Velyaminov, B. A.: 1960, *Atlas of Interacting Galaxies* (Moscow).

Stars with Helium-Rich Cores.

R. KIPPENHAHN

Max-Planck-Institut für Physik und Astrophysik - Munich

Introduction.

In 1938 OEPIK first mentioned that inhomogeneous stellar models presumably can explain the giant stars, and at the present time we are quite convinced that we can explain at least some of the features of the H-R-diagrams of galactic and of globular clusters as evolutionary effects which are caused by the exhaustion of hydrogen in the interior of the stars. In my lecture therefore I will speak about this phase of stellar evolution in which the stars leave the main sequence and become giants or supergiants because of the exhaustion of hydrogen in the central region.

In the first chapter we will investigate whether the helium which comes from hydrogen burning in a star will be stored in the central region or whether it will be mixed with the matter in the outer regions. In the latter case the star would follow the evolutionary track of a completely mixed star, as it has been described by Dr. SANDAGE. In the next chapter we will discuss the physical problems of nonmixed evolution. We will see that for stars which are not very hot the outer boundary conditions are very important. In the last chapter therefore we will discuss the problem of the outer layers of these stars, especially the theory of the convective regions near the surface.

1. – Mixing due to meridional circulation.

1·1. *Nonspherical distortions.* – We write the equations of stellar structure in the case of radiative energy transport in vector form:

$$\nabla \Pi = -\varrho \nabla \Phi + \boldsymbol{c}\,, \tag{1}$$

$$\nabla^2 \Phi = 4\pi G \varrho\,, \tag{2}$$

$$f = -\frac{4ac}{3\varkappa\varrho} T^3 \nabla T \,, \tag{3}$$

$$\nabla \cdot f = \varepsilon\varrho \,, \tag{4}$$

(Π pressure, ϱ density, Φ gravitational potential, G gravitational constant, f vector of radiative energy flux, a constant of radiation pressure, c velocity of light, $\varkappa$ opacity, T temperature, ε rate of energy production). To make the system complete we have to assume that ϱ is given as a function of Π and T by an equation of state and that $\varkappa$ and ε are known functions of Π and T. The vector $\boldsymbol{c}$ may be a disturbing force (dyne per cm^3) which may produce a deviation from spherical symmetry in the star. As examples of the disturbing force we take rotational distortion, tidal distortion, and the distortions due to a toroidal and a meridional magnetic field.

a) Rotational distortion. Here $\boldsymbol{c}$ is the centrifugal force. With a spherically-symmetric distribution of the angular velocity $\omega = \omega(r)$, we have in spherical co-ordinates r, ϑ, φ:

$$c_r = \varrho\omega^2 r \sin^2\vartheta = \tfrac{2}{3}\varrho\omega^2 r(1 - P_2) \,, \tag{5}$$

$$c_\vartheta = \varrho\omega^2 r \sin\vartheta \cos\vartheta = -\frac{1}{3}\varrho\omega^2 r \frac{dP_2}{d\vartheta} \,, \tag{6}$$

$$c_\varphi = 0 \,, \tag{7}$$

with the spherical harmonic $P_2 = (3\cos^2\vartheta - 1)/2$, where $\vartheta = 0$ is the axis of rotation. We assume for simplicity that ω is a function of r only, but the following can also be done for $\omega = \omega(r, \vartheta)$.

b) Tidal distortion. The star we are considering may be star I of a double-star system with circular orbits. The center of star I may be the origin of a system of spherical co-ordinates rotating with the orbital motion. The center of the second star (star II with the mass M_{II}) may be on the axis $\vartheta = 0$ all the time. The distance of the centers of the two stars may be D. In this rotating system of co-ordinates star I may be at rest, which means, that star I rotates like a solid body with a rotational period which equals the orbital period. Therefore one has two types of distortion acting simultaneously on star I: the gravitational distortion caused by star II and the centrifugal force due to the rotation of star I around an axis perpendicular to the plane of the orbit. If we restrict ourselves to small distortions, both effects can be superposed and the rotational distortion gives a force as given in *a*) and the gravitational distortion is (as long as $r/D \ll 1$) given by

$$c_r = \frac{2}{D^3} G M_{\mathrm{II}} \varrho r P_2 , \tag{8}$$

$$c_\vartheta = \frac{1}{D^3} G M_{\mathrm{II}} \varrho r \frac{\mathrm{d} P_2}{\mathrm{d}\vartheta} , \tag{9}$$

$$c_\varphi = 0 . \tag{10}$$

This disturbing force has the potential

$$V(r, \vartheta) = -\frac{1}{D^3} G M_{\mathrm{II}} r^2 P_2 . \tag{11}$$

If one wants to add the centrifugal force given in eqs. (5), (6), (7) to the tidal force, one has to keep in mind, that ϑ has a different meaning in both cases.

c) Toroidal magnetic field. Nonspherical distortions can also be produced by a magnetic field. A simple toroidal field is given in spherical co-ordinates r, ϑ, φ by

$$\boldsymbol{B} = (0, 0, \psi(r) \sin\vartheta) , \tag{12}$$

with an arbitrary function $\psi(r)$, which may fulfil certain boundary conditions (for instance $\psi = 0$ at the surface to avoid sheet currents on the surface of the star) which are of no interest in this connection. For the disturbing force we get

$$\left\{ \begin{aligned} \boldsymbol{c} &= -\frac{1}{4\pi} [\boldsymbol{B} \times [\nabla \times \boldsymbol{B}]] \\ &= -\frac{1}{4\pi} \left[\frac{2}{3r} \frac{\mathrm{d}}{\mathrm{d}r} (r\psi) \psi (1 - P_2) , \; -\frac{6}{r} \psi^2 \frac{\mathrm{d}P_2}{\mathrm{d}\vartheta} , \; 0 \right] . \end{aligned} \right. \tag{13}$$

d) Meridional magnetic field. A simple meridional magnetic field is given by

$$\boldsymbol{B} = \left(\frac{2}{r^2} \psi \cos\vartheta , \; -\frac{1}{r} \frac{\mathrm{d}\psi}{\mathrm{d}r} \sin\vartheta , \; 0 \right) , \tag{14}$$

with an arbitrary function $\psi(r)$, which may fulfil certain boundary conditions which are of no interest in this connection. The disturbing force is then given by

$$\boldsymbol{c} = -\frac{1}{4\pi} \left[\frac{2}{3r^2} \frac{\mathrm{d}\psi}{\mathrm{d}r} \gamma (1 - P_2) , \; -\frac{1}{3r^2} \psi\gamma \frac{\mathrm{d}P_2}{\mathrm{d}\vartheta} , \; 0 \right] , \tag{15}$$

with $\gamma = \mathrm{d}^2\psi/\mathrm{d}r^2 - 2\psi/r^2$.

All four examples discussed here have in common, that c_r has the form $A(r)+B(r)P_2$ and c_ϑ the form $C(r)\,\mathrm{d}P_2/\mathrm{d}\vartheta$ and $c_\varphi=0$. In the next paragraph therefore we will assume $\boldsymbol{c}$ to have this form without specifying whether rotation or magnetic distortion or anything else causes the deviation from spherical symmetry.

1·2. *The proof of existence of circulations.* – For each of the examples *a)*–*d)* in the foregoing paragraph we can now write an expression for $\boldsymbol{c}$ in eq. (1). For small distortions we can assume that the solutions of the eqs. (1)–(4) with distortion are not very different from the solutions of the undisturbed spherically symmetric configuration. Therefore we write

$$\Pi(r,\vartheta)=\Pi_0(r)+\Pi_2(r)P_2\,, \tag{16}$$

$$\Phi(r,\vartheta)=\Phi_0(r)+\Phi_2(r)P_2\,, \tag{17}$$

$$T(r,\vartheta)=T_0(r)+T_2(r)P_2\,, \tag{18}$$

where Π_0, Π_2, Φ_0, Φ_2, T_0, T_2 are functions of r only. The functions Π_2, Φ_2, ... are small compared to Π_0, Φ_0 .... The functions Π_0, Φ_0 ... are in general *not* identical with the functions of the undisturbed star since in all our examples but in the case of tidal distortion the disturbing force has a radial part of spherical symmetry. But the functions Π_0, Φ_0 ... are not very different from the undisturbed case. We neglect second- and higher-order terms like $\Pi_2\cdot T_2$ and use the equation of state for simplicity with $\mu=\text{const}$ (μ means molecular weight); the chemical composition is assumed to be homogeneous everywhere in the region of interest and ionization neglected. Finally we have again to keep in mind, that ϱ, $\varkappa$, ε in the eqs. (1)–(4) can be considered as abbreviations for known functions of Π and T.

From eq. (3) it follows that the r-component of $\boldsymbol{f}$ has the form

$$f_r=f_{r0}(r)+f_{r2}(r)P_2 \tag{19}$$

and that there is a ϑ-component f_ϑ of $\boldsymbol{f}$ which has the form

$$f_\vartheta=f_{\vartheta 2}(r)\,\frac{\mathrm{d}P_2}{\mathrm{d}\vartheta}\,. \tag{20}$$

Separating in eqs. (1)–(4) the terms which only depend on r or are proportional to P_2 respectively $\mathrm{d}P_2/\mathrm{d}\vartheta$ we get linear ordinary differential equations for the functions with the subscript 2 with coefficients which depend on the spherically-symmetrical model only. In the coefficients we can for simplicity replace the functions Π_0, Φ_0 ... by the functions of the undisturbed model. The error is of second order. Table I gives the number of ordinary differential equations, which can be obtained. Every vector equation gives three ordinary equations: two from the r-component, and one from the ϑ-component. Every scalar equation yields two ordinary differential equations, one from the terms which depend on r only and one from the terms which are proportional to P_2. The four columns of the Table correspond to the four eqs. (1)–(4). The numbers in the lines indicate whether a certain equation has terms which are of spherical symmetry, are proportional to P_2 or to $\mathrm{d}P_2/\mathrm{d}\vartheta$. The number 1 in the table means that there are terms of this type in the equation, which therefore give one ordinary differential equation. The number 0 means that there are no such terms.

TABLE I.

	Equation No.			
	(1)	(2)	(3)	(4)
function of r only:	1	1	1	1
proportional: P_2	1	1	1	1
proportional: $\mathrm{d}P_2/\mathrm{d}\vartheta$	1	0	1	0

Table I shows, that we have 6 ordinary differential equations for the quantities with the subscript 2. On the other hand, the unknown variables with the subscript 2 (when ϱ, $\varkappa$, ε are replaced by Π and T) are:

$$\Pi_2\,,\quad \Phi_2\,,\quad T_2\,,\quad f_{r2}\,,\quad f_{\vartheta 2}\,.$$

We therefore have 5 unknown functions and 6 (linear) equations to determine these functions.

This means that if one for instance determines the variables in such a way that eqs. (1)–(3) and the spherically symmetrical part of eq. (4) are fulfilled then one cannot fulfil the other part of eq. (4). Therefore, the radiation flux cannot carry away all the energy. This means that there would be regions which are heated or cooled. This gives rise to meridional circulation which will transport energy in such a way that energy conservation is fulfilled at every point in the star. Therefore, we have to replace eq. (4) by a new equation which takes account of this additional energy transport (see for instance SCHWARZSCHILD, 1958 or KIPPENHAHN, 1958):

$$\nabla\cdot\boldsymbol{f}=\varepsilon\varrho+\frac{GM_r\varrho}{r^2}\,\frac{c_p\mu}{\mathfrak{R}}\left[\frac{\mathrm{d}\ln T}{\mathrm{d}\ln\Pi}-\left(\frac{\mathrm{d}\ln T}{\mathrm{d}\ln\Pi}\right)_{\mathrm{ad}}\right]v_r\,. \tag{21}$$

Here v_r is the radial component of the meridional velocity. For the factor in front of v_r one can take the quantities from the undisturbed model since v_r is already of first order. Since the nonvanishing part of $\nabla \cdot \boldsymbol{f} - \varepsilon \varrho$ is proportional to P_2 we have the same angular dependence for v_r:

$$v_r = v_{r2}(r) P_2 \,. \tag{22}$$

The equation of continuity than gives a ϑ-component

$$v_\vartheta = v_{\vartheta 2}(r) \frac{\mathrm{d}P_2}{\mathrm{d}\vartheta} \,. \tag{23}$$

We therefore introduce two new variables v_{r2}, $v_{\vartheta 2}$ and one new equation, the continuity equation, and have now seven equations and seven unknown functions, which in principle can be determined.

The meridional circulations in the case of rotational distortion have been found by VOGT (1925) and EDDINGTON (1926) both using von Zeipel's formalism (1924). The circulations in the case of tidal distortion have been investigated by GRATTON (1945).

The foregoing conclusions from the number of equations and the variables are only valid under the assumption that the equations are independent. By choosing a special law of rotation, $\omega = \omega^*$ (or in the magnetic case by choosing a special magnetic field $\boldsymbol{B} = \boldsymbol{B}^*$) one can force the system of the 6 linear differential equations for the five variables

$$\Pi_2\,, \quad \Phi_2\,, \quad T_2\,, \quad f_{r2}\,, \quad f_{\vartheta 2}$$

to have a linear dependence and then there is no need to introduce meridional circulations in a first-order theory as we are using here. In the case of Eddington's standard model (with $\varkappa = \text{const}$, $\varepsilon = \text{const}$), SCHWARZSCHILD (1942) has shown that this special law of rotation is the solid-body rotation $\omega = \omega^* = \text{const}$. For more realistic stellar models the particular angular velocity distribution ω^* is more complicated.

I think one of the most important problems of the theory of stellar rotation is to determine whether a star can actually have such a special angular velocity distribution. Suppose there is a given initial distribution of ω at time $t = 0$. From that will result a meridional circulation velocity. Now Coriolis forces begin to act and change the initial ω-distribution. This will also change the circulations. Can a star then adjust rotation in such a way, that finally $\omega = \omega^*$ and the circulations are zero in the first order? There is no solution of this problem at the moment. We do not know whether the star will adjust its ω in such a way that the circulations are at minimum. But a necessary condition for this at least would be that the state of zero circulation is a *stable* one in the theory which is of first order. Let us take a simple example. For the standard model the special ω is $\omega^* = \text{const}$. If we assume a standard model to rotate uniformly, the first-order theory will give no meridional circulation. Now take a small arbitrary deviation from $\omega = \text{const}$ which will cause circulations. Will the circulations re-establish the special $\omega = \omega^* = \text{const}$ or will they make the deviations bigger? One can formulate the same problem also for more realistic stellar models and their distinguished angular velocity distribution. A similar discussion holds for the magnetic case but not for the case of tidal distortions since in this case the circulations do not change the disturbing force. Therefore the case of tidal distortion is much simpler than the cases of rotational or magnetic distortion.

In the foregoing considerations we assumed that the molecular weight μ is constant. But our conclusions are also correct when μ varies due to ionization. In this case μ is a function of Π and T and we can write $\mu = \mu_0(r) + \mu_2(r)P_2$ with

$$\mu_2 = \left(\frac{\partial \mu}{\partial \Pi}\right)_0 \Pi_2 + \left(\frac{\partial \mu}{\partial T}\right)_0 T_2 \,, \tag{24}$$

where the coefficients with the subscript 0 are known from the undisturbed model. Equation (24) shows how we have to eliminate μ_2 by Π_2 and T_2. One therefore does not increase the number of unknown functions by including ionization effects and therefore meridional circulations will be present also in this case.

1·3. *Properties of the meridional circulations.* – To find the meridional circulation for a given star with a given distribution of ω one has to solve the system of linear differential equations *with* meridional velocities described in Section **1·2**. For solid body rotation this problem was first solved by SWEET (1950) for the Cowling model. It turns out that the meridional velocities are very slow and for many stars the circulation will not finish a revolution within the life-time of the star. A very rough estimation of the velocity v_r in the stellar interior is given by (SWEET (1950), BAKER-KIPPENHAHN (1959))

$$v_r \approx \frac{L}{4\pi G M \varrho R}\,\frac{\nabla_{\text{ad}}}{\nabla - \nabla_{\text{ad}}}\,\chi \,, \qquad \left(\nabla_i = \frac{\text{d}\ln \Pi}{\text{d}\ln T}\right). \tag{25}$$

χ is of the order of the ratio of the disturbing force to the gravitational force and therefore in the case of rotation, which we are considering here one has in a region defined by r

$$\chi_{\text{rot}} = \frac{\omega^2 r^3}{G M_r} \,. \tag{26}$$

A formula similar to (25) also holds for tidal and magnetic distortions, in these two cases χ is defined by

$$\chi_{\text{tid}} = \frac{M_{\text{II}}}{M_r} \frac{r^3}{D^3}, \tag{27}$$

$$\chi_{\text{magn}} = \frac{\boldsymbol{B}^2 r}{4\pi G M_r \varrho}. \tag{28}$$

Sweet found in the case of rotation, that for all stars earlier in type than F the meridional circulation is fast enough to transport within the star's life-time on the main sequence matter from the central region to the surface and mix the star completely. This result can be obtained by putting reasonable numerical values for the characteristics of the star and for the angular velocities into eq. (25). The result could be very serious for the theory of stellar evolution. It would mean that most of the evolutionary tracks which are published in Schwarzschild's book (1958) must be replaced by evolutionary tracks for completely mixed evolution which seem to be in contradiction with the observation. Fortunately shortly after Sweet's paper MESTEL (1954) found a way out of this dilemma. But before discussing Mestel's results we will investigate the properties of the circulations more in detail.

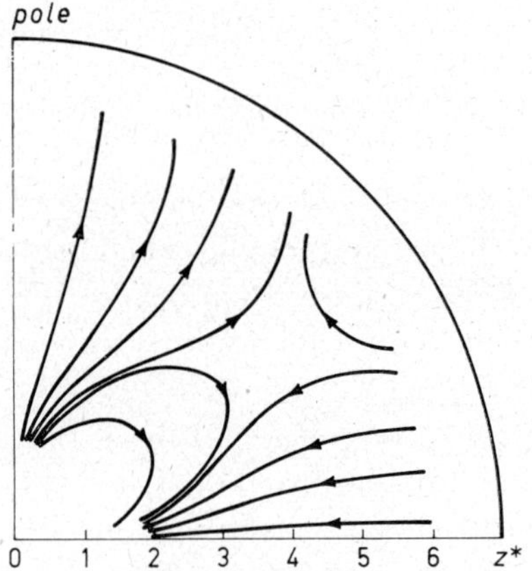

Fig. 1. – The meridional circulations in the standard model for $\omega \sim 1/r^2$. The units are Emden units, Z^* corresponds to the radius.

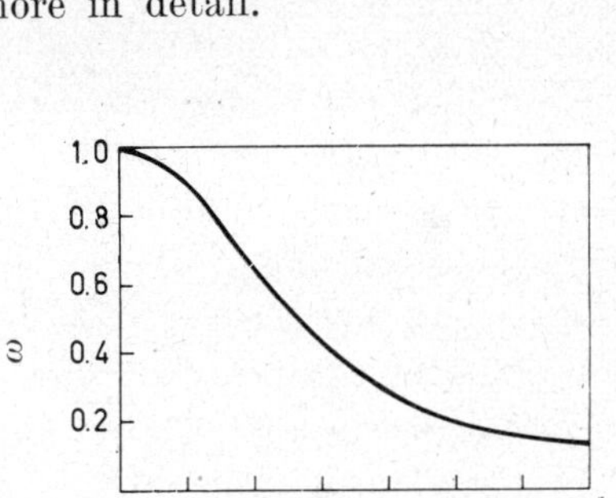

Fig. 2*a*. – Distribution of angular velocity for which in Fig. 2*b* the meridional circulations are given for the standard model. ω is the normalized angular velocity. The abscissa like in Fig. 1.

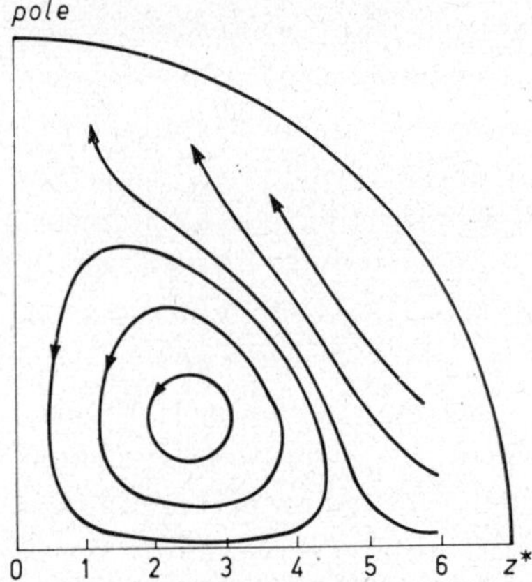

Fig. 2*b*. – The meridional circulations in the standard model for the angular velocity distribution of Fig. 2*a*.

Sweet's circulations go in such a way that they rise at the poles and descend at the equator. This is a general property of solid body rotation, as has been shown by BAKER and KIPPENHAHN (1959). But for nonuniform rotation there may also be regions where the sense of the circulation is just in the other direction. This for instance can be seen in Fig. 1, 2 for nonuniform distribution of angular velocity for the standard model (KIPPENHAHN, 1958). In these examples we also see that the stream lines of the meridional circulations do not necessarily connect the surface regions with the deep interior. The circulations may divide the meridional plane in several subregions which are not connected by stream lines.

For tidal distortions the circulation always goes down at the line between the centers of the two stars and rises in the plane perpendicular to this axis.

1·4. *The μ-currents.* – In Section **1·2** we assumed that $\mu=\mu(\Pi, T)$ or $\mu=\mu(\varrho, T)$. But our result, which was derived by counting the number of unknowns and the number of equations does not hold, when μ is assumed to be a free function which can be adjusted. Then we have one free function more and the system is not overdetermined any more. Then there is no reason to introduce meridional velocities as new free functions as has been done in Section **1·2**.

The meaning of this is that the chemical elements in an inhomogeneous star can be distributed in a nonsymmetrical way which gives a new degree of freedom, and the function μ together with the other functions which appear in the system will give just the right number of unknowns. We therefore define for a given angular-velocity distribution ω a particular μ-distribution μ^* which describes a distribution of molecular weight in such a way that no meridional circulations can appear. As with the function ω^*, we do not know whether in an actual case the star will really approach this μ^*. If the interior of a rotating star becomes richer in helium than the outer envelope, than the meridional circulation will carry helium into outer regions and, since this necessarily must happen in a (spherically) unsymmetric way, a nonspherical μ-distribution is built up. It may happen, that the star then will approach the particular distribution μ^* and the circulation will die out.

The structure of stars with nonsymmetric distributions of molecular weight has been studied by MESTEL (1954). Following his procedure we restrict ourselves to first-order effects. This has the advantage, that one can investigate circulations, which are produced by rotation and those produced by an asymmetric μ-distribution separately and superpose both. The distinguished μ^* then has to produce meridional circulation which cancels the circulation due to rotation everywhere. We start with a given asymmetric distribution of chemical composition in a nonrotating star

$$X = X_0(r) + X_2(r)P_2\,, \qquad \mu = \mu_0(r) + \mu_2(r)P_2\,. \tag{29}$$

From eq. (1) with $\boldsymbol{c} = 0$ (we are considering a *nonrotating* star with asymmetric μ-distribution) we obtain

$$\varrho = \varrho(\Phi)\,, \qquad \Pi = \Pi(\Phi) \tag{30}$$

and this together with eq. (2) shows that the surfaces $\Phi = \text{const}$ are spheres, according to a theorem of potential theory. Therefore

$$\Phi_2 = 0\,, \qquad \varrho_2 = 0\,, \qquad \Pi_2 = 0 \tag{31}$$

and

$$\frac{\mu_2}{\mu_0} = \frac{T_2}{T_0}\,. \tag{32}$$

The star therefore has a pressure and density distribution which is spherically symmetric and the asymmetry in μ is compensated by an asymmetry of the temperature distribution.

For simplicity we assume that $\varepsilon = 0$, which is valid in the radiative envelope of the Cowling model. If we assume that the absorption is due to scattering, and if we neglect elements heavier than helium we have

$$\varkappa = 0.19(1 + X)\,, \tag{33}$$

$$\mu = \frac{4}{3 + 5X}\,, \tag{34}$$

$$\frac{\mu_2}{\mu_0} = -\frac{5X_0}{3 + 5X_0}\frac{X_2}{X_0}\,, \tag{35}$$

and therefore

$$\frac{\varkappa_2}{\varkappa_0} = -C\frac{\mu_2}{\mu_0} \qquad \text{with} \qquad C = \frac{3 + 5X_0}{5(1 + X_0)}\,. \tag{36}$$

Now we must derive $(\boldsymbol{\nabla}\cdot f)_2$, the P_2-part of the left-hand side of eq. (4) which gives rise to meridional circulation caused by the asymmetry in μ. From

$$f = -\frac{4ac}{3\varkappa\varrho}\nabla T\,, \tag{37}$$

with $\alpha = \mu_2/\mu_0$ one gets (primes denote derivatives with respect to r)

$$f_{r2} = f_{r0}\left[(4 + C)\alpha + \frac{T_0}{T_0'}\alpha'\right], \tag{38}$$

$$f_{\vartheta 2} = -\alpha f_{r0} \tag{39}$$

and finally

$$(\nabla \cdot \boldsymbol{f})_2 = f_{r0} \left\{ \left[(4 + C)\alpha + \frac{T_0}{T_0'} \alpha' \right]' - \frac{6}{r} \alpha \right\}. \tag{40}$$

This term causes now circulations, Mestel's μ-currents. The r-component v_r^μ of the μ-currents is given by a formula similar to eq. (21). The star starts with homogeneous chemical composition and burns hydrogen. The rotation gives rise to circulations v_r^ω which change the μ-distribution (and due to the Coriolis forces also the ω-distribution). A time step Δt later, when hydrogen has been burned into helium and has been transported by the ω-circulation into the radiative envelope, there will occur an asymmetric μ-distribution in the star:

$$\alpha = - v_{r2}^\omega \frac{\mathrm{d} \ln \mu_0}{\mathrm{d}r} \Delta t \,. \tag{41}$$

This means that additional μ-currents v_r^μ appear according to eq. (40). Then in the next step Δt the change of the μ-distribution is given by

$$\frac{\mathrm{d}\alpha}{\mathrm{d}t} = - (v_{r2}^\omega + v_{r2}^\mu) \frac{\mathrm{d} \ln \mu_0}{\mathrm{d}r} \,. \tag{42}$$

To reach a steady state one would require

$$v_{2r}^\omega + v_{2r}^\mu = 0 \,. \tag{43}$$

We have seen that in principle a steady state is possible in which the circulations due to rotation are stopped by an asymmetric μ-distribution but it has not yet been shown that in an actual star the steady state will be reached. We do not even know whether the μ-currents (caused by a μ-asymmetry produced by ω-currents according to eq. (41)) go in the opposite direction to the ω-currents. The reason is that in formula (40), which determines the sign of v_r^μ, higher derivatives of α appear. One should follow this problem in detailed calculations. As a first step the problem of tidal, instead of rotational distortions would be simpler and should be attacked first.

But let us now forget the difficulties mentioned above and assume, that a steady state can be reached. Then one can find a *necessary* condition for reaching a steady state without circulations which mix the star. The time-scale of the ω-circulation must be not be too short compared with the time-scale of the change of μ due to nuclear reactions in the core. Otherwise the matter of the rising part of a stream line will have almost the same composition as the matter on the falling part of the line and all differences in chemical composition will be almost completely smeared out and the asymmetry of μ will be too small to produce big enough μ-currents.

The time the ω-circulation needs to make one revolution is roughly given by

$$\tau \approx \frac{R}{v_r^\omega}, \tag{44}$$

(R radius of the star). If we assume, that $\partial \mu / \partial t$ is constant in time in the core, then

$$\mu_2 \approx \left(\frac{\partial \mu}{\partial t}\right)_{\text{core}} \tau . \tag{45}$$

If we assume now that the ω-circulation is stopped by an asymmetry in μ then μ_2 should be roughly $\chi \mu_0$. This is the case for all functions which appear in the linearized eqs. (1)–(4): $\Pi_2 \approx \chi \pi_0$, $T_2 \approx \chi T_0$, $f_{r2} \approx \chi f_{r0}$ (KIPPENHAHN (1958)). Therefore

$$\chi \approx \frac{\mu_2}{\mu_0} \approx \frac{R}{v_r^\omega} \frac{\partial \ln \mu}{\partial t} . \tag{46}$$

If M_c is the mass of the convective core and L the luminosity of the star, differentiating eq. (34) we get

$$\frac{\partial \ln \mu}{\partial t} = - \frac{5X_0}{3+5X_0} \frac{\partial \ln X_0}{dt} \approx \frac{\partial \ln X_0}{\partial t} = 10^{-18.8} \frac{L}{M_c} . \tag{47}$$

The velocity of the ω-circulation is according to eq. (25)

$$v_r^\omega \approx \frac{L}{3gM} \frac{\nabla_{\text{ad}}}{\nabla - \nabla_{\text{ad}}} \chi \approx \frac{L}{gM} \chi , \qquad \left(g = \frac{GM_r}{r^2}\right). \tag{48}$$

Here we put $\nabla_{\text{ad}}/3 \cdot (\nabla - \nabla_{\text{ad}}) \approx 1$. Equation (46) then gives

$$\chi^2 \approx 10^{-18.8} Rg \frac{M}{M_c} . \tag{49}$$

This is the case, in which $\alpha \sim \chi$. The μ-circulations can be stopped, when χ is not bigger than the value given by eq. (49), since then an appreciably asymmetry in μ can build up. If χ in the region of interest is bigger then the value given in (49) then the μ-circulation cannot be stopped.

For Kushwaha's model (1957) with $M = 10\, M_\odot$ one gets for the region right above the convective core from eq. (49) $\chi \approx 10^{-1.6}$. If we assume, that stars rotate roughly like solid bodies, than χ increases considerably if one goes from the core outwards since the centrifugal acceleration will increase and the gravitational acceleration will decrease. For main sequence stars with convective cores χ may be about 20 to 30 times bigger at the surface then on top of the convective core. Therefore even if the star is almost rotationally unstable at the equator ($\chi \approx 1$), at the boundary of the inner convective core χ will be smaller than the critical value given by eq. (49) and therefore sufficiently high asymmetries in μ can be built up. But this only holds for solid body rotation.

1.5. *Nonuniform rotation during evolution.* – In the following we consider a star of $1.2\,M_\odot$. In its early evolutionary phases the star may have come through a Hayashi phase in which the star was fully convective. The enormous turbulent friction caused by convection gave the star a kind of steady-state distribution of angular velocity which no longer depended on the initial conditions of the star. Then the star may have approached the main sequence, in an evolutionary phase in which almost all the mass of the star became stable and its temperature distribution was governed by radiative energy transport. Then the only sources of viscosity were molecular and radiative friction. But since the friction is very small the time scale in which friction will change the distribution of angular velocity is larger then the lifetime of the star. One therefore should assume, that in the radiative parts of the star every mass element should conserve its angular momentum during evolution.

Let us assume that the star, after it has reached the main sequence, may rotate with an uniform angular velocity and may have an equatorial velocity of 100 km/s, which is a little bit above the mean rotational velocity of F0 stars. After hydrogen exhaustion at the center, the star gets an isothermal helium core and moves to the right. Before helium burning starts some 40 or 50 % of its mass has contracted into the helium core of high density. The outer layers have expanded and the total radius increased: the star becomes a red giant. Figure 3 gives $\log r$ *vs.* M_r for the star on the main sequence *a*) and in a later phase, *b*) where 23 % of the star's mass are in the core from calculations by HENYEY and collaborators (1962). In the interior (left-hand side of the diagram) the contraction decreases the central distance of a mass-element by a factor of 10.

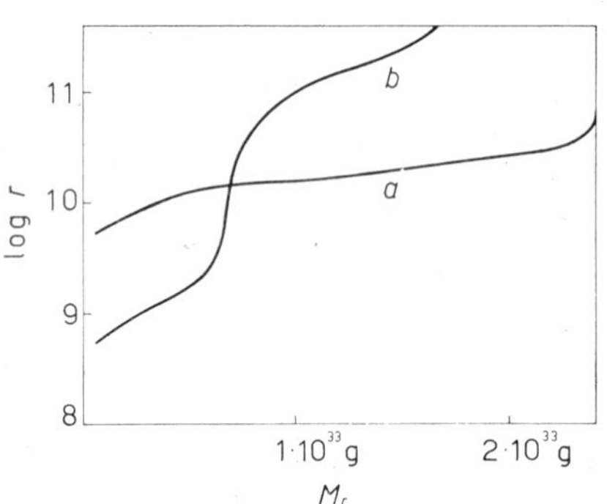

Fig. 3. – The change in mass distribution during evolution of a $1.2\,M_\odot$ star of population II after HENYEY and collaborators (1962). The curve *a* corresponds to the star at the main sequence ($M_{\text{bol}} = +2.494$, $T_e = 7701°$) and curve *b* to a later state ($M_{\text{bol}} = +0.876$, $T_e = 5272°$), where 23% of the star's mass are in the isothermal helium core.

From our assumption of conservation of angular momentum for each mass-element it follows that, because of the contraction of the interior, the helium core will rotate about 100 times faster than the same element in the main sequence phase.

From $\chi \sim \omega^2 r^3 / M_r$ and $\omega r^2 = \text{const}$ we conclude that during evolution, the value χ at a certain depth, characterized by M_r changes according to the relation

$$\chi \sim \frac{1}{r_M}, \tag{50}$$

where r_M is the radius of the sphere which contains the mass M_r. In the interior of our star χ therefore will increase by a factor of 10 during the evolution which leads from a to b in Fig. 3.

On the main sequence for uniform rotation with an equatorial velocity of 100 km/s on a sphere which is characterized by $M_r = 0.2\,M$ the value for χ is $10^{-2.8}$. During evolution χ increases by a factor of 10 and one has $\chi \sim 10^{-1.8}$ in the region $M_r = 0.2\,M$. But there are F0 stars with equatorial velocities of 200 km/s or more, in which a rapidly spinning core may evolve which has even higher values for χ. Furthermore our estimation is based on the assumption that on the main sequence the star rotates like a solid body. If for main-sequence stars the angular velocity increases inwards, the star may come to values of χ not to far from 1 $M_r = 0.2\,M$, and then mixing may occur. Therefore it may well be possible that the most rapidly rotating F0 stars are not able to follow the complete evolutionary track. At a certain time, namely when the core rotates too quickly, they may get mixed and more back to the left in the HR-diagram to the main sequence of mixed stars. This may give an explanation for the main sequence stars above the turn-off points of globular clusters mentioned by Dr. SANDAGE.

Our argument does not hold if a magnetic field imposes a distribution of angular momentum over the whole star. The uniform rotation along magnetic field lines can be maintained and the core never can rotate considerably faster than the outer regions.

2. – The theory of inhomogeneous stars.

In this Section is given a brief summary of some of the physical problems which one encounters if one tries to calculate theoretical evolutionary tracks for stars which leave the main sequence due to hydrogen exhaustion. The reader should keep in mind that this shall not be a Handbuch article but a lecture. Therefore, it is complete neither in mentioning the problems nor in the list of references.

2·1. *The Chandrasekhar-Schönberg limit.* – After a star has settled onto the main sequence, nuclear burning converts the hydrogen in the core to helium. The star will then enter a phase in which it has an isothermal helium core and a hydrogen envelope, the two being separated by a thin shell in which hydrogen is being burned, and which eats its way outwards. We will consider the fitting procedure for models of such stars.

We begin defining two parameters, U and V (since from now on Legendre polynomials will not be used any more, we will use the letter P for the pressure):

$$U = \frac{\mathrm{d}\ln M_r}{\mathrm{d}\ln r} = \frac{4\pi r^3 \varrho}{M_r}, \qquad V = -\frac{\mathrm{d}\ln P}{\mathrm{d}\ln r} = \frac{G\varrho M_r}{rP}. \tag{51}$$

Every integration of the equations of stellar structure will give a line in the U-V-plane. Solutions which are homologous are represented by the same line. All integrations of isothermal gaseous spheres for a perfect gas which fulfil the central boundary conditions are represented by a single curve, the solid line in Fig. 4, which starts at $U=3$, $V=0$ (center) and spirals in the outer regions. The outer, radiative layers, if one assumes a simple opacity law, are represented by a one-dimensional set of lines (– – – – – –) in the diagram. For this purpose one can for instance use envelope integrations for the Cowling model. The parameter which characterizes this set of lines is

$$C = \text{const}\,\frac{LR^{0.5}}{M^{5.5}}. \tag{52}$$

The constant in eq. (52) depends on the chemical composition of the envelope. (For the theory of the Cowling model see for instance WRUBEL (1958).) The dotted lines finally are lines of constant $q = M_r/M$ for the envelope. At the boundary between the core and the envelope, P, M_r, r and T must be continuous, which leads to the conditions

$$\frac{U_i}{U_e} = \frac{V_i}{V_e} = \frac{\varrho_i}{\varrho_e} = \frac{\mu_i}{\mu_e}. \tag{53}$$

(The subscript e refers to the envelope and i to the core at the interface). The eqs. (53) are necessary conditions for the fit. Let us at first try to put an isothermal core into the envelope without discontinuity in μ. Then the stellar model is represented by one of the dotted lines in Fig. 4 which intersects the solid line and from the point of intersection by the part of the solid line which goes to the center ($U=3$, $V=0$). The fraction of stellar mass which is in the isothermal core is given by the value of q at the point of intersection and from Fig. 4 one can see that such a value of q can never be bigger than about 0.35. Therefore we must conclude that there are no models which have more than some 35% of their total mass in an isothermal core. If we assume that there is a discontinuity in the molecular weight then the limit

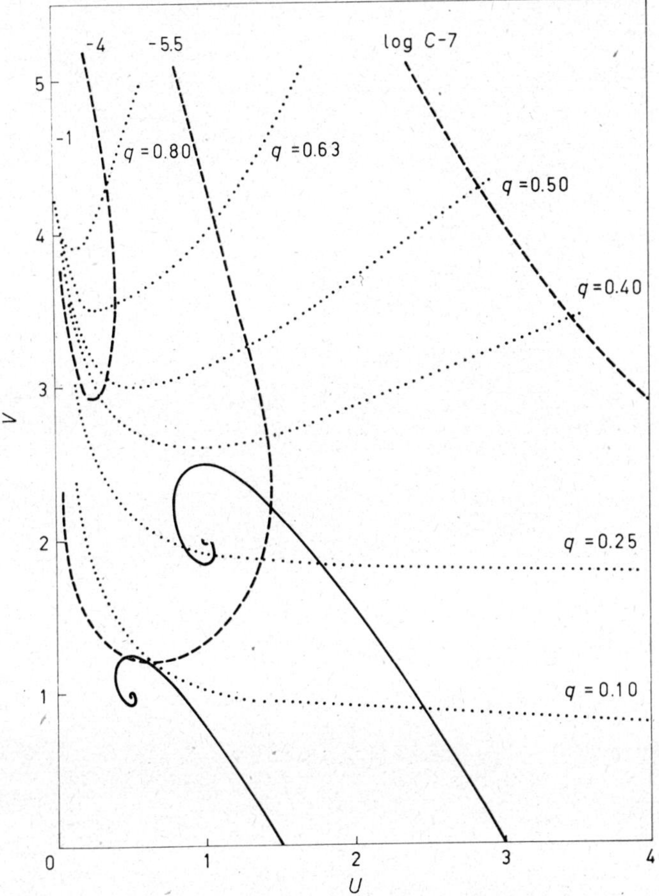

Fig. 4. – The U-V-diagram for the fitting of a radiative envelope with an isothermal core.

mass for the core is even lower. Let us assume, that $\mu_e/\mu_i = 2$. Then eqs. (53) give

$$\frac{U_i}{U_e} = \frac{V_i}{V_e} = \frac{1}{2}, \tag{54}$$

for the fitting conditions. This means that to fit the envelope with the core one may construct an auxiliary curve in the diagram for the interior (thin solid line) by replacing every point U, V of the isothermal solution curve by the point $U/2$, $V/2$ and then find intersection points between the envelope solutions and the new curve. One can see that these intersection points never can have q-values bigger than 0.1. (SCHÖNBERG-CHANDRASEKHAR (1942)).

The result that there is a limiting percentage of mass which can be in an isothermal core of a star means, that if a star burns its helium in the interior and builds up an isothermal helium core with a hydrogen-burning shell then the shell can only grow in mass until the limit mass is reached. There are two ways a star can escape the difficulty which then arises. Less massive stars may get a degenerate interior (our mass limit consideration was based in the assumption that the gas is perfect). More massive stars which cannot have degenerate interiors must contract. This will release gravitational energy in the core which then cannot be isothermal since the energy has to be carried away by a finite temperature gradient. We will see that both possibilities are realized in nature.

2.2. *The early evolution of very massive stars.* – We start with the evolution of stars with masses greater than $10\,M_\odot$. Such stars have on the main sequence a radiative envelope and a convective core. In the main sequences all the energy is generated in the core. Radiation pressure has to be taken into account in these stars; then

$$P=\frac{\mathfrak{R}}{\mu}\varrho T+\frac{a}{3}T^4\,, \tag{55}$$

where $\mathfrak{R}=10^{7.92}$ (cgs-units) is the gas constant and $a=10^{-14.12}$ the radiation pressure constant. For these stars one can assume simple surface conditions,

$$P=0\,,\qquad T=0 \tag{56}$$

from which we can also derive $\varrho=0$ by a power series expansion at the surface. Eventual convective zones near the surface are very thin and ineffective, and therefore can be neglected. The opacity almost everywhere in the star is given by scattering

$$\varkappa=0.19\,(1+X) \tag{57}$$

for a hydrogen-helium mixture with heavier elements negligible in weight ($Z\ll 1$). It may even suffice to assume the opacity formula (57) to hold everywhere in the star, since the thin surface region where this introduces an error may not affect the structure of the interior considerably.

The boundary between radiative envelope and convective core is determined by the stability criterion $\nabla_{\rm r}<\nabla_{\rm ad}$ where ∇_r and $\nabla_{\rm ad}$ are the radiative and the adiabatic double logarithmic gradients:

$$\nabla_{\rm r}=\left(\frac{{\rm d}\ln T}{{\rm d}\ln P}\right)_{\rm radiative}\,,\qquad \nabla_{\rm ad}=\left(\frac{{\rm d}\ln T}{{\rm d}\ln P}\right)_{\rm adiabatic}\,.$$

In ∇_{ad} one has to include the radiation pressure:

$$\nabla_{ad} = \frac{2}{5}\,\frac{1 + 3(1-\beta)}{1 + 6(1-\beta) - \frac{3}{5}(1-\beta)^2} = \frac{4 - 3\beta}{16 - 12\beta - \frac{3}{2}\beta^2}, \tag{58}$$

where β is the ratio of the gas pressure to the total pressure. When $\beta = 0$, $\nabla_{ad} = 0.4$ and when $\beta = 0.9$ then $\nabla_{ad} = 0.326$, therefore the radiation pressure decreases the value of ∇_{ad}.

Because of the discontinuity of composition at the interface between the core and the envelope, a difficulty arises in fitting according to the stability criterion at this point, since one can derive from eq. (3)

$$\nabla_r = \frac{\varkappa L_r}{16\pi c G(1-\beta) M_r}. \tag{59}$$

From eq. (59) we can see that the radiative gradient will change discontinuously at the interface since it is proportional to $\varkappa$, in our case being higher in the envelope than in the core. Let us examine the situation naively as we move outward from the center in a diagram where ∇_r and ∇_{ad} are plotted *vs.* M_r. On the right-hand side of Fig. 5 is the interior with the helium-rich material of the convective core. In this region the chemical composition is homogeneous since the material is well-mixed by convection. In this region is $\nabla_r < \nabla_{ad}$. The more one approaches the boundary of the convective core from the inside the closer ∇_r and ∇_{ad} come together and the boundary of the core ($M = M^*$) should be the layer with $\nabla_r = \nabla_{ad}$. But then at this layer the chemical composition should have a discontinuity since outside the material must have the original hydrogen-rich composition. But the sudden increase of X in outward direction will produce a discontinuity in $\varkappa$ so that ∇_r goes up. But then the region above the convective zone should be convective too. But if it were convective, then there would be no discontinuity and then it should be stable. The same difficulty of course would arise, if we would go inwards in the radiative envelope until we reach and try to continue with a hydrogen-poor convective core.

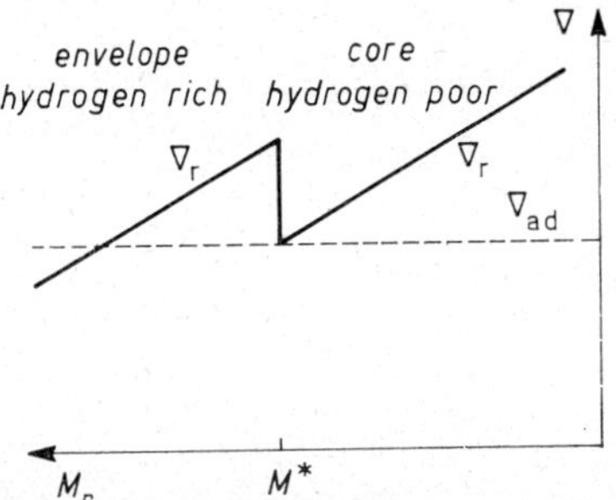

Fig. 5. – The difficulty which occurs if one tries to fit a radiative envelope with a convective helium-enriched core when the opacity is determined by scattering.

To get out of this dilemma one has to follow the evolution of a convective core with decreasing hydrogen content. For simplicity we assume at first that the mass in the convective core remains the same during evolution.

For the homogeneous star the situation is simple. Figure 6*a* shows the adiabatic and the radiative gradients and the actual gradient which will determine the actual temperature stratification and which is equal to the radiative gradient in the stable region and equal to the convective one in the unstable region.

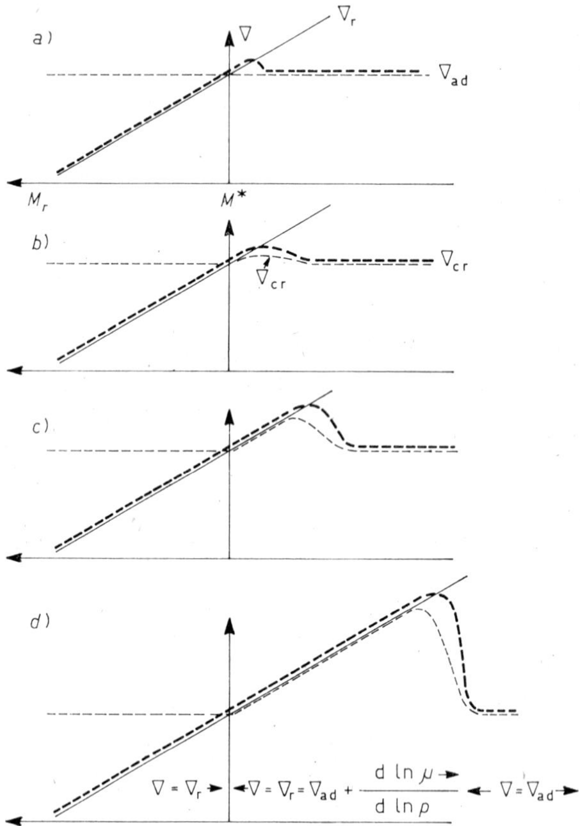

Fig. 6. – How an intermediate zone of marginal stability is built up during evolution. For simplicity the opacity is assumed to be independent of chemical composition. To make the drawings clear lines which should coincide are drawn parallel and closely together. ——— the radiative gradient; - - - - the critical gradient; **- - -** the actual gradient which will be built up.

But there is a thin layer right below $M = M^*$ where convective energy transport cannot transport all the energy since there the convective velocity (which must go to zero at $M = M^*$) is too small. In this thin layer the actual gradient will lie between the radiative and the adiabatic gradient. The thickness of this region may be of the order of one mixing length. Now in regions which are chemically inhomogeneous one has to keep in mind that the stability criterion has to be modified. A region where the heavier element is at the bottom may be stable even if the radiative gradient exceeds the adiabatic one since a certain amount of energy is required to lift the heavier element. It is not difficult to repeat the considerations which lead to the normal stability criterion taking into account this effect. The result is, that the quantity

$$\nabla_{cr} = \nabla_{ad} + \frac{d \ln \mu}{d \ln P}, \tag{60}$$

has to be compared with ∇_r. $\nabla_r < \nabla_{cr}$ gives stability, $\nabla_r > \nabla_{cr}$ gives instability.

Now let us make another simplification and assume, that $\varkappa$ does not depend on the chemical composition and be simply constant, and investigate the behavior of the region between envelope and core during evolution. The initial situation is given in Fig. 6*a*. Now let us take a later time. The molecular weight in the envelope increased a little bit. The helium has been equally distributed through the convective core, but since the convective velocity cannot go to zero at $M = M^*$ discontinuously, there must be a thin but finite transition layer right below $M = M^*$ where the convective velocity was too small to mix the helium completely into this layer within the time we are considering. Therefore there is no real discontinuity of μ but a continuous decrease within a thin layer, which may have a thickness of one mixing length. The variation in μ changes the ∇_{cr} in this region (see Fig. 6*b*) and during the evolution ∇_{cr} will increase there till $\nabla_{cr} = \nabla_r$. Then no mixing can occur since an increase of the helium content in the transition layer would make the layer more stable and no convection and therefore no change in chemical composition can occur. After some time the helium content of the core has again increased and the situation of Fig. 6*c* will be approached. One has right below a region in which the chemical composition is such that $\nabla_r = \nabla_{cr}$ in a finite region. In this region the matter is just in the state of marginal stability. On the bottom of this layer the convective velocity must be zero and therefore below this layer there must be a thin region in which the actual gradient lies between the radiative and the adiabatic gradient and in which the mixing is incomplete. A further increase of the helium content in the core will produce a finite gradient and the region of marginal stability will increase in mass more and more as it is indicated in Fig. 6*d*.

The next step now is to include the effect of varying chemical composition on $\varkappa$ and ∇_r. In principle the result is the same as in the case discussed above. The only difference is, that in the transition layer the ∇_r-curve will bend down simultaneously when the ∇_{cr}-curve goes upwards (the curve of the actual gradient always between both) until both curves meet and coincide in a finite region (Fig. 7).

Finally we have to remember that the convective core increases in mass.

In our $\nabla - M_r$-diagram this means that the whole ∇_r-curve is lifted up during evolution. Therefore every marginal instability region will become slightly unstable after some time and therefore a slight mixing will occur which increases the ∇_{cr} curve a little bit and lowers the ∇_r curve until both coincide again. The result again is an intermediate region which is marginally stable and in which the chemical elements are distributed by convection in such a way, that

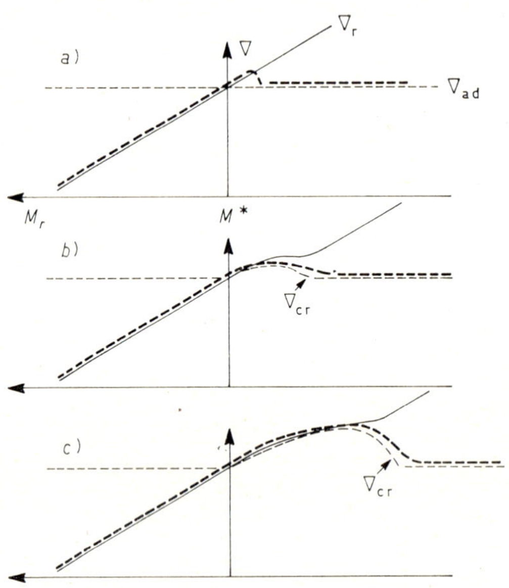

Fig. 7. – The same as in Fig. 6 but now including the effect of chemical composition on the opacity and therefore also on the radiative gradient.

$$\nabla_r = \nabla_{cr} . \tag{61}$$

The thin transition layer between a convective core and a radiative envelope has been investigated by LEDOUX (1947).

The finite intermediate regions with marginal convection which appear in massive stars have been investigated by SCHWARZSCHILD and HÄRM (1958) who gave it the name « semi-convection zone » and by SAKASHITA, ONO, HAYASHI (1959). For other absorption mechanisms the opacity in the helium-rich region is higher than outside and no difficulty arises. But since the mass of the convective core decreases a stable layer of varying chemical composition will be built up right above the convective core as it is the case for instance in Kushwaha's models (1957).

The evolution of massive stars has been investigated by SCHWARZSCHILD and HÄRM (1958) for stars between 216 and 28 solar masses and initial composition $X_e = 0.73$ and $Y_e = 0.22$ up to the point where hydrogen in the core is exhausted and the core consists of isothermal helium, before contraction begins according to the Chandrasekhar-Schönberg mass limit. However, they simply took $\nabla_r = \nabla_{ad}$ for the determination of the distribution of μ in the semi convective zone instead of the condition (61). Their evolutionary tracks are given in Fig. 8. The transition regions with variable μ are for massive stars not at all negligible, they may contain 30 % or more of the total mass of the star.

2·3. *Further evolution of massive stars.* – After the central convective zone has burnt out, the helium core contracts, releasing gravitational energy. The core then goes over from an isothermal state to one with temperature increasing toward the center and since the core is not isothermal any more it can increase in mass beyond the Chandrasekhar-Schönberg limit. The central temperature will continue to increase with time until helium burning begins.

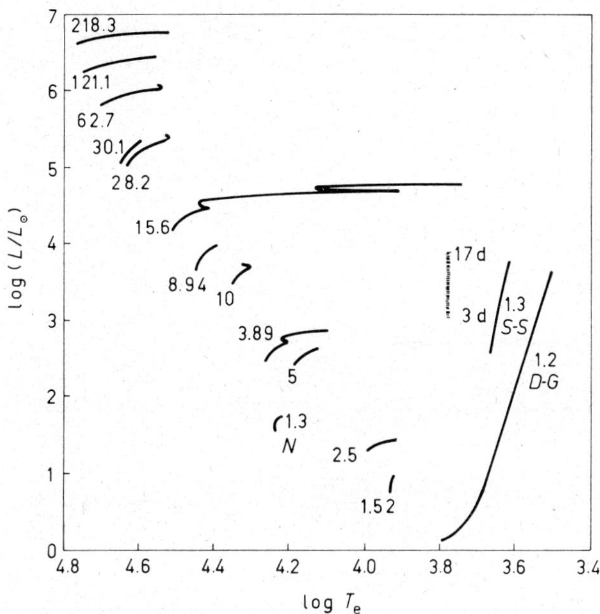

Fig. 8. – Same typical theoretical evolutionary tracks (see Table II). The hatched region gives the position of cepheids in galactic clusters.

An evolutionary track which goes from the main sequence to the most advanced evolutionary state has been calculated by Sakashita, Ono and Hayashi (1959), Hayashi and R. C. Cameron (1962) for $M = 15.6\,M_\odot$. From their papers we learn about the features of the evolutionary track which may be typical for all stars of massive type. The evolutionary track is given in Fig. 8. We will discuss the main features of the evolution in Fig. 9 and 10. The stage from A to C represents hydrogen burning (A to B) and exhaustion of hydrogen (B to C) in the convective core. In this phase the mass of the convective core decreases from 0.41 to about 0.18, releasing a zone of varying chemical composition; then the core is almost pure helium. During this phase the nuclear energy generation is gradually replaced by release of gravitational energy and at the same time a hydrogen-burning shell is built up. Then a very short phase sets in between C and D in Fig. 9: more gravitational energy is released and when point D in the HR-diagram is reached the central temperature is high enough so that helium burning releases as much energy as contraction. In the following phase the star has a hydrogen-rich envelope

TABLE II. – *The evolutionary tracks of Fig.* 8.

$M/M_\odot$	X	Y	Z	Author:
1.2	0.999	0.000	0.001	DEMARQUE, GEISLER (1962)
1.3	0.9	0.099	0.001	SCHWARZSCHILD, SELBERG (1962)
1.3	0.543	0.456	0.001	NISHIDA (1962)
1.52 389 8.94 30.1	0.75	0.23	0.02	HOYLE (1960)
2.5 5.0 10.0	0.90	009	0.01	KUSHWAHA (1957)
15.6	0.90	0.08	0.02	SAKASHITA, ONO, HAYASHI (1959) HAYASHI, CAMERON (1962)
28.2 67.2 121.1 218.5	0.75	0.22	0.03	SCHWARZSCHILD, HÄRM (1958)

(initial composition) and a radiative envelope which contains matter which has been in the convective core in earlier phases and therefore is enriched in helium. At the bottom of this layer is the hydrogen-burning shell and below this zone is the helium core, consisting of a radiative outer part and a central convective core in which helium burning goes on and which therefore increases its carbon content with time. When the point E in the HR-diagram is reached the helium in the convective core is exhausted and a new contraction phase sets in which may carry the star over the Hertzsprung gap to the right to a region where carbon burning may set in.

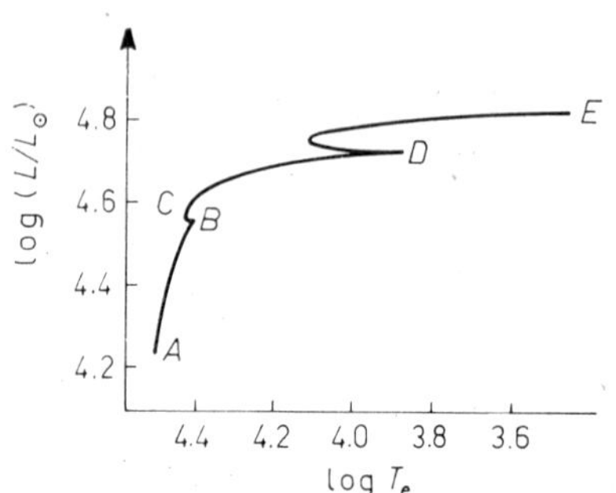

Fig. 9. – The evolutionary track obtained by HAYASHI and CAMERON (1962).

2·4. *Evolution of moderate masses.* – For masses between 2 and 15 solar masses evolutionary tracks have been calculated which cover the phases immediately after the stars have left the main sequence. The evolutionary track which goes to the most advanced stage is that of 3.89 solar masses by HOYLE (1960) which shows the turn-off-point where the convective zone disappears and an isothermal helium core is built (which corresponds to the points B, C in Fig. 9). Hoyle's evolutionary track ends where the helium-burning phase starts. The convective cores of all stars with masses smaller than $10\,M_\odot$ decrease in mass and therefore leave an intermediate region of variable chemical composition in the stable radiative region right above. The fitting procedure for models of this property has been described by SCHWARZSCHILD (1958).

As we can see in Fig. 8 the calculated evolutionary tracks are far away from the *Cepheid* region but an extension of the work of HAYASHI and CAMERON, described in the preceeding Section, to lower masses would give evolutionary tracks which cross the *Cepheid* region and would give theoretical masses and therefore a theoretical mass-luminosity relation for cepheids. As soon as one has evolutionary tracks through the cepheid region one could determine the pulsation period of these models and their stability behavior by methods described in Dr. Baker's lecture. A check of the results with observations would give a new test for the theory of stellar evolution. Evolutionary tracks into more advanced phases of evolution for moderate masses may be a little bit more difficult to obtain than a track for $15.6\,M_\odot$ since degeneracy may become more important but in the helium-burning phase

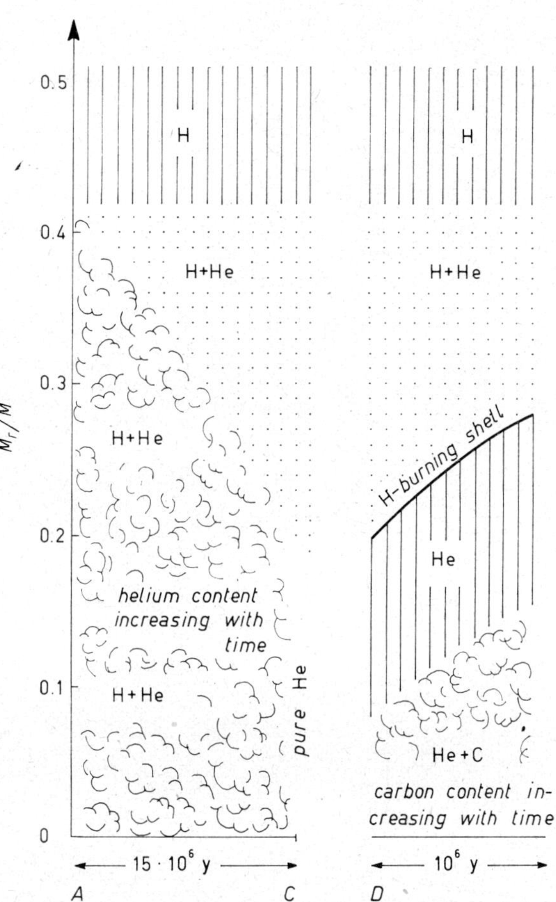

Fig. 10. – The change of thei nterior structure of a star of $16.6M$ according to HAYASHI and CAMERON (1962). The letters at the bottom correspond to the letters in Fig. 9. [radiative region symbol] radiative region; [intermediate region symbol] intermediate region of variable chemical composition; [convective region symbol] convective regon; [degenerate region symbol] degenerate region with electron conduction.

it may not produce such difficulties as the helium flash for lower masses (see Section **2·6**).

2·5. *The evolution of* 1.2 ÷ 1.3 *solar masses of population II.* – Since these stars do not have a convective core on the main sequence, a hydrogen-burning shell around an isothermal helium core appears when the star leaves the main sequence. The central region is degenerate and therefore there is no Chandrasekhar-Schönberg limit for the mass of the isothermal core which therefore can increase without considerable release of gravitational energy. A fitting procedure for these models is described by SCHWARZSCHILD (1958). Another

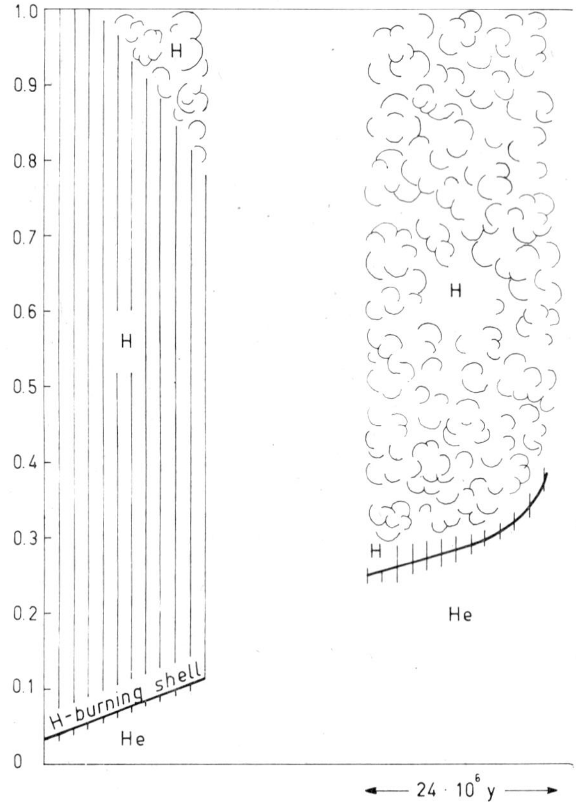

Fig. 11. – The change of the interior structure of a star of $1.2 \div 1.3 M_{\odot}$: Left: right after the star has left the main sequence (very schematically); Right: immediately before helium burning sets in (SCHWARZSCHILD, SELBERG, 1962).

procedure has been used by KIPPENHAHN, TEMESVÁRY and BIERMANN (1958) and in a slight variation by DEMARQUE, GEISLER (1962). Recently (1962) HENYEY and his collaborators started an evolutionary track for 1.2 $M_{\odot}$ using the Henyey method for integration. This method turned out to replace all the old fitting procedures perfectly and since it is a procedure which can be done automatically it is the best method for big computers—apart from its other advantages.

The first evolutionary track has been calculated by HOYLE, SCHWARZSCHILD (1955) and since this time this track has been repeated by several authors with different improvements of the models. In Fig. 8 we give the track by DEMARQUE, GEISLER (1962). We also give there an evolutionary track for the phase right before helium burning starts (SCHWARZSCHILD, SELBERG (1962)). In Fig. 11 we give schematically the change of internal structure during this phase of evolution. On the left the star after it has left the main sequence and on the right shortly before helium burning starts.

The evolutionary tracks go into a relatively cool region of the HR-diagram and therefore the outer convective zones become important. It turns out that the uncertainties which normally come into the models by the determination of the entropy in the adiabatic part of the zones make the evolutionary track quite uncertain. If one uses a mixing length theory for instance, slight changes in the definition of the mixing length produce a considerable change in the slope of the evolutionary track (see Section 3·5). The slope of the evolutionary track is also affected by the metal content: the lower the metal content, the steeper the increase to the red super giant region. It seems that there is no way to determine the right parameters for the outer convective layers until the problem of convection is solved.

Fortunately one can overcome this difficulty by adjusting the uncertain parameters in the mixing length theory (or whatever other dubious theory is used in the convective region) in such a way that the observed HR-diagrams of globular clusters can be explained. After fixing the free parameters by observation, theory is able to study the further evolution.

For one of the very advanced stages of the evolution SCHWARZSCHILD and STOTHERS (1961) studied the eigenperiod which seemed to be in good agreement with the period of *Mira* stars in globular clusters. This may indicate that the theoretical models are not too bad—even with the uncertain outer convective regions.

2·6. *The helium flash.* – SCHWARZSCHILD and HÄRM (1962) followed the evolutionary track for $1.3\,M_\odot$ into the supergiant region after making some assumptions for the uncertain outer layers which make the evolutionary tracks agree with observation. In their models the helium core, which increases in mass and decreases in radius, releases gravitational energy so that a finite temperature gradient is necessary to transport energy into the outer regions. The matter is degenerate in the core and the energy is transported by conduction. During the evolution the central temperature increases until the temperature for helium-burning is reached. When helium-burning starts new energy is released, but since for a degenerate gas the pressure does not increase with temperature the core cannot expand and all the new energy must be used to increase the temperature even more, which magnifies the energy generation rate of helium burning. Therefore the helium burning will begin very rapidly and will be connected with an increase of temperature which continues until the central region becomes nondegenerate. This happens at $3\cdot10^8$ degrees. Then the conductivity decreases and an increase in the temperature gradient will produce convection. How the structure of a star changes during the helium flash is shown in Fig. 12.

These violent changes in the deep interior of the star make it very difficult to calculate models for subsequent time steps and sometimes it turns out that the evolution of the stars goes faster than the calculation of this phase with a fast computer.

These stars have deep outer convective zones and between them and the convective core after the helium flash there is only a very thin radiative region. SCHWARZSCHILD therefore pointed out that it may well be possible, that the outer and the inner convective zones come so close together that mixing may occur. Then the star may move very rapidly to the left-hand side in the HR diagram and settle on the main sequence for mixed models at a much higher luminosity than it had when it left the main sequence.

2·7. *Nishida's mixed models.* – If one assumes that mixing occurs during or right after the helium flash the star would go to the mixed main sequence with a much higher luminosity than it had when it left the main sequence. This is due because its composition has changed meanwhile and the star has a homogeneous helium-rich composition ($X = 0.4 \div 0.5$). On the main sequence for mixed stars the star would start again with hydrogen burning in its central region. Recently NISHIDA (1962) has calculated helium-rich main-sequence models of $M = 1.3\, M_\odot$, which have convective cores in which the helium content increases even more during evolution. They again become inhomogeneous models and leave the main sequence, now the second time during the life-time of the star. The time scale for the exhaustion of hydrogen in the star is very short, since the star has a very high luminosity compared to its mass. It is of the order of 10^7 years. The convective core decreases its mass during evolution leaving an intermediate zone with varying chemical composition in the radiative layer above. NISHIDA calculated three evolutionary tracks for the same mass but with different initial chemical composition. The track for an initial value of $X = 0.543$ is plotted in Fig. 8.

Such an evolution—as SCHWARZSCHILD suggested—may be responsible for the horizontal branch of the HR diagrams of globular clusters. It therefore may be of interest to follow the evolutionary tracks into the region of the

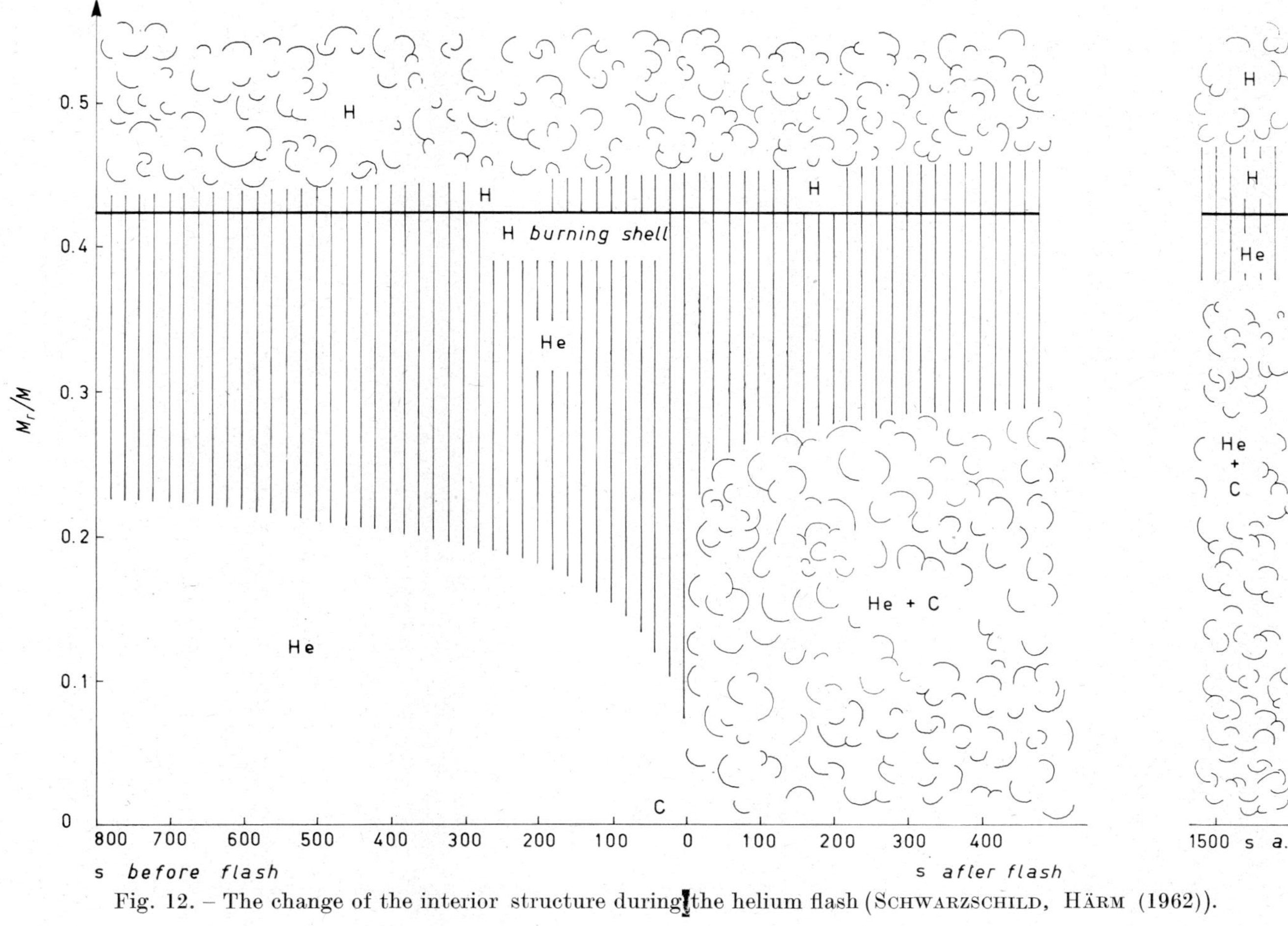

Fig. 12. – The change of the interior structure during the helium flash (SCHWARZSCHILD, HÄRM (1962)).

RR Lyrae stars and determine the pulsational frequences of the models. This may give a check for this part of evolution.

3. – The outer layers of stars with convective envelopes.

Some months ago I had a discussion with Dr. SCHWARZSCHILD about astronomical books. He made the statement that books on astrophysics should be printed on such a kind of cheap paper that they could not last longer than the validity of the ideas in the book. I am a little embarrassed when I think that the following chapter shall appear in the volumes the Academic Press is publishing from this Summer School since I have seen the kind of paper they use. If one wants to take Dr. Schwarszchild's words literally and if one thinks about the state of the theory of convection then one should print parts of this chapter on a kind of paper which disintegrates already during the printing process.

3·1. *The photospheric boudary conditions.* – For hot stars with negligible convective zones one can use the simple boundary conditions $P=0$, $T=0$ at the surface of the star without introducing too large errors. But for stars with convective envelopes one has to be more careful. In these cases one should start with a more or less simplified model atmosphere. If $\bar{\tau}$ is the mean optical depth, the theory of stellar atmospheres provides us with a formula for the temperature-optical depth-dependence:

$$T = T(\bar{\tau}) \,. \tag{62}$$

Integration of the equation of hydrostatic equilibrium

$$\frac{\mathrm{d}P}{\mathrm{d}\bar{\tau}} = \frac{g}{\varkappa} \qquad (\varkappa = \varkappa(P, T))\,, \tag{63}$$

then gives P as a function of $\bar{\tau}$. If P and T are known functions of $\bar{\tau}$ one can define a photosphere (or another layer from which on inwards one wants to integrate the equations of stellar structure). At this point P and T are known from the model atmosphere and one then has the necessary initial values to carry out an inward integration.

3·2. *The cause of the instability.* – In the outer layers of the stars the opacity normally is a very complicated function of pressure and temperature. In Fig. 13 we have a three-dimensional model of the Rosseland mean of the absorption coefficient (values from E. VITENSE (1951)) for the outer regions of a population I star. The steep rise in $\varkappa$ at the right is due to H^- absorption; the long ridge running diagonally is the region of hydrogen and helium ionization; the flat area at lower left is the region where electron scattering is the most important absorption process. For any point in the star there is a given pair of values $\log P$, $\log T$, and a corresponding point on the $\varkappa$-surface. The structure of a stellar model corresponds to a line on the $\varkappa$-surface. On the model we plotted the lines for the outer layers of the sun (——) and of a star near the Cepheid region (----). Both lines go into the steep H^- region as one goes into the stars. This steep increase in opacity can be easily explained. The opacity for H^- is given by

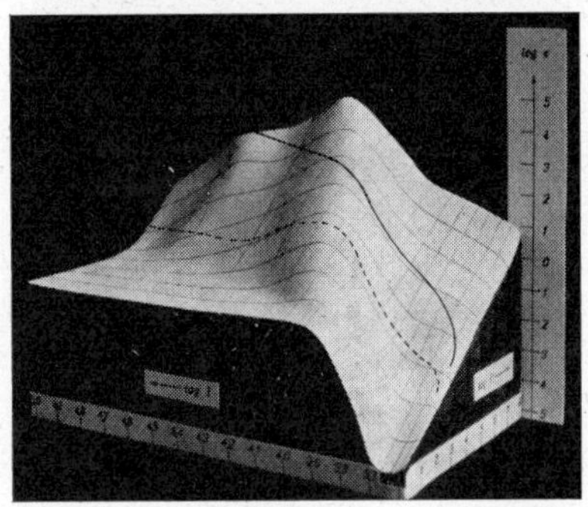

Fig 13. – A three-dimensional model of $\log \varkappa$ as a function of $\log P$ and $\log T$.

$$\varkappa_{H^-} = x(1-x)P_e f(T) , \tag{64}$$

where $f(T)$ is a relatively slow varying function of T. $\varkappa$ is the degree of ionization of hydrogen. The electron pressure P_e is given by

$$P_e = (x + A^{-1})\mu XP , \tag{65}$$

where A is the number of hydrogen atoms per metal atom. All metals are here assumed to be ionized. If one goes from the outer layer into the star $\varkappa$ increases very rapidly as soon as it becomes bigger than A^{-1} since then an increase of temperature increases P_e and therefore the opacity. The temperature as well as ∇_r increases more rapidly with $\varkappa$ increasing the opacity further and so on. With the increasing ∇_R the radiative gradient overcomes ∇_{ad} and the layer becomes unstable. At the same time the adiabatic gradient decreases due to ionization effects which again makes the layer unstable. In hot stars the sharp increase of $\varkappa$ occurs in the atmospheres where it does no harm since in these optically thin regions ∇_r is very low. Therefore hot stars have very thin and ineffective convective zones. In the convective layer a certain fraction of the total energy flux is carried by convection and therefore the temperature gradient depends on the convective transport mechanism.

3·3. *Mixing length theory.* – Let us consider a thermally convective layer in a quasi-stationary state that is in a state of motion in which mean values, taken over a sufficient long time interval Δ at the time t, are independent of t. Then let us take two points P', P'' separated by a distance d. At both points we can measure a certain velocity component, say the x-component v_x in a cartesian system of co-ordinates simultaneously. The velocities may be v'_x and v''_x. Then we take the time mean value $\overline{v'_x v''_x}$ (over a sufficiently long time interval Δ). If we change the position of point P'', which may change the distance d then we can consider $\overline{v'_x v''_x} = f(d)$ as a function of d. With a normalization $f(0) = 1$ we would get a function as given in Fig. 14-*a*. The closer the two points, the closer is the correlation between v'_x and v''_x. The farther the two points the less correlated are the two velocities and therefore the smaller the mean value will be since in the time interval in which the mean is taken the function $v'_x v''_x$ will be as often positive as negative. From the function given in Fig. 14-*a* we can derive a mean distance l in which the velocities are correlated. The same can be done by measuring the velocity v'_x at P' at time t and measuring the velocity v'''_x at the same point P' but at the later time $t+\delta$. The product $\overline{v_x v'''_x}$ considered as a function $g(\delta)$ of δ gives a mean value, taken over a sufficiently long time interval Δ around t. For $g(\delta)$ plotted as a function of δ we get a similar dependence as in Fig. 14-*a* which is now given in Fig. 14-*b*. Again one can define a mean time interval τ for the correlation.

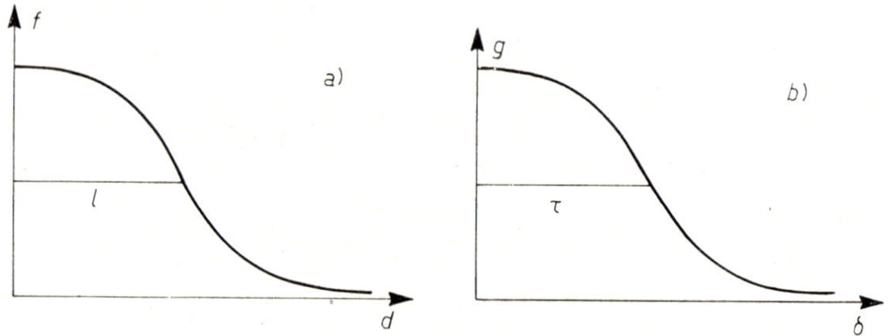

Fig. 14. – The characteristic length l and the characteristic time τ in a convective velocity field.

Now let us consider a number of gas blobs of roughly the same size with random motion in a volume. All matter within one blob may have almost the same velocity which is the velocity of the blob. If we repeated our procedure of determining l and τ, we then would find that l is roughly the diameter of the blobs and l/τ their mean velocity in x-direction. In the mixing length theory we replace the convective gas by a gas of blobs with the diameter l and the mean velocity l/τ. We do not assume that the blobs keep their identity all the time but that they disappear within the time τ and mix with the surrounding matter in which new blobs (convective elements) are born. l is called the *mixing length*.

This picture is very rough and one cannot believe that it will give a good description of the truth but if it is accepted one can derive the forces which act on these blobs, study their physical behavior and find numerical values for the mean velocity and for the energy transport.

We therefore assume that there are rising and falling elements of the size of l and with a « mean free path » of the order of l. The elements have the same pressure as the surrounding matter but a different temperature which causes buoyancy forces and drives the elements. After an element has moved one mixing length l it may dissolve and mix with the surrounding matter.

Let us first assume that the rising and falling elements do not lose energy during their motion along the mixing length and therefore undergo an *adiabatic* change. The (mean) temperature gradient ∇ which will determine the temperature-pressure relation in the convective layer will be smaller than the radiative gradient ∇_r (since not all the energy is transported by radiation) and bigger than the adiabatic gradient ∇_{ad} (since in an exactly adiabatic layer no energy can be transported by adiabatically changing elements). Therefore we have $\nabla_r > \nabla > \nabla_{ad}$.

Now we assume with E. VITENSE (1953) that the rising hot elements will lose some energy to the surrounding matter by radiation. Then the temperature change δT of the element will not be related to its pressure change δP by ∇_{ad} but by a gradient ∇' which is smaller than the gradient ∇ in the surrounding matter and bigger than the adiabatic gradient. We therefore have $\nabla_r > \nabla > \nabla' > \nabla_{ad}$.

The formulae of the energy transport by convection have been derived by BIERMANN (1945) and improved by BÖHM-VITENSE (1958).

They are

$$F_c + F_r = \frac{16\sigma T^4}{3\varkappa\varrho H_p}\nabla_r , \tag{66}$$

$$F_r = \frac{16\sigma T^4}{3\varkappa\varrho H_p}\nabla , \tag{67}$$

$$F_c = c_p\varrho T v(\nabla - \nabla')\frac{l}{2H_p} , \tag{68}$$

$$v^2 = \frac{gl^2}{8H_p}(\nabla - \nabla') , \tag{69}$$

$$\frac{c_p\varrho^2 lTv\varkappa}{24\sigma T^4} = \frac{\nabla - \nabla'}{\nabla' - \nabla_{ad}} . \tag{70}$$

F_r and F_c are the radiative and convection fluxes. σ is the Stefan-Boltzmann constant, v gives the mean velocity, $H_p = \Re T/g\mu$ the pressure scale height, g the gravity acceleration and c_p the specific heat. We do not try to derive these equations since they are derived for instance in Unsöld's book (1955) and (with some slight changes) by E. BÖHM-VITENSE (1958). We used the equations of the latter author and differ with her only slightly in eq. (69) where we distinguish between l and H_p whereas she puts $l = H_p$:

The physical meaning of the five equations is the following:

eq. (66): convective and radiative flux together must carry all the energy which goes through the outer layer;

eq. (67): the radiative transport is given by the same formula as without convection but the gradient on the right-hand side is the gradient which is actually in this layer (and not the radiative one which is defined so that the radiative transport carries all the energy);

eq. (68): the convective energy flux is given. It depends on the temperature difference between the moving element and the mean temperature of the layer;

eq. (69): the convective velocity is determined by estimating the buoyancy forces which act on each element;

eq. (70): the loss of energy each element suffers during its life-time due to radiation loss is estimated.

The first two equations are exact, the other three are derived under several assumptions which cannot be proved, but which seem to be reasonable, as long as one keeps the uncertainty in mind and does not take the whole theory too seriously.

The equations can be simplified by introducing new quantities U, ξ which are defined by

$$U = \frac{24\sigma T^3}{c_p \varrho^2 l^2 \varkappa}\sqrt{\frac{2H_p}{q}}, \tag{71}$$

$$\xi = \sqrt{U^2 + \nabla - \nabla_{\text{ad}}},$$

which allows us to replace the whole set of eqs. (66) to (70) by the single one:

$$\tfrac{9}{8}(\xi - U)^3 + U\xi^2 - U^3 - (\nabla_r - \nabla_{\text{ad}})U = 0 \tag{72}$$

which is an algebraic equation for ξ and can be solved easily. ξ then gives ∇ (eq. (71)).

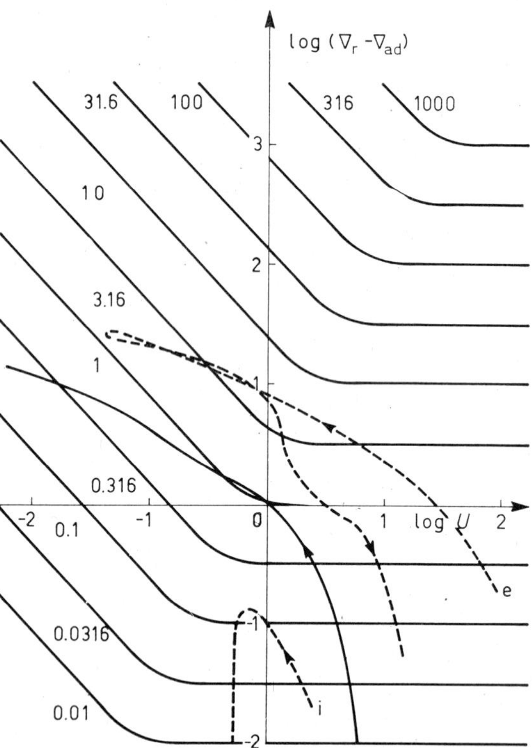

Fig. 15. – A diagram for the solution of eq. (72) with line $\nabla - \nabla_{\text{ad}} = $ const.

Some properties of the solutions of eq. (72) for given values of ∇_r, ∇_{ad} and U can be seen in Fig. 15 where in a diagram of $\log U$ and $\log(\nabla_r - \nabla_{ad})$ the quantity $\nabla - \nabla_{ad}$ is plotted. There is a similarity relation eq. (72) so that if ∇ is a solution for a given triplet ∇_r, ∇, U_{ad} then $\tau^2\nabla$ is a solution for the triplet $\tau^2\nabla_r$, $\tau^2\nabla_{ad}$, τU for any value of τ. Due to this similarity relation the lines $\nabla - \nabla_{ad} = \text{const}$ can be generated by shifting one of these lines parallel along the line

$$\log(\nabla_r - \nabla_{ad}) = 2 \log U . \tag{73}$$

As it is shown in Fig. 16 the diagram can be divided into three regions: The region in which ∇ is very close to the radiative gradient, the intermediate region and the region where the gradient ∇ is almost equal to the adiabatic one. In Fig. 17 the solid line characterizes the top of the convective zone obtained by the mixing-length theory with the mixing-length equal pressure-scale height of the sun. The arrow indicates the direction into the sun. We see that the outermost layer is more radiative and then layers become more and more adiabatic the deeper one goes into the sun.

E. Böhm-Vitense found that there are stars which have two separate convective

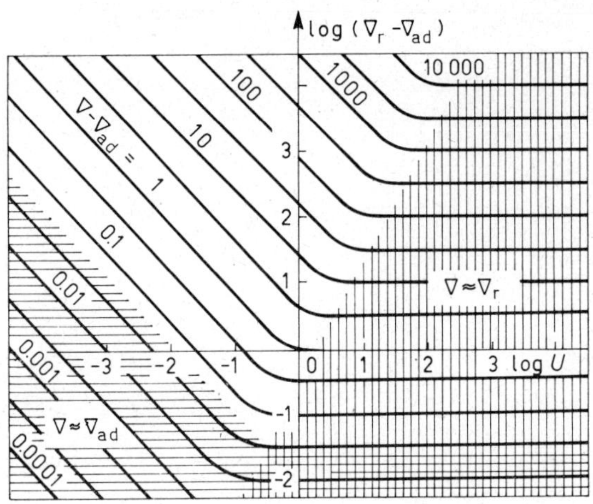

Fig. 16. – The regions in the ∇-U-diagram where $\nabla \approx \nabla_r$ and where $\nabla \approx \nabla_{ad}$.

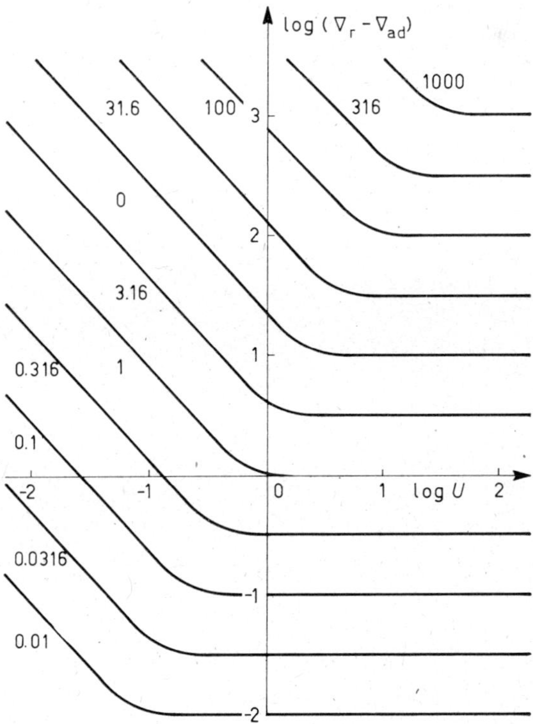

Fig. 17. – The sun (———) and a star with two separated convection zones (– – –) in the ∇-U-diagram.

zones in their outer layers. One is due to the hydrogen ionization and its effects on $\varkappa$ and ∇_{ad}. A deeper one is due to the second helium ionization.

In Fig. 17 the behavior in the $\nabla - U$ diagram is given for such a star ($M = 11.5\,M_\odot$, $L = 5\,000\,L_\odot$, $T_e = 5\,390°$). The outer zone has parts in the intermediate region while the helium convective zone has a radiative stratification.

The fact that the eqs. (66)–(70) can be replaced by an equation which only has ∇_r, ∇_{ad} and one other parameter has been found by TEMESVÁRY (1959).

3·5. *What is the right mixing length?* – The mixing length is a quantity which is not determined by the eqs. (66)–(70). It is therefore a free parameter in the theory. It is not the only free parameter. Even if we could determine l exactly the mixing-length theory would not give the right solutions because of the simplified picture of the velocity field (existence of elements) and because of the uncertainties in eqs. (68)–(70) in which many free parameters should appear which are replaced by plausible values. For instance, one appears at the point where the radiation loss of a rising element is estimated. Therefore, we really have several free parameters in the mixing-length theory and for all but one we assumed plausible values but the last free parameter, l, therefore, is not really the mixing length any more but also stands for all the errors which have been introduced by fixing the other free parameters. It seems therefore to be only of limited value to try to determine the mixing length too well since it is not the true mixing length any more.

That the mixing-length theory has at the very end at least one free parameter is quite natural. Any theory of convection which is not based on an exact solution of the hydrodynamic equations must have somewhere an assumption where something is simplified and this normally means that for some unknown things plausible values have been taken. This means that some free parameters have been fixed artificially. Only the exact solution is allowed to have no free parameters. Any other theory which does not show at least one free parameter must have somewhere in the derivation a place where free parameters have been fixed tacitly.

In the last years several authors (for instance BÖHM-VITENSE (1958), KIPPENHAHN, TEMESVÁRY, BIERMANN (1958), DEMARQUE, GEISSLER (1962)) took l equal to H_p or a multiple of H_p. In the paper by KTB the Hoyle-Schwarzschild evolutionary track for $M = 1.2\,M_\odot$ has been repeated with a temperature-pressure dependence in the outer layers fixed by the mixing-length theory. Their results showed that the evolutionary track in the HR diagram depends sensitively on the choice of l. The tracks for $l = H_p$ and $l = 2H_p$ are given in Fig. 18. They differ considerably but the mixing-length theory cannot decide between them.

For practical use, therefore, one has to take values for l which give tracks which agree with the observation instead of predicting the evolutionary tracks theoretically.

SCHWARZSCHILD (1962) pointed out that the density should determine the mixing length more accurately than the pressure.

Indeed one can think of an experiment with an incompressible fluid in which the mixing length is bigger than the pressure-scale height and in which the pressure-scale height turns out to be irrelevant. In an incompressible fluid the density ϱ is practically independent of the pressure but decreases slowly with increasing temperature. We now consider a fluid forming a layer upon an infinite horizontal plane in a homogeneous gravity field. The thickness of the layer may be h. On top of the layer a constant pressure P_0 may be applied on the free surface. The pressure at the bottom then is $P_1 = P_0 + g\varrho h$. The pressure-scale height in the depth x below the upper surface is

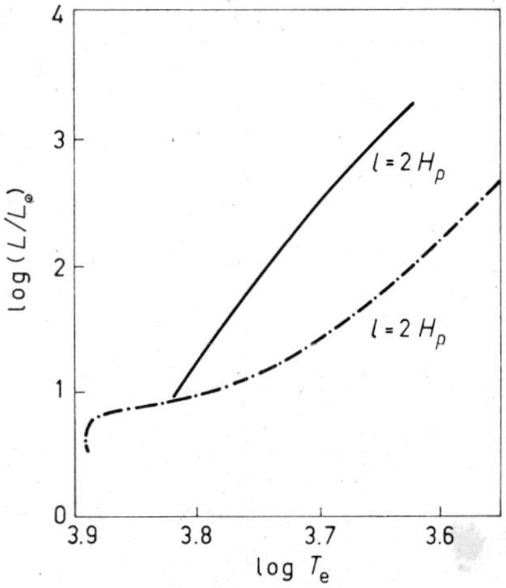

$$H_p = \left|\frac{dr}{d\ln P}\right| = \left|P\frac{dr}{dP}\right| = \frac{P_0}{g\varrho} + x\,. \tag{74}$$

Fig. 18. – Theoretical evolutionary tracks for $M = 1.2 M_\odot$ with $l = H_p$ and $l = 2H_p$.

We now assume that the thickness h is bigger than the maximum value of H_p in the layer. If we now generate convection in the layer by heating the bottom we can determine the characteristic length l of the velocity field at every point in the layer by measurements described in Section 3·3. Since the Navier-Stokes equations, the continuity equation and the equation of conservation of energy (thermal equation) only contain dP/dx, but not P the velocity field cannot change if we change the pressure P_0 on top of the layer which does not change dP/dx. But by changing P_0 we change H_p. Therefore a change of P_0 does not have any effect on the velocity field and therefore on l. Therefore H_p cannot determine the mixing length. In our problem the mixing length will be determined by the thickness h of the layer and in some respect the thickness of the layer is something like a density-scale height since the top of the layer is defined as the region in which ϱ drops to zero.

It seems difficult to prove the same for compressible media. But let us think about the consequences which will follow if we introduce the density-scale height as the mixing length. From the equation of state we derive the relation between H_p and H_ϱ where H_ϱ is defined by $H_\varrho = dr/d\ln\varrho$:

$$H_\varrho = \frac{H_p}{1 - (1 - \partial\ln\mu/\partial\ln T)\nabla + \partial\ln\mu/\partial\ln P}\,. \tag{75}$$

For simplicity we neglect ionization effects which cause changes in μ. Then we have

$$H_\varrho = \frac{H_p}{1-\nabla}. \tag{76}$$

In the outer layers of a hydrogen convective zone which is calculated with the gradient, ∇ is bigger than one which causes a density inversion (*i.e.* a decrease of density with depth), since

$$\frac{\mathrm{d}\ln\varrho}{\mathrm{d}\ln P} = 1-\nabla, \tag{77}$$

and therefore $\mathrm{d}\ln\varrho/\mathrm{d}\ln P$ becomes negative when $\nabla > 1$. This cannot happen if $l \sim H_\varrho$ since if we integrate inwards and when ∇ approaches one, then l increases according to eq. (76). But the bigger is l the smaller U and the closer ∇ will approach ∇_{ad}. Since ∇_{ad} is 0.4 the gradient will be kept below 1 and no density inversion can appear. Figures 19, 20 give the P-T-relation and the ϱ-P-relation of some models of the outer layers of the sun with a mixing length connected with H_p and with H_ϱ (BAKER (1962)). The H_ϱ-models have no density inversion any more. But despite this fact it seems that they do not differ very much from models with $l \sim H_p$.

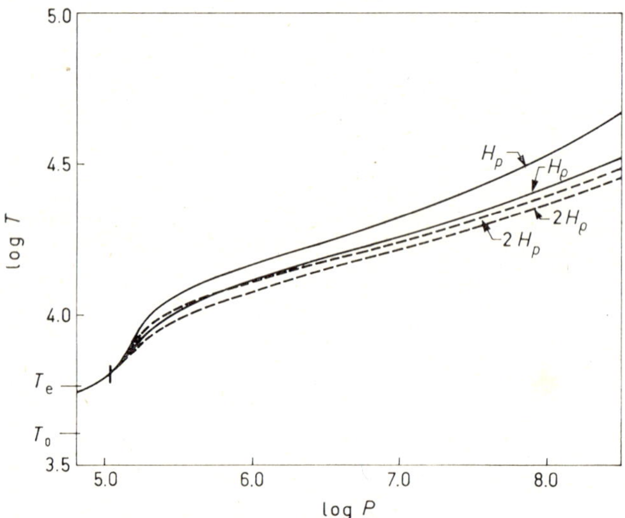

Fig. 19. – The temperature stratification for models with $l = H_p$, $2H_p$, H_ϱ, $2H_\varrho$.

The density-scale height and the pressure-scale height depend both only on local quantities. But the elements which transport the energy at a certain point of the zone will also have properties which are determined by the physical quantities at their origin one scale height deeper or higher. Therefore SCHWARZSCHILD (1962) proposed to take a *nonlocal* density-scale height as mixing length instead of the *local* density-scale given in eq. (75). This would mean that the mixing length in a layer is equal to the distance (outwards or inwards or a mean of both distances) in which the actual density changes by a factor e. In the local theory we took the distance in which the density extrapolated with the local gradient changes by a factor e. Models for the nonlocal theory have to be calculated by an iteration process since the nonlocal density-scale height can only be determined when the structure of the layer is already known.

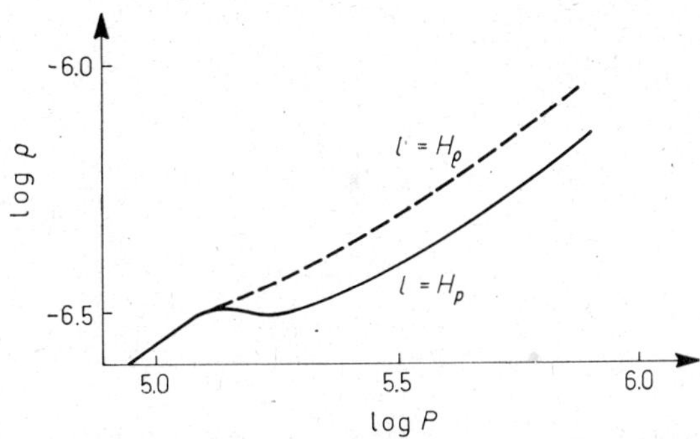

Fig. 20. – The density stratification for models with $l = H_p$, $l = H_\varrho$. The latter has no density inversion.

I think it is not necessary to emphasize that all attempts to get rid of the uncertainties in the mixing-length theory (as well as all attempts to replace the mixing-length theory by something else) cannot be very successful at the very end if one does not go back to the basic equations, the Navier-Stokes equations with mass and energy conservation. Only their solution can give the clue. The time may not be too far away when available computer facilities may encourage the astrophysicists to tackle the problem from this side.

* * *

It is a pleasure for me to thank all my collegues who kindly made their recent yet unpublished results accessible to me: Dr. N. BAKER, Drs. P. DEMARQUE and J. E. GEISLER, Dr. L. HENYEY and J. FORBES, Dr. M. NISHIDA, Dr. M. SCHWARZSCHILD.

BIBLIOGRAPHY

BAKER, N. and KIPPENHAHN, R.: 1959, *Z. f. Astrophys.*, **48**, 140.
BAKER, N.: 1962, private communication.
BIERMANN, L.: 1945, *Erg. d. exakt. Naturwiss.*, **21**, 2.
BÖHM-VITENSE, E.: 1958, *Z. f. Astrophys.*, **46**, 108.

Demarque, P. and Geisler, J. E.: 1926, preprint.
Eddington, A.: 1929, *M. N.*, **90**, 54.
Gratton, L.: 1945, *Mem. della Soc. Astr. Ital.*, vol. xvii-2.
Hayasha, C. and Cameron, R. C.: 1962, *Ap. J.*, **136**, 166.
Henyey, L.: 1962, private communication.
Hoyle, F. and Schwarzschild, M.: 1955, *Ap. J. Suppl.* II, No. 13.
Hoyle, F.: 1960, *M. N.*, **120**, 22.
Kippenhahn, R.: 1958, *Z. f. Astrophys.*, **46**, 26.
Kippenhahn, R., Temesváry, S. and Biermann, L.: 1958, *Z. f. Astrophys.*, **46**, 257.
Kushwaha, R. S.: 1957, *Ap. J.*, **125**, 242.
Ledoux, P.: 1947, *Ap. J.*, **105**, 305.
Mestel, L.: 1953, *M. N.*, **113**, 716.
Nishida, M.: 1962, private communication.
Oepik, E.: 1938, *Publ. de l'Obs. Astr. de l'Université de Tartu.*
Sakashita, S., Ono, Y. and Hayashi, C.: 1959, *Progr. Theor. Phys.*, **21**, 315.
Schönberg, M. and Chandrasekhar, S.: 1942, *Ap. J.*, **96**, 161.
Schwarzschild, M.: 1942, *Ap. J.*, **95**, 441.
Schwarzschild, M.: 1958, *Structure and Evolution of the Stars*, Princeton, N. J.
Schwarzschild, M. and Härm, R.: 1958, *Ap. J.*, **128**, 348.
Schwarzschild, M. and Härm, R.: 1962, *Ap. J.*, **136**, 158.
Schwarzschild, M. and Selberg, H.: 1962, *Ap. J.*, **136**, 150.
Schwarzschild, M.: 1962, private communication.
Stothers, R. B. and Schwarzschild, M.: 1961, *Ap. J.*, **133**, 343.
Sweet, P. A.: 1950, *M. N.*, **110**, 548.
Temesváry, St.: 1960, *Modèls d'Etoiles et Evolution Stellaire*, Liège.
Unsöld A.: 1955, *Physik der Sternatmosphären*, Berlin.
Vitense, E.: 1951, *Z. f. Astrophys.*, **28**, 81.
Vitense, E.: 1953, *Z. f. Astrophys.*, **32**, 135.
Vogt, H.: 1925, *Astr. N.*, **223**, 229.
Wrubel, M.: 1958, *Handbuch der Physik*, vol. li, 1.
Zeipel, V.: 1924, *M. N.*, **84**, 665, 684, 702.

Theoretical Models for the Pulsations of Cepheids.

N. Baker (*)

Goddard Institute for Space Studies
National Aeronautics and Space Administration - New York, N. Y.

Introduction.

The subject that I am going to discuss is related to topics that two of the other lecturers have talked about in this course.

The theories that I shall describe are a special case of the general stability investigations which Ledoux (1958, 1962) has been discussing. We apply a type of stability analysis to certain stellar models in order to try to understand some of the observations of intrinsic variables which Sandage (1962) has discussed. I shall not attempt to present the entire theory of pulsational instability. Rather, I wish to discuss some of the general aspects of this theory, and then to present some of the physical ideas which are behind the recent treatments of cepheid instability. The work that I shall describe is essentially an attempt to interpret the pulsations of some of the intrinsic variable stars as being a phenomenon of vibrational instability. Professor Ledoux has discussed this type of instability in general and I shall confine myself only to some very specific cases. At the end of this talk I want to present some of the most recent results that have been obtained, and to indicate their possible significance and range of validity. Our object in doing this type of stability analysis is to try to understand the origin of the instability and thus to explain why the variable stars appear in their particular regions of the Hertzsprung-Russell diagram. In particular, we hope to be able to understand why the classical cepheids and the *RR-Lyrae* stars occupy a very narrow, almost vertical region in the H-R diagram and, hopefully, to lay the basis for a theoretical understanding of the observationally important period-luminosity relationship. An additional reason for undertaking this type of investigation is the hope that an understanding of the cepheids will enable us to get a better

(*) Present address: Institute for Advanced Study, Princeton, New Jersey, U.S.A.

picture of the structure of giant and supergiant stars and will perhaps provide some information about the evolution of stars in regions of the H-R diagram far away from the main sequence.

1. – The pulsation equations.

We must first discuss briefly the equations which are used in the study at hand. I shall give only a very general discussion, but a complete discussion of the equations may be found in the fundamental review article of LEDOUX and WALRAVEN (1958). A discussion of the equations in the form that I shall use may also be found in BAKER and KIPPENHAHN (1962) (quoted as BK). We begin by writing down the general equations for the structure of a spherically symmetric star. We explicitly allow the possibility that the various quantities may change with time,

$$\frac{\partial^2 r}{\partial t^2} = -4\pi r^2 \frac{\partial P}{\partial M_r} - \frac{GM_r}{r^2}, \tag{1}$$

$$\frac{\partial r}{\partial M_r} = \frac{1}{4\pi r^2 \varrho}, \tag{2}$$

$$\frac{\partial T}{\partial M_r} = -\frac{3\varkappa L_r}{64\pi^2 acr^4 T^3}, \tag{3}$$

$$\frac{\partial L_r}{\partial M_r} = \varepsilon - c_p \frac{\partial T}{\partial t} + \frac{1}{\varrho}\frac{\partial P}{\partial t}. \tag{4}$$

The equation of state used is that for a perfect gas, including radiation pressure where necessary.

Equation (1) is just the generalization of the equation of hydrostatic equilibrium. The left-hand side is the inertial term which must be included when we are considering a hydrodynamic situation. Equation (2) expresses the continuity of mass. The third is the equation of radiative transport, as it is generally written in stellar structure problems. This is a diffusion equation, which, of course, implies a short mean free path for radiation. The last equation expresses energy conservation. The luminosity gradient here is due not only to the nuclear energy source term ε, but also to the last two terms on the right-hand side, which represent the time derivative of the heat contained in a unit element of gas. These two terms come then from the thermodynamics and they are related to the thermal time-behavior of the gas. The four differential equations are all coupled with each other, but we may notice at this point that the first two equations refer primarily to the mechanical (acoustical) behavior of the gas while the last two equations essentially describe the thermal properties of the gas and the heat transport.

There are several comments to be made about this set of equations. First, we see that they are all written in Lagrangian form; that is, the independent variable, M_r, which is the total mass inside a sphere of radius r, is a variable which does not change during the pulsation. The Lagrangian form of the equations is very convenient when we are discussing this type of time-dependent behavior. Furthermore, we have not included the dissipative effects of viscosity and we shall assume that they are small. The probable effect of including the viscosity terms would be simply to increase the stability of the stars in question.

The use of eq. (3) implies that we are assuming radiative transport throughout the star. For regions of a star in which convection is important, this equation must be replaced by a more general one which describes the additional effect of convective transport. In fact, convection has been taken into account to some extent in the models that I shall discuss, and I shall indicate later how this was done. It should also be noted that we are considering only radial displacements. We have not written vector equations here and the assumption will be that only radial pulsations occur. Furthermore, we shall neglect the nuclear energy source term ε because we shall, for the most part, be discussing only outer regions of stars in which no nuclear energy generation occurs. Thus we are explicitly neglecting any influence that the nuclear energy generation may have on the stability.

The first step in our analysis of these time-dependent equations is to linearize them. This is a common procedure in this type of stability calculation and, of course, ignores the possibility of nonlinear oscillations whose instability depends upon the amplitude. In order to carry out the linearization, we write

$$T = T_0(M_r) + T_1(M_r, t)\,, \qquad \text{(similarly for } P,\ r, \text{ and } L_r\text{)};$$

where the T_0, etc., are functions only of the equilibrium configuration of the star. The time-dependent part is written

$$T_1 = T_0 t \exp[i\omega t]\,, \quad \text{etc.},$$

and the linearization consists in ignoring quadratic and bilinear terms in the t, etc. The harmonic time-dependence is used in order that we may carry out the usual sort of normal-mode analysis which is applicable to linear problems. By the means described here, eqs. (1)–(4) are written in linearized form. The linearized forms of these four equations in terms of the nondimensional variables, p, t, x, l, $(x = (r_1/r_0)\exp[i\omega t])$ are:

$$p' = -p - (4 + c_3\sigma^2)x\,, \tag{5}$$

$$x' = c_4(3x + c_5 p - c_6 t)\,, \tag{6}$$

$$t' = c_7(l - 4x + c_8 p - c_9 t)\,, \tag{7}$$

$$l' = i\sigma c_1(c_2 t - p)\,, \tag{8}$$

where the prime denotes differentiation with respect to $\ln P_0$, which has been taken as the independent variable. The dimensionless coefficients c_1 through c_9 depend only upon the equilibrium parameters of the star. They are given in full in BK. σ is a dimensionless frequency, the counterpart of ω. Again we note that the first-order eqs. (5) and (6) refer primarily to fluctuations of the mechanical qualities of the star, while eqs. (7) and (8) describe fluctuations in the thermal properties. We now consider some of the applications of the above set of equations.

2. – Linear adiabatic theory.

For purposes of orientation, we first discuss the simplest form of the pulsation theory. Equations (5)–(8) describe the mechanical oscillations of the star as they are coupled to the thermodynamic properties. If, however, we assume for the moment that the mechanical energy of pulsations is conserved during the pulsations, we need consider only adiabatic acoustical oscillations. Then we replace the eqs. (7) and (8) with the single nondifferential relation

$$\frac{P_1}{P_0} = \gamma\,\frac{\varrho_1}{\varrho_0}\,.$$

This leads to the following second-order differential equation (cf. LEDOUX and WALRAVEN 1958, Section **58**)

$$\frac{\mathrm{d}^2 x}{\mathrm{d}r_0^2} + \left(\frac{4}{r_0} - \frac{GM_r\varrho}{r_0^2 p}\right)\frac{\mathrm{d}x}{\mathrm{d}r_0} + \frac{\varrho}{\gamma p}\left[(4-3\gamma)\,\frac{GM_r}{r_0^3} + \omega^2\right]x = 0\,.$$

The (Lagrangian) indipendent variable here is $r_0(M_r)$.

We may write this equation in nondimensional form in terms of the homology-invariant variables:

$$z = \frac{r_0}{R}\,,\qquad q = \frac{M_r}{M}\,,\qquad \tilde{p} = \frac{P}{n_1(M, R)}\,,\qquad \tilde{t} = \frac{T}{n_2(M, R)}\,.$$

The equation then takes the form

$$\frac{d^2x}{dz^2} + f_1(z)\frac{dx}{dz} + f_2(z)[f_3(z) + \sigma^2]x = 0 ,$$

which is invariant under homology transformations which keep the quantities z, q, $\tilde{p}$, $\tilde{t}$ invariant. The nondimensional eigenfrequency

$$\sigma = \frac{1}{\sqrt{G\bar{\varrho}}\,\Pi},$$

is thus the same for all the stars of a homologous series (Π is the period); that is,

$$\Pi\sqrt{\bar{\varrho}} = \text{constant}$$

for homologous stars. This relation is often written in the form

$$\Pi_{(\text{days})}\sqrt{\frac{\bar{\varrho}}{\bar{\varrho}_\odot}} = Q .$$

It turns out that according to model calculations, the cepheids are not a strictly homologous group and so Q is not strictly constant. However, we can use this relationship in an approximate way to make some preliminary arguments regarding the periods of cepheids.

The period-mean density relationship provides us with a first step toward a period-luminosity relationship. Thus, we have

$$\Pi = \Pi(M, R) .$$

Now, if we also assume that there exists a mass-luminosity relationship (which must be obtained from evolutionary considerations),

$$M = M(L) ,$$

then we require only a relation of the form

$$R = R(L)$$

in order to have a single equation relating period and luminosity. This requires a study of the mechanism of pulsation, which in turn necessitates consideration of a nonadiabatic theory. This argument can be put in a slightly different form, as SANDAGE has done in his lectures:

In Fig. 1 is indicated schematically the region of the classical cepheids and *RR-Lyrae* stars in the H-R diagran. The oblique lines running through this region are lines of constant period, assuming Q = constant. From this picture we see that for a given luminosity there will be only a small range of possible periods in the instability region. This is clearly a consequence of the narrow width of the instability strip. If the strip were very much wider in the horizontal direction, there would be a wide range of periods for a given value of L. Thus, we see again that the existence of a period-luminosity relationship depends upon the fact that there is a rather unique relationship between radius or effective temperature and luminosity for unstable stars. To investigate this relationship, we now turn to a discussion of the nonadiabatic theory.

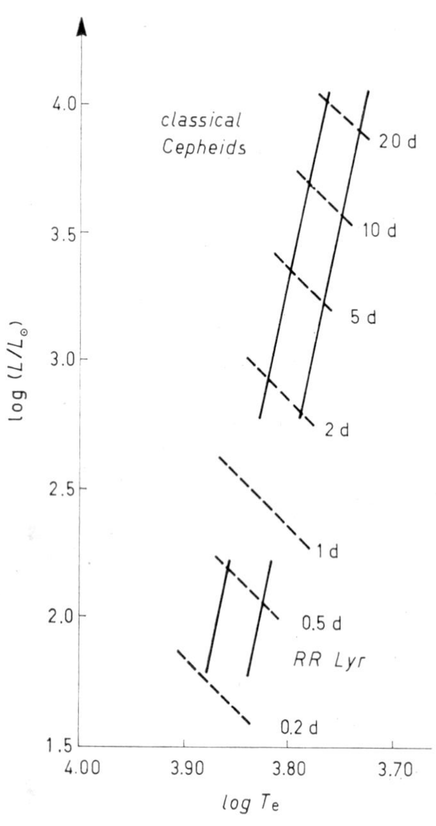

Fig. 1. – Approximate location of the classical cepheid and *RR Lyrae* regions in the luminosity-effective temperature diagram. Dashed lines are lines of constant period, assuming $Q = 0.04$.

3. – Linear non-adiabatic theory.

3·1. *The quasi-adiabatic approximation.* – Although the pulsational energy is strictly conserved in the adiabatic theory, we may use the adiabatic approximation in order to make a first estimate of the radiative damping of a pulsating star. The linearized form of eq. (3) may be written as follows:

$$L_1 = f_L(r_1, P_1, T_1, T_1') ,$$

where f_L is a linear function of the variables. (Primes here denote differentiation with respect to r_0.) If we put in the values of the arguments which are obtained from the adiabatic solution, we get a first-order approximation to L_1. On the other hand, conservation of energy gives us

$$L_1' = 4\pi r^2 \varrho \frac{\partial Q}{\partial t},$$

where Q is the total heat content of a unit mass of gas. Now the total heat leakage per cycle has been given by EDDINGTON (1926):

$$\Delta W \sim -\oint \mathrm{d}Q \frac{T_1}{T_0} \sim -\oint \frac{T_1}{T_0} L_1' \mathrm{d}t\,.$$

If this quantity is then integrated over the entire star, we have an expression for the total amount of energy loss from pulsations during one cycle. In this way one may estimate the damping time. EDDINGTON (1926) for *δ-Cephei* got a time of 8000 years, using $\gamma = 1.374$. More recently ZHEVAKIN (1953), with the modern value $\gamma = 1.667$, obtains a value of 300 years. This shows us the necessity of considering a mechanism for maintenance of the pulsations. First, however, we can gain some physical insight into the mechanisms involved by considering the range of validity of the quasi-adiabatic approximation.

3·2. *The validity of the quasi-adiabatic approximation.* – The reason for the nonadiabatic nature of the situation we are considering is the fact that a given volume of gas, when, during a pulsation, it is hotter than its mean value, tends to radiate more energy to cooler regions; and when it is cooler than average, tends to absorb radiant energy. If, however, the amount of energy radiated or absorbed during the course of a pulsation is very small compared to the total energy content of the particular volume of gas we are considering, then the deviation from the adiabatic situation will be very small and can be neglected. Put in another way, the quasi-adiabatic approximation will be a good one if the characteristic distance for heat diffusion during one cycle is small compared to the distance over which a physical variable, like the temperature fluctuation, varies. In order to make a rough estimate, let us imagine an infinitesimally thin shell source of heat at some radius r in the interior, which varies periodically with frequency ω. Then we have the usual equation for heat diffusion

$$\frac{\partial^2 T}{\partial r^2} = \frac{1}{K}\frac{\partial T}{\partial t},$$

where the radiative conductivity K has a dependence of the form

$$K \sim \frac{T^3}{\varkappa \varrho^2 c_p}\,.$$

In such a formulation, the characteristic length for heat diffusion is given by

$$\lambda_r \approx \sqrt{\frac{K}{\omega}}\,.$$

We can say roughly that the quasi-adiabatic approximation will be good so long as this quantity is small compared to the wave length of the acoustic oscillations,

$$\lambda_s \approx c_s/2\pi\omega\,.$$

Thus the condition is

$$\lambda_r \ll \lambda_s\,, \qquad \text{or} \qquad \sqrt{K/\omega} \ll c_s/2\pi\omega$$

for the approximation to be good. There are two things to note about this formula. First, as one goes deep into the star, K becomes small and thus the radiative diffusion distance decreases. On the other hand, λ_s becomes large because $c_s \sim \sqrt{T}$. Thus the quasi-adiabatic approximation becomes increasingly good in the interior of the star, and we may speak of the adiabatic interior which can provide neither positive nor negative dissipation by means of interaction with radiation.

The second thing to note is that the approximation becomes poorer with increasing ω. Thus, the adiabatic zone is smaller for the higher frequencies.

3·3. *Mechanisms for negative dissipation.* – We have seen how the pulsation of a star may be damped due to radiative diffusion. We must now consider possible mechanisms for overcoming this damping. We shall mention here two possibilities (EDDINGTON 1926). The first mechanism depends on the extreme temperature-sensitivity of the nuclear reactions in the center of the star. During contraction, the central temperature increases and thus the rate of nuclear-energy generation increases. When the star expands, less energy is produced, due to the somewhat lower temperature, and this sort of a mechanism, analogous to a diesel motor, injects energy into the pulsation in the proper phase, so as to contribute to the driving. This is considered in more detail in the lectures of LEDOUX, and here we mention only that although this process always produces a destabilizing effect, the magnitude of this effect in normal stars seems to be too small to overcome the radiative damping. Thus, we must look for another mechanism to provide the necessary negative dissipation.

To illustrate the second mechanism we consider, as a physical analog, a piston with a gas inside. A constant flux of radiation enters the piston through the bottom. The walls are not completely opaque, and a certain amount of energy may be radiated through the walls. This energy loss will be greatest during compression when the gas is hottest, and least during expansion. This is the analog of the radiative damping considered above. In the top of the piston there is a valve which, when opened, will let radiation through. When it is closed, the entire top of the piston is opaque. If the valve is closed when the piston is compressed, the radiation will be trapped in the piston, the gas will start to heat up, and the piston will expand. If then, at the time of maximum expansion, the valve is opened, the gas will begin to cool off and the piston will contract. It is seen, then, that in order properly to modulate the radiation fiux, so that part of this flux is transformed into mechanical energy, it is only necessary that the valve be opened and closed in the proper phase. Such a process could actually take place in a star, if there were a region in the star where the opacity increases during contraction and decreases during expansion. Of course, this is not true in most regions of a star. If we write the opacity in the form

$$\varkappa = \varkappa_0 \varrho^m T^n ,$$

then we know that, in general, m is positive and approximately equal to one; whereas n, for a Kramer's opacity law (which holds in most of the interior) is about -3.5. Thus, in general, the opacity decreases in compression and increases in expansion, exactly the opposite of the behavior needed to produce driving; indeed this will contribute a positive damping to the star. What we need is a region where n and m are both positive. It will also help if c_p is large, for then the heating during expansion will be smaller, and the amount of radiative damping will be decreased.

It turns out that both of these circumstances are found in regions of the star where abundant elements are being ionized. In the hydrogen and helium ionization zones, m and n are positive (see BK, Fig. 1 and 2) and c_p is large. Thus, the suggestion was made (see, *e.g.*, EDDINGTON 1926 and ZHEVAKIN 1953) that these ionization zones might be able to provide sufficient negative dissipation to overcome the radiative damping. The opacity effect is particularly great in the hydrogen ionization zone, but EDDINGTON and later ZHEVAKIN noted that the total amount of mass in the hydrogen ionization zone, in stars with surface temperatures like those of cepheids, is extremely small. Since there is not much gas in this zone, the total amount of energy that can be injected into this gas or dissipated from it is very small and so this extreme outer region is very inefficient in either driving or damping.

This led ZHEVAKIN (1953) to suggest that the He II ionization zone ($He^+ \rightarrow He^{++}$) might be the region which is responsible for cepheid instability. The behavior of the absorption coefficient in this zone is not as pronounced as it is in the hydrogen zone, but the zone lies considerably deeper in the star because of the higher ionization potential, and the total amount of gas which may take part in the excitation mechanism is correspondingly much greater. This suggestion has received considerable attention in recent years (see, *e.g.*, Cox 1958, 1960, 1962 *a* and BK). Our most recent calculations seem to suggest, however, that the cepheids may have lower surface temperatures than has been thought from observations and, as a consequence of this, the ionizations of H and He I seem to play almost as great a role as does that of He II in the destabilizing mechanism. We shall discuss the details of this later on.

3·4. *Reason for the instability region.* – We are now able to see qualitatively why, for a given luminosity, only stars in a certain small range of effective temperatures are unstable. We consider the star schematically as being made up of three regions. The very outer surface region contains very little mass and therefore very little total heat capacity and this region can have very little effect on the total energy balance in the pulsations. The innermost part of the star is the adiabatic region, which contributes neither to the damping nor driving as long as we are considering only radiative mechanisms. Hence our attention is centered on the intermediate nonadiabatic region, which usually produces a damping effect but might, under proper circumstances, be able to produce a negative dissipation. Let us consider, following a discussion of J. P. Cox (1961), what happens in a series of models, all of which are assumed to have the same luminosity and the same mass. These models lie in a horizontal line in the H-R diagram, and we ask what happens as we go from left to right along this line.

At the high-temperature end, the hydrogen and both helium ionization zones fall in the atmosphere of the star. Beneath the surface layers these elements are completely ionized, and even as we go to slightly cooler stars, all of these ionization zones are still found in the thin outer region. As the surface temperature is decreased still more, however, the He II ionization zone, and later the H–He I ionization zones, fall in the intermediate nonadiabatic region. In this small range of intermediate temperatures, there is the possibility of sufficient negative dissipation to drive the star. At still lower effective temperatures, the ionization zones find themselves still farther inside the star until they are fully in the adiabatic region, and then they can no longer contribute to the driving. Thus, we expect that both the hottest and the coolest stars will be stable, while there will be a less stable intermediate temperature region. This provides a T_e-L, or R-L, relation, which is what we need.

This qualitative picture appears at the present time to be basically correct, although, at least for the cooler classical cepheids, there may be other important effects.

At the right-hand side of the classical cepheid region, it seems that the stars have such low surface temperatures that the outer convection zone associated with the ionization of H and He becomes very deep. So far the problem of treating pulsations of a region of a star where convection is the dominant energy transport mechanism has proved very difficult to solve. It seems clear, however, that if convection is the major transport mechanism in the ionization zones, then the negative dissipation mechanism which we have described, and which depends upon radiative energy transport, cannot be effective. Our recent calculations indicate that convection becomes very important at the cooler edge of the cepheid strip, and there seems still to be merit in the suggestion (BK) that existence of an extensive and effective convection zone in stars of lower effective temperature is the reason for the right-hand boundary of the cepheid zone. This problem, however, has not yet been satisfactorily resolved. We have not yet produced accurate models for *RR-Lyrae* stars, but it seems probable that the influence of convection in these is not so great as it is in the classical cepheids. The smaller mass and higher effective temperature of the *RR-Lyrae* stars may reduce the effectiveness of the convection in the ionization zones. It seems possible that in these stars one may completely neglect convection; in that case the simple picture outlined above may be entirely sufficient.

4. – Numerical models of Cepheids.

In recent years a number of workers have constructed numerical models of pulsating stars. Zhevakin, following his original suggestion of the He II mechanism has constructed many models which are discussed in a series of papers. He uses polytropic envelopes as equilibrium models and carries out nonadiabatic calculations using a number of shells or zones to simplify the integration. The most extensive series of calculations has been carried out by J. P. Cox (1958, 1960 and 1962 *a*). His equilibrium models are also polytropic envelopes and he has constructed models of classical cepheids, *RR-Lyrae* stars and other types of variables. Although his equilibrium models are simplified ones, the absorption coefficient and its dependence upon density and temperature in the region of He II ionization, as well as the specific heat in this zone, are included explicitly. Convection is not taken into account. In general Cox's results are in good qualitative agreement with the results to be discussed below.

Recently there have been some interesting attempts to study the nonlinear effects of pulsations. Whitney (1962) and his collaborators have discussed nonlinear effects in envelopes with particular reference to the *W-Virginis* stars, the observations of which are indicative of shock waves in the atmospheres.

Recently CHRISTY (1962) has studied nonlinear effects in Cepheids with extensive convection zones. He has considered the interaction of convection with the pulsations, particularly with a view toward trying to understand the observed phase shift between the radius and luminosity fluctuations. It is not entirely clear at present what role the convection must play in the classical cepheids and *RR-Lyrae* stars. However, it is certain that any discussion of cooler pulsating stars, like the *Mira* stars (Long-Period Variables), must take full account of convection, since the outer convection zones in these stars are very extensive and the energy transport in outer parts of the stars is nearly 100 % convective.

I shall devote the rest of this talk to a discussion of certain cepheid models constructed in the past several years by R. KIPPENHAHN and myself.

4·1. *Equilibrium models.* – It is not possible at this time to construct complete equilibrium models of supergiant stars, since evolutionary calculations have not yet been pushed into this region of the H-R diagram. We can, however, construct good envelope models, and as we shall see, this is sufficient for our purposes. The models are constructed essentially as described in BK. In order to begin, we must have values of mass, luminosity, and effective temperature. The luminosities and effective temperatures are taken in a range suggested by observations. The masses are taken from SANDAGE's (1958) semi-empirical mass-luminosity relation, although it may prove possible later to determine the masses on the basis of the observed periods. The opacities are obtained from opacity tables which have been calculated with the help of the codes developed by the Los Alamos group (COX, 1962 *b*). These are probably the best opacities which are presently available; perhaps their most serious deficiency is the lack of corrections for spectral lines. These corrections might change the opacities by a factor of 2 or 3 in the innermost regions of our envelopes (COX and STEWART, 1961).

In equilibrium models where convection is important, the mixing-length theory, as applied by BÖHM-VITENSE (1958), has been used. In most of the calculations I shall discuss, a mixing length equal to 1.5 times the local pressure scale height has been used. As mentioned above, the convection does seem to play an important role in the cooler classical Cepheids. In general, the integrations are carried from the surface of the star down to the point where $M_r = M/2$. The assumption is thereby made that the star is chemically homogeneous in the outer half of its mass, and that no energy production occurs in this region. It is actually not necessary to carry our envelope integrations this deep, but it is convenient for the determination of the periods, as I shall discuss below. The equations used are eqs. (5)–(8), which describe the fully nonadiabatic radial pulsations. These equations form a fourth-order complex homogeneous system of equations.

4·2. *The boundary conditions.* – We must now discuss briefly the boundary conditions.

a) Outer boundary conditions. The luminosity is of course related to radius and effective temperature through the equation:

$$L = 4\pi\sigma R^2 T_e^4 .$$

By linearizing this formula in the way described above, we obtain one boundary condition. Actually, the condition is slightly modified in practice to take account of the change in optical depth during a pulsation (see BK).

The second boundary condition is a condition on the mechanical variables. It is obtained in the following way. Let us assume an infinite plane atmosphere on the star, with adiabatic or isothermal standing or running waves. A discussion of this may be found in LEDOUX and WALRAVEN (1958, Section **68**). If we consider this atmosphere to be an isothermal layer, we find that there is a characteristic frequency, given by

$$\omega_0 = c_s/2H ,$$

where H is the scale height. In the case that

$$\omega < \omega_0 ,$$

the relation between p and x is real and the pulsations are pure standing waves. If, on the other hand,

$$\omega > \omega_0 ,$$

the relation between p and x is complex and there is a component of running waves. If we consider a two-layer atmosphere, *i.e.*, one isothermal layer underlying a hotter isothermal layer (which might represent a hot corona) we can also get progressive waves (SCHATZMAN, 1956). Such progressive waves imply a loss of mechanical energy through the surface of the star, and preliminary calculations have shown that even a relatively small component of progressive waves can provide a rather considerable mechanical damping. In some cases these progressive waves probably become shock waves as they go through the atmosphere.

b) Inner boundary conditions. If we had a complete stellar model we could of course apply the conditions of regularity at the center of the star. Since, however, we have only an envelope model, such a requirement is meaningless. Thus, at the inner boundary of the envelope we use the condition

$$p = \frac{c_p \mu}{\Re} t ,$$

which is the relation between p and t for adiabatic pulsations. The effect of this boundary condition is to select only those solutions of the equation which become truly adiabatic in the interior. The final condition, since we are dealing with homogeneous equations, is a normalization. It has proved convenient to normalize by requiring that the relative radius fluctuation x be equal to 1 and real at the inner boundary of the envelope.

4·3. *The eigenvalue problem.* – The boundary conditions given above are not yet sufficient to determine our problem because the eigenvalue ω must still be obtained. ω is complex and in order to determine both the real and imaginary parts we require in fact two more conditions. The real part of ω, which gives the period, is obtained by requiring that the solution have the proper number of nodes. The way this is done in practice may be seen from Fig. 2. If the frequency chosen is too low, the function begins to increase as the integration is carried toward the center because of the singularities there. If, on the other hand, the frequency is somewhat too large, one will find a node appearing already in the fundamental mode.

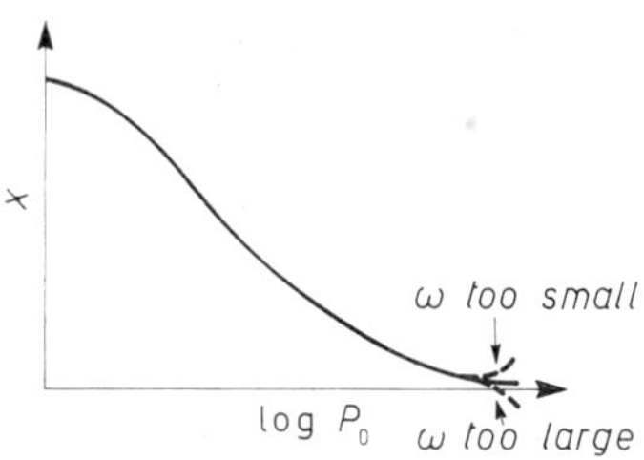

Fig. 2. – Method of determining the frequency ω of the pulsation

It turns out that if the envelope contains half the mass of the star, one can, by this method, determine ω_R to many more decimal places than is necessary. In fact, it seems that one needs only about the outer 30% of the star (in radius) in order to determine the period to a few percent, as long as it may be assumed that the mass distribution in the interior is not too different from that of an $n=3$ or $n=4$ polytrope. The imaginary part of the eigenvalue tells us whether the net dissipation in the envelope is positive or negative. $\omega_I > 0$ means that the star is damped and $\omega_I < 0$ means that the amplitude of the oscillations will increase. The increase or decrease of the pulsational energy appears then as a change in amplitude of the pulsations. The condition we should apply in order to get ω_I is that

$$\oint_{\text{cycle}} P \, \mathrm{d}V = 0 ,$$

when evaluated at the inner boundary. This will be valid so long as the inner boundary is in the adiabatic region.

In practice a different but equivalent method turns out to be more convenient for evaluating ω_I. We imagine the envelope to be driven by a piston at its inner boundary in such a way that the pulsations are strictly periodic ($\omega_I = 0$). Then energy flows through the inner boundary, being exchanged with the piston. At any spherical surface in the star the quantity

$$W = \oint P \, \mathrm{d}V ,$$

where V is the entire volume inside the surface, is equal to the net work done per cycle by the entire mass of the star *outside* of the surface considered, upon that part of the star which is inside.

Thus the quantity

$$W_{\text{in}} = \oint P \, \mathrm{d}V \big|_{r=R_{\text{in}}} ,$$

is the integrated driving or damping of the entire envelope. Then it is quite straightforward to show that

$$\frac{\omega_I}{\omega_R} \approx \frac{W_{\text{in}}}{\frac{1}{2}\int_0^M |\dot{r}_1|^2 \, \mathrm{d}M_r} .$$

If the quantity W is plotted as a function of the independent variable ($\log P_0$) one can see the damping or driving effect of the various regions. When W is increasing it means that that region of the model is contributing to the negative dissipation. When W decreases it means that the region is producing damping. In Fig. 3*a* we see the situation in a star which is hot enough so that the ionization regions all fall outside of, or in the outer part of, the nonadiabatic region. Here the ionization regions do provide some negative dissipation but this is cancelled by the positive dissipation farther inside the nonadiabatic region. Figure 3*b* shows W for a quite unstable star. We

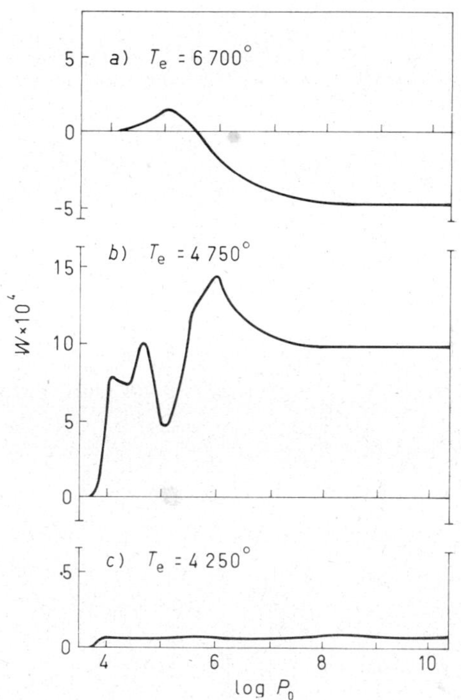

Fig. 3. – Regions of positive and negative dissipation for three model envelopes.

can see the negative dissipation due to the H, He I and He II regions. In this case the inner boundary of the He II region extends almost into the adiabatic region and there is very little positive dissipation internal to the He II ionization zone. Figure 3*c* shows a still cooler star where convection is transporting nearly all of the energy in the ionization zones. W remains practically constant and very small, which is a consequence of our assumption that the convection does not interact in any way with the pulsations.

4·4. *Some preliminary results.* – Although the work that I am describing is not yet complete, it is possible to discuss some preliminary results. For a given mass and luminosity we calculate a series of models having a certain range of effective temperatures. For each model we find the appropriate period as described above and the quantity

$$\varepsilon = -\frac{\omega_I}{\omega_R}.$$

Then ε may be plotted as a function of effective temperature for the particular L and M that we are considering. Examples of such curves are shown in Fig. 4 and 5. One notes in Fig. 4 that for the hotter stars the function ε

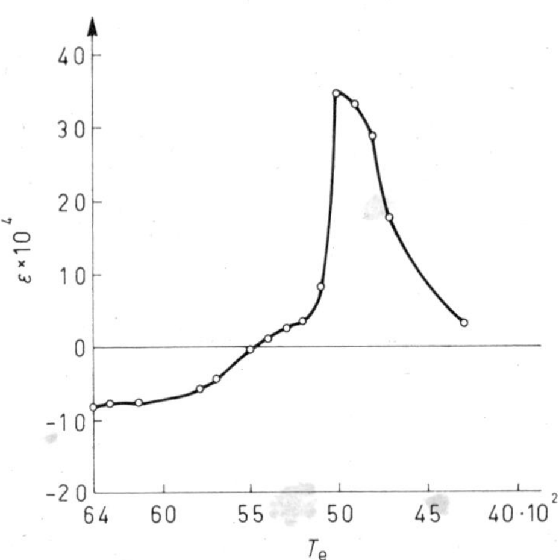

Fig. 4. – The instability parameter $\varepsilon = -\omega_I/\omega_R$ as a function of effective temperature for a series of models. $M/M_\odot = 7.68$, $L/L_\odot = 4100$, $X = 0.602$; $Y = 0{,}354$, $Z = 0.044$. Fundamental mode.

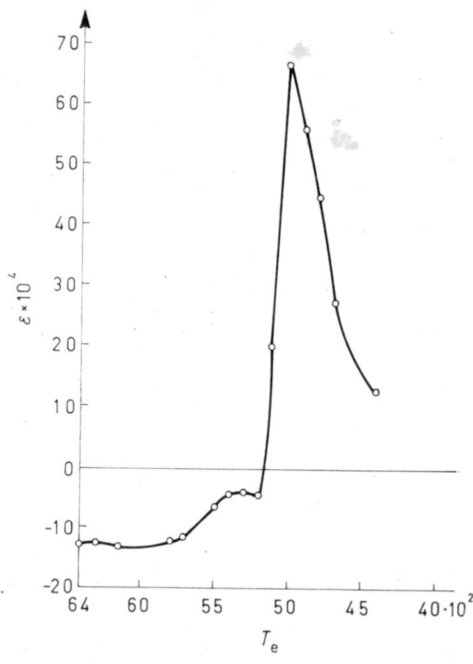

Fig. 5. – The instability parameter $\varepsilon = -\omega_I/\omega_R$ as a function of effective temperature for a series of models. $M/M_\odot = 7.68$, $L/L_\odot = 4100$; $X = 0.602$, $Y = 0.354$, $Z = 0.044$. First overtone.

is definitely negative, and that as the models become cooler ε increases and becomes very strongly positive in the range near 5 000°. As we move to still cooler models, the ε decreases again, partly because the He II region is now moving into the adiabatic zone, but primarily because of the increasing influence of convection. Because of our assumption that the pulsations do not interact with the convection, a model with a large convection zone will have an almost neutral stability. The approach to this neutral stability is seen as we move to the right in Fig. 4 and 5. In Fig. 4, one may also note that, with the number of models which we have available here, there is an uncertainty of several hundred degrees as to where the actual peak of the curve should be. Of course this could be determined more accurately by more integrations, although it is not clear how meaningful this would be. In any case, this uncertainty leads to a certain range of possible models, all of which may be unstable, and this will be reflected when we try to plot these data in an H-R diagram, as well as in the attempt to get a period-luminosity relationship.

Figure 5 shows the same series of models where the data now refer to the first overtone rather than to the fundamental. It should be noted that the peak comes about at the same temperature. At the maximum, however, the function is considerably greater. This does not necessarily mean that in this region the overtones are more strongly excited than the fundamental, since there are sources of dissipation that we have not considered, and most types of damping will tend to damp overtones more strongly. (For example, progressive waves at the surface dissipate energy more efficiently if their frequency is higher, and the higher modes have large amplitudes at the surface.) We may also note that the width of the peak of this function, which is of the same order of magnitude as the uncertainty in the peak, is about 300 or 400 degrees—roughly the same as the observed width of the instability region. In

TABLE I. – *Properties of cepheid models.*

Observed			Semi-empirical $M/M_\odot$	Calculated		
$L/L_\odot$	T_e (°K)	Π (day)		Z	T_e (at peak) (°K)	Π (at peak) (day)
2080	6000	5.3	6.11	0.022	5340±100	6.7±0.5
				0.044	4960±100	8.4±0.7
2950	5920	7.2	6.81	0.022	5270±150	8.7±1.0
				0.044	5080±100	10.3±0.8
4410	5770	10.8	7.68	0.022	5140±100	12.4±0.9
				0.044	5000±100	13.1±1.2

our work so far, we have calculated models with three different luminosities and corresponding masses, as shown in Table I. Each of the models was calculated for two different chemical mixtures which differ primarily in the fraction of heavy elements Z. Both models have $X \approx 0.6$, $Y \approx 0.4$. In one of the mixtures $Z = 0.022$, in the other $Z = 0.044$. When these data are plotted in the H-R diagram (Fig. 6) the most striking thing is that the cepheid region, according to these calculations, lies considerably to the right of the region given by the most recent observations (KRAFT, 1961). It turns out to be very difficult indeed to change any parameter enough to cause a considerable shift in the temperature of the instability strip. The periods which we find are all systematically larger than the observed ones, as may be seen from Table I. This, however depends on the mass-luminosity relation one takes, and we might at this point turn the argument around and use the observed periods to obtain masses and thereby a mass-luminosity relation.

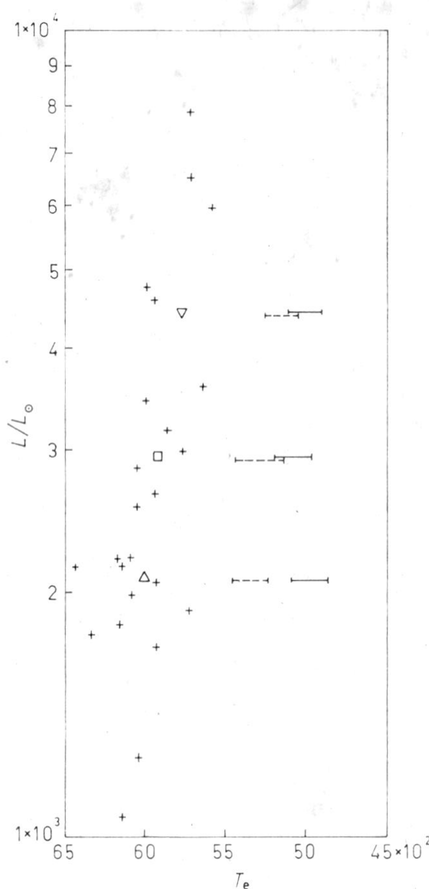

Fig. 6. – Temperature-luminosity diagram for cepheids. +'s represent observed cepheids from KRAFT (1961). Specifically indicated are *δ Cep* (△), *η Aql* (□), and *Z Lac* (▽). Horizontal bars denote range of instability for calculated models: ——— $Z = 0.044$, - - - $Z = 0.022$.

We may also note that the models with $Z = 0.022$ lie, in general, at somewhat higher temperatures than the others and that the slope of the instability strip for $Z = 0.022$ is somewhat greater. It also turns out that this mixture gives a period-luminosity relationship closer to the observed one (see below). On the other hand, the models with $Z = 0.022$ are systematically less strongly excited, *i.e.*, ε at the peak is a factor of about two smaller than it is for the models with $Z = 0.044$. The fact that the models with lower metal content lie farther to the left in the H-R diagram is due to the smaller absorption from H^- in the photosphere. This has the consequence that these models have a lower temperature at the same pressure in the photosphere, and this shows some effect also deeper in the envelope. The result is that as far as the ionization zones and their negative dissipation are concerned, these models behave like cooler models of higher metal content.

Finally, we may indicate a preliminary result for the period-luminosity relation. If we write this relation in the form

$$M_v = A - B \log \Pi ,$$

then it is of particular interest to calculate the coefficient B. The results are shown in Table II. It is observed (SANDAGE, 1962) that the slopes of the period-luminosity relationship for $M31$, the small Magellanic cloud and IC 1613 are all the same, while there is a definite deviation for the large Magellanic cloud. In our calculations, B is rather too large, for both of the chemical mixtures

TABLE II. – *Observed and calculated values of B, the slope of the period-luminosity relation.*

	Object		B
Observed	M31 and SMC		2.47
	LMC		2.82
	Z	B (fundamental)	B (lst overtone)
Calculated	0.022	3.1	3.1
$(\pm \sim 30\%)$	0.044	4.3	3.7

we have taken. It is interesting, however, that B is about the same whether calculated on the basis of the fundamental mode or of the first overtone; and it is also interesting that the difference for the two mixtures is of the order of 30 %, roughly the same as that which is observed as the discrepancy between the large and small Magellanic clouds. We do not know how the period-luminosity relation may be changed if we take a different mass-luminosity relation, or what would happen if a different helium-hydrogen ratio were used. In general, then, one may say that there are a number of parameters that might be varied in order to bring the period-luminosity relation into agreement with the observed one. On the other hand, it would seem very difficult to change the temperature at which the instability strip occurs by any change in the parameters. These are some of the points which we hope to investigate further in the near future.

BIBLIOGRAPHY

BAKER, N. and KIPPENHAHN, R.: 1962, *Z. f. Astrophysik*, **54**, 114.
BÖHM-VITENSE, E: 1958, *Z. f. Astrophysik*, **46**, 108.
CHRISTY, R. F.: 1962, *Ap. J.* **136**, 887.
COX, J. P.: 1958, *Ap. J.*, **127**, 194.
COX, J. P.: 1960, *Ap. J.*, **132**, 594.
COX, J. P.: 1961, *I.A.U. 11th General Assembly*, Berkeley, Comm. 35.
COX, J. P.: 1962, To be published.
COX, A. N.: 1962*b*, preprint, *Stellar Absorption Coefficients and Opacities*; and private communications.
COX, A. N. and STEWART, J. N.: 1962, *Astron. J.* **67**, 113.
EDDINGTON, A. S.: 1926, *The Internal Constitution of the Stars*, Cambridge.
KRAFT, R. P.: 1961, *Ap. J.*, **134**, 616.
LEDOUX, P.: 1958, *Handbuch der Physik*, **51**, 605; *Stellar Stability.*
LEDOUX, P.: 1962, This volume, p. 394.
LEDOUX, P. and WALRAVEN, TH.: 1958, *Handbuch der Physik*, **51**, 353; *Variable Stars.*
SANDAGE, A.: 1958, *Ap. J.*, **127**, 513.
SANDAGE, A. and GRATTON L.: 1962, This volume p. 11.
SCHATZMAN, E.: 1956, *Ann. d'Astrophysique*, **19**, 45.
WHITNEY, C. A.: 1962, *Astron. J.* **67**, 286.
ZHEVAKIN, S. A.: 1953, *Russ. Astrophys. J.*, **30**, 161.

White Dwarfs and Type I Supernovae.

E. Schatzman

Institut d'Astrophysique - Paris

We are going to study first the equation of state of matter at high densities or temperatures and the rate of thermonuclear reactions at high densities.

1. – Equation of state.

In the $\log \varrho - \log T$ diagram (Fig. 1), we shall represent the different regions corresponding to the various properties of the matter. The classical equation of state of dense matter for free electrons without any interactions, either with each other or with the nuclei, is the classical equation of state given by Chandrasekhar (1938, Chapter 10, eqs. (198) and (204)), and reproduced in Schatzman's *White Dwarfs* (1958). We also can draw the limit of degeneracies, in the $\log \varrho$-$\log T$ diagram. In a recent paper, Salpeter (1961) has carefully included all possible corrections: electrostatic corrections, quantum effects, correlation energy. He has also considered inverse β-decay due to capture of electrons by the nuclei, and has calculated, for zero temperature, the decrease of Z as a function of the number of reactions per cubic centimeter, improving greatly calculations by Schatzman. For densities larger than 10^{29} electrons cm^{-3}, all corrections are small, and the main effect is electron capture by nuclei. The usual technique for computing this effect is to maximize the quantity

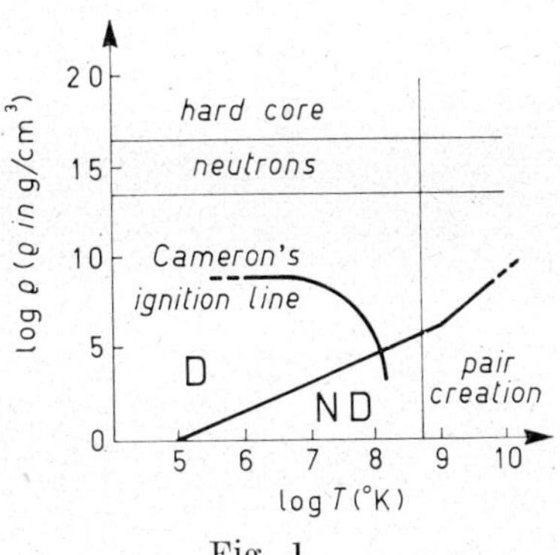

Fig. 1.

$$b^* = \frac{B(Z, A) - ZE'_f}{A},$$

where B is the total binding energy of the nucleus, and E'_f is the Fermi energy of the electron gas minus the neutron-hydrogen rest mass energy difference.

It should be noticed that if we include in E'_f the contribution of the temperature due to the fact that the gas is not actually at zero temperature, we increase E'_f, and when maximizing b^*, we find a charge of the nucleus smaller than the charge at zero temperature. No work has yet been done to take this effect into account.

At zero temperature, SALPETER has calculated the change in Z for various nuclear species, and for the equilibrium atom which corresponds to the chemical composition of lowest energy. When the density increases, the number Z decreases so much that finally b^* becomes negative. In such a case, we have no more nuclei, but rather free neutrons. This occurs for ϱ larger than $10^{11.5}$. At higher densities, we have to deal with the equation of state of a neutron gas. This problem has been considered by CAMERON (1959) using Skyrme's (1959) results. The same subject has also been studied recently by AMBARZOUMIAN and co-workers (1959, 1960). An effect similar to the so-called excluded volume effects becomes important at densities larger then 10^{15} and above $10^{16.5}$ the « hard core » model of the neutron gas seems to be a fairly good representation. In the series of papers by AMBARZOUMIAN, a very detailed discussion of the various phases which appear as a result of the production of hyperons can be found.

On the side of high temperatures, the electron-pair production becomes an important effect, and CHIU (1961) has given extended tables for the pressure in case of pair creation due to interactions with photons. This effect has no real importance below $5 \cdot 10^8$ degrees.

2. – Thermonuclear reactions.

An analysis of the screening effect of the electrons can be found in many places. Let us mention especially a course given in Varenna by SCHATZMAN in 1958. In a recent paper, CAMERON (1959) has studied to a great extent the increased rate factor in α-reactions at high densities. Considering for example the $3\alpha \to {}^{12}C$ reaction, we can draw a curve in the $\log T - \log \varrho$ plane which limits the region where the rate of this reaction is negligible.

3. – Zero-temperature white dwarfs.

HAMADA and SALPETER, in their 1961 paper, have given the mass-radius relation for various models of white dwarfs, including those states which are partially made of neutrons. This curve can be reasonably fitted to Cameron's

(1959) curve for the neutron state. As it will be noticed in Fig. 2, the consequence of the change of Z as a function of density is the existence of a maximum of mass for a finite value of the radius. It can be checked by studying the pulsations of a star *near to* the maximum mass that it becomes dynamically unstable *at* maximum mass. The physical interpretation of this fact is obvious: for a small contraction the capture of electrons by nuclei will reduce the gas pressure in such a way that the gas will no longer be able to resist the weight of the star.

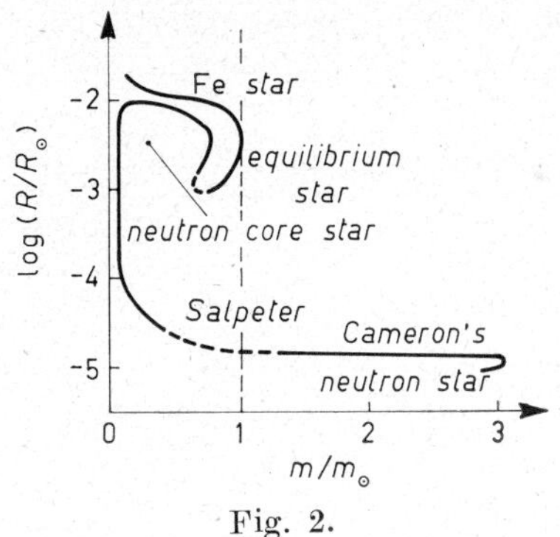

Fig. 2.

4. – Nonzero-temperature white dwarfs.

As mentioned by HOYLE and FOWLER (1960) the effect of the temperature being to increase the pressure, the matter at the same density will be able to resist a greater stellar weight, and for the same radius, a nonzero-temperature white dwarf will have a larger mass, the higher its temperature. As a consequence, the cooling of a white dwarf would ordinarily lead to a minor increase in density. Things are different above maximum mass. There the contraction of a star will increase both the density and the temperature as the contribution of the fully degenerate gas to the pressure will not be sufficient to resist the weight of the star. The exact place of the maximum of mass of the nonzero temperature white dwarf is a little difficult to guess, as on one hand the pressure increase for a given number of electrons will increase the mass, while on the other hand the decrease in Z will decrease the number of free electrons per nucleon. These two effects acting in opposite directions, it is not possible, without detailed calculation, to predict what is the exact place of the mass-radius relation with respect to the $T=0$ mass-radius relation. Anyhow, the lines of constant T certainly intersect so that for masses lower then the maximum mass the radius decreases when the temperatures decreases, and above maximum mass, the radius decreases with rising temperature.

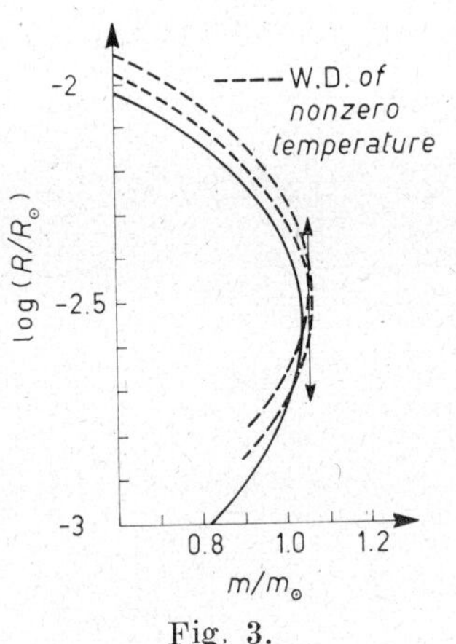

Fig. 3.

5. – Type I supernovae.

If we represent in the $\log T - \log \varrho$ diagram the model of white dwarfs, we find a line which goes to the degenerate region with rising temperature, and then, as a consequence of the high conductivity of the degenerate electron gas, stays at constant temperature (Fig. 4). We see that for a dense white dwarf, this line crosses the region of fast reactions studied by CAMERON. A white dwarf could exist only if its chemical composition were such that no fuel exists in the dangerous region. But a white dwarf of mass larger than the maximum mass will evolve during contraction towards higher temperature. The line representing the model will shift towards the region of higher temperatures and the outer layers will enter the dangerous region. Then let us consider the exponents of density and temperature which are effective for the vibrational instability effects. As can be found in *White Dwarfs* (p. 131) it is the combination $6\mu + 4\nu$ which is important, and ν_{eff} is given by the approximate relation

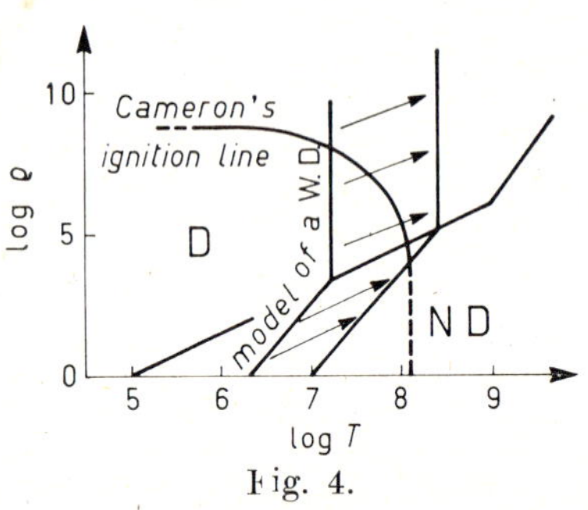

Fig. 4.

$$\nu_{\text{eff}} = \varepsilon\nu/\bar{\varepsilon} ,$$

where ε is the rate of energy generation by the nuclear reactions we consider and $\bar{\varepsilon}$ is the average rate of energy generation of the star. In the dangerous region, ε becames comparable to $\bar{\varepsilon}$, and ν_{eff} and μ_{eff} approach ν and μ. For the 3α reaction, ν is of the order of 7, and μ of the order of 12, at low temperature. For surface energy generation $6\mu + 4\nu$ should be smaller than 16, a condition which is obviously not fulfilled.

We can conclude that when the matter of the outer layers enters the dangerous region, the star becomes highly vibrationally unstable, and can explode. As can be seen in Hoyle and Fowler's paper, the energy available exceeds the energy needed for ejection of 1/100 of the stellar mass, or even for disruption of the star. A possibility seems to be that a small amount of mass is ejected, and that the remnant part of the star being left on the lower branch of the mass-radius relation, collapses until it becomes a neutron star (Fig. 2).

The accretion of interstellar gas, though not a very efficient process, could shift a white dwarf from below to above the critical mass and then be the cause of a type I supernova explosion. A wild hypothesis consists of assuming that the extragalactic radio sources are due to the simultaneous outburst of 10^6 supernovae. The sudden beginning of accretion could trigger this fantastic sequence of explosions.

BIBLIOGRAPHY

SCHATZMAN, E.: 1958, *White Dwarfs.*
SALPETER, E. E.: 1961, *Ap. J.*, **134**, 669.
CAMERON, A. G. W.: 1959, *Ap. J.* **130**, 884.
SKYRMO, T. H. R.: 1959, *Nuclear Phys.*, **9**, 615.
AMBARZOUMIAN, V. A.: 1959-1960.
CHIU, H. Y.: 1961, *Phys. Rev.*, **123**, 1040.
CAMERON, A. G. W.: 1959, *Ap. J.*, **130**, 916.
HAMADA, T. and SALPTER, E. E.: 1961, *Ap. J.*, **134**, 683.
HOYLE, F. and FOWLER, W.: 1960, *Ap. J.*, **132**, 565.
CHANDRASEKHAR, S.: 1938, *Stellar Structure*, Chicago.

Stellar Stability and Stellar Evolution.

P. Ledoux

Institut d'Astrophysique de l'Université - Liège

Introduction.

It is well known, at least qualitatively, that some phases of the evolution of a star may be strongly affected by the instabilities arising from a rotation (through the conservation of angular momentum) or from the existence of an external gravitational field due to a close companion (tidal effects and Roche limit; unstable Lagrangian points in the orbital plane). The effects of a magnetic field are not quite so straightforward but they certainly affect the formation and early evolution of stars and they may present interesting interactions with the instabilities of the first two types.

However, as these factors will be discussed in other lectures, we shall mainly concentrate here on purely spherical stars devoid of rotation, magnetic fields or companions. But even then, at least in a strict sense, the problems of stellar evolution and stellar stability are intimately related since a configuration which would be rigorously stable would not evolve.

However in many phases of its life, the evolution of a star is a very slow process which can be depicted as a succession of quasi-equilibrium states (*i.e.* second-order time-derivatives can be neglected but first-order derivatives must be retained) or even proper equilibrium states (*i.e.* first-order time-derivatives themselves can be neglected). As an example of the first case, we may take the ordinary phases of gravitational contraction characterized by a time-scale

$$\tau_G = -\frac{V}{2\bar{L}}, \tag{I.1}$$

where V represents the gravitational potential of the star in its present state and $\bar{L}$, its average luminosity in the course of the preceding contraction. On the other hand, the second case is generally considered a good approximation during the part of a star-life controlled by a given thermonuclear reaction since its duration

$$\tau_N \simeq \frac{E_n}{L}, \tag{I.2}$$

where E_N is the total energy available from the considered nuclear reaction, is generally much longer than the time defined in (1.1).

In those circumstances, it seems permissible to replace the direct and much more difficult study of the stability of the slow evolutionary motion by a discussion of the stability of equilibrium configurations. Of course, in doing this, we neglect the small lack of balance between energy radiated and energy generated which results in the slight instability which causes evolution. However as long as the time-scales of the motions studied in the stability problem are short as compared to (1.1) or (1.2), this is certainly justified. Anyway, this is the only approach which, up to now, has been the object of detailed discussions and has yielded significant results. One may also foresee that the general direct approach which will lead to a complicated problem in partial differential equations will raise serious difficulties even if tackled numerically on very large electronic computers as it was once suggested (COX and BROWNLEE, 1960) because of the very different time-scales involved and we shall not discuss it any further.

Of course, since the problem consists in studying the stability of the successive equilibrium configurations of a system depending on one or more slowly varying parameters, one might attempt to extend to it the theory of the linear series with loss or exchange of stability at their bifurcation points as initiated by POINCARÉ to discuss the stability of the equilibrium configurations of incompressible fluids in rigid rotation.

Actually since we neglect here any extraneous factors such as rotation, magnetic field or external gravitational fields, the problem is geometrically simpler, the equilibrium shape remaining spherical at all times. But, unfortunately, compared to the mass of perfect fluid, of the classical investigations recalled above, any realistic stellar model is a very complicated system with important thermodynamical properties and parameters.

While in the classical problem, the slowly varying parameter is a simple mechanical quantity such as the total angular momentum $|\mathscr{H}|$ or the angular velocity $|\boldsymbol{\omega}|$ of the rigid rotation, here its choice is much less evident. Since, in any evolutionary change, the total entropy S is increasing, one might want to take it as a parameter and in some phases of evolution this might be the only possible choice. But then the star must be treated *explicitly* as a dissipative system and this is likely to affect the method considerably. Furthermore, it is likely that the total entropy would not characterize the configuration or its evolution in a univoque manner and one might have to evaluate it at each point in the star.

During phases controlled by thermonuclear reactions, one might take the mean molecular weight $\bar{\mu}$ as the parameter but here again, in general, its complete distribution in the star must be given since it is the only way to characterize completely the changes in chemical composition resulting from the thermonuclear reactions and causing the evolutionary motion. In some cases, however, the situation may be simpler. For instance in the evolution of a star with a well-mixed convective core outside which the nuclear reactions are negligible, the value of $\bar{\mu}$ in the core could serve indeed as a single-valued parameter.

In these cases, the most obvious extrapolation of Poincaré's method would consist in linearizing the equilibrium equations governing the internal structure of the star for the general lagrangian perturbation associated with the variation of the chemical composition resulting, in a time interval Δt, from the known distribution of the nuclear sources of energy in the star at the time considered. Any point in the course of the evolution where the resulting system of linear equations ceases to admit a unique solution or has no solution at all should by analogy correspond to a bifurcation point and the manifestation of some kind of instability. However, the connection between this system of linear differential equations and the existence or not of a true minimum for the *total* potential energy which comprises here the gravitational potential energy plus the available thermal energy (cf. P. LEDOUX, 1958, Section **10**) is not as straightforward as in the case of a purely mechanical system.

This is probably the reason why, up to now, while one could recover in this way the usual condition of secular stability in a particular case (GARCET, 1961), it has been impossible for instance to show that the establishment of a convection zone or its disappearance is associated with the impossibility of solving the linear system referred to above although physically this should correspond to a bifurcation point between two linear series. The same may be said of the appearance of a degenerate core.

Of course, in these last two cases, we have at our disposal local criteria which seem quite sufficient to solve the problem in practice. Nevertheless it might by worthwhile spending some further efforts on this generalization of Poincaré's method as it might provide a particularly adequate tool to disclose automatically any critical point in an evolutionary sequence built by methods similar to those introduced by HENYEY (HENYEY *et al.*, 1959).

It might also throw some light on some of the well known hydrostatic incompatibilities such as the one leading to the existence of an upper limit to the relative mass of an isothermal core. However, one should also be aware of some of the limitations of this method, namely that it is not always clear off hand what kind of instability (dynamical, vibrational, local, secular) occurs at a bifurcation point and, in many cases, a detailed discussion of the stability would still remain necessary.

1. – The problem of stellar stability.

1·1. *Definition and methods.* – According to the remarks in the Introduction, we shall be concerned here with the stability of a star as a *whole* when it has reached a state of pure hydrostatic equilibrium or is, at least, very close to it.

We shall not go into the delicate question of the best possible and most embracing definition of stability for a continuous system but anyone interested may be referred to a previous article (LEDOUX, 1958, Section 2) where other references are given and to the first chapter of Chandrasekhar's recent book (CHANDRASEKHAR, 1961). Here, we shall say that a given state is stable provided that, ε_1 being an arbitrary small quantity, it be always possible to define another small quantity ε_2 such that any initial disturbance corresponding to variations of the relevant variables at any point in the system smaller in modulus than ε_2, gives rise to motions in which all « displacements » taken in absolute values remain always smaller than ε_1.

One should be aware that this leaves out some types of instability such as metastability or the kind that gives rise to hard self-excited oscillations and which manifest themselves only for finite perturbations greater than certain critical values. To the best of my knowledge, the discussion of this aspect has never been tackled in the stellar case and it is likely to prove very hard although it might be significant at least for some types of variable stars.

In the frame of the above definition, two methods are available. In the *energy method*, a generalized potential (including the effects of thermodynamical or other nonmechanical factors) must go through an absolute minimum (*i.e.* min. along *all* independent variables) in any stable equilibrium state. By the way, this method should provide the natural approach to the generalization of Poincaré's theory referred to in the Introduction. However the discussions by THOMAS (1930) and TOLMAN (1939) based on this method have remained isolated. An account of them can be found in the present author's article in the *Handbuch der Physik* (LEDOUX, 1958) where the main results concerning stellar stability were, as far as possible, presented from this point of view.

In the *small perturbation method*, a small disturbance is applied to the equilibrium configuration and its further behaviour under the natural forces of the system is studied. It is assumed that if the initial perturbation is small enough, the solutions of the *linearized* equations will reveal the trend of this motion in a small region around the equilibrium state and thus will permit to decide whether or not the general definition above is satisfied.

In practice, both methods have their own advantages but if fully developed they should probably prove strictly equivalent (for an example, cf. LEDOUX, 1958, Section **12**). Most of the advantages of the energy method may be recovered, in the small perturbation method, by the variational interpretation of the equations and since most of the known results have been obtained by the second method we shall adopt it here.

1·2. *General equations.* – Since we are dealing with continuous systems, the language of hydrodynamics seems appropriate and we know that we have two representations at our disposal.

In the Eulerian approach, the spatial co-ordinates ($\boldsymbol{r}$; x, y, z) are independent geometrical variables, the velocity field $\boldsymbol{v}$ being defined as a function of $\boldsymbol{r}$ and t. In the Lagrangian representation, the spatial co-ordinates ($\boldsymbol{r}$; x, y, z) are attached to the particles and are dependent mechanical variables, functions of t and parameters characterizing the particle considered, for instance, the initial values ($\boldsymbol{r}_0$; x_0, y_0, z_0) of its co-ordinates.

The equations governing a small perturbation of the system are derived by linearizing (*i.e.* taking the first variation with respect to the dependent variables) the general nonlinear equations expressing the conservation of mass, momentum and energy, the equation of state and those defining the different fields of force acting on the system.

In the Eulerian representation these first variations which we shall denote by a *prime* are taken locally and all space operators are invariant towards this operation. In the Lagrangian representation the variations denoted by δ are taken following the motion and the space operators are no longer invariant. When more than one space co-ordinate is involved, this last circumstance complicates somewhat the equations and it is generally easier to work in Eulerian co-ordinates. One should note however that, usually, the energy equation and the boundary conditions are simpler in the Lagrangian representation.

Of course, it is always possible to pass from one representation to the other and in particular the following relations between the Lagrangian (denoted by a dot) and the local Eulerian time-derivatives or the Lagrangian and Eulerian perturbations should be noted:

$$\dot{f} = \frac{\partial f}{\partial t} + \boldsymbol{v}\cdot\nabla f; \qquad \delta f = f' + \delta\boldsymbol{r}\cdot\nabla f\,. \tag{1.1}$$

The general equations have been discussed at length in the article on intrinsic variable stars by LEDOUX and WALRAVEN (1958) and we shall simply recall them here, using Eulerian co-ordinates, and keeping only the terms which have proven significant up to now.

The present problem differs from the usual hydrodynamical ones in two ways:

a) factors such as radiative stresses and energy, ionization and the liberation of subatomic energy alter somewhat the hydrodynamical and thermodynamical equations;

b) gravity which usually is the dominant field of force is altered by the modifications of the system and, in general, its perturbations are not negligible as it is the case in the laboratory or even in the terrestrial atmosphere or oceans.

1·2.1. Conservation of mass. – Up to now, conditions encountered in stellar models are such that relativistic corrections to the mass of material particles other than electrons are negligible as well as the inertia of the radiation energy even for temperatures as high as 10^9 to 10^{10} °K since, as far as we know, these occur simultaneously with very high material densities of the order of 10^8 to 10^9 g cm^{-3}. Concurrently annihilation or materialization of heavy particles (barring the recent superdense models of Ambartzumian) are irrelevant and, in all other types of nuclear reactions, the mass loss is so small as to be negligible.

In these conditions and for any inertial frame of reference (x^i), the hydrodynamical continuity equation may be written

$$\frac{\partial \varrho}{\partial t} + \nabla\cdot(\varrho \boldsymbol{v}) = 0 \qquad \text{or} \qquad \frac{1}{\varrho}\frac{d\varrho}{dt} + \nabla\cdot\boldsymbol{v} = 0\,, \tag{1.2}$$

where ϱ is the classical material density.

1·2.2. Conservation of momentum. – With respect to any inertial frame, we have for the k-component of momentum per unit volume

$$\frac{\partial(\varrho v^k)}{\partial t} + \sum_i \Delta_i(\varrho v^i v^k - P^{ik}) = \varrho X^k\,, \tag{1.3}$$

where X^k is the k-component of the total force acting per unit mass, Δ_i denotes the covariant derivative which in cartesian co-ordinates reduces simply to $\partial/\partial x^i$ and P^{ik} is the total stress tensor.

In absence of turbulence and if the radiation stresses are negligible, the latter reduces to its usual expression for a gas

$$P^{ik} = -\, p_G g^{ik} + \mathscr{P}^{ik}_G \tag{1.4}$$

with

$$\mathscr{P}^{ik}_G = -\tfrac{2}{3}\,\eta_G\, g^{ik} \sum_j \Delta_j v^j + \eta_G \sum_j (g^{ij}\Delta_j v^k + g^{kj}\Delta_j v^i)\,. \tag{1.5}$$

or, in cartesian co-ordinates,

$$(1.6)\qquad \begin{cases} \mathscr{P}^{ii}_{G} = -\dfrac{2}{3}\,\eta_G \boldsymbol{\nabla}\cdot\boldsymbol{v} + 2\eta_G \dfrac{\partial v^i}{\partial x^i}\,, \\ \mathscr{P}^{ik}_{G} = \eta_G \left(\dfrac{\partial v^k}{\partial x^i} + \dfrac{\partial v^i}{\partial x^k}\right) \qquad i \neq k\,. \end{cases}$$

The dynamical coefficient of viscosity and the gaseous pressure are given, in a first approximation, by

$$(1.7)\qquad \eta_G = \frac{1}{3}\,\varrho \bar{c} \bar{l}\,, \qquad p_G = nkT = \frac{\mathscr{R}\varrho T}{\bar{\mu}}\,,$$

where $\bar{c}$ is the root mean square thermal velocity and $\bar{l}$, the molecular mean free path.

The general expressions of the radiative stresses are rather complicated (cf. THOMAS, 1930, HAZLEHURST and SARGENT, 1959; SIMON, 1962) but, for our present purpose, we may limit ourselves to the first order terms in $|\boldsymbol{v}|/c$ where c is the velocity of light. We shall also assume local thermodynamical equilibrium and adopt for the solution of Schwarschild's equation for the intensity of radiation, the first approximation which implies that the variations of $\log T$ and of (v^k/c) along a mean free path $\bar{l}_R$ are much smaller than unity. If, furthermore, we admit that the time derivative of any macroscopic quantity divided by c is small as compared to its space derivatives, we may decompose P^{ki}_R exactly as in (1.4) writing

$$(1.8)\qquad P^{ik}_R = -\,p_R g^{ik} + \mathscr{P}^{ik}_R\,,$$

where $\mathscr{P}^{ik}_R$ has exactly the same form as (1.5) with η_R substituted for η_G. The value of the former,

$$(1.9)\qquad \eta_R = \frac{4aT^4}{15\varkappa\varrho c}\,,$$

where T is the temperature, may be derived from (1.7) by replacing $\bar{c}$ by c, ϱ by aT^4/c^2 and $\bar{l}$ by

$$(1.10)\qquad \bar{l}_R = \frac{1}{\varkappa\varrho}\,,$$

where the coefficient of opacity per unit mass $\varkappa$ shall be assumed to be of the form

$$(1.11)\qquad \varkappa = \varkappa_0 \varrho^m T^{-n}\,.$$

However the decomposition (1.8), implies for the pressure of radiation, the following definition

$$p_R = \tfrac{1}{3} a T^4 - \tfrac{5}{3} \eta_R \nabla \cdot \boldsymbol{v} . \tag{1.12}$$

The radiative viscous stresses are small, usually at most of the order of the gaseous viscous stresses which, in our present problem, are practically always negligible. Thus, most of the time we may reduce (1.12) to its usual form

$$p_R = \tfrac{1}{3} a T^4 . \tag{1.13}$$

NOTE. – *In all this and the following all quantities referring to the radiation are the* proper *values as defined in a frame of reference instantaneously at rest with respect to the matter at the point considered.*

Many stellar models contain convection zones and it is likely that in cosmical bodies, convection is always turbulent. Anyway, up to now, our only way to deal with thermal convection in the interior of a star is to treat it as turbulence using some simple picture based on the analogy with the kinetic theory of gas (cf. COWLING, 1936; LEDOUX, 1940; WASIUTYNSKI, 1946, also LEDOUX and WALRAVEN, 1958).

In that case and as far as the conservations of mass and momentum are concerned, eqs. (1.2) and (1.3) remain essentially valid with the following conventions:

a) the state variables ϱ and T must now be interpreted as mean values taken on a sufficiently large fraction of space or on a sufficiently long interval of time and $\boldsymbol{v}$ is a mean velocity defined in such a way that its product by the mean density is equal to the mean momentum ($\overline{\varrho v^i} = \bar{\varrho}\, \bar{v}^i$);

b) furthermore in eq. (1.3), the Reynolds stresses

$$P_t^{ik} = \overline{\varrho V^i V^k} ,$$

where V denotes the turbulent component of the instantaneous velocity ($v^i = \bar{v}^i + V^i$, $\overline{\varrho V^i} = 0$), must be included in the general stress tensor. In a first approximation, we may again write as in (1.4) and (1.5)

$$P_t^{ik} = -p_t g^{ik} + \mathscr{P}_t^{ik} \tag{1.14}$$

and

$$\mathscr{P}_t^{ik} = \tfrac{2}{3} \eta_t' g^{ik} \sum_j \Delta_j v^j + \eta_t \Big(\sum_j g^{ij} \Delta_j v^k + g^{kj} \Delta_j v^i \Big) \tag{1.15}$$

with

$$p_t = \tfrac{1}{3} \varrho \bar{C}_t^2 , \qquad \eta_t = \tfrac{1}{3} \varrho \bar{l}_t \bar{C}_t , \tag{1.16}$$

where all macroscopic quantities are supposed to take their mean values and $\bar{C}_t$ is a mean absolute velocity of turbulence of the same order as the root mean square velocity and $\bar{l}_t$ is the mean free path of the turbulent elements.

Although we do not dispose of a complete theory of turbulence which would fix the value of $\bar{l}_t$, the latter is in general many orders of magnitude larger than $\bar{l}_G$ or $\bar{l}_R$ and, as a consequence, while the effects of ordinary (molecular or radiative) viscosity on the *mean* motion are usually negligible, the effects of turbulent viscosity may become large.

Let us note also that ordinary viscosity may be important at the lower scale of the turbulent motions because of the large values of $\overline{\partial V^i/\partial x^j}$ which occur there and that, together with conductivity (radiative and molecular), it is the corresponding dissipation which insures the establishment of a *stationary* state of turbulent convection in presence of superadiabatic gradients.

Anyway, we see that we can always use eqs. (1.2) and (1.3) provided that, if necessary, ϱ, $\boldsymbol{v}$ be interpreted as appropriate averages and that the total stresses: gaseous, radiative and turbulent be included. Equation (1.3) may also be written more explicitly

$$(1.17)\qquad \frac{\partial(\varrho v^k)}{\partial t}+\sum_i \Delta_i(\varrho v^k v^i)=\varrho X^k-\sum_i g^{ik}\,\Delta_i(p+p_t)+\sum_i \Delta_i(\mathscr{P}^{ik}+\mathscr{P}^{ik}_t)\,,$$

or using the equation of continuity and the definition (1.1)

$$(1.18)\qquad \frac{\mathrm{d}\boldsymbol{v}}{\mathrm{d}t}=\boldsymbol{X}-\frac{1}{\varrho}\Delta(p+p_t)+\frac{1}{\varrho}\nabla\cdot\mathfrak{P}\,,$$

where, in first approximation,

$$(1.19)\qquad p=p_G+p_R=\frac{\mathscr{R}\varrho T}{\bar{\mu}}+\frac{1}{3}aT^4\,,$$

and $\mathfrak{P}$ is the tensor of the total viscous stresses. If $\eta=(\eta_G+\eta_R+\eta_t)$ is constant, this last equation reduces to the usual Navier-Stokes equation

$$\frac{\mathrm{d}\boldsymbol{v}}{\mathrm{d}t}=\boldsymbol{X}-\frac{1}{\varrho}\nabla(p+p_t)+\frac{1}{3}\eta\,\mathrm{grad}\,\mathrm{div}\,\boldsymbol{v}+\eta\,\nabla^2\boldsymbol{v}\,.$$

Generally, gravity is largely predominant in a star and, in most of our discussions, we shall admit that it is the only force present so that

$$(1.20)\qquad \boldsymbol{X}=-\,\mathrm{grad}\,\varphi\,,$$

where the gravitational potential φ satisfies Poisson's equation

$$\nabla^2\varphi = 4\pi G\varrho \,. \tag{1.21}$$

1'2.3. Conservation of mechanical energy. – In absence of turbulence, multiplying (1.18) scalarly by $\boldsymbol{v}$, we obtain

$$\frac{\mathrm{d}}{\mathrm{d}t}\left(\frac{1}{2}v^2\right) = \boldsymbol{X}\cdot\boldsymbol{v} - \frac{\boldsymbol{v}}{\varrho}\cdot\nabla p + \frac{\boldsymbol{v}}{\varrho}\cdot(\nabla\cdot\mathfrak{P})\,, \tag{1.22}$$

which expresses the conservation of mechanical energy per unit mass. Per unit volume, one finds

$$\frac{\partial}{\partial t}\left(\frac{1}{2}\varrho v^2\right) + \nabla\cdot\left(\frac{1}{2}\varrho v^2\boldsymbol{v}\right) = \varrho\boldsymbol{X}\cdot\boldsymbol{v} - \boldsymbol{v}\cdot\nabla p + \boldsymbol{v}\cdot(\nabla\cdot\mathfrak{P})\,. \tag{1.23}$$

In presence of turbulence, eq. (1.23) is still essentially valid for the mean motion ($\bar{\varrho}$, $\overline{v_i}$, etc.) but p must now include the turbulent pressure p_t defined by (1.16) and $\mathfrak{P}$ may be reduced to $\mathfrak{P}_t$ because of the large predominance of the effects of turbulent viscosity on the mean motion.

However, in this case, there is an independent equation corresponding to the conservation of the turbulent energy which may be written, keeping only the largest terms,

$$\frac{\partial}{\partial t}\left(\overline{\frac{1}{2}\varrho V^2}\right) + \nabla\cdot\left(\frac{1}{2}\varrho V^2\boldsymbol{v}\right) + p_t\nabla\cdot\boldsymbol{v} = \varrho\varepsilon_4 - \varrho\varepsilon_2 - \nabla\cdot\boldsymbol{F}_2 - \boldsymbol{V}\cdot\nabla p\,, \tag{1.24}$$

where ε_2 which is of the order of

$$\varepsilon_2 \simeq \frac{\bar{C}_t^3}{\bar{l}_t}\,, \tag{1.25}$$

represents the total rate of dissipation per unit mass of the kinetic energy of turbulence due to molecular and radiative viscosity; ε_4 represents the rate of transformation of the kinetic energy of mean motion into turbulent energy due to the turbulent stresses

$$\varepsilon_4 \simeq \frac{1}{\bar{\varrho}}\sum_{i,k}\mathscr{P}_t^{ik}(\boldsymbol{v})\,\Delta_i v_k\,, \tag{1.26}$$

and $\boldsymbol{F}_2$ corresponds to the convective flux of turbulent energy

$$\boldsymbol{F}_2 = \frac{1}{2}\,\overline{\varrho V^2\boldsymbol{V}}\,. \tag{1.27}$$

The last term in (1.24) corresponds to the transformation of gravitational potential energy into energy of turbulence.

1·2.4. Conservation of the total energy. – If we neglect the excitation energy, the total thermal energy per unit mass may be written

$$U = \frac{3}{2}\frac{\mathcal{R}T}{\bar{\mu}} + I + \frac{aT^4}{\varrho}, \tag{1.28}$$

where $\bar{\mu}$ is the mean molecular weight given by

$$\bar{\mu} = \frac{1}{Nm_H}, \tag{1.29}$$

if N is the total number of material particles per unit mass and I is the ionization energy. Let us denote by E_S the « subatomic » energy associated with the proper mass of the particles, and by ε_ν, the energy of the neutrinos created per unit mass and unit time. We shall admit further that, except for the emission of neutrinos, all energy liberated by nuclear processes can be considered as transformed instantaneously into thermal energy in equilibrium at the temperature of the medium.

Balancing the variation following the motion of the total energy of a mass limited by a surface S bound to the particles and the subatomic energy released inside it against the negative total flux across this surface (positive normal outward) plus the work of the external forces $\boldsymbol{X}$ on the whole mass and that of the tensions on the surface, one obtains

$$\frac{\mathrm{d}}{\mathrm{d}t}\left(U + E_S + \frac{v^2}{2}\right) + \varepsilon_\nu = \boldsymbol{X}\cdot\boldsymbol{v} - \frac{1}{\varrho}\nabla\cdot(\boldsymbol{F}_R + \boldsymbol{F}_c) - \frac{1}{\varrho}\nabla\cdot(p\boldsymbol{v}) + \frac{1}{\varrho}\nabla\cdot[\boldsymbol{v}\cdot\mathfrak{P}(\boldsymbol{v})]\,. \tag{1.30}$$

The radiative and conductive flux denoted respectively by $\boldsymbol{F}_R$ and $\boldsymbol{F}_c$ are both related to the temperature gradient by a formula of the type

$$\boldsymbol{F} = -\lambda\nabla T, \tag{1.31}$$

where λ is a coefficient of conductivity. In the case of radiation, up to the same approximation as the one adopted in Section **1**·2.2, $\lambda = (4acT^3/3\varkappa\varrho)$ so that

$$\boldsymbol{F}_R = -\frac{4acT^3}{3\varkappa\varrho}\nabla T\,. \tag{1.32}$$

1'2.5. Conservation of thermal energy. – If we substract (1.22) form (1.30), we get

$$(1.33)\qquad \frac{dU}{dt}+\frac{p}{\varrho}\nabla\cdot\boldsymbol{v}=-\frac{dE_S}{dt}-\varepsilon_\nu-\frac{1}{\varrho}\nabla\cdot(\boldsymbol{F}_R+\boldsymbol{F}_c)+\frac{1}{\varrho}\sum_{i,k}\mathscr{P}^{ik}\Delta_i v_k\,.$$

This is nothing more than the explicit expression of the first principle of thermodynamics where the second member represents the rate at which « external » energy is given to the system.

Of course, apart from the fluxes which represent really the exchange of energy with the exterior, it is normal to count the heat liberated by friction (last term in (1.33)) as energy furnished to the system since it comes from the kinetic energy of mass motions which is outside the description of U. In the same way, the emission of neutrinos represents a real sink of energy for the system since practically they do not react with it. As far as the term $(-dE_S/dt)$ is concerned, we could as well have kept it on the left-hand side of the equation since it is mainly a question of convention whether we count E in the internal energy or not.

However, in most stars, the variations of E_S are due to thermonuclear reactions and dE_S/dt is usually very small compared to U. In the same way, the rates of variation of $\bar{\mu}$ or of the total number of particles N due to these reactions which are given by (*)

$$(1.34)\qquad \frac{1}{N}\frac{dN}{dt}=-\frac{1}{\bar{\mu}}\frac{d\bar{\mu}}{dt}\simeq\frac{5}{4}\frac{\bar{\mu}}{e_N}\frac{dE_S}{dt}\,,$$

where e_N is the total energy liberable per unit mass by the relevant reaction, are extremely small. Furthermore, dE_S/dt is a function only of ϱ and T and not of their time variations. In these conditions, our convention in (1.33) is quite appropriate.

Let us note also that the losses of energy to neutrinos emitted in the usual thermonuclear reactions are never very large and we may as well discount these losses directly and define the rate of energy generation due to thermonuclear reactions by

$$(1.35)\qquad \varepsilon_N=-\frac{dE_S}{dt}-\varepsilon_\nu\,.$$

In that case, eq. (1.33) becomes, eliminating also $\nabla\cdot\boldsymbol{v}$ by the continuity

(*) The particular value $\bar{\mu}$ in (1.34) corresponds to the common reactions in which $4H\rightarrow He$.

equation

$$\frac{dU}{dt} - \frac{p}{\varrho^2}\frac{d\varrho}{dt} = \varepsilon_N - \frac{1}{\varrho}\nabla\cdot(\boldsymbol{F}_R + \boldsymbol{F}_c) + \frac{1}{\varrho}\sum_{i,k}\mathscr{P}^{ik}\Delta_i v_k\,, \tag{1.36}$$

On the other hand, for extreme temperatures and (or) densities, the nuclear processes tend to reach equilibrium conditions and dE_S/dt becomes a function of dT/dt and $d\varrho/dt$ rather than a function of ϱ and T. In those cases, the energy liberated by the displacement of the equilibrium may become very large as well as the change is N or $\bar{\mu}$ and it is more appropriate there to combine directly E_S with U writing (1.33) as

$$\frac{d(U+E_S)}{dt} - \frac{p}{\varrho^2}\frac{d\varrho}{dt} = -\varepsilon_\nu - \frac{1}{\varrho}\nabla\cdot(\boldsymbol{F}_R + \boldsymbol{F}_c) + \frac{1}{\varrho}\sum_{i,k}\mathscr{P}^{ik}\Delta_i v_k\,. \tag{1.37}$$

In presence of turbulence, multiplying (1.33) by ϱ and rewriting then the first member of the equation as

$$\frac{\partial(\varrho U)}{\partial t} + \nabla\cdot(\varrho U\boldsymbol{v}) + p\nabla\cdot\boldsymbol{v}\,,$$

we get, after substituting $v^i = \overline{v^i} + V^i$, averaging and neglecting the effects of ordinary viscosity on the mean motion

$$\bar{\varrho}\frac{d}{dt}\left(\frac{\overline{\varrho U}}{\bar{\varrho}}\right) + \bar{p}\nabla\cdot\boldsymbol{v} = \overline{\varrho\varepsilon_N} + \bar{\varrho}\varepsilon_2 - \bar{\varrho}\varepsilon_3 - \nabla\cdot(\overline{\boldsymbol{F}}_R + \overline{\boldsymbol{F}}_c + \overline{\boldsymbol{F}}_t)\,, \tag{1.38}$$

where we have kept the bar averaging-symbol notation for clarity and assumed that the nuclear processes reduce to thermonuclear reactions.

In this expression, $\boldsymbol{F}_t$ represents the mean turbulent flux of thermal energy

$$\boldsymbol{F}_t = \overline{\varrho U \boldsymbol{V}}\,, \tag{1.39}$$

and ε_3 is defined by

$$\bar{\varrho}\varepsilon_3 = \overline{p\nabla\cdot\boldsymbol{V}}\,, \tag{1.40}$$

while ε_2 is given as before by (1.25).

1·2.6. Special forms of the equation of conservation of thermal energy – Let us consider first, eq. (1.36). It is useful to eliminate U in terms of two of the three state variables p, ϱ, T introducing generalized adiabatic coefficients Γ_1, Γ_2, Γ_3 defined by

$$\Gamma_1 = \frac{C_p}{C_v}\left\{-\frac{\tau}{p}\left(\frac{dp}{d\tau}\right)_T\right\} = \frac{\tau}{p}\frac{[(\partial U/\partial\tau)_p + p]}{(\partial U/\partial p)_\tau}, \tag{1.41}$$

$$C_v = \left(\frac{\partial U}{\partial T}\right)_\tau, \qquad C_p = \left(\frac{\partial H}{\partial T}\right)_p = \left(\frac{\partial H}{\partial \tau}\right)_T\left(\frac{\partial \tau}{\partial T}\right)_p + \left(\frac{\partial H}{\partial T}\right)_\tau, \tag{1.42}$$

$$\frac{\Gamma_2}{\Gamma_2 - 1} = \frac{(\partial U/\partial T)_\tau \Gamma_1 T}{[(\partial U/\partial \tau)_T + p]\tau} = -\frac{C_p T(\partial p/\partial\tau)_T}{p[(\partial U/\partial\tau)_T + p]}, \tag{1.43}$$

$$\Gamma_3 - 1 = \frac{(\Gamma_2 - 1)\Gamma_1}{\Gamma_2} = \frac{\tau[(\partial U/\partial\tau)_T + p]}{C_v T}, \tag{1.44}$$

where τ is the specific volume ($\tau = 1/\varrho$), H is the enthalpy ($H = U + p\tau$) and C_p and C_v represent generalized specific heats defined by (1.42). Equation (1.36), may then be written

$$\frac{dp}{dt} - \frac{\Gamma_1 p}{\varrho}\frac{d\varrho}{dt} = (\Gamma_3 - 1)\varrho\left[\varepsilon_N - \frac{1}{\varrho}\nabla\cdot(\boldsymbol{F}_R + \boldsymbol{F}_c) + \frac{1}{\varrho}\sum \mathscr{P}^{ik}\Delta_i v_k\right], \tag{1.45}$$

or

$$\frac{1}{T}\frac{dT}{dt} - \frac{\Gamma_3 - 1}{\varrho}\frac{d\varrho}{dt} = \frac{1}{C_v T}\left[\varepsilon_N - \frac{1}{\varrho}\nabla\cdot(\boldsymbol{F}_R + \boldsymbol{F}_c) + \frac{1}{\varrho}\sum \mathscr{P}^{ik}\Delta_i v_k\right]. \tag{1.46}$$

In presence of turbulence, eq. (1.38) becomes in the same way

$$\frac{dp}{dt} - \frac{\Gamma_1 p}{\varrho}\frac{d\varrho}{dt} = (\Gamma_3 - 1)\varrho\left\{\varepsilon_N + \varepsilon_2 - \varepsilon_3 - \frac{1}{\varrho}\nabla\cdot(\boldsymbol{F}_R + \boldsymbol{F}_c + \boldsymbol{F}_t)\right\}. \tag{1.47}$$

where all quantities are appropriate means.

Note that eq. (1.24) expressing the conservation of kinetic turbulent energy may also be written in a similar form. If we assume that turbulence is isotropic and use (1.16), we get

$$\frac{dp_t}{dt} - \frac{5}{3}\frac{p_t}{\varrho}\frac{d\varrho}{dt} = \frac{2}{3}\varrho\left(\varepsilon_4 - \varepsilon_2 - \frac{1}{\varrho}\nabla\cdot\boldsymbol{F}_2 - \frac{1}{\varrho}\boldsymbol{V}\cdot\nabla p\right), \tag{1.48}$$

which, in a way, means that in our treatment of turbulence, the turbulent elements behave as the particles of a monatomic gas ($\Gamma_1 = \Gamma_3 = \frac{5}{3}$).

1·3. *Values of* Γ_1, Γ_2, Γ_3.

1·3.1. Effects of radiation. – If the contribution of radiation and ionization to U is negligible, the stellar gas can be treated as monatomic

$$\Gamma_1 = \Gamma_2 = \Gamma_3 = \gamma = \tfrac{5}{3}. \tag{1.49}$$

If radiation is not negligible (large stellar masses) but the variations of the ionization energy are still very small one gets the usual values

$$(1.50)\qquad \begin{cases} \Gamma_1 = \beta + \dfrac{(4-3\beta)^2(\gamma-1)}{\beta+12(\gamma-1)(1-\beta)}\,, \\ \Gamma_3 - 1 = \dfrac{(4-3\beta)(\gamma-1)}{\beta+12(\gamma-1)(1-\beta)}\,, \qquad \Gamma_2 = \dfrac{\Gamma_1}{1+(\Gamma_1-\Gamma_3)}\,, \end{cases}$$

where

$$(1.51)\qquad \beta = \frac{p_G}{p}\,,$$

and the ratio of specific heats for the gas, γ, can be put equal to $\frac{5}{3}$.

1·3.2. Effects of ionization. – If we are in conditions where the ionization energy changes rapidly with T and ϱ we have to take I explitictly into account in the definition (1.28) of U and use Saha's equation to estimate its variations in terms of those of ϱ and T. With the present large abundances of H and He this type of situation is encountered only in the external layers (say $T<10^5$ °K) where the effects of the ionizations of H I, He I and He II must be taken into account. However, as those ionizations occur in fairly well separated ranges, it is often sufficient to consider that only one ion at a time is in a critical stage of ionization. In that case (cf. LEDOUX and WALRAVEN, 1958, Section **53**; also BAKER and KIPPENHAHN, 1962) let us define the following quantities:

a) X_i is the abundance by weight of the element in a critical stage of ionization;

b) x is the relative abundance by numbers of all electrons with respect to all ions;

c) x_i^r is the fraction of the atoms i which have lost r electrons (the critical stage of ionization is between the two states having lost r and $(r+1)$ electrons);

$$(1.52)\qquad d)\qquad B = \frac{X_i x x_i^{r+1}(1-x_i^{r+1})}{x(1+x)+X_i x_i^{r+1}(1-x_i^{r+1})}\,,$$

$$(1.53)\qquad e)\qquad \begin{cases} A_{\varrho,T} = \left(\dfrac{\partial \ln N}{\partial \ln \varrho}\right)_T = -\dfrac{B}{1+B}\,, \\ A_{T,p} = \left(\dfrac{\partial \ln N}{\partial \ln T}\right)_p = B\left[\dfrac{5}{2}+\dfrac{\chi_i^r}{kT}+\dfrac{4(1-\beta)}{\beta}\right], \\ A_{T,\varrho} = \left(\dfrac{\partial \ln N}{\partial \ln T}\right)_\varrho = \dfrac{B}{1+B}\left(\dfrac{3}{2}+\dfrac{\chi_i^r}{kT}\right), \end{cases}$$

where χ_i^r is the ionization potential of the ion having lost r electrons so that

$$(1.54)\qquad \mathrm{d}I = -\chi_i^r\,\mathrm{d}M_i^r = \chi_i^r\,\mathrm{d}N_e = \chi_i^r\,\mathrm{d}N\,.$$

where M_i^r, N_e and N are respectively the total numbers of ions (i, r), of electrons and of all particles per unit mass.

With these definitions and noting that apart from the ionization considered the gas can be treated as monatomic, one finds

$$(1.55)\qquad \Gamma_1 = \frac{\{16-12\beta-1.5\beta^2+\beta[4-1.5\beta+\beta\chi_i^r/kT]A_{T,p}\}[(1+A_{\varrho,T})}{12-10.5\beta+\beta\,(1.5+\chi_i^r/kT)A_{T,\varrho}}\,,$$

$$(1.56)\qquad \Gamma_3-1 = \frac{\Gamma_2-1}{\Gamma_2}\,\Gamma_1 = \frac{4-3\beta+\beta A_{T,\varrho}}{12-10.5\beta+\beta(1.5+\chi_i^r/kT)A_{T,\varrho}} = \frac{\Gamma_1-\beta(1+A_{\varrho,T})}{(4-3\beta)+\beta A_{T,\varrho}}\,.$$

The corresponding expressions of the generalized specific heats are

$$(1.57)\qquad C_v = c_v\left[1+\frac{8(1-\beta)}{\beta}-\frac{2}{3}\frac{A_{T,\varrho}^2}{A_{\varrho,T}}\right],$$

and

$$(1.58)\qquad C_p = c_v\left\{\frac{5}{3}+\frac{8(1-\beta)}{3\beta}\left[5+\frac{4(1-\beta)}{\beta}\right]+\frac{2}{3}\frac{A_{T,p}^2}{B}\right\},$$

where c_v is the specific heat at constant volume of the gas.

In regions where H or He are in a critical stage of ionization, these formula yield values of C_v which are much larger than c_v and values of Γ_1, Γ_2 and Γ_3 which tend towards unity and can, in any case, be appreciably smaller than $\frac{4}{3}$ because of the large value of the χ_i^r as compared to kT in these regions. But in the main interior of most stars, these effects of ionization cease to be relevant and the values (1.50) of Γ_1 and Γ_3 are the significant ones.

1·3.3. Effects of nuclear processes. – If we go to the extreme conditions for which the appropriate form of the equation of conservation of energy is (1.37), we may expect that the value of C_v and C_p and of the Γ's will also be strongly affected by the nuclear equilibrium which tends to get established. In these conditions, the effects of ionization are negligible and it is easy to verify that, in general, Γ_1 and Γ_3, will be given by

$$(1.59)\qquad \Gamma_1 = \frac{16-12\beta-1.5\beta^2+6\beta(1-\beta)A_{\varrho,p}-\beta(4-3\beta)(\varrho/NkT)(\partial E_S/\partial\varrho)_p}{12-10.5\beta-6\beta(1-\beta)A_{p,\varrho}+\beta(4-3\beta)(p/NkT)(\partial E_S/\partial p)_\varrho}\,,$$

and

$$(1.60)\qquad \Gamma_3-1 = \frac{4-3\beta-1.5\beta A_{\varrho,T}-(\beta\varrho/NkT)(\partial E_S/\partial\varrho)_T}{12-10.5\beta+1.5\beta A_{T,\varrho}+(\beta T/NkT)(\partial E_S/\partial T)_\varrho}\,,$$

where the A's have the same general meaning as in (1.53) but cannot, of course, be defined in terms of B and χ_i^r. If ϱ and T are to be preferred as independent variables, Γ_1 may also be expressed in terms of the partial derivatives with respect to them by means of the relations:

$$(1.61)\quad \left\{\begin{aligned} A_{p,\varrho} &= A_{T,\varrho}[(4-3\beta)+\beta A_{T,\varrho}]^{-1}\,, \\ A_{\varrho,p} &= [(4-3\beta)A_{\varrho,T}-\beta A_{T,\varrho}][(4-3\beta)+\beta A_{T,\varrho}]^{-1}\,, \\ \left(\frac{\partial E_S}{\partial \varrho}\right)_p &= \left(\frac{\partial E_S}{\partial \varrho}\right)_T - \left(\frac{\partial E_S}{\partial T}\right)_\varrho \frac{T}{\varrho}\,\frac{\beta(1+A_{\varrho,T})}{[(4-3\beta)+\beta A_{T,\varrho}]}\,, \\ \left(\frac{\partial E_S}{\partial p}\right)_\varrho &= \left(\frac{\partial E_S}{\partial T}\right)_\varrho \frac{T}{p}\,\frac{1}{[(4-3\beta)+A_{T,\varrho}]}\,. \end{aligned}\right.$$

The integrability condition for the entropy yields also

$$(1.62)\qquad A_{T,\varrho} = -\frac{3}{2}A_{\varrho,T} - \frac{\varrho}{NkT}\left(\frac{\partial E_S}{\partial \varrho}\right)_T\,.$$

One of the first « nuclear » equilibria which we are likely to encounter when T is rising corresponds to the creation and annihilation of electron pairs

$$(1.63)\qquad e^+ + e^- \rightleftarrows (h\nu)$$

the main contribution in stars corresponding probably to the two photon annihilation. Some computations have been made in this case by SOUFFRIN (1960) and more recently by MARTIN (1962). It is found, that at low densities of the order of 1 g cm^{-3}, the Γ's are already reduced, on this account, to values smaller than $\frac{4}{3}$ for $T \geqslant 5\cdot 10^8$ °K while for higher densities this occurs only for higher T: if

$$\varrho \simeq 10^3 \text{ g cm}^{-3}\,, \qquad \Gamma_1 < \tfrac{4}{3} \qquad \text{for} \qquad T \geqslant 10^9\,^\circ\text{K}$$

$$\varrho \simeq 10^6 \text{ g cm}^{-3}\,, \qquad \Gamma_1 < \tfrac{4}{3} \qquad \text{for} \qquad T \geqslant 5\cdot 10^9\,^\circ\text{K}.$$

However for densities of the order of 10^6 to 10^8 and temperatures in the range 2 to $4\cdot 10^9$ °K, the nuclear reactions become very fast and tend to set up a statistical equilibrium between the different nuclei which would have to be taken into account. In fact, it has been suggested first by HOYLE (1946) that, in the course of stellar evolution, the conditions in the central part of a star reach these extreme values and that the resulting statistical equilibrium is a fundamental step in the theory of the formation of the chemical elements (cf. also BURBIDGE, BURBIDGE, FOWLER and HOYLE, 1957 and CAMERON, 1960).

Unfortunately the situation is a complicated one and a reliable quantitative evaluation of the Γ's seems difficult.

But a purely qualitative approach may already yield some useful indication. One may assume that, as in the case of ionization equilibrium (cf. eq. (1.54)), the variations of E and N are related by some general relation say

$$\mathrm{d}E_s = \chi\,\mathrm{d}N$$

which will permit to eliminate the partial derivatives of E_s in (1.59), in terms of the A's defined previously. Let us assume furthermore that β is still close to 1, then using the relation (1.61) and assuming $\chi/kT > 1$, the expression of Γ_1 reduces to

$$\Gamma_1 = \frac{5 + (7 + 2\chi/kT)A_{T,\varrho}}{3 + (3 + 2\chi/kT)A_{T,\varrho}} \simeq \frac{5 + (2\chi/kT)A_{T,\varrho}}{3 + (2\chi/kT)A_{T,\varrho}}. \tag{1.64}$$

In the lower part of the range of temperatures of interest (2 to $3\cdot 10^9$ °K) when the elements of the iron peak are being built, $\mathrm{d}E_s$ and $\mathrm{d}N$ are both negative for $\mathrm{d}T$ positive (χ:+, $A_{T,\varrho}$:−) and provided the product $(2\chi/kT)\cdot A_{T,\varrho}$ be in absolute value < 3, values of Γ_1 appreciably larger than $\frac{5}{3}$ would seem possible.

In the upper part of the range ($T > 4\cdot 10^9$ °K) when the iron gets converted into He again with rising temperatures, $\mathrm{d}E_s$ and $\mathrm{d}N$ are both positive (χ:+, $A_{T,\varrho}$:+) and Γ_1 may certainly become much smaller than $\frac{5}{3}$ and probably fairly close to 1.

However this approach may be partly invalidated by the emission of neutrinos and antineutrinos which is probably important in the reactions around the iron peak as photobeta decay rates may be large (cf. CAMERON, 1960). The emission of neutrinos and antineutrinos introduces an irreversible process which invalidates somewhat the notion of a true equilibrium, the importance of its effects depending essentially on the order of magnitude of the corresponding rate of energy loss ε_ν appearing in (1.37) which, with the previous definitions of Γ_1 and Γ_3, may still be written,

$$\frac{\mathrm{d}p}{\mathrm{d}t} - \frac{\Gamma_1 p}{\varrho}\frac{\mathrm{d}\varrho}{\mathrm{d}t} = (\Gamma_3 - 1)\varrho\left[-\varepsilon_\nu - \frac{1}{\varrho}\nabla\cdot(\boldsymbol{F}_R + \boldsymbol{F}_c) + \frac{1}{\varrho}\sum_{i,k}\mathscr{P}^{ik}\Delta_i v_k\right]. \tag{1.65}$$

It should be kept in mind that, while the last two terms in the bracket are usually small, ε_ν may be large.

According to CHIU (1961), large neutrinos and antineutrinos emission could even take place well before the iron peak is reached and be so large as perhaps to prevent the star ever reaching the conditions for the formation of this peak. The process advocated by CHIU is the annihilation of a pair of electrons by emission of a neutrino and an antineutrino so that, in appropriate conditions, the equilibrium (1.63) discussed earlier should be completed as follows

$$(1.63') \qquad h\nu \rightleftarrows e^+ + e^- \rightarrow \nu + \bar{\nu} .$$

Here again ε_ν is very large in the vicinity of $T = 10^9\,^\circ\mathrm{K}$, of the order of

$$(1.66) \qquad \varepsilon_\nu \simeq \frac{4.3 \cdot 10^5}{\varrho} T_9^9 \ \mathrm{erg/g\,s} ,$$

where T_9 is T expressed in $10^9\,^\circ\mathrm{K}$. ε_ν should be kept explicitly in eq. (1.65) which, in those conditions, neglecting the last two small terms in the bracket could be written as

$$\frac{\mathrm{d}p}{\mathrm{d}t} - \frac{\Gamma_1 p}{\varrho}\frac{\mathrm{d}\varrho}{\mathrm{d}t} = -(\Gamma_3 - 1)\varrho\varepsilon_\nu = -(\Gamma_3 - 1)4.3 \cdot 10^5\, T_9^9 ,$$

or

$$(1.67) \qquad \frac{\mathrm{d}}{\mathrm{d}t}\left[\lg\left(\frac{p}{\varrho^{\Gamma_1}}\right)\right] = -(\Gamma_3 - 1)\alpha\frac{p^8}{\varrho^9} ,$$

where α is a numerical constant. This implies that, in such a system, p and ϱ are explicitly functions of the time which already points to some kind of secular instability which however may have a very short time-scale considering the large absolute value of ε_ν.

1·4. *Small motions and linearized equations.* – Suppose that we know a solution f_0, of the general equations discussed in the preceding sections, f being any of the dependent variables entering these equations. Let us denote by f' a small Eulerian perturbation of these dependent variables. To study the evolution of the perturbation f' we reintroduce $f = f_0 + f'$ in the equations, take into account the fact that f_0 is itself a solution of the equations and develop the remaining terms in powers of f'. If we limit ourselves to the first-order terms in f', we obtain the linearized equations of motion which should at least reveal the trend of the evolution of the perturbation and permit, in general, to decide whether the solution f_0 is stable or not.

Actually since these equations are linear, we can always represent the time-dependence by a factor of the type e^{st} where s may be complex. It is clear that if all the characteristic values s have negative real parts, the equilibrium is certainly stable. On the other hand, if at least one of the characteristic values s has a positive real part, the configuration is certainly unstable. However, if one or more of the real parts of the characteristic values are rigorously zero, the situation becomes more complicated and the discussion must be pushed one step further to include at least the first nonlinear terms before a decision as to the stability can be reached unless the principle of the exchange of stability is applicable, which essentially will be true if it can be shown that s is a pure real number in the vicinity of $s=0$.

The linearized equations can also be obtained by taking the first variation of the general equations with respect to the dependent variables. Of course in the Eulerian representation that we are using, the space co-ordinates are independent variables and the corresponding operators ∇, ∇^2... are invariant but one must remember that all absolute time derivatives ($\mathrm{d}/\mathrm{d}t$) should be expanded as in eq. (1.1) before the first variation is taken.

Here we shall discuss only the perturbations of an equilibrium state and, for the time being anyway, we shall admit that the only nuclear processes of interest are thermonuclear reactions so that the appropriate form of the energy equation is (1.47) or (1.45) depending on whether convection is present or not.

In that case, the equilibrium state is characterized by:

$$(1.68)\quad \begin{cases} & \boldsymbol{v}=0\,, \\ (a) & \dfrac{1}{\varrho}\nabla(p+p_t)=-\nabla\varphi\,, \\ (b) & \nabla^2\varphi=4\pi G\varrho\,, \\ (c) & \varepsilon_N+\varepsilon_2-\varepsilon_3=\dfrac{1}{\varrho}\nabla\cdot(\boldsymbol{F}_R+\boldsymbol{F}_t)\,, \\ (d) & \varepsilon_2+\dfrac{1}{\varrho}\boldsymbol{V}\cdot\nabla\cdot p=\nabla\cdot\boldsymbol{F}_2\,. \end{cases}$$

If turbulent convection is absent, $\boldsymbol{V}$, ε_2, ε_3, $\boldsymbol{F}_t$ and $\boldsymbol{F}_2$ vanish so that the last equation which corresponds to the conservation of turbulent energy vanishes also. We have also neglected the conduction flux $\boldsymbol{F}_c$ which in ordinary stars is negligible.

In white dwarfs where it is predominant, we simply replace $\boldsymbol{F}_R$ by $\boldsymbol{F}_c$.

Since, in these circumstances, the equilibrium shape of the star is spherical, the only significant components of the vectorial quantities in (1.68) are the radial ones. In that case, in particular, the mean radial component of $\boldsymbol{V}$ and $\boldsymbol{F}_t$ are given, with the notations (1.53), by

$$(1.69)\qquad \bar{V}_r=\bar{l}_t C_t\left(\frac{4-3\beta}{\beta}+A_{T,p}\right)D\,,$$

and

$$(1.70)\qquad F_{t,r}=\varrho\bar{l}_t C_t c_v T\left\{1+\frac{2}{3}\frac{A^2_{T,p}}{B}-\frac{2}{3}\frac{A_{T,p}}{\beta}+\frac{2(1-\beta)}{\beta}\left[5+\frac{4(1-\beta)}{\beta}\right]\right\}D\,,$$

where c_v is the specific heat of the gas. D is defined by

$$(1.17)\qquad D=\frac{\beta}{4-3\beta}A=\frac{\beta}{4-3\beta}\left(\frac{1}{\varrho}\frac{\mathrm{d}\varrho}{\mathrm{d}r}-\frac{1}{\Gamma_1 p}\frac{\mathrm{d}p}{\mathrm{d}r}\right)=-\left(\frac{1}{T}\frac{\mathrm{d}T}{\mathrm{d}r}-\frac{\Gamma_2-1}{\Gamma_2}\frac{1}{p}\frac{\mathrm{d}p}{\mathrm{d}r}\right)=$$

$$=\frac{1}{T}\left|\frac{\mathrm{d}T}{\mathrm{d}r}\right|-\frac{\Gamma_2-1}{\Gamma_2}\frac{1}{p}\left|\frac{\mathrm{d}p}{\mathrm{d}r}\right|,$$

since, in all cases of interest, $\mathrm{d}T/\mathrm{d}r$ and $\mathrm{d}p/\mathrm{d}r$ are simultaneously negative. Thus the condition for convective instability is

$$(1.72)\qquad A>0 \quad \text{or} \quad D>0\,.$$

Instead of $F_{t,r}$, one may also introduce a generalized turbulent flux $F^*_{t,r}$ which includes the rate of the pressure work on the turbulent elements:

$$(1.73)\qquad F^*_{t,r}=F_{t,r}+p\bar{V}_r=\varrho\bar{l}_t\bar{C}_t C_p D\,,$$

where C_p is the generalized specific heat defined by (1.58). In terms of the definition (1.73), the 4th eq. (1.68) becomes

$$(1.74)\qquad \varepsilon_N=\frac{1}{\varrho}\nabla\cdot(\boldsymbol{F}_R+\boldsymbol{F}_t+\boldsymbol{F}_2)\,.$$

Applying the procedure outlined at the beginning of this section, we obtain for the Eulerian perturbation of the equilibrium state, the following equations:

$$(1.75)\qquad \frac{\partial\varrho'}{\partial t}+\nabla\cdot(\varrho\boldsymbol{v}')=0\,,$$

$$(1.76)\qquad \frac{\partial\boldsymbol{v}'}{\partial t}=-\nabla\varphi'+\frac{\varrho'}{\varrho^2}\nabla(p+p_t)-\frac{1}{\varrho}(p'+p_t')-$$

$$-\frac{\varrho'}{\varrho^2}\nabla\cdot\mathfrak{P}(\boldsymbol{V})+\frac{1}{\varrho}\nabla\cdot[\mathfrak{P}_t(\boldsymbol{v}')+\mathfrak{P}(\boldsymbol{V}')]\,,$$

$$(1.77)\qquad \nabla^2\varphi'=4\pi G\varrho',$$

$$(1.78)\qquad \frac{\partial p'}{\partial t}+\boldsymbol{v}'\cdot\nabla p-\frac{\Gamma_1 p}{\varrho}\left(\frac{\partial\varrho'}{\partial t}+\boldsymbol{v}'\cdot\nabla\varrho\right)=$$

$$=(\Gamma_3-1)\varrho\left\{\varepsilon_N+\varepsilon_2-\varepsilon_3-\frac{1}{\varrho}\nabla\cdot(\boldsymbol{F}_R+\boldsymbol{F}_t)\right\}',$$

$$(1.79)\qquad \frac{\partial p_t'}{\partial t}+\boldsymbol{v}'\cdot\nabla p_t-\frac{5}{3}\frac{p_t}{\varrho}\left(\frac{\partial\varrho'}{\partial t}+\boldsymbol{v}'\cdot\nabla\varrho\right)=\frac{2}{3}\varrho\left[\varepsilon_4-\varepsilon_2-\frac{1}{\varrho}\nabla\cdot\boldsymbol{F}_2-\frac{1}{\varrho}\boldsymbol{V}\cdot\nabla p\right]',$$

where

$$v' = \frac{\mathrm{d}\,\delta\boldsymbol{r}}{\mathrm{d}t}\,, \tag{1.80}$$

is the small velocity associated with the perturbation.

According to eq. (1.1), we may interchange at will the Eulerian and Lagrangian perturbation f' and δf of any quantity f which, in the equilibrium state, is identically zero or for which grad $f = 0$. As this is the case for the second members of (1.78) and (1.79), one may also write them

$$\frac{\mathrm{d}\,\delta p}{\mathrm{d}t} - \frac{\Gamma_1 p}{\varrho}\frac{\mathrm{d}\,\delta\varrho}{\mathrm{d}t} = (\Gamma_3 - 1)\varrho\delta\{\}\,, \tag{1.81}$$

$$\frac{\mathrm{d}\,\delta p_t}{\mathrm{d}t} - \frac{5}{3}\frac{p_t}{\varrho}\frac{\mathrm{d}\,\delta\varrho}{\mathrm{d}t} = \frac{2}{3}\varrho\delta[\,]\,, \tag{1.82}$$

if we take (1.80) into account and note from eqs. (1.1) that, for any perturbation of an equilibrium state ($\boldsymbol{v} = 0$), the total and partial time-derivatives are identical in the linear approximation.

Of course, one may also linearize eq. (1.46) which becomes

$$\frac{1}{T}\frac{\mathrm{d}\,\delta T}{\mathrm{d}t} - (\Gamma_3 - 1)\frac{1}{\varrho}\frac{\mathrm{d}\,\delta\varrho}{\mathrm{d}t} = \frac{1}{C_v T}\,\delta\left\{\varepsilon_N + \varepsilon_2 - \varepsilon_3 - \frac{1}{\varrho}\nabla\cdot(\boldsymbol{F}_R + \boldsymbol{F}_c + \boldsymbol{F}_t)\right\}. \tag{1.83}$$

An « adiabatic » perturbation is one in which the second members of (1.81), (1.82), (1.83) are negligibly small and can be put equal to zero.

2. – Purely radial perturbation.

2·1. *Definitions and remarks.* – The case of purely radial motions in which the star keeps its spherical symmetry is by far the simplest one and it already permits to isolate the main types of possible stellar instabilities.

Furthermore, to avoid too much algebraic complications, we shall, from now on, disregard turbulence and viscosity and be content with some remarks on their effects when needed.

2·2. *The general third-order equation for radial motion.* – The elimination of ϱ' and p' by means of (1.75) and (1.78) from (1.76) requires taking a time-derivative of the last equation and if we eliminate $\boldsymbol{v}'$ by means of (1.80), the general equation in $\delta\boldsymbol{r}$ will be, as already pointed out by JEANS (1928), of the third order in the time. Quite generally, this suggests that our discussion will fall into three parts corresponding to motions on three different time-scales.

For purely radial motions and with the hypothesis of Section 2·1, the relevant eqs. (1.75), (1.76), (1.77) and (1.78) become

$$\frac{\partial \varrho'}{\partial t} + \frac{1}{r^2}\frac{\partial}{\partial r}\left(r^2 \varrho \frac{\partial\, \delta r}{\partial t}\right) = 0 \,, \tag{2.1}$$

$$\frac{\partial^2 \delta r}{\partial t^2} = -\frac{\partial \varphi'}{\partial r} + \frac{\varrho'}{\varrho^2}\frac{\partial p}{\partial r} - \frac{1}{\varrho}\frac{\partial p'}{\partial r} \,, \tag{2.2}$$

$$\frac{1}{r^2}\frac{\partial}{\partial r}\left(r^2 \frac{\partial \varphi'}{\partial r}\right) = 4\pi G \varrho' \,, \tag{2.3}$$

$$\frac{\partial p'}{\partial t} + \frac{\partial\, \delta r}{\partial t}\cdot\frac{\partial p}{\partial r} - \frac{\Gamma_1 p}{\varrho}\left(\frac{\partial \varrho'}{\partial t} + \frac{\partial\, \delta r}{\partial t}\cdot\frac{\partial \varrho}{\partial r}\right) = (\Gamma_3 - 1)\varrho\left\{\delta\varepsilon_N - \frac{\mathrm{d}\, \delta L(r)}{\mathrm{d}m}\right\} , \tag{2.4}$$

where $L(r) = 4\pi r^2 F_R(r)$ is the total flux across the sphere of radius r and $m(r)$, the mass contained in it; m is invariant following the motion.

After taking the time-derivative of (2.3) and using (2.1) to eliminate $\partial\varrho'/\partial t$, Poisson's equation may be integrated once with respect to r, yielding

$$\frac{\partial}{\partial r}\left(\frac{\partial \varphi'}{\partial t}\right) = -4\pi G \varrho \frac{\partial\, \delta r}{\partial t} \,. \tag{2.5}$$

Eliminating $\partial\varrho'/\partial t$, $\partial p'/\partial t$, $\partial\varphi'/\partial t$ from (2.2) derived once with respect to the time, we obtain the following equation:

$$\frac{\partial^3 \delta r}{\partial t^3} = \frac{1}{\varrho}\frac{\partial}{\partial r}\left[\frac{\Gamma_1 p}{r^2}\frac{\partial}{\partial r}\left(r^2 \frac{\partial\, \delta r}{\partial r}\right)\right] - \frac{4}{\varrho r}\frac{\partial p}{\partial r}\frac{\partial\, \delta r}{\partial t} - \frac{1}{\varrho}\frac{\partial}{\partial r}\left[(\Gamma_3 - 1)\varrho\left\{\delta\varepsilon_N - \frac{\mathrm{d}\, \delta L}{\mathrm{d}m}\right\}\right] . \tag{2.6}$$

Let us define the relative displacement and assume that (2.6) is separable in t and r, writing

$$\frac{\delta r}{r} = \xi(r) \exp[st] \,. \tag{2.7}$$

In all generality, both ξ and s may be complex; however, usually, the imaginary part $I(\xi)$ of ξ is very small except perhaps very close to the surface. Using (2.7), eq. (2.6) becomes

$$s^3 r\xi = \frac{s}{\varrho}\frac{1}{r^3}\left\{\frac{\mathrm{d}}{\mathrm{d}r}\left(\Gamma_1 p r^4 \frac{\mathrm{d}\xi}{\mathrm{d}r}\right) + r^3 \xi \frac{\mathrm{d}}{\mathrm{d}r}[(3\Gamma_1 - 4)p]\right\} - \frac{1}{\varrho}\frac{\mathrm{d}}{\mathrm{d}r}\left[(\Gamma_3 - 1)\varrho\left\{\delta\varepsilon - \frac{\mathrm{d}\, \delta L}{\mathrm{d}m}\right\}\right] . \tag{2.8}$$

Solutions of this equation must satisfy the boundary conditions:

$$\delta r = 0 \qquad \text{at } r = 0\,, \tag{2.9}$$

$$\delta p = -\Gamma_1 p\left(3\xi + r\frac{\mathrm{d}\xi}{\mathrm{d}r}\right) + \frac{(\Gamma_3 - 1)}{s}\varrho\left[\delta\varepsilon_N - \frac{\mathrm{d}\,\delta L}{\mathrm{d}m}\right] = 0\,, \quad \text{at } r=R\,(p=0)\,, \tag{2.10}$$

which imply that ξ and its first derivative must remain finite everywhere.

If we multiply eq. (2.8) by $4\pi\varrho r^3\xi^*\,\mathrm{d}r$ and integrate over the whole star, we get an average equation for s which can be written

$$s^3 + s(A - B) + C = 0 \tag{2.11}$$

with

$$\begin{cases} A = \dfrac{1}{J}\displaystyle\int_0^R 4\pi\Gamma_1 p r^4 \left|\frac{\mathrm{d}\xi}{\mathrm{d}r}\right|^2 \mathrm{d}r\,, \\[2ex] B = \dfrac{1}{J}\displaystyle\int_0^R |\xi|^2 \frac{\mathrm{d}}{\mathrm{d}r}[(3\Gamma_1 - 4)p]\,4\pi r^3\,\mathrm{d}r\,, \\[2ex] C = \dfrac{1}{J}\displaystyle\int_0^R 4\pi r^3\xi^*\frac{\mathrm{d}}{\mathrm{d}r}\left[(\Gamma_3 - 1)\varrho\left\{\delta\varepsilon_N - \frac{\mathrm{d}\,\delta L}{\mathrm{d}m}\right\}\right]\mathrm{d}r\,, \end{cases} \tag{2.12}$$

where an asterisk denotes complex conjugate quantities. The normalization factor J is given by

$$J = \int_0^M |\xi|^2 r^2\,\mathrm{d}m\,. \tag{2.13}$$

The constants A, B and J are always real while C may be complex: $C = C' + iC''$. Very generally both $|C'|$ and $|C''|$ are much smaller then $|A|$ or $|B|$ and $|C''|$ is smaller than $|C'|$.

After an integration by parts, taking the boundary conditions (2.10) into account and noting that, after separation of the time, (2.1) may be written

$$\frac{\delta\varrho}{\varrho} = -\frac{1}{r^2}\frac{\mathrm{d}}{\mathrm{d}r}(r^3\xi)\,, \tag{2.14}$$

we find that the last expression for C becomes

$$C = \frac{1}{J}\int_0^R \left(\frac{\delta T}{T}\right)_a^* \left(\delta\varepsilon_N - \frac{\mathrm{d}\,\delta L}{\mathrm{d}m}\right)\mathrm{d}m\,, \tag{2.15}$$

where the adiabatic relative variation of the temperature T, denoted here by an index a, is related to $\delta\varrho/\varrho$ by the relation (1.83) with the second member neglected.

Equation (2.11) which admits a root s_1 proportional to C can be written

$$(s^2 + 2Ps + Q)(s - s_1) = 0\,,$$

with

$$s_1 = -\frac{C}{Q}\,, \qquad Q = \mathscr{A} + s_1^2\,, \qquad \mathscr{A} = A - B\,, \qquad P = \frac{s_1}{2} = -\frac{C}{2Q}\,,$$

The two other roots are given by

$$s_{2,3} = -P \pm \sqrt{P^2 - Q}\,. \tag{2.16}$$

If we neglect completely the dissipation ($C' \simeq 0$, $C'' \simeq 0$, $P \simeq 0$: *adiabatic approximation*) the roots (2.16) become

$$s_a = \pm i\sqrt{Q_a}\,, \qquad Q_a = \mathscr{A}_a\,,$$

or

$$-s_a^2 = +\sigma_a^2 = +Q_a = \tag{2.17}$$

$$= +\frac{1}{J_a}\left[\int_0^R 4\pi\Gamma_1 p r^4 \left(\frac{\mathrm{d}\xi_a}{\mathrm{d}r}\right)^2 \mathrm{d}r - \int_0^R \xi_a^2 \frac{\mathrm{d}}{\mathrm{d}r}[(3\Gamma_1 - 4)p]4\pi r^3\,\mathrm{d}r\right],$$

where of course ξ_a should be the corresponding real adiabatic solution. This, in fact, is exactly the expression of σ_a^2 yielded by the variational interpretation of the equation governing the adiabatic radial oscillations of a gaseous sphere (LEDOUX and PEKERIS, 1941).

If we treat ξ_a as a constant, formula (2.17) yields the following fairly reasonable approximation for the fundamental mode (no node in $0 \leqslant r \leqslant R$) of radial oscillation:

$$\sigma_a^2 = \overline{(3\Gamma_1 - 4)}\,\frac{\displaystyle\int_0^M \frac{Gm(r)\,\mathrm{d}m^2}{r}}{\displaystyle\int_0^M r^2\,\mathrm{d}m} = -\overline{(3\Gamma_1 - 4)}\cdot\frac{V}{I}\,, \tag{2.18}$$

where V is the gravitational potential and I the moment of inertia with respect to the centre.

The motion will be either a pure oscillation (σ_a^2: +) or an exponentially increasing motion (σ_a^2: —) with a time scale defined essentially by

$$\tau_0 = \frac{1}{\sigma_a} \simeq \left(\frac{I}{-V}\right)^{\frac{1}{2}} \simeq \left(\frac{R^3}{GM}\right)^{\frac{1}{2}} \simeq \left(\frac{3}{4\pi G\bar{\varrho}}\right)^{\frac{1}{2}}, \tag{2.19}$$

where $\bar{\varrho}$ is the mean density and which is very much shorter than the gravitational time-scale τ_G as defined by (I.1). For a star like the sun, τ_0 is of the order of one hour while τ_G is of the order of 10^7 years or 10^6 years if, as suggested by HAYASHI, convective equilibrium prevails in the contracting phases.

The question of determining whether σ^2 is positive (stability) or negative (instability) is the problem of the *dynamical stability* of the star and, in this respect, the main result is already apparent from (2.18): the star will be dynamically unstable only if an appropriate mean value of Γ_1 over the whole star is smaller than $\frac{4}{3}$.

In the usual case of well-marked dynamical stability the roots (2.16) may be noted

$$s_{2,3} = \pm i\sigma_a - \sigma' = \pm i\sqrt{\mathscr{A}} - P$$

introducing a damping coefficient

$$\sigma' = P = -\frac{C}{2\mathscr{A}} \simeq -\frac{C'_a}{2\mathscr{A}} = -\frac{1}{2\sigma_a^2 J_a}\int_0^M \left(\frac{\delta T}{T}\right)\left\{\delta\varepsilon_N - \frac{\mathrm{d}\,\delta L}{\mathrm{d}m}\right\}_a \mathrm{d}m\,, \tag{2.20}$$

which may be evaluated by means of the adiabatic solution as noted in the last expression by a subscript a to all the terms. The amplitude of the adiabatic oscillation will decrease or increase depending on whether σ' is positive or negative and, correspondingly the star will be said to be *vibrationally stable* or *unstable* (overstable in Eddington's terminology). Actually (2.20) is the usual approximation for σ' as obtained by the perturbation method (ROSSELAND 1931, 1932; cf. also LEDOUX and WALRAVEN, 1958) or by thermodynamical arguments (EDDINGTON, 1926 and 1941).

Using the approximation (2.18) for σ_a^2 and noting that J_a is of the order of MR^2 while the integral in (2.20) is essentially proportional to L, one sees immediately that σ' is associated with a time-scale

$$\tau_d \simeq \alpha\left(\frac{-V}{L}\right) \simeq 2\alpha\tau_G\,. \tag{2.21}$$

Although α may vary in a large range, this shows that, in general, the time-scale associated with vibrational stability or instability will be much larger than τ_0.

As to the last root s_1, it is proportional to the small dissipation term $-C$ contrarily to the real part of $s_{2,3}$ which is proportional to $+C$. Of course this time, C must correspond to a solution of (2.8) associated directly with s_1 or, in other words, with a very long time-scale $\tau_s = 1/s_1$. This is, in general, so long that deviations from adiabacy in the corresponding motion are very large and the adiabatic solution ξ_a can certainly not be used to evaluate C as it was done in (2.20). Rather in this case, it is the acceleration which may be neglected so that the appropriate solution must be found from (2.8) in which, while the left-hand member may be dropped, the last term on the right must be expressed explicitly in terms of ξ and its derivatives. This is the problem of *secular stability* and, according to the process outlined above, it must, in all generality, lead to a complicated differential equation.

On the other hand, if the acceleration terms are neglected, the solution ξ_1, appropriate to s_1 *may* be of the nature of the relative displacement between two quasi-static equilibrium configurations. But among these, there is a well-known family corresponding to a homology transformation

$$r = r_0\eta\,, \qquad \varrho = \varrho_0\eta^{-3}\,, \qquad p = p_0\eta^{-4}\,, \qquad T = T_0\eta^{-1} \tag{2.22}$$

which verifies the continuity eq. (2.14), the hydrostatic eq. (1.68,a), the equation of state (1.19) and which, for a star in radiative equilibrium with a law of opacity of the form (1.11), implies through eq. (1.32) that $L(r)$ transforms according to

$$L(r) = L_0(r_0)\,\eta^{-(n-3m)}\,. \tag{2.23}$$

We know also that, for thermonuclear reactions, we may, at least in not too long intervals of ϱ and T, represent ε_N by

$$\varepsilon_N = \varepsilon_0\varrho^\mu T^\nu, \tag{2.24}$$

which yields the transformation

$$\varepsilon_N = (\varepsilon_R)_0\eta^{-(3\mu+\nu)}\,. \tag{2.24}$$

The relative variations going from one of these configurations to another one close to it, are

$$\left\{\begin{aligned} &\frac{\delta r}{r} = \xi_1 = \frac{\delta\eta}{\eta}\,, &&\frac{\delta\varrho}{\varrho} = -\,3\xi_1\,, &&\frac{\delta T}{T} = -\,\xi_1\,, \\ &\frac{\delta L}{L} = -\,(n-3m)\xi_1\,, &&\frac{\delta\varepsilon_N}{\varepsilon_N} = -\,(3\mu+\nu)\xi_1\,, && \end{aligned}\right. \tag{2.26}$$

where ξ_1 is a constant throughout the star.

Substituting these values in

$$s_1 = 2P_1 = -\frac{C_1}{Q} = -\frac{1}{\sigma_a^2 J_a}\int_0^M (\Gamma_3 - 1)\left(\frac{\delta\varrho}{\varrho}\right)_1\left(\delta\varepsilon_N - \frac{\mathrm{d}\,\delta L}{\mathrm{d}m}\right)_1 \mathrm{d}m\,,$$

we find immediately, if the Γ's are constant and using (2.18) for σ_a^2,

$$s_1 = \frac{3(\Gamma_3 - 1)}{(3\Gamma_1 - 4)}[(n - 3m) - (3\mu + \nu)]\frac{L}{(-V)}\,, \tag{2.27}$$

which shows that the appropriate time-scale τ_1 is generally of the order of τ_G and is thus very long. In this approximation, the condition of secular stability for a gaseous star, $s_1 < 0$, becomes

$$n - 3m < 3\mu + \nu \tag{2.28}$$

a condition already written down by JEANS (1928) and which has been used extensively.

As long as $|\mathscr{A}| \gg |C'|$ and $|C''|$, it is easy to write successive approximations to the roots and, up to the second-order terms, one gets

$$s_1 = -\left(\frac{C' + iC''}{\mathscr{A}}\right)_1\,, \tag{2.29}$$

$$s_{2,3} = i\left(\pm\mathscr{A}^{\frac{1}{2}} + \frac{C''}{2\mathscr{A}} \mp \frac{3}{8}\frac{C''^2 - C'^2}{\mathscr{A}^{\frac{3}{2}}}\right) + \frac{C'}{2\mathscr{A}} \mp \frac{3}{4}\frac{C'C''}{\mathscr{A}^{\frac{5}{2}}}\,, \tag{2.30}$$

where $\mathscr{A}$, C' and C'' must, in each case, be computed by means of the appropriate solutions ξ evaluated to the same order. This shows that, in principle at least, the slow secular motion (cf. (2.29)) may have an oscillatory component but of very long period.

As far as the dynamical motion is concerned (cf. (2.30)), we see that the first-order correction to the frequency ($\mathscr{A} > 0$) is proportional to C'' while, as we already found in (2.20), the dominating term in the real part of $s_{2,3}(\sigma')$ is of the order of C'.

It may be noticed that the presence of an imaginary part C'' in C leads to the splitting of the radial frequencies into two very close frequencies. This might find some application in the theory of intrinsic variable stars presenting long beats.

If $\mathscr{A}$ vanishes (vanishing of the dynamical stability in the previous sense: $\sigma_a = 0$), the roots become

$$s_{1,2,3} = \varrho^{\frac{1}{3}} \exp\left[i\frac{\theta}{3}\right]; \qquad \exp\left[i\,\frac{2\pi + \theta}{3}\right]; \qquad \exp\left[i\,\frac{(4\pi + \theta)}{3}\right], \tag{2.31}$$

where $\varrho^2 = (C'^2 + C''^2)$, and θ is defined by

$$\sin\theta = -C''/\varrho\,, \qquad \cos\theta = -C'/\varrho\,.$$

Thus, strictly speaking, the three solutions remain periodic when $\mathscr{A} = 0$, true dynamical instability appearing only for some small but finite negative value of $\mathscr{A}$. Of course, our way of writing (2.31) implies that ϱ or $|C|$, and consequently the time-scales, are of the same order for all the solutions which will be true only if ξ has the same behaviour. But this seems possible in this critical case since, with σ_a tending to zero, ξ_a tends to become constant throughout the star just as the solution ξ_1, used above for the discussion of secular stability.

In particular, if $C'' = 0$, the solutions (2.31) become

$$\text{(2.32)} \qquad s_{1,2,3} = |C'|^{\frac{1}{3}}\left(1, -\frac{1}{2} \pm \frac{i\sqrt{3}}{2}\right) \qquad \text{or} \qquad s_{1,2,3} = |C'|^{\frac{1}{3}}\left(-1, \frac{1}{2} \pm \frac{i\sqrt{3}}{2}\right),$$

depending on whether C' is negative or positive. In the first case, (positive dissipation), the oscillatory solutions are damped in a few periods ($\tau_d \simeq \tau_0$) while the solution associated with the first root is continuously amplified (secular instability). In the second case (negative dissipation), the situation is reversed and the instability would appear as oscillations of increasing amplitudes.

This first survey has isolated the three types of stability problems arising in connection with stellar structure and has already defined some of their important characteristics such as their respective time-scales, at least for gaseous stars in which the energy generation is due to thermonuclear reactions. We shall now turn to a somewhat more detailed account of each, trying especially to underline their significance for stellar evolution.

Let us note, however, that the explicit elimination of ($\delta\varepsilon : -\mathrm{d}\,\delta L/\mathrm{d}m$) in terms of ξ and its derivatives in the general eq. (2.8) may raise the order of this equation with respect to s, especially if the full equations governing the perturbations of the abundances of the elements entering the nuclear reactions are taken into account. This may introduce new time-scales characterizing new types of stable or unstable motions. However, up to now, this aspect of the problem has received little attention.

3. – Stability and stellar evolution.

3·1. *Dynamical stability towards radial perturbation.* – We have seen that, in this case, we may limit our investigation to « *adiabatic* » perturbations *i.e.*, in the energy equation, we may neglect the right-hand member unless neutrino emission constitutes a large dissipation (cf. eq. (1.65)).

For radial oscillations, the problem then reduces to the resolution of the following eigenvalue problem:

$$
(3.1)\qquad \begin{cases} \dfrac{\mathrm{d}}{\mathrm{d}r}\left(r^4 \Gamma_1 p \dfrac{\mathrm{d}\xi_a}{\mathrm{d}r}\right) + \xi_a \left\{\sigma_a^2 \varrho r^4 + r^3 \dfrac{\mathrm{d}}{\mathrm{d}r}[(3\Gamma_1 - 4)p]\right\} = 0\,, \\ \delta r = 0 \quad \text{at} \quad r = 0\,, \\ \delta p_a = p' + \delta r \dfrac{\mathrm{d}p}{\mathrm{d}r} = -\Gamma_1 p \left(3\xi + r\dfrac{\mathrm{d}\xi}{\mathrm{d}r}\right) = 0 \qquad \text{at } r = R\,. \end{cases}
$$

The differential equation has regular singularities at $r=0$ and $r=R$ and the solutions have the same general properties as those of a classical Sturm-Liouville problem. There exists an infinite discrete set of eigenvalues which, when ordered by increasing values, will be denoted by σ_{a0}^2, σ_{a1}^2, σ_{a2}^2, ... When all the σ_{ai}^2 are positive (stability), the associated eigensolutions $\xi_{a,0}$, $\xi_{a,1}$, ... which form a complete set of orthogonal functions

$$
(3.2)\qquad \int \xi_{ai}\xi_{ak}\varrho r^4 \,\mathrm{d}r = 0 \qquad\qquad i \neq k\,,
$$

correspond to the successive modes of oscillations: ξ_{a0} has no node in $0 \leqslant r \leqslant R$, ξ_{a1} has one node, ξ_{a2} has two nodes, etc. If they are normalized in such a way that $|\xi_a| = 1$ at the centre, $|\xi|_a$ reaches higher and higher values at the surface as the order of the mode increases. However the intermediate maxima of $|\xi_a|$ in the main part of the star tend to become smaller and smaller and, for very high modes, they can be much smaller than 1 in a large fraction of the mass (cf. BOURY et Mme HUSTIN-BRETON, 1961; LEDOUX, 1962).

Since $\sigma_{a,0}^2$ is the smallest eigenvalue, it is through the fundamental mode that dynamical instability ($\sigma_a^2 < 0$) will first arise. But we have already seen than $\sigma_{a,0}^2$ is given by (2.17) which, from a variational point of view, can also be interpreted by saying that if ξ_a is any regular function at $0 \leqslant r \leqslant R$, the right-hand member of (2.17) goes through a minimum equal to $\sigma_{a,0}^2$ when ξ_a is varied and the particular function $\xi_{a,0}$ corresponding to this minimum is the fundamental mode.

If Γ_1 is constant throughout the star, the second member of (2,17) is definite positive since $\mathrm{d}p/\mathrm{d}r$ is always negative and $\sigma_{a,0}^2$ will be positive if $\Gamma_1 > \frac{4}{3}$ which is the well-known condition. If Γ_1 varies, it is an appropriate average Γ_1 which must be greater than $\frac{4}{3}$ for stability. TOLMAN (1939) and LEDOUX (1949) have shown that, in general, the variations of Γ_1 occuring around the middle of the star radius have the strongest influence and that, all conditions in this region being equal, the most favourable case for instability occurs when $(3\Gamma_1 - 4)$ has the deepest possible minimum both at the centre and close to the surface.

Let us now consider the factors which affect Γ_1. According to formula (1.50), Γ_1 decreases as the pressure of radiation increases (β decreases). Consequently the dynamical stability decreases as we go to larger and larger masses; however, the limiting value of $\Gamma_1(\beta \to 0,\ M \to \infty)$ is $\frac{4}{3}$ and cannot be reached for finite masses. Thus, radiation cannot by itself cause dynamical instability but, of course, it can help in combination with other factors.

Among these, the ionization of an abundant element may, as we have seen in discussing eqs. (1.55) to (1.58), lower Γ_1 considerably below $\frac{4}{3}$. However, with the large predominance of H and He, the corresponding zones of low Γ_1 in normal stars occur very close to the surface where the pressure is very small and they affect the average $\bar{\Gamma}_1$ very little so that dynamical instability cannot occur on this account in ordinary gaseous stars.

Let us now turn to objects or stars which, to be in extreme physical conditions, are nevertheless interesting and important for stellar evolution and let us consider first the early phases of gravitational contraction. At a time when the general predominance of H and He was not yet established BIERMANN and COWLING (1939) showed that for masses of the order of 1 to 10 $M_\odot$ there were phases corresponding to radius between 30 and 100 $R_\odot$ where the contracting star would become dynamically unstable due to the ionization of the heavy elements ($\bar{\Gamma}_1 < \frac{4}{3}$) unless it contained a fairly large amount of H and He already completely ionized at this stage.

Later (cf. LEDOUX, 1958, Section **16**), it was pointed out that if the star is composed predominantly of H and He, this instability would probably not be avoided but would simply occur for a larger radius of the order of 1000 $R_\odot$ or 5 A.U. But, as CAMERON (1962) has recently suggested, there is still an earlier phase where H would be mainly in the form of H_2 molecules and he estimates that a phase of dynamical instability, due to the dissociation of H_2 would start for values of R of the order of 100 A.U. and would merge directly into the phases of dynamical instability due to the ionization of H, He I and He II bringing a protostar of a solar mass very rapidly to a radius of the order of 40 $R_\odot$. In the same article, CAMERON points out that this fast contraction combined with the conservation of angular momentum may have very important cosmological consequences such as the shedding of practically all the original mass in a flattened nebula. In any case, a detailed study of these fast contraction phases would be very interesting.

The other case of interest concerns the opposite extreme of stellar evolution. As we already recalled in Section **1**.3.3, temperature and (or) densities may reach such high values in those phases that nuclear equilibria tend to get established. The simplest one consists in the equilibrium between pairs of electrons and radiation and we have seen that it can lower the Γ's below $\frac{4}{3}$. However it is difficult to judge the possible effect without actual stellar models to evaluate the relative mass of the regions affected. For instance, assuming that $\Gamma_1 = 1$ in a central region and $\Gamma_1 = \frac{5}{3}$ in the external part, one finds, for the standard model, that the relative mass m_c/M of the central core has to be of the order of 0.35 ($r_c/R = 0.23$) before the whole configuration becomes unstable.

If a star can reach conditions passed the iron peak where iron is converted again into He ($^{56}Fe \rightarrow 13\,^4He + 4n$), there is little doubt that Γ_1 will become very close to 1 probably in a fairly large region and the resulting violent instability (with a very short time-scale of the order of a few seconds: $\bar{\varrho} \simeq 10^6$) could very well be, as it was first suggested by HOYLE (1946), the cause of the supernovae explosions.

On the other hand, the discussion of eq. (1.64) in Section 1·3.3 suggested that while the iron peak is forming, the Γ's might take values greater than $\frac{5}{3}$. In other words, in this range of physical conditions, the compressibility would be decreased at least for adiabatic modifications. If this would remain true during the secular changes of the configuration, its effects combined with those of the conservation of angular momentum might be interesting since, according to JEANS, a compressible mass with $\Gamma_1 > 2.2$ in presence of an increasing angular velocity, would behave more like an incompressible configuration and we know that this implies major changes in the possible equilibrium configurations and the nature of the ultimate rotational instability (Jacobi's ellipsoids and dynamically unstable pear-shaped configurations rather than equatorial shedding).

Of course, this is rather speculative and the approximate expression (1.64) might not even give a reasonable order of magnitude of Γ_1 during those phases where a complete study of the effects of the different nuclear equilibria is bound to be very complicated. It would also be essential to check that these nuclear equilibria can in fact follow the very rapid adiabatic modifications considered here or, in other words, that the time constants of the different reactions leading to equilibrium are short or at most of the same order as the period. Otherwise, phase-delays will appear and it seems likely that, if they are long, these equilibrium reactions will no longer affect the dynamical stability but rather the vibrational stability of the star (Γ_1 would acquire an imaginary part and the corresponding term in the first member of (1.45) would be of the same nature as those in the right-hand member).

Furthermore, it is just during those phases (building up of the iron group) that Chiu's process (emission of neutrinos and antineutrinos) might be most efficient. If we compare eq. (1.65) appropriate for this case to eq. (1.45) which we have used in the general discussion of Section 2·2, it seems that the only necessary change is to replace ε_N by $-\varepsilon_\nu$ and since the other dissipation terms due to conduction are comparatively small, we may neglect them, writing for C

$$(3.3) \qquad C = \frac{1}{J}\int_0^M \frac{\delta T}{T}(-\,\delta\varepsilon_\nu)\,\mathrm{d}m\,,$$

which will be large and negative and, qualitatively, the solutions should be of the same nature as those corresponding to the first group of solutions (2.32) *i.e.* large secular motion, rapidly damped oscillatory motions.

Of course, this also implies no phase delays in the emission of ν and $\tilde{\nu}$ or in the way the nuclear equilibrium adapts itself to the rapid motion considered. Otherwise, as already mentioned above, these would introduce imaginary parts in ε_ν as well as in Γ_1 and partial interchange of the rôles of the two processes with respect to dynamical and vibrational stability. Without a detailed study of all the relevant processes and the mean life-time of the elements involved it is quite impossible to evaluate even qualitatively the resulting effects. There is however no doubt that Chiu's process could start a violent secular instability. But when the rate of the corresponding motion becomes fast, we would again have to take into account the possible phase-delays.

Another nuclear process which, at least at very high densities and namely in white dwarfs, should be taken into consideration, is the equilibrium between electron captures and β-radioactivity. For instance, if we limit our discussion to white dwarfs assuming complete degeneracy, it has been shown (SAUVENIER-GOFFIN, 1949, 1950) that although Γ_1 tends to $\frac{4}{3}$ as complete relativistic degeneracy tends to get established, σ_a^2 remains finite so that dynamical stability persists. However, SCHATZMAN (1958) has shown that, for high enough densities, the above nuclear equilibrium reduces Γ_1 to values below $\frac{4}{3}$ leading to dynamical instability.

3·2. *Dynamical stability towards nonradial oscillations.* – In this case the displacement $\delta\boldsymbol{r}$ has three distinct components δr, $r\,\delta\theta$, $r\sin\theta\,\delta\varphi$ of which δr like ϱ', p', φ' may be represented by expressions of the form

$$(3.4) \qquad \alpha(r,\theta,\varphi,t) = \alpha(r)\,Y_l^m(\theta,\varphi)\exp[i\sigma t] = \\ = \alpha(r)\,P_l^m(\cos\theta)\exp[im\varphi]\exp[i\sigma t]\,, \qquad -l\leqslant m\leqslant l\,,$$

while $r\delta\theta$ and $r\sin\theta\,\delta\varphi$ are respectively proportional to $\partial Y_l^m/\partial\theta$ and $\partial Y_l^m/\partial\varphi$.

The problem becomes appreciably more complicated in this case and, for details, the reader is referred to the review article by LEDOUX and WALRAVEN (1958, Chap. IV) and original papers by ROSSELAND (1932), PEKERIS (1938),

COWLING (1941), LEDOUX (1949, Chap. II), KOPAL (1949), SAUVENIER-GOFFIN (1951), SKUMANICH (1955), OTTELET (1960), etc.

A complete analytical solution (PEKERIS, 1938) is available only for the homogeneous compressible model for which the differential problem is of the second order. The finite expressions for the discrete eigenvalues show that:

1) in absence of perturbing forces (rotation, magnetic fields, tidal effects), the problem is degenerate $\sigma_{l,m} = \sigma_l$ for $-l \leqslant m \leqslant l$.

2) for any value of l, the discrete spectrum falls into two parts: one starting from a minimum positive value and tending to infinity (stable modes or, with Cowling's definition, p-modes) and another one starting from a minimum negative value and converging towards zero (unstable modes related to Cowling's g-modes).

One finds also that $\Gamma_1 = \frac{4}{3}$ is no longer a critical value for stability and, actually it does not seem that there is any critical value of Γ_1 directly significant for the stability here. The origin of the instability manifested in the second part of the spectrum ($\sigma_g^2 < 0$) can be traced back to the fact that, in this model, the condition for convective instability $A > 0$ (cf. (1.72)) which reduces here to $-(1/\Gamma_1 p)(\partial p/\partial r) > 0$ is realized at all points. Note that for purely radial oscillations, the value of A has no influence at all.

As far as the amplitudes of ϱ', p', T' are concerned, they all vanish at the centre so that nonradial oscillations always have a loop as well as a node at that point. For a given mode (fixed number of nodes of δr in $0 \leqslant r \leqslant R$), $|\sigma^2|$ increases with l and, in particular for the unstable modes, the dynamical instability increases when the horizontal dimensions of the perturbation decrease. On the other hand, for a given l, the two sets of σ^2 (σ_p^2 and σ_g^2) increase algebraically with the order of the mode so that, in particular, the instability decreases with the vertical wave-length.

Qualitatively these seem to be general properties which have been confirmed by all the models studied (polytropes: COWLING, 1941; Roche model: OTTELET, 1960; LEDOUX, in press) In fact, for any physical model, if we neglect φ' (which appears to be a fairly good approximation as soon as l is equal to a few units) one can, for modes of sufficiently high orders, write the following expressions for the frequencies

$$(3.5)\qquad \sigma_p^2 = \frac{\displaystyle\int_0^R \frac{\Gamma_1}{r^2 p^{(2/\Gamma_1)-1}} \left(\frac{\partial v}{\partial r}\right)^2 \mathrm{d}r - \int_0^R \frac{Ag\varrho}{r^2 p^{2/\Gamma_1}}\, v^2\, \mathrm{d}r}{\displaystyle\int_0^R \frac{\varrho v^2}{r^2 p^{2/\Gamma_1}}\, \mathrm{d}r},$$

and

$$\sigma_g^2 = -\frac{l(l+1)\int\limits_0^R \frac{Ag\varrho v^2}{r^2 p^{2/\Gamma_1}}\,\mathrm{d}r}{l(l+1)\int\limits_0^R \frac{\varrho v^2}{r^2 p^{2/\Gamma_1}}\,\mathrm{d}r + \int\limits_0^R \frac{\varrho}{p^{2/\Gamma_1}}\left(\frac{\mathrm{d}v}{\mathrm{d}r}\right)^2 \mathrm{d}r}, \tag{3.6}$$

where

$$v = r^2 p^{1/\Gamma_1}\,\delta r\,.$$

These expressions show that if A is everywhere negative (subadiabatic gradient everywhere) σ_p^2 and σ_g^2 are always positive and any form of dynamical instability is excluded.

If A is positive everywhere (superadiabatic gradient everywhere) all the σ_g^2 become negative leading to dynamical instability. The question is more difficult to decide for the p-modes but since $(\partial v/\partial r)$ increases with the order of the mode, we may surmise that p-modes of high enough orders will remain stable.

If A is positive in some part of the star only, the occurence of negative σ_g^2's depends on the relative extent of these superadiabatic regions. In all cases, the absolute value of σ_g^2 decreases when the order of the mode increases $[(\partial v/\partial r)^2$ increases$]$ while $|\sigma_g^2|$ increases with l. Thus, as long as one neglects all dissipation, dynamical instability if present will be the stronger, the smaller the horizontal wave-length and the larger the vertical wave-length.

One may also note that the integral in the numerator of (3.6) is simply the total work of the buoyancy force (counted positively in a superadiabatic region and negatively elsewhere). Thus, essentially, the present instability depends on whether for a given mode of displacement the gravitational potential energy liberable in the superadiabatic regions is larger than the necessary work to be spent against gravitation in the subadiabatic regions.

If, by somewhat artificial boundary conditions, motions could be limited to the superadiabatic region ($\delta r = 0$ outside of it), some instability according to (3.6) would always manifest itself locally leading to a mixing of this zone and a lowering of the gradient close to its adiabatic value, *i.e.* convection.

If, in the course of stellar evolution, superadiabatic gradients develop it seems that they must do so very gradually starting first in narrow zones growing in extent very slowly. From the preceding, one would then expect that convection would always follow very closely, preventing the building up of extensive regions with appreciable superadiabatic gradients. If it were not for this, the existence of large superadiabatic regions would provide a nice source of dynamical instability which, following the arguments of UNSÖLD (1930) and BIERMANN (1939), could very well have been the cause of ordinary novae.

However, convection does not reduce the gradients exactly to their adiabatic values but the subsisting superadiabatic excess is generally extremely small. Nevertheless, in a star with a large convection zone involving most of its mass as in the most massive of the stellar models built by SCHWARZSCHILD and HÄRM (1958) or in the very cool stars of very small masses (LIMBER, 1958) it might be interesting to assess the existence and the type of nonradial perturbation leading to the kind of dynamical instability discussed here.

Of course, even if such modes exist, the star would not get shattered into pieces but the fortuitous building up of a suitable perturbation might lead to the ejection of matter in the form of local jets. More important perhaps and, at least as likely, it might result in a general re-mixing of the whole star involving the stable as well as the unstable regions, the star settling down afterwards in a state of lower total potential energy. In some extreme models, with large convection zones depleted of nuclear fuels, such a global re-mixing, even if the total mass of the stable region containing fresh nuclear fuels is small, could have spectatular consequences.

3·3. *Vibrational stability towards radial pulsation.*

3·3.1. Some general remarks. – The expression (2.20) for the damping coefficient σ' of the fundamental mode may be generalized to any mode simply by adding an extra subscript, $k = 0, 1, 2, \ldots$, corresponding to the order of the mode. However, as we shall be mainly concerned with the fundamental mode ($k = 0$) we shall not use this extra subscript unless explicitly needed. Neglecting convection and viscous forces as in Section **2**·2, we have

$$\sigma' = -\frac{1}{2\sigma_a^2 J_a}\int_0^M \left(\frac{\delta T}{T}\right)_a \left\{\delta\varepsilon_N - \frac{\mathrm{d}\,\delta L}{\mathrm{d}m}\right\} \mathrm{d}m\,. \tag{3.7}$$

Some care should be taken in evaluating this integral. First we must remember that this form of σ' has been obtained by transforming C from its expression (2.12) to the expression (2.15) using the boundary condition (2.10). In particular, since ε_N (and $\delta\varepsilon_N$) becomes quite negligible well before reaching the surface and since p vanishes there, the second boundary condition may be written

$$\frac{\mathrm{d}\,\delta L(r)}{\mathrm{d}r} = 0 \qquad \text{at } r = R\,, \tag{3.8}$$

Furthermore, with $\sigma^2 = -s^2$ the general eq. (2.8) may be written

$$\text{(3.9)} \quad \frac{d^2\xi}{dr^2} + \frac{d\xi}{dr}\left[\frac{4}{r} + \frac{1}{\Gamma_1 p}\frac{d(\Gamma_1 p)}{dr}\right] + \xi\left\{\frac{\sigma^2\varrho}{\Gamma_1 p} + \frac{1}{r\Gamma_1 p}\frac{d}{dr}[(3\Gamma_1 - 4)p]\right\} = = \frac{1}{i\sigma\Gamma_1 p}\frac{d}{dr}\left[(\Gamma_3 - 1)\varrho\left(\delta\varepsilon - \frac{d\,\delta L}{dm}\right)\right],$$

and close to the surface keeping only the terms of the order of $1/p$ and noting that at the limit the Γ's may be treated as constants, it becomes

$$\text{(3.10)} \quad \varrho\left[-\Gamma_1 g\frac{d\xi}{dr} + \xi\left(\sigma^2 - \frac{(3\Gamma_1 - 4)}{r}g\right)\right] = -\frac{(\Gamma_3 - 1)}{i\sigma r}\frac{d}{dr}\left[\frac{1}{4\pi r^2}\frac{d\,\delta L}{dr}\right],$$

where g denotes the gravity. As $\varrho \to 0$ when $r \to R$, we also have

$$\text{(3.11)} \quad \frac{d^2\,\delta L}{dr^2} = 0 \qquad \text{at } r = R\,.$$

Thus very close to the surface, δL must be practically constant and the integral in (3.7) should really be extended only up to a radius R^* (*i.e.* a mass M^*) above which the heat capacity, the pressure and the density are so small that differential motions in the corresponding superficial layer cannot pratically affect δL. This level R^* is usually close to the photosphere and one may admit that practically the atmosphere above it adapts itself, at each instant, to a flux $(L+\delta L)$ constant through the layer and imposed by the interior.

On the other hand, when we come from the centre, the terms in the right-hand member of (3.9) remain very small compared to those on the left for a very large fraction of the mass. However, the right-hand member increases towards the surfaces as p^{-1}, while the dominating terms in the first member increase only as ϱp^{-1} (cf. also eq. (3.10)) and if we denote by R_a (or M_a) the level at which the two members become of the same order, we may say that the adiabatic solution (which determines the dynamical stability of the star and which should be used in evaluating (3.7) is a good approximation in the region $0 \leqslant r \leqslant R_a$ (or $0 \leqslant m \leqslant M_a$). However above R_a, the deviations with respect to adiabacy may become large and, in that case, the curly brackett in (3.7) should no longer be evaluated by means of the adiabatic solution in the corresponding region $R_a \leqslant r \leqslant R^*$.

However, in many stars, especially those of the main sequence, that region has already a very small heat capacity and its contribution to the integral (3.7) is negligible and we may as well take M_a as the upper limit of the integral. The only exception is the case where an important ionization (H or He) occurs in the region $R_a \leqslant r \leqslant R^*$ and we shall come back to it later but for the time being we want first to discuss (3.7) when the upper limit M of the integral is replaced by M_a.

3'3.2. *The effects of radiation and thermonuclear reactions: critical mass.* – The term due to the energy generation (cf. (2.24) where we may put $\mu=1$)

$$(3.12)\qquad \delta\varepsilon_N = \varepsilon_N\left(\frac{\delta\varrho}{\varrho}+\nu\,\frac{\delta T}{T}\right)_a = \varepsilon_N\left(\frac{1}{\Gamma_3-1}+\nu\right)\left(\frac{\delta T}{T}\right)_a,$$

is always of the same sign as $(\delta T/T)_a$, and so always contributes a negative part to σ' and thus favours vibrational instability.

In the second term, the variation of the total radiation flux is, according to (1.32) and for an opacity law of the form (1.11),

$$(3.13)\qquad \delta L(r) = L(r)\left[4\xi + (4+n)\,\frac{\delta T}{T} - m\,\frac{\delta\varrho}{\varrho} + \frac{(\mathrm{d}/\mathrm{d}r)(\delta T/T)}{(1/T)(\mathrm{d}T/\mathrm{d}r)}\right],$$

and, for usual values of m and n, it contributes positively to σ' and thus favours stability.

In all gaseous stars of ordinary masses and comprising an important radiative zone, ξ_a, $(\delta\varrho/\varrho)_a$ and $(\delta T/T)_a$ increase fairly rapidly from $r=0$ to $r=R$ (for instance in the standard model: $(\xi_R/\xi_c)\simeq 20$, $(\delta\varrho/\varrho)_R/(\delta\varrho/\varrho)_c\simeq 60$).

But since ε_N is appreciable only close to the centre, the corresponding distabilizing term in (3.7) is weighted by a small factor as compared to the stabilizing term which can be large in the external layers. As pointed out first by COWLING (1936), this renders ordinary stars very stable and, if one computes the critical value of ν, say ν_c for which vibrational instability enters, one finds that they are much larger (say 10^3 to 10^4) than the actual values of ν.

However, for larger and larger masses, $(1-\beta)=p_R/p$ increases and the Γ's decrease tending towards $\frac{4}{3}$, a limiting case for which ξ_a, $(\delta\varrho/\varrho)_a$ and $(\delta T/T)_a$ become constant throughout the star so that the weighting with respect to $(\delta T/T)_a^2$ so favourable to stability in small mass stars, ceases to be efficient here and ν_c becomes of the order of 1. This is now so small, that for any given thermonuclear reactions, there must exist a critical mass M_c above which stars become vibrationally unstable and have to get rid of the excess mass fairly rapidly. Thus the criterion of vibrational instability introduces an upper limit to the possible masses of gaseous stars.

Using the standard model and Kramer's law of opacity ($m=1$, $n=3.5$), I found that, for the carbon cycle, the critical mass was approximately $M_c = 100\,M_\odot\bar{\mu}^{-2}$ (LEDOUX, 1941). Recently, SCHWARZSCHILD and HÄRM (1959) using their refined models for massive stars with very extensive convective cores and an opacity due to electron scattering throughout have reduced the upper limit of stable masses to $M=60\,M_\odot$. The reduction is due to the change in opacity and to the fact that, for a given value of $\bar{\Gamma}_1$, the large convective cores reduce the central condensation of the model and the rate of increase of ξ_a from $r=0$ to $r=R$ as compared to the corresponding standard model.

For « primeval stars », formed of pure H and taking only the p-p chain into account, LEDOUX and BOURY (1960) using again the standard model found a critical mass of the order of $1200\,M_\odot$. But BOURY 1963 has shown that the use of better models again reduces this limit to something like $300\,M_\odot$. As these stars are the seat of a continuous production of ^{12}C (EZER, 1961) through the 3α reaction and since the carbon cycle plays an important rôle in the energy generation, these stars are also the seat of a special kind of secular instability which makes them evolve rather rapidly away from the main sequence with *decreasing* central temperatures. As this goes on, the sensitivity of the carbon cycle to the temperature T which is very small at the very high initial temperature (in fact it is essentially controlled by the two radioactive processes) increases again and it would be very interesting to try to follow this evolution further, as the masses in the range $60\,M_\odot < M < 300\,M_\odot$ must ultimately become vibrationally unstable.

In fact, it is only in the frame of a theory of stellar evolution that the real significance of vibrational instability will appear. For instance, the above critical masses refer to stars in equilibrium on the original main sequence. As SCHWARZSCHILD and HÄRM (1959) remarked, if in the course of the subsequent evolution, the vibrational instability vanishes before it is able to bring the velocity of pulsation in the external layers up to a value of the order of the velocity of escape, it is clear that it has very little importance. In this way, they were able to raise the limiting mass from 60 to $65\,M_\odot$.

Conversely, if a mass greater than M_c starts contracting, vibrational instability will manifest itself before the star reaches the main sequence as soon as a large enough fraction of the energy radiated is due to nuclear reactions but it will start as a very mild instability. As the star goes on contracting, the amplitude of any induced pulsation will increase until its velocity at the surface reaches the velocity of escape and shock waves start pushing shells of material away from the star. It is then very likely that the process will go on until the mass is reduced to M_c and the star can resume its normal evolution.

3·3.3. Effects of the position in the star of the energy-generating region. – Let us note first, that in presence of different types of nuclear reactions in the same star, only those contributing an appreciable fraction of the total flux L can have some influence on its vibrational stability. For instance, all nuclear processes taking place under special conditions (magnetic fields, etc.) in the external layers, although they may be important for the observed relative abundances of the elements, can be neglected in the present discussion.

One of the most obvious way to favour vibrational instability is to move the generation of energy from the centre towards the exterior. This occurs in the course of evolution, when a He-isothermal core gets formed and the nuclear reactions take place in a spherical shell outside it. But this shell is still far too close to the centre when the isothermal core reaches the Chandrasekhar-Schönberg limit to have an appreciable effect. Even for small masses, with a degenerate He-core which may contain a large fraction of the mass (HOYLE and SCHWARZSCHILD, 1955), the conditions don't seem too favourable but no detailed discussion is available.

3'3.4. Effects of the central condensation and of convection. – As far as the internal distribution of mass in the model is concerned, one may discuss its influence roughly in terms of the central condensation $\varrho_c/\bar{\varrho}$. As this ratio increases, ξ (and $\delta\varrho/\varrho$, $\delta T/T$ etc.) increases more and more rapidly from the centre to the surface and this enhances vibrational stability. Inversely, the smaller $\varrho_c/\bar{\varrho}$, the easier it is to bring about vibrational instability.

For ordinary gaseous stars, the physical model with the smallest central condensation is the polytrope $n=\frac{3}{2}$ which can be used to represent a star of fairly small mass ($\beta \to 1$, $\Gamma_1=\gamma=\frac{5}{3}$) in convective equilibrium throughout. BIERMANN (1935) suggested that such stars should be vibrationally unstable however COWLING (1938), taking into account the external radiative layer, found that this was not quite the case, but that the critical ν was reduced nevertheless to a very small value of the order of 9. Thus in presence of the carbon-cycle, ($\nu \simeq 16$), such models would be unstable but all the evidence is that they apply best to small cool stars in which the p-p reaction with a ν of the order of 5 is responsible for the energy generation.

One should note however, that even in that case, the margin of stability is very small and new detailed computations would be welcome. First, they should take into account oui better understanding of the effects of the very external radiative layer which, in fact, might not be able to store any appreciable energy at all and thus might play a lesser rôle than the one estimated by COWLING.

On the other hand, the evaluation of the effects of convection itself on the damping is rather delicate. There are no formal difficulties in using the general eqs. (1.81) and (1.82) if one admits that $(\delta\bar{l}_t/\bar{l}_t)$ is of the order of $\delta r/r$. Actually, for these small cool stars ($\Gamma_1=\gamma=\frac{5}{3}$), one can show (cf. LEDOUX, and WALRAVEN, 1958, Sections **65**, **67**) that the influence of convection in the main interior reduces to adding to the expression (3.7) of σ', a term

$$\frac{1}{2\sigma_a^2 J_a}\int\limits_{M_t}\left(\frac{\delta T}{T}\right)_a\left(\frac{\mathrm{d}\,\delta L_t^*}{\mathrm{d}m}\right)_a \mathrm{d}m = -\,\frac{1}{2\sigma_a^2 J_a} = -\,\frac{1}{2\sigma_a^2 J_a}\int\limits_{M_t}\delta L_t^*\,\frac{\mathrm{d}}{\mathrm{d}m}\left(\frac{\delta T}{T}\right)_a \mathrm{d}m\,,$$

since δL_t^*, as L_t^*, vanishes at the boundaries of the mass M_t in convective equilibrium. But

$$\frac{\delta L_t^*}{L_t^*} = (1-3\gamma)\xi - \left(\gamma + \frac{4}{3}\right) r \frac{\mathrm{d}\xi}{\mathrm{d}r},$$

is positive at compression and thus this extra term always enhances the instability.

However, the validity of this result implies that convection adjusts itself practically instantaneously to the pulsation. This is reasonable as long as the relaxation time of convection is small compared to the period of pulsation. But this is rather unlikely in the kind of stars considered since the period (which is twice the time taken by an acoustic wave to move from the centre to the surface) will rather be shorter than the time taken by a turbulent element to travel a distance of the order of R or an appreciable fraction thereof. In that case, it is usually admitted that the resulting phase-delays in the adjustment of convection during the pulsation will reduce greatly the effects of the former and that it might just be as well to neglect them completely. However even if $\bar{C}_t$ and $\bar{l}_t$ do not adapt themselves to the pulsation, the effects of the other factors in δL_t^* and $\delta\varepsilon_2$ may still be important.

Some preliminary work in Liège (MUKENDI, 1960) suggests that, even in the presence of the p-p reaction, all these factors might reduce the margin of stability of a wholly convective star to very little or even push it on the instability side and we might have to face a rather difficult problem, in the case of the small cool stars, in trying to reconcile our present views on their internal structure and the requirements of vibrational stability.

A last remark should be made concerning convection. We have seen that the coefficient of turbulent viscosity η_t, is much larger than η_G or η_R. While the influence of the last two is quite negligible at least for the first few modes, the damping due to η_t which adds to σ' (eq. (3.7)), a term

$$\text{(3.14)} \qquad \frac{2}{3J_a}\int_{M_t} \frac{\eta_t r^2}{\varrho}\left(\frac{\mathrm{d}\xi}{\mathrm{d}r}\right)^2 \mathrm{d}m\,,$$

might be considerable even for the fundamental mode (cf. COUNSON, LEDOUX and SIMON, 1956) and might stabilize the star. However, here again the strict application of this notation of turbulent viscosity is somewhat delicate.

However, if it applies, one should not loose sight of the fact that, apart from the pure dissipation (3.14), viscosity also transfers momentum from the external layers with relatively large velocities (large ξ) to the central regions (small ξ) tending to make ξ a constant throughout the star, a limiting case in which the dissipation (3.14) vanishes. But this tendency to reduce the variation of ξ from the centre to the surface, modifies the weighting of the different terms in the expression (3.7) of σ' and enhances again the distabilizing rôle of the energy generation by thermonuclear reactions (LONGE, 1956).

3'3.5. The case of white dwarfs. – In the case of white dwarfs, the problem remains essentially the same except that conduction due to the degenerate electrons plays the major part in the transfer of energy. Taking this into account, one can see easily that the combination of the two factors discussed above: $\varrho_c/\bar{\varrho}$ and Γ is very favourable to the setting in of vibrational instability in presence of thermonuclear reactions (LEDOUX and SAUVENIER-GOFFIN, 1950). In the nonrelativistic phases of degeneracy, the effective Γ_1 is still close to $\frac{5}{3}$ but $\varrho_c/\bar{\varrho}$ is low, of the same order as for the convective model (polytrope $n=\frac{3}{2}$). For models of higher and higher mean densities, relativistic degeneracy spreads from the centre outwards and $\varrho_c/\bar{\varrho}$ increases but the mean value of Γ_1 decreases and tends towards $\frac{4}{3}$. Furthermore, in any given configuration, Γ_1 increases from the centre outwards and this also tends to lower further the ratio ξ_R/ξ_c. This always reduces the positive dissipation due to heat conduction relatively to the amplification (negative dissipation) due to the energy generation.

On this basis alone, it appears that thermonuclear reactions are practically incompatible with the stable existence of white dwarfs. Considerations relating to the secular instability of white dwarfs (MESTEL, 1952) also seem to indicate that nuclear reactions must be absent in their interior. Phases immediately preceding the white-dwarf stage are unfortunately little known but it is likely that most of the hydrogen must already have been burnt. According to SCHATZMAN (1945), the rest will diffuse into the external part where the temperature may become high enough to permit nuclear reactions while the interior becomes at least partially degenerate. In these circumstances, it is rather likely that vibrational instability will arise and it may play a rôle in allowing the star to get rid of its last reserve of hydrogen and perhaps even of any mass in excess of Chandrasekhar's critical mass.

3'3.6. Effects of the external layers. – As we have already noted, the influence of the nonadiabatic layer $(M_a - M^*)$ may be important if its heat capacity is large, which is the case when the ionization of H and He takes place there. Then we must use the general expression (2.15) of C to evaluate the contribution say σ'_{Ex} of this region to σ' and one finds

$$\sigma'_{Ex} = \frac{1}{2\sigma_a^2 J_a}\int_{M_a}^{M^*}\left(\frac{\delta T}{T}\right)_a \frac{\mathrm{d}\,\delta L}{\mathrm{d}m}\,\mathrm{d}m\,,$$

since ε_N and $\delta\varepsilon_N$ vanish there and since the denominator is practically not affected by these external layers of very small mass. Of course, δL should now be computed using the general nonadiabatic solution. Rigorously, the latter can only be obtained by solving the complete problem obtained in expressing explicitly δL in (2.4) ($\delta\varepsilon_N = 0$), by means, for instance, of (3.13) if radiative equilibrium prevails there. This leads in general to a fourth-order differential system in complex variables so that altogether a real system of the eighth order has to be solved. However with electronic computers the easiest way is perhaps the direct attack followed recently by COX (1961) and by BAKER and KIPPENHAHN (1962). To be significant, this also requires a fairly realistic equilibrium model for these external layers and this is already a problem in itself. Here we shall only try to stress the qualitative features of the problem.

First, let us recall that the study of vibrational stability was started by EDDINGTON (1926) to find a mechanism susceptible to excite and to maintain the radial pulsations to which the observed variations of the Cepheids are attributed. He immediately recognized that an increase in opacity at contraction in some region of the star (valve mechanism) should be just as efficient to produce instability as an increase in the energy generation and this is true whether it occurs in the adiabatic or the nonadiabatic region. However as we have already seen, the variation of opacity is usually just in the opposite direction throughout the main interior.

On the other hand, the masses of the Cepheids are well below the critical mass M_c above which thermonuclear reactions render the star vibrationally unstable. For the present at least, it is difficult to think of any reasonable model in which the effects of thermonuclear reactions could be enhanced sufficiently to bring about instability. In fact, central condensation should be high in Cepheids and, for such a model, COX (1955) found that it would take $2 \cdot 10^9$ years for the nuclear reactions to amplify the amplitude by a factor e while the damping time due to radiative conductivity in the interior is only of the order of 10 days.

Thus it seems that the source of the instability at the origin of the Cepheids' pulsation must be looked for in the external layers. After the large predominance of H had been established, EDDINGTON (1941, 1942) recognized that layers in which its ionization was in progress would have a large heat capacity (high values of C_v and low values of the Γ's). But, for H, this layer is very external (say towards the top of the region: $R_a - R^*$) and the nonadiabatic terms are likely to be largely dominant. In that case, and using drastic simplifications (for a critical discussion cf. LEDOUX and WHITNEY, 1961), EDDINGTON came to the conclusion that the main effect would be a phase-shift by about 90° of δL through the considered region. This would have two consequences. First, it would just about make all dissipation due to radiative conductivity vanish so that the star would be unstable due to the thermonuclear reactions (but let us recall that EDDINGTON dit not realize at the time how weak this effect is in Cepheids). Secondly, as the corresponding phase-shift in ξ is very small, this would at the same time explain the observed phase relation between the light and velocity curves. Physically the mechanism is quite sound since heat is accumulated in these layers in the form of ionization energy during the contraction. Unfortunately, it seems that the ionization zone of H is really a little too far out, to possess the required heat capacity (for some estimates cf. LEDOUX and WALRAVEN, 1958, Section **69**).

However, since around 1950, ZHEVAKHIN has been extending these considerations to the ionizations of He. As these occur deeper in the star, the non-adiabatic terms are somewhat less important but the Γ's being very low, the values of $(\delta T/T)_a$ corresponding to a given value of $\delta\varrho/\varrho$ are rather small and this tends, even in the case of Kramers law, to reverse the sign of the opacity variations in the course of the pulsation providing a « valve mechanism » much more similar to the original one advocated by EDDINGTON. Furthermore, these ionizations modify the opacity which now tends to increase with both ϱ and T in the critical region and this reinforces strongly the efficiency of the « valve ». In particular, ZHEVAKHIN found, on the basis of a rather crude model, that, if the transport of energy in the layer where the ionization of He takes place is still mainly by radiation, this effect could bring about an appreciable instability and this has been confirmed by the much more elaborated discussions of COX (1961) and of BAKER and KIPPENHAHN (1962).

Let us note first that this instability could never have drastic consequences for the star since, if the contraction goes as far as ionizing completely all He and H in $(R_a - R^*)$ the source of instability itself vanishes. This means that the excited pulsation will have a finite limit and, of course, this is exactly what we need to explain variable stars. Naturally, other factors may also play a rôle in the limitation of the amplitude.

Furthermore, the papers referred to above suggest that this effect would be especially important in the region occupied by the Cepheids in the Hertzsprung-Russell diagram.

However there is a difficulty in pushing the source of instability in the very external layers. Indeed, we have seen that if ξ is normalized to unity at the centre, it reaches higher and higher values near the surface as we go to higher and higher modes. Thus one should expect that the first few higher modes should be excited at least as strongly as the fundamental one. One may think of other damping factors (viscosity-progressive waves) which tend to affect these higher modes more strongly than the fundamental one. However I doubt that these differences could really eliminate completely the higher modes which apparently are not at all present in ordinary Cepheids.

3'3.7. The case of very high modes. – Talking of higher modes, the behaviour of the amplitude of very high modes described in Section 3'1 may also have some interesting implications. There is even a remote possibility that these modes could become excited by the energy generation at the centre alone and this probability is greatly increased if the source of instability occurs in the very external regions. But a detailed discussion is certainly needed to settle this. However it suggests certainly that the variations of energy generation due to any short-period perturbation of the central region have a tendency to get dissipated only close to the surface and thus, there may be a closer correlation between short-lived fluctuations at the surface and deep-seated phenomena than has been usually assumed.

3'3.8. Phase-delays in energy generation. – Phase-delays in energy generation (cf. ROSSELAND, 1949; LEDOUX and WALRAVEN, 1958) during the pulsation don't have very important consequences in many of the most usual cases. However, in general, the coefficients μ_e and ν_e in

$$\frac{\delta\varepsilon}{\varepsilon} = \mu_e \frac{\delta\varrho}{\varrho} + \nu_e \frac{\delta T}{T}, \tag{3.15}$$

are not necessarily equal to the exponents of ϱ and T in (2.24) at equilibrium. In some critical cases, the determination of the exact values of μ_e and ν_e may be a somewhat lengthy process (for a detailed example, cf., for instance, LEDOUX and BOURY, 1960) however the following remarks are helpful to decide whether or not a detailed discussion is needed. If the mean lifes of all elements entering the reaction are short as compared to the period of the pulsation considered, μ_e and ν_e can be taken equal to their values in the static model. On the other hand, if the mean-life of an element is long compared to the period, the variation of its abundance will exhibit a phase-shift of the order of $\pi/2$ which does not affect σ' (but rather σ: very little) and we may treat its abundance as constant. If this is true of all the elements entering the considered chain of reactions $\mu_e = \mu$ and ν_e is a mean value of the different ν_{ij} weighted with respect to the energy Q_{ij} released by the reaction (i, j). This is the case, for instance, of all collision processes in the C-N cycle in ordinary stars. Since the two disintegrations liberate a relatively small fraction of the total energy and since the ν_{ij} corresponding to the different collision processes are not very different from each other μ_e and ν_e remain close to their equilibrium values. It is only at very high temperature that a detailed discussion may prove necessary. On the other hand, in the p-p chain, the mean-life of ^{2}H is very short, in general, compared to any pulsation period and its abundance will vary in phase with the pulsation making ν_e somewhat greater than ν. In exceptional circumstances (SCHATZMAN, 1951), $(\nu_e - \nu)$ and $(\mu_e - \mu)$ may become important in this chain.

The 3α-reaction deserves a special discussion since the first step in the reaction is practically in equilibrium while the second step is determined by a resonance.

3·3.9. Effects of progressive waves. – The exact structure of the external layers of a star is a very complex problem and it may happen that, for instance due to the presence of a chromosphere or a corona, it does not insure the perfect reflection which we have postulated up to now (boundary conditions (2.10) violated). In that case, the oscillations will take there a more or less marked progressive character and the wave energy escaping or destroyed in the surrounding medium will contribute to the damping of the pulsation.

The usual linearized treatment of infinite idealized atmospheres (ROSSELAND, 1949, Section 6·4; SIMON, 1959 *a*; LEDOUX and WALRAVEN, 1958, Section **68**) is not too satisfactory. Schatzman's model (1956) of a finite isothermal layer at temperature T supporting an infinite isothermal atmosphere at temperature T' $(T' > T)$ is perhaps one of the most useful. Let us denote by ϱ and ϱ' the densities on both sides of the interface. Then the correction to σ' due to the progressive waves leaking across this interface is

$$\sigma'_{\text{progr}} = -\frac{1}{2}\frac{\Delta K}{K_M},$$

where

$$\Delta K = -2\pi R^2 c' \left[\tfrac{1}{2}\varrho'\sigma^2(\delta r)^2_R\right],$$

represents the average loss of energy per second, K_M, the maximum kinetic energy of the pulsation

$$K_M = \frac{\sigma^2}{2}\int_0^M (\delta r)^2 \, \mathrm{d}m \,, \tag{3.16}$$

and c' the velocity of sound in the external region.

If, as SCHWARZSCHILD and HÄRM (1959), we replace in (3.16) ϱ' by the photospheric density, we obtain a rather high upper limit to the value of σ'_{progr}. There is no doubt that if vibrational instability is present, this effect may become important when the amplitude increases and nonlinear terms must be taken into account.

3·4. *Vibrational stability towards nonradial oscillation.* – The coefficient of vibrational stability relative to the k-mode of the oscillation represented by the spherical harmonic of degree l and order m (SIMON, 1957 *b*) is independent of m and can be written, neglecting viscosity and convection.

$$\sigma'_{k,l} = -\frac{1}{2(\sigma_{k,l})^2}\,\frac{\int(\delta T/T)_{k,l}\{\varepsilon_N - (1/\varrho)\operatorname{div}\boldsymbol{F}_R\}'_{k,l}\,\varrho r^2 \sin\theta \,\mathrm{d}r\,\mathrm{d}\theta\,\mathrm{d}\varphi}{\int(\delta\boldsymbol{r})_{k,l}\cdot(\delta\boldsymbol{r})_{k,l}\,\varrho r^2\sin\theta\,\mathrm{d}r\,\mathrm{d}\theta\,\mathrm{d}\varphi}, \tag{3.17}$$

where all the terms are to be evaluated by means of the corresponding adiabatic solution. It is advantageous here to keep to the Eulerian variation, denoted by a prime, for the bracket.

As we have mentioned earlier (Section **3**·2), ϱ' and T' (or $\delta\varrho$ and δT) vanish at the centre for nonradial oscillations and one must expect that, as long as the energy generation is concentrated there, its effects on the vibrational stability will be smaller than for radial pulsations. This has been confirmed quantitatively by SIMON who studied numerically the case of the p-mode, $l=2$, $k=0$, for the standard model. I don't remember any other detailed application. Such applications also to g-modes and to models with energy-generation at some distance of the centre would be very welcome.

3·5. *Secular stability.* – As we have seen, in the case of dynamical and vibrational stability, we have at our disposal clear-cut criteria and many of the factors affecting these criteria have already been the subject of extensive discussions. But the study of secular stability has been rather neglected up to now.

Its study seems to have been initiated by a remark of RUSSELL at a time when very little was known on nuclear reactions in stars except that they should provide a much longer life-time for ordinary stars than that allowed by the Helmholtz-Kelvin contraction scheme. RUSSELL pointed out that these reactions should depend on ϱ and T in such a way that any excess (defect) of the luminosity with respect to the energy generated can be compensated by a small contraction (expansion) and the resulting increase (decrease) in ϱ and T. If this condition is not verified, then, although all the static equilibrium equations might be satisfied, any small perturbation will *start* the star contracting or expanding at a rate defined by the gravitational time-scale (I.1).

The criterion (2.28) which expresses this condition for a gaseous star in the case of an homologous perturbation was already established by RUSSELL and JEANS and, as far as I know, we are still practically at the same point.

However there does not seem to be any obvious reason why the radial perturbation considered should be homologous. Not only could the relative amplitude $(\delta r/r)$ vary from one point to the next but it could also change sign in the star. In fact, as it has been recalled on many occasions in this course, some of the most interesting phases in stellar evolution according to our present views, are those in which the core contracts and the envelope expands and it would certainly be natural to test secular stability for this type of motions as well.

Unfortunately, the problem seems difficult. We have already mentioned in the Introduction the line of approach which would consist in generalizing Poincare's theory of linear series but we have also pointed out there some of the difficulties encountered in that direction.

On the other hand, if one follows THOMAS who started from the energy principle using the entropy as the dependent variable one is led to a rather difficult integro-differential problem which was reviewed in some detail in my paper on *Stellar Stability* (LEDOUX, 1958). It was also suggested there that progress in the solution of this problem might be possible by means of series development in terms of the eigen-solutions $\xi_{a,i}$ of the adiabatic pulsation problem (3.1). However since then, a few attempts in that direction have proved rather fruitless as the amount and the complications of the implied algebraic and numerical work becomes rapidly prohibitive.

One may also go back to the general problem (2.1) to (2.8) in which the second and higher-time derivatives may now be neglected. Then eq. (2.8) in which the first member is neglected may be integrated once from any level r to the surface. Taking the second boundary condition (2.10) into account, this yields (LEDOUX, 1960)

$$s_1\left[\frac{\Gamma_1 p}{r^2}\frac{\mathrm{d}}{\mathrm{d}r}\left(\frac{r^3}{p_r}\frac{\mathrm{d}\alpha}{\mathrm{d}r}\right)-4\alpha\right]=(\Gamma_3-1)\varrho\left(\delta\varepsilon_N-\frac{\mathrm{d}\,\delta L}{\mathrm{d}m}\right), \tag{3.18}$$

where

$$\alpha=-\int_r^R \xi\,\frac{\mathrm{d}p}{\mathrm{d}r}\,\mathrm{d}r\,,$$

and p_r denotes $\mathrm{d}p/\mathrm{d}r=-Gm(r)\varrho/r^2$.

Equation (2.2), in which the first member is again neglected, may be written, using (2.1), (2.5) and (2.7)

$$\frac{1}{\varrho}\frac{\mathrm{d}\,\delta p}{\mathrm{d}r}-4\xi\,\frac{Gm(r)}{r^2}=0\,, \tag{3.19}$$

after reintroducing δp instead of p' by means of (1.1). From the equation of state (1.19) and the definitions of $\beta=p_G/p$, we have:

$$\frac{\delta p}{p}=\beta\,\frac{\delta\varrho}{\varrho}+(4-3\beta)\,\frac{\delta T}{T}\,. \tag{3.20}$$

Eliminating δp in (3.19) by means of (3.20) and integrating from any r to R, we get

$$\frac{\delta T}{T}=-\frac{\beta}{4-3\beta}\,\frac{\delta\varrho}{\varrho}-\frac{4\alpha}{(4-3\beta)p}\,,$$

while (2.1) gives immediately

$$\frac{\delta\varrho}{\varrho} = -\frac{1}{r^2}\frac{\mathrm{d}}{\mathrm{d}r}(r^3\xi)\,.$$

These values may now be substituted in the expressions (3.12) and (3.13) of $\delta\varepsilon_N$ and δL to evaluate the second member of (3.18). This yields a fourth-order differential equation for α

$$(3.21)\qquad s_1 p\left[\frac{\Gamma_1}{r^2}\frac{\mathrm{d}}{\mathrm{d}r}\left(\frac{r^3}{p_r}\frac{\mathrm{d}\alpha}{\mathrm{d}r}\right)-\frac{4\alpha}{p}\right]=$$

$$= (\Gamma_3-1)\varrho\varepsilon_N\left[\left(\frac{\beta\nu}{4-3\beta}-\mu\right)\frac{1}{r^2}\frac{\mathrm{d}}{\mathrm{d}r}\left(\frac{r^3}{p_r}\frac{\mathrm{d}\alpha}{\mathrm{d}r}\right)-\frac{4\nu\alpha}{(4-3\beta)p}\right]-$$

$$-\frac{(\Gamma_3-1)}{4\pi r^2}\frac{\mathrm{d}}{\mathrm{d}r}\left\{\frac{4L(r)}{p_r}\frac{\mathrm{d}\alpha}{\mathrm{d}r}+L(r)\left[\left(\frac{(4+n)\beta}{4-3\beta}+m\right)\frac{1}{r^2}\frac{\mathrm{d}}{\mathrm{d}r}\left(\frac{r^3}{p_r}\frac{\mathrm{d}\alpha}{\mathrm{d}r}\right)-\frac{4(4+n)}{4-3\beta}\frac{\alpha}{p}\right]+\right.$$

$$\left.+\frac{L(r)T}{(\mathrm{d}T/\mathrm{d}r)}\frac{\mathrm{d}}{\mathrm{d}r}\left[\frac{\beta}{4-3\beta}\frac{1}{r^2}\frac{\mathrm{d}}{\mathrm{d}r}\left(\frac{r^3}{p_r}\frac{\mathrm{d}\alpha}{\mathrm{d}r}\right)-\frac{4\alpha}{(4-3\beta)p}\right]\right\},$$

which should be solved subject to the boundary conditions

$$\frac{\mathrm{d}\alpha}{\mathrm{d}r}=0 \qquad \text{at } r=0\,,$$

$$\alpha=0\,,\qquad \frac{\mathrm{d}\alpha}{\mathrm{d}r}=0\,,\qquad \text{and}\qquad \frac{\mathrm{d}\,\delta L}{\mathrm{d}r}=0\qquad \text{at } r=R\,.$$

The general properties of this boundary-value problem have not been discussed and it is difficult to say anything of its solutions. It is possible to express s_1 as an integral over the whole star but this does not seem very useful in the general case. However if one assumes again an homologous perturbation (cf. (2.26)), one recovers immediately the usual condition (2.28). However the meaning of this criterion can only be very limited since for $\xi=$ ct., eq. (3.21) reduces to

$$s_1(3\Gamma_1-4)p=(\Gamma_3-1)\varrho\varepsilon_N[(n-3m)-(3\mu+\nu)],$$

which, apart from the critical case $s=0$, $(n-3m)=(3\mu+\nu)$ everywhere, cannot be verified simultaneously at each point in the star.

There are cases too where a type of secular instability may not manifest itself most directly in the form of motions. For instance, in the case of white dwarfs, MESTEL (1952) has pointed out that, as the pressure which is mainly due to the degenerate electrons depends very little on the temperature, the regulating mechanism between energy generation due to nuclear reactions and energy radiated is lacking. In this case, if at any time the energy generation exceeds the luminosity, the temperature rises accelerating further the reaction rate while the pressure, at least in a first approximation, does not change. This must lead to a temperature increasing extremely fast unless some new mode of energy transfer comes into play. However the dynamical effects will remain negligible until T becomes large enough to lift the degeneracy. This is usually taken as excluding the possible existence of nuclear fuel in stable white dwarfs. The same mechanism appears to be responsible for the very fast and difficult phases of stellar evolution occuring towards the top of the red giant sequence in Schwarzschild and Hoyle's theory (SCHWARZSCHILD and HÄRM, 1962).

Neutrino emission, as we have mentioned at the end of Section (**1**.3), also induces strong instabilities which at the very beginning at least, must be of the secular type. However the rate of energy loss may be so large that certainly the usual approach adopted here (neglect of higher time-derivatives) is no longer justified.

On the other hand, it is obvious that, in presence of very violent sources of instability, the linear methods will never be able to describe the phenomena very far, but coupled with the careful study of evolutionary sequences of stellar models, and applied at the phases where these instabilities just appear, they might still provide the best guide to the real trend of further developments.

BIBLIOGRAPHY

BAKER, N. and KIPPENHAHN, R.: 1962, *Z. f. Astrophys*, **54**, 114.

BIERMANN, L.: 1953, *Astr. Nach.*, **257**, 269.

BIERMANN, L.: 1939, *Z. f. Astrophys.*, **18**, 344.

BIERMANN, L. and COWLING, T. G.: 1939, *Z. f. Astrophys.*, **19**, 1.

BOURY, A. and Mme HUSTIN-BRETON,: 1961, *Bull. Ac. Roy. de Belgique, Cl. des Sc.*, 5ème série, **47**, 543.

BOURY, A.: 1963, *Ann. d'Astrophys.*, **26**, 354.

BURBIDGE, E. M., BURBIDGE, G. R., FOWLER, W. A. and HOYLE, F.: 1957, *Rev. of Mod. Phys.*, **29**, 547.

CAMERON, A G. W.: 1960, *IXème Colloque Internat. d'Astrophys. de Liège, Mém. Soc. Roy. des Sc. de Liège*, 5ème sér., **3**, 163.

CAMERON, A. G. W.: 1926, *Icarus*, **1**, 13.

CHANDRASEKHAR, S.: 1961, *Hydrodynamic and Hydromagnetic Stability*, London.

CHIU, H. Y.: 1961, *Phys. Rev.*, **123**, 1040.

COUNSON, J., LEDOUX, P. and SIMON, R.: 1956, *Bull. Soc. Roy. Sc. de Liège*, **25**, 144.

Cox, J. P.: 1955, *Ap. J.*, **122**, 286.
Cox, J. P.: 1961, *Ap. J.*, **132**, 594.
Cox, A. N. and Brownlee, R. R.: 1960, *IXème Colloque Internat. d'Astrophys. de Liège, Mém. Soc. Roy. Sc. de Liège*, 5ème série, **3**, 469.
Cowling, T. G.: 1936, *Month. Not. R.A.S.*, **96**, 42.
Cowling, T. G.: 1938, *Month. Not. R.A.S.*, **98**, 528.
Cowling, T. G.: 1941, *Month. Not. R.A.S.*, **101**, 367.
Eddington, A. S.: 1926, *The Internal Constitution of the Star*, sect. 132, Cambridge.
Eddington, A. S.: 1941, *Month. Not. R.A.S.*, **101**, 182.
Eddington, A. S.: 1942, *Month. Not. R.A.S.*, **102**, 154.
Ezer, D.: 1961, *Ap. J.*, **133**, 159.
Garcet, M.: 1961, *Mémoire de Licence*, Université de Liège.
Härm, R. and Schwarzschild, M.: 1958, *Ap. J.*, **128**, 348.
Hazlehurst, J. and Sargent, W. L. W.: 1960, *Ap. J.*, **130**, 276.
Henyey, L. G., Wilets, L., Böhm, K. H., Le Levier, R. and Levee, R. D.: 1959, *Ap. J.*, **129**, 628.
Hoyle, F., 1946, *Month. Not. R.A.S.*, **106**, 343.
Hoyle, F. and Schwarzschild, M.: 1955, *Ap. J. Suppl.*, n. 13, **2**, 1.
Jeans, J.: 1928, *Astronomy and Cosmology*, par. 108, Cambridge.
Kopal, Z.: 1949, *Ap. J.*, **109**, 509.
Ledoux, P.: 1940, *Astrophys. Norv.*, **3**, 193.
Ledoux, P. and Pekeris, C. L.: 1941, *Ap. J.*, **94**, 124.
Ledoux, P.: 1941, *Ap. J.*, **94**, 537.
Ledoux, P.: 1949, *Contribution à l'étude de la structure interne des étoiles*, in *Mém. Soc. Roy. Sc. de Liège*, **9**, Chap. IV and V.
Ledoux, P. and Sauvenier-Goffin, E.: 1950, *Ap. J.*, **111**, 611.
Ledoux, P. and Walraven, Th.: 1958, *Handbuch der Physik* (ed. S. Flügge), **51**, 353
Ledoux, P.: 1958, *Stellar Stability, Hdb. der Phys.*, (ed. S. Flügge), **51**, 605.
Ledoux, P. and Boury, A.: 1960, *IXème Colloque Internat. d'Astrophys. de Liège, Mém. Soc. Roy. Sc. de Liège*, 5ème série, **3**, 298.
Ledoux, P.: 1960, *Bull. Acad. Roy. de Belgique, Cl. des Sc.*, 5ème série, **46**, 429.
Ledoux, P. and Whitney, C. A.: 1961, n° 1, *Suppl. Nuovo Cimento*, **22**, 131.
Ledoux, P.: 1962, *Bull. Acad. Roy. de Belgique, Cl. des Sc.*, 5ème sér., **48**, 240.
Limber, D. N.: 1958, *Ap. J.*, **127**, 363, 387.
Longe, P.: 1956, *Bull. Soc. Roy. Sc. de Liège*, **25**, 541.
Martin, Y.: 1962, *Mémoire de Licence*, Université de Liège.
Mestel, L.: 1952, *Month. Not. R.A.S.*, **112**, 583.
Ottelet, I.: 1960, *Ann. d'Astrophys.*, **23**, 218.
Pekeris, C. L.: 1938, *Ap. J.*, **88**, 189.
Rosseland, S.: 1931, *Univ. Obs. Oslo*, Publ. n. 1.
Rosseland, S.: 1932, *Univ. Obs. Oslo*, Publ. n. 2.
Sauvenier-Goffin, E.: 1949, *Mém. Soc. Roy. Sc. de Liège*, 4ème série, **10**, 1.
Sauvenier-Goffin, E.: 1950, *Bull. Soc. Roy. Sc. de Liège*, **19**, 47.
Sauvenier-Goffin, E.: 1951, *Bull. Soc. Roy. Sc. de Liège*, **20**, 20.
Schatzman, E.: 1951, *Ann. d'Astrophys.*, **14**, 305.
Schatzman, E.: 1956, *Ann. d'Astrophys.*, **19**, 45.
Schatzman, E.: 1958, *White Dwarfs*, Amsterdam.
Schwarzschild, M. and Härm, R.: 1959, *Ap. J.*, **129**, 637.
Schwarzschild, M. and Härm, R.: 1962, *Ap. J.*, **136**, 158.
Skumanich, A.: 1955, *Ap. J.*, **121**, 408.

SIMON, R.: 1957*a*, *Bull. Acad. Roy. de Belgique, Cl. des Sc.*, 5ème série, **43**, 471.
SIMON, R.: 1957*b*, *Bull. Acad. Roy. de Belgique, Cl. des Sc.*, 5ème série, **43**, 610.
SIMON, R.: 1962, A & ES Report 62-1, Purdue University.
SOUFFRIN, P.: 1960, *IXème Colloque Internat. d'Astrophys. de Liège, Mém. Soc. Roy. des Sc. de Liège*, 5ème série, **3**, 245.
THOMAS, L. H.: 1930*a*, *M.N.R.A.S.*, **91**, 122 and 619.
THOMAS, L. H.: 1930*b*, *Quart. J. of Math.*, **1**, 239.
TOLMAN, R. C.: 1939, *Ap. J.*, **90**, 541, 568.
UNSÖLD, A.: 1930, *Z. f. Astrophys.*, **1**, 138.
WASIUTYNSKI, J.: 1946, *Astrophys. Norv.*, 4 (chap. I).
ZHEVAKHIN, S. A.: 1960, *Astron. Journ. U.S.S.R.*, **36**, 268, 394, 996; many references are given there to its extensive earlier work on the subject.

Stellar Hydromagnetics.

I. W. Roxburgh

Department of Applied Mathematics and Theoretical Physics
University of Cambridge - Cambridge, England

1. - Introduction.

No detailed discussion of the strength and structure of the magnetic fields to be expected inside stars is yet available. The interstellar field is of order 10^{-6} G and if a contracting proto-star traps this field when its radius is of order a parsec, then the resulting field will be of order 10^8 G, with magnetic and gravitational energies comparable. Whether the freezing of the field into the material of the contracting proto-star is exact, or not, is still a matter of debate, although the work of Mestel and Spitzer (1956) suggests that freezing may not occur until the proto-star is much less than a parsec in diameter. Even so, a newly formed star may be expected to possess a strong large-scale magnetic field, as a relic of the intergalactic magnetic field.

A difficulty of this « fossil » theory of the origin of stellar magnetic fieldsy is the recently discovered Hayashi (1961) phase of contraction, where the star becomes completely convective. The turbulence may so tangle the lines of force that the field quickly decays to zero. Alternatively, a turbulent dynamo may operate (Batchelor, 1950) maintaining a strong magnetic field, with a large-scale component of an unknown structure. Again, even if a dynamo does not operate, the turbulence may only push the lines of force to the surface, in the way envisaged by Spitzer (1957), and leave the external dipole component unchanged; with the subsequent decay of the Hayashi phase, the field lines may diffuse from the surface layers into the rest of the star. In the absence of any definite discussion of the interaction of magnetic fields and turbulence, no conclusion can be reached.

In stellar conditions, the decay time of a large-scale magnetic field in the radiative regions of stars is of order 10^9 to 10^{10} years (Wrubel, 1952; Cowling, 1953), so any field left as a result of the stars formation, either a « fossil field » or a « turbulent dynamo field » will be present for at least the main-sequence stage of stellar evolution, and may therefore have important effects on the internal structure of stars.

Another possibility is that magnetic fields may be maintained by a convective dynamo in the core of a star, by a mechanism similar to the one proposed by BULLARD for the maintenance of the earth's magnetic field (BULLARD and GELLMAN, 1954). No discussion of this possibility is yet available.

There remains the mechanism proposed by BIERMANN (1950), that the partial pressure of the electron gas in a rotating star acts like a « battery », driving electric currents which maintain a magnetic field around the axis of rotation. BIERMANN found that large magnetic fields could be maintained by this mechanism; even in the sun, which is rotating very slowly, a field of 500 G could easily be maintained. This « battery » effect of the electron partial pressure has recently been re-examined by MESTEL and ROXBURGH (1962), and they find that if the star originally possesses a weak poloidal field, as a relic of its formation, then the « battery » is ineffective. A poloidal field of 10^{-4} G is sufficient to reduce the field in the sun from 500 to 10^{-3} G.

Let us assume that stars possess internal magnetic fields and examine their effect on the internal structure.

2. – Structure of magnetic stars.

We shall confine our attention to the radiative zones of an upper main sequence star, and assume that all the energy generation takes place in the convective core. The star is assumed to be rotating with angular velocity Ω, and to have a magnetic field $\boldsymbol{H}$.

The centrifugal and magnetic fields produce nonspherical forces which will, in general, drive meridian circulation currents in the manner already described by KIPPENHAHN (1962). In the absence of circulation, the angular velocity and magnetic fields interact, tending to set up isorotation (FERRARO, 1937). If there is a maintained meridional circulation, the moving elements convect angular momentum, causing a strong variation along a poloidal field line, and so generating an extra toroidal component of field. The toroidal magnetic force alters the angular momentum of the circulating elements, and the change in the toroidal field depends not only on the nonuniform rotation along a poloidal field line, but also on the convection of the toroidal field by the circulation. A discussion of the interaction process is very difficult and we shall assume that a steady state has been reached, and examine the distribution of the angular velocity and magnetic fields.

If we consider the material of the star to be a simple perfect conductor, then the velocity $\boldsymbol{V}$, the magnetic field $\boldsymbol{H}$, and the electric field $\boldsymbol{E}$, are related by the equation:

$$E + \frac{V \wedge H}{c} = 0 \,. \tag{1}$$

If we write the magnetic and velocity fields as:

$$H = H_p + H_t \,, \qquad V = V_p + \Omega \varpi t \,, \tag{2}$$

where ϖ is the distance from the axis of symmetry of the star, t the unit toroidal vector, the subscript p and t denoting the poloidal and toroidal components respectively, then in a steady state the electric field is irrotational and eq. (1) gives

$$\nabla \wedge (\Omega \varpi t \wedge H_p + V_p \wedge H_t + V_p \wedge H_p) = 0 \,. \tag{3}$$

From our assumption of axial symmetry, and the condition that the magnetic field is solenoidal, eq. (3) gives (MESTEL, 1961):

$$V_p = k H_p \,, \qquad \Omega - \frac{k H_t}{\varpi} = \alpha \,, \tag{4}$$

where α is constant along a line of H_p, and k is a scalar. If we substitute the expression for V_p in the equation of continuity

$$\nabla \cdot (\varrho V_p) = 0, \tag{5}$$

we obtain

$$\varrho k = \eta \,, \tag{6}$$

where η is again constant along a field line of H_p.

Moreover the condition for a steady-state solution is that there is no toroidal component of force, the Coriolis force is balanced by the magnetic torque:

$$\frac{1}{\varpi} V_p \cdot \nabla \Omega \varpi^2 = \frac{j_p \wedge H_p}{c \varrho} \,, \tag{7}$$

which after manipulation yields:

$$H_t - 4\pi \varrho k \Omega \varpi^2 = \beta \,, \tag{8}$$

where again β is constant along a line of H_p. Combining eqs. (4)–(8) we obtain:

$$\Omega = \frac{(\alpha + \eta\beta/\varrho\varpi^2)}{(1 - 4\pi\eta^2/\varrho)}, \qquad H_t = \frac{\beta/\varpi + 4\pi\eta\alpha\varpi}{(1 - 4\pi\eta^2/\varrho)} \,. \tag{9}$$

Equations (9) determine the variation of Ω and H_t around a poloidal field line. A detailed derivation is given by MESTEL (1961).

3. – A detailed model.

So far we have ignored the problem of the generation of the circulation currents $\boldsymbol{V}_p$. In principle, we could introduce a stream function S for the unknown circulation; prescribe the functions $\alpha(S)$, $\beta(S)$, $\eta(S)$ (so ensuring that they are constants along the stream lines); and then introduce the expression for $\boldsymbol{H}_p$, $\boldsymbol{H}_t$ and Ω into the equations of thermal equilibrium, as explained by KIPPENHAHN in a previous lecture, the resulting set of simultaneous equations could then be solved for S. In practice this is not feasible and we can only obtain consistent solutions under simplifying assumptions (ROXBURGH, in press).

Let us consider the case of a star with just a toroidal component of magnetic field $\boldsymbol{H}_t$. Equation (3) must now be replaced by

$$\nabla\wedge(\boldsymbol{V}_p\wedge\boldsymbol{H}_t) = 0 \,, \tag{10}$$

which, on using the equation of continuity (5), yields:

$$\boldsymbol{V}_p\cdot\nabla\left(\frac{H_t}{\varrho\varpi}\right) = 0 \,, \tag{11}$$

and hence

$$H_t = \gamma\varrho\varpi \,, \tag{12}$$

with γ constant along stream lines of the circulation. The condition that $\boldsymbol{H}_p$ and Ω be zero is, of course, unnecessarily strong, all we require is that the energy-density of the toroidal component of field be much greater than that of the poloidal components and the centrifugal field. If the poloidal component of field is sufficiently small so that

$$\frac{4\pi\eta^2}{\varrho} = \frac{4\pi\varrho V_p^2}{H_p^2} = \frac{V_p^2}{(V_{\text{Alfvén}})^2} \gg 1 \,, \tag{13}$$

(*i.e.* the flow is everywhere « super Alfvén »), then with β sufficiently small eq. (9) reduces to eq. (12), with γ constant along the coinciding stream lines and poloidal field lines.

If we further simplify by taking γ constant over the whole star, and so *a fortiori* constant on stream lines whatever the form of the stream lines, then the magnetic-body force is derivable from a potential

$$(14)\qquad \frac{\boldsymbol{j}_p \wedge \boldsymbol{H}_t}{c\varrho} = -\nabla\left(\frac{\gamma^2 \varrho \varpi^2}{4\pi}\right).$$

If we now assume that the star is not grossly perturbed from the zero-order spherically symmetric state, then the density in the expression (14) can be replaced by the zero-order density, and the potential of the magnetic force can be expressed as

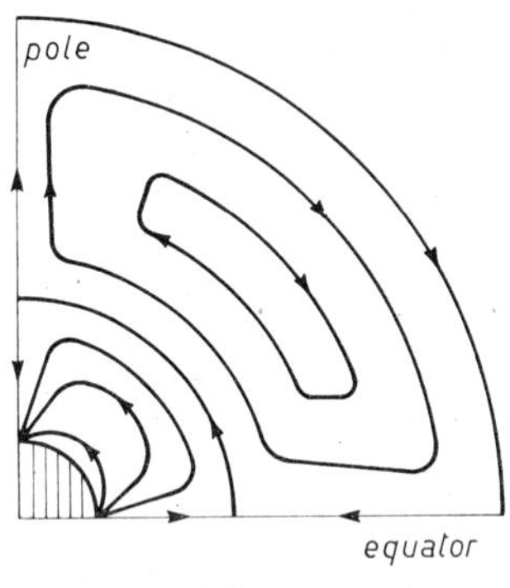

Fig. 1.

$$(15)\qquad \Phi = a_0(r) + a_2(r)\, P_2(\cos\theta)\ ;$$

where $P_2(\cos\theta)$ is the second Legendre polynomial, and (r, θ, φ) are spherical polar co-ordinates with origin at the centre of star and $\theta = 0$ the rotation axis. The circulation currents due to such a perturbing potential can be easily evaluated by a first order perturbation technique (BAKER and KIPPENHAHN, 1959) and detailed calculations have been made, taking a Kushwaha model of 2.5 solar masses in initial main sequence state as the unperturbed star (KUSHWAHA, 1957; ROXBURGH, in press). The resulting stream lines are shown in Fig. 1. The radial component of velocity is given by:

$$(16)\qquad V_r = \varepsilon \frac{LR^2}{GM^2} P_2(\cos\theta)\, V(x)\,,$$

where $V(x)$ is a tabulated function of order 5 over most of the star, but becoming infinite near the surface, and

$$(17)\qquad \varepsilon = \frac{3\gamma^2}{16\pi^2 G}\,,$$

is of order the ratio of the magnetic force to the gravitational force. For the model star used in the integration we find

$$(18)\qquad V_r \sim 10^{-2}\varepsilon \text{ cm/s}\,.$$

4. – Mixing between the core and envelope.

If we accept Mestel's estimate for the efficiency of mixing, including the effects of the inhomogeneities of molecular weight (MESTEL, 1953), then the ratio of the perturbing force to gravity must be greater than 1/30 at the boundary between the convective core and the radiative envelope. With the magnetic field (12) the approximate value of this ratio is given by the ratio of the two potentials, this is

$$\delta \sim U\varepsilon \,, \tag{19}$$

where U is the homology variable $4\pi\varrho r^3/M_r$, and this decreases from 3 at the centre to zero at the surface. Consequently, imposing a condition on δ at the core surface does not cause difficulties at any other point of the star; in contrast with the case of solid body rotation, where the centrifugal and gravitational force must become comparable at the surface, before the ratio of the two forces is greater than 1/30 at the core.

We conclude by remarking that the field (12) vanishes at the surface of a star ($\varrho = 0$), and so cannot be directly observed, and therefore cannot be eliminated on observational grounds. If a star possesses a toroidal magnetic field, with energy-density much greater than that of both the poloidal component and the centrifugal field, than it is possible that a homogeneous, rather than an inhomogeneous, model of evolution may be the required one.

BIBLIOGRAPHY

BACHELOR, G. K.: 1950, *Proc. Roy. Soc.* **201**, 405.
BAKER, N. and KIPPENHAHN, R.: *Zs. f. Ast.*, 1959, **48**, 140.
BIERMANN, L.: 1950, *Zs. f. Nat.*, **5** a, 65.
BULLARD, E. and GELLMAN, H.: 1954, *Phil. Trans. Roy. Soc.*, **247**, 213.
COWLING, T. G.: 1953, *Solar Electrodynamics*, in the Sun, ed. KUIPER, Chicago.
FERRARO, V. A. C.: 1937, **97**, 458.
HAYASHI, C.: 1961, *Pub. Ast. Soc. Japan*, **13**, 450.
KIPPENHAHN, R.: 1962, This volume p. 330.
KUSHWAHA, R.: 1957, *Ap. J.*, **125**, 242.
MESTEL, L.: 1953, *M. N.*, **113**, 716.
MESTEL, L.: 1961, *M. N.*, **122**, 473.
MESTEL, L. and ROXBURGH, I. W.: 1962, *Ap. J.*, **136**, 615.
MESTEL, L. and SPITZER, L.: 1956, *M. N.*, **116**, 503.
ROXBURGH, I. W.: *M. N.* (in press).
SPITZER, L.: 1957, *Ap. J.*, **125**, 525.
WRUBEL, M.: 1952, *Ap. J.*, **116**, 291.

The Evolution of Close Binary Systems.

M. Hack

Osservatorio astronomico - Merate

I. - Observational Data.

1. – Introduction.

About fifty per cent of the stars in our galaxy are binary stars and a large percentage of double stars are close binaries. By close binaries we mean a pair of stars which revolve at a distance of the same order as the sum of their radii; that is $(R_1+R_2)/a$ ranges between about 1 and 0.05. Close double stars manifest themselves as spectroscopic or eclipsing binaries. According to Plaskett and Pearce (1931) at least 30 % of stars of spectral type O and B are close doubles. According to Harper (1937) only 2 or 5 % of A-type stars are close binaries, but the percentage of close binaries becomes high again among late F and G, K stars. In a recent study C. and M. Jaschek (1957) found that the percentage of spectroscopic binaries along the main sequence is approximately constant and equal to 30 %. This agrees with the results found by Kuiper (1935) for the visual binaries.

The close eclipsing binaries are about 0.2 % of the total stellar population. These systems are the most interesting because the combination of photometric and spectroscopic data gives masses, dimensions, temperatures and hence luminosities of the stars and their distance in km. Moreover the two stars are often distorted because of mutual gravitational attraction and they assume an elliptical shape. Now two spherical stars act as though their masses were concentrated at the center-of-mass; but this is not true of ellipsoidal stars; they revolve in a potential field which departs from the simple inverse square law. A primary consequence of this is that the entire orbit precesses and the rate of rotation of the line of the apsides is a function of the distribution of mass in the stars. If the stars are highly concentrated toward the center the apsidal motion is less than if the density gradient is smaller. So the apsidal motion gives information about the distribution of density.

However all these data can be determined only if the light curve and the radial velocity curves for both stars are available. At present we know about 3 000 eclipsing binaries, but only 100 are in this happy situation.

Close binaries are population I objects; they are found in galactic clusters and associations; they are rare among the stars of the galactic center and absent in globular clusters. This fact may be a consequence of the conditions at the time of their formation, or the effect of disrupting forces acting over a 10^{10} period of years.

The main factors governing stellar evolution are the mass and the initial chemical composition of the stars. H-R diagrams of clusters show the evolutionary dispersion of stars of the same age and chemical composition but different initial masses. Binary systems can provide the opportunity to test our ideas on stellar evolution by studying the effects of differential evolution of stars having exactly the same age and chemical composition and only different initial masses. Visual binaries and wide pairs of spectroscopic and eclipsing binaries are a good test of theories on stellar evolution. Their components have positions in the H-R diagrams which are interpretable by the theory of stellar evolution. But in the case of close binaries the situation is more complicated and difficult to understand. In fact the theory permits us to foresee, and the observations confirm, that transfer of mass from one component to the other and loss of mass from the system takes place. Therefore the evolution of these pairs of stars depends not only upon the mass but also upon the interaction between the two components and upon the rate at which mass is lost to outer space or transferred to the other component.

What the evolutionary pattern of close pairs may be is an open question. We shall try to give first an account of the more important observational facts, and then a review of the more important ideas concerning the evolution of these systems.

2. – Equipotential surfaces in the vicinity of a double system.

The equipotential surfaces for a single star are spheres with the center in the center of the star. The equipotential surfaces for a binary system are given by consideration of centrifugal and gravitational forces in a co-ordinate system revolving with the system itself. Two potential wells surround the two stars but they merge into each other for a value of the potential known as Roche limit (Roche's model: the masses are concentrated in the center of the star). The equipotential surfaces for a binary system are given by

$$-\Omega = \frac{m_1}{r_1} + \frac{m_2}{r_2} + \frac{x^2 + y^2}{2}\,,$$

where Ω is the negative potential, r_1 and r_2 the distances of a point $P(x, y, z)$ to the mass centers. The units are: $m_1 + m_2 = 1$; $a = 1$ (where a is the distance of the two centers of the two stars); angular velocity $= 1$. For a value $-\Omega = C_1$ the two equipotential surfaces touch each other in a point L_1. Only the matter which is enclosed inside one of the two surfaces can belong permanently to the respective star. For values of C less than C_1 we have a single equipotential surface, which for $-\Omega = C_2$ presents another singular point L_2 (Fig. 1). The matter inside this surface belongs permanently to the system but not to one single star. The matter outside of this surface does not belong any longer to the system. The two limiting surfaces defined by L_1 and by L_2 depend only on the mass ratio m_2/m_1, the radii of the equipotential lobes being given in units of the distance a. KUIPER (1941) has computed the inner and outer contact surfaces for different mass ratios.

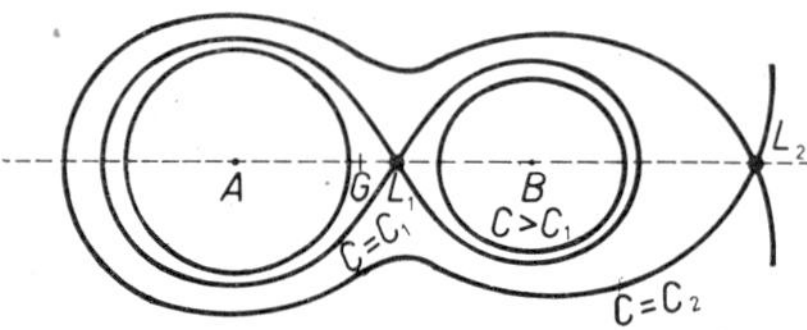

Fig. 1. – Equipotential surfaces for a double system. A and B are the two centers of mass, G is the center-of-gravity of the system.

Now we know several close systems the two components of which have masses and radii which are the same as those of single stars occupying the same position in the H-R diagram. Therefore we infer that the internal constitution of the components of these close systems is the same as that of a single star, and also the first evolutionary stages will be the same; when the more massive star has depleted its core of hydrogen it will begin to expand. But the expansion of the component of a close binary will be limited by the size of the maximum close equipotential surface. When this size is reached, no further expansion can take place, but only a transfer of matter to the companion through the lagrangian point L_1. This fact is probably the main reason why the evolution of a component of a close binary is different from the evolution of a single star of the same initial mass and of the same age.

3. – Classification of close eclipsing binaries.

Several systems of classification of eclipsing binaries have been proposed. The importance of a classification consists in collecting under a same class all the types physically similar and in finding the transitions among different classes; such transitions may give some hints concerning the evolutionary patterns followed by the close binary components. One classification adopted for many years consists in dividing the eclipsing binaries according to their period and spectral type. These types are named after their prototypes:

a) *β Lyrae type*- both the components have spectral type between O and B9, the mass ratio is close to 1 and the periods range between 1 and 12 days. Their separation is of the same order as the sum of the radii.

b) *Algol type*- one component, generally the primary, is a main sequence star, and the secondary is a subgiant. We point out that subgiants are rather rare among single stars; however this type of close systems is very frequent. Another fact on which I would call attention is that the less massive star is out of the main sequence, and hence is probably the most evolved, and not the primary, as we might expect. Periods range between 1 and 10 days, and the mass ratio is about 0.5.

c) *W UMa type*- the two components have spectral types ranging between F5 and M. Both components have about the same spectral type and the difference of magnitude is small; however the mass ratio is about 0.5. The two components are in contact with each other through L_1. The periods are less than one day.

A more detailed classification is given by KRAT (1944). He introduces seven classes, A, B, C, D, E, F, G.

Group A includes systems both components of which belong to types ranging from O to B9. Examples are *β Lyrae*, 29 *CMa*, A0 *Cas*. The radii are comparable with the semi-axis of their orbits. These systems show therefore an ellipticity effect. To this group belongs a subgroup which has relatively smaller stars, and less pronounced ellipticity effects (examples *Y Cygni*, *VV Orionis*). The orbital period is a few days.

Group B includes systems with the fainter components belonging to spectral classes A0 to F5. Examples *CG Cygni* ($P \sim$ few tenths of day) or *λ Tauri* ($P \sim$ several days).

Group C has at least one component a subgiant. The ellipticity effect is small. The period is longer than one day to 10^d or more. Examples *β Per*, *U Cep*, *δ Lib*. A large number of eclipsing variables belong to this group,

Group *D* includes systems with both components on the main sequence. The spectrum of the secondary is later than F5. Periods from a fraction of day to several days. Small ellipticity effect. Examples *α CrB*, *R CMa*, *RT Per*.

Group E consists of pairs similar to *W UMa*. The spectral types of the two components are between F5 and M.

Group F includes systems where one or both components are degenerate. We know few systems of this kind: *UX UMa*, *Nova DQ Her* are some examples. *Nova WZ Sagittae* differs from the other two systems in possessing a conspicuous secondary minimum (KRZEMINSKI, 1962).

Group G one or both components are giant or supergiant. It includes *ζ Aur*, *ε Aur*, 31 *Cygni*, 32 *Cygni*, *VV Cephei*. The periods are very long, several years.

Sahade (1960) has recently proposed another classification related to the evolutionary ages of the different systems. Group 1: includes systems with components still gravitationally contracting. Group 2: systems with both components on the main sequence. Group 3: systems with one giant or subgiant component. Group 4: systems with two giant or subgiant components. Group 5: systems with components below the main sequence.

Kopal (1955) has proposed another classification which differs from the previous ones by being based upon the relative size of the radii of the stars and the radii of their respective maximum close equipotential surfaces, instead of upon the spectral type:

a) when the stars do not fill the singular equipotential surfaces we have a *detached system*;

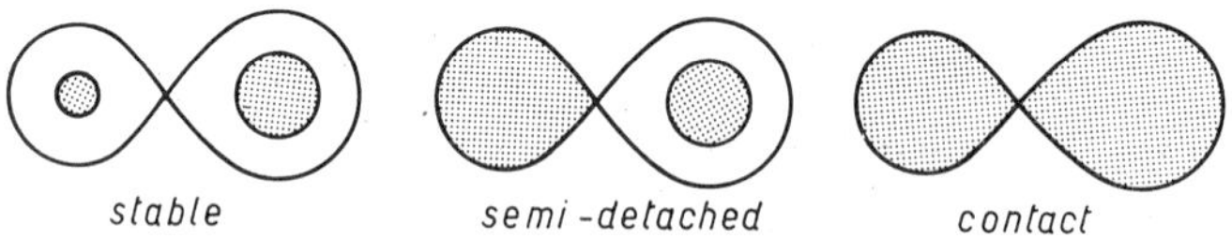

Fig. 2. – Detached, semidetached and contact systems.

b) when one of the two components fills its singular equipotential surface we have a *semidetached system*;

c) when both the two components fill their singular equipotential surfaces we have a *contact system* (Fig. 2).

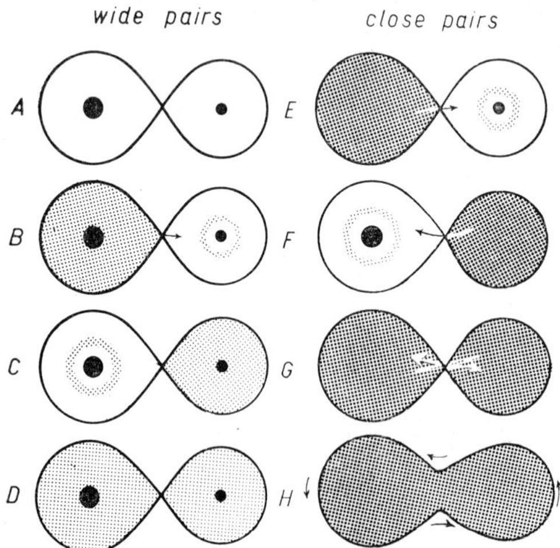

Fig. 3. – Eight possible configurations of close binary systems.

Struve (1957) describes eight possible configurations of double stars all existing in nature (Fig. 3):

A: no interaction between the components. Visual binaries and many spectroscopic binaries belong to this group;

B: a gaseous cloud surrounds the more massive component. The gas escaping through L_1 can form a ring around the less massive component. ε *Aur* probably belongs to this group. If the shell is quiescent all of it is retained gravitationally by the star; but a small disturbance, due to turbulence or thermal agitation will permit some of the atoms to fly outside the equipotential surface. A very small disturbance at the point L_1 will propel the atoms into the empty space surrounding the less massive star;

C: the reverse of *B*: the less massive star has a tenuous atmosphere filling the equipotential surface. The gas escaping through L_1 can form a ring around the primary;

D: both stars have extended atmospheres;

E: this is a very unusual configuration; only a very few systems are known to belong to this class, for example *UX Mon*, composed of a giant massive star G2p IV filling its equipotential surface and of an A6 less massive star, and *SZ Psc* (Kl III + A).

F: this configuration (the opposite of *E*: is the one most frequently observed;

G: *β Lyrae* is a representative of this configuration;

H: *W UMa* stars are representative of this configuration, which is very frequently observed.

4. – Spectrographic observations of close binary systems.

Several close binary systems display peculiarities which prove the existence of interchange of matter between the two components. These peculiarities are more often observed in spectroscopic binaries which are also eclipsing variables. This fact suggests that the phenomena involved generally occur close to the orbital plane. Often emission lines are visible in the spectra of such stars. In general the binaries show the same emission lines as those observed in single stars: H, He II, He I in binaries of early spectral type, O, B; H lines in binaries of spectral type A; Ca II lines in binaries of late type.

STRUVE (1958) divides the binaries into five classes according to the emission line present in their spectra:

1) Wolf-Rayet binaries;

2) Spectroscopic binaries of types Of to A and F; in which the emission lines of H are double and are explained in terms of gaseous streams.

3) Spectroscopic binaries of later spectral types in which the emission lines of Ca II are produced in the tidal bulges of the deformed stellar bodies.

4) *W UMa* systems in which the emission lines of Ca II may be produced in a current of gas flowing around the contact binaries.

5) Several kinds of systems in which permitted and forbidden emission lines are produced in enormous gaseous shells surrounding the double star.

Systems I and II often show the vast shell characteristic of class V. For example *V 444 Cyg* (class I) and *β Lyr* (class II). Characteristic of class V is *W Ser.* We know little about systems of class III and IV. The emission lines present in binaries of F5 or later spectral type are usually those of Ca II. They are dynamically associated with one of the two components, or we have two emissions, each associated with one of the two components, as for example in BD + 74° 493. In the early type binaries, on the contrary, the emission lines are generally associated with streams flowing from one star to another, or with concentration near to the lagrangian point L_1. The Ca II emissions resemble those often found in single G, K and M type stars, but they are usually stronger in binary systems suggesting that the binary nature of the object favors the conditions for the formation of the emission.

We can summarize here some of the most important spectroscopic phenomena which are observed in close binaries, *i.e.* evidence of: *a*) gaseous rings, *b*) concentration of matter at lagrangian points, *c*) absorption effects of gaseous streams and rings, *d*) expanding shells.

a) A characteristic example of a system with a gaseous ring is offered by *RW Tau.* The primary is a giant B9, the companion a subgiant K0, the spectrum of which is observed only during primary eclipse. The existence of a gaseous ring is suggested by the following facts: at the precise moment in which the B9 star is completely hidden by the K star, red-shifted emission lines of hydrogen become visible. The Doppler shift indicates a velocity of about +350 km/s. During the totality only the K0 spectrum is visible. Before the B9 star becomes again visible, we see emission lines again, but now they are violet-shifted, indicating a velocity of about − 350 km/s.

In all the known systems showing similar characteristics the ring has been found to rotate in the same direction as that of the orbital motion. Struve (1946) found empirically that the velocities of rotation are related to the period by the Keplerian relation $V^3_{\text{em}} \propto P^{-1}$. In many systems there are variations in character and intensity which suggest that the motions in the ring are complicated, and that these formations are not permanent nor of uniform density

b) Evidence of concentrations of matter at lagrangian points is found, for example, in systems such as *β Per, V 444 Cyg, HD* 47 and *UX Mon.* In *β Per* the emission lines Hα, Hβ, Hγ appear only during the elongations. This means that at the conjunctions the gaseous matter is eclipsed by either star of the system. At the elongation which follows the primary minimum the emission is red-shifted relative to the absorption line, and at the other it is violet-shifted, hence the motion of the gas is retrograde. In the case of *UX Mon* we observe double emission lines of hydrogen at the time of the two elongations. The components are sharp, showing that the gas has a density too low to produce Stark broadening. We infer that when the two stars are observed side by side there are in the space between them masses of gas (prominences) whose motions are predominantly of approach, and other masses of prominences whose motions are predominantly of recession. In front of neither star are there any large masses moving at right angles to the line of sight since we do not observe strong undisplaced absorption lines.

c) The matter streaming out from one component to form the ring often produces absorption effects when it is seen projected upon the surface of the star whose spectrum is observed. These effects consist in distorting the motions of the orbital velocity curve due to the motions of the gas, or in the presence of satellite lines due to gas streaming from one lobe to the other (for example β *Lyr*, or v *Sgr*) or to absorption of the rotating ring when it is projected upon the surface of the brighter star (for example ε *Aur*) showing strong effect of dilution of radiation.

d) Expanding shells. The observations show the existence of large shells surrounding several binary systems. Often these shells, which produce absorption and emission lines, have negative radial velocities indicating that they are expanding and therefore matter is escaping from the system. For example β *Lyrae* has a shell expanding at a velocity of -170 km/s, HD 47129 at -700 km/s, the Wolf-Rayet binary γ^2 *Vel* at -1300 km/s, *W Ser* at -60 km/s. Strong effects of dilution of radiation are always present and sometimes forbidden emission lines are present, as in the case of *W Ser*.

The spectra of the *W UMa* stars show evidence of an envelope of gas surrounding the whole system, thick enough to give continuous radiation. Actually when the lines are double (at the elongations) the intensities are much less than we would expect from the filling of the continuum of the companion, suggesting that the lines are filled with continuous radiation of the envelope; at the conjuctions the lines are deeper because of the envelope absorption.

5. – Mechanism of mass loss.

Kopal (1956) and Gould (1957) have computed the trajectories that particles of gas emitted under different initial conditions can describe. The problem is treated as a restricted three-body problem. The main results are the following: the direction of the particles will depend upon whether the angular velocity of rotation is larger or smaller than the orbital velocity. In the first case the direction will be the same as that of the revolution of the system (direct); in the second case the direction will be opposite (retrograde). In both cases the velocity of the particles would be the difference between the velocity of rotation and the velocity of the orbital revolution. PRENDERGAST (1960) observes that the problem cannot be treated simply as a restricted three-body problem. Since the mean free path of the particles is several orders of magnitude less than the sizes of the system, an hydrodynamical treatment is necessary. In fact the density of the streams is of the order of 10^{10} cm^{-3}; for a collision cross-section of the order of 10^{-15} or 10^{-16} cm^2 the mean free path is $l = 1/N\sigma \simeq 10^5$ or 10^6 cm. Therefore there will be interaction among different parts of the stream and the stream motions must be considered as motions of large masses of gas rather than individual particles. Prendergast finds that rings and strong concentrations of gas at L_1 (but not at L_2) are possible.

However prominences and violent ejections of gas are necessary to explain the observed high expansion velocities with loss of mass to outer space.

6. – Photometric evidence of mass loss in close binary systems, and intrinsic variability in brightness of these systems.

The asymmetry of light curves observed in some eclipsing variables can possibly be explained by the presence of gaseous streams (MERGENTALER, 1950). Large masses of cool gas might cover the photosphere of one star decreasing its brightness. The asymmetry of the stream can explain the asymmetry of the light curve. MARTYNOV (1957) observes that fluctuations of the photometric periods are often observed and they correspond to the actual fluctuations of the sideral periods of orbital motion, which is possible only if changes of the rotational momentum of the system are taking place. Indeed fluctuations of the period might be explained (WOOD, 1950) by transitory ejections of very large masses of gas from a system. The observed fluctuations of the periods of *RZ Cas*, *U Cep*, *AO Cas*, *AR Lac* and *U Sge* require an ejection of the order of 10^{-7} or 10^{-6} solar masses. But this ejection would give strong spectroscopic effects which are not observed in those eclipsing binaries. For example the *P Cygni* stars lose about 10^{-6} or 10^{-7} solar masses per year. Perhaps it may be some kind of corpuscular emission nonobservable spectroscopically.

Secular increase of period, such as observed for *W Ser* or *β Lyrae* can be caused by a continuous ejection of mass. For example the period of *W Ser* seems to increase at a present rate of about 20 seconds per year (SAHADE and STRUVE, 1957).

Often close binary systems have components whose light irregularly fluctuates. For example *UX Mon* (WOOD, 1957) sometimes presents large fluctuations; on about a two and a half hour interval, variations of almost 0.2

magnitudes in the blue and 0.1 in the yellow were observed; irregular fluctuations occur in the supergiants ε *Aur*, 32 *Cyg*, *VV Cep*, ζ *Aur*, while 31 *Cyg* so far appears to be constant.

Almost all the *W UMa* stars fluctuate irregularly.

7. – Statistical relations between the physical parameters of the close binary systems.

KOPAL (1955, 1956) has collected all the existing data concerning masses, spectral types and absolute magnitudes for close binary systems. Let us examine the mass-luminosity relation for detached (*d*), semidetached (*sd*) and contact (*c*) systems, and for the particular group of the *R CMa* stars. This last group includes systems in which both components are overluminous (*i.e.* too luminous for their masses) by a larger extent than any other stellar object so far known, and these systems are also characterized by a small value of the mass ratio m_2/m_1. They are detached systems but they differ in their statistical properties from the normal detached systems and form a group by themselves.

As Fig. 4 shows the *d* systems have a mass-luminosity relation which is practically indistinguishable from that derived for the visual systems.

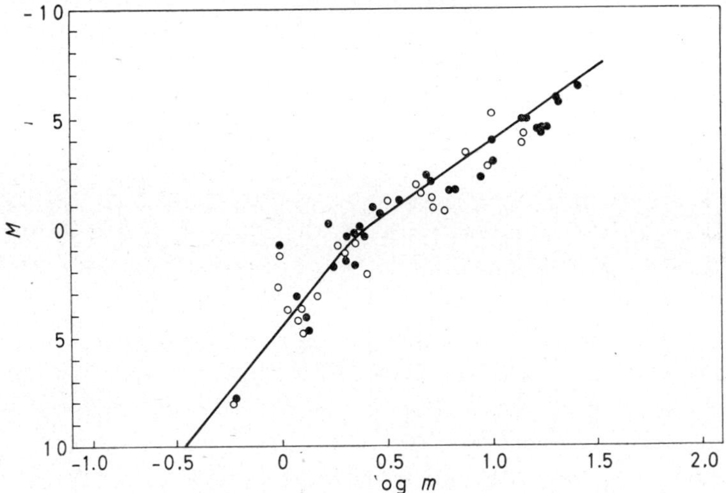

Fig. 4. – Mass-luminosity relation for *d*-systems (● pr., ○ sec.).

The more massive and brighter component of the *sd* systems (Fig. 5) is generally close enough to the average mass-luminosity relation, but the companion on the contrary is very often overluminous. The effect is strong: sometimes the companion is 7 or 8 magnitudes brighter than expected from its mass.

Among the *c* systems the three pairs of early spectral type are close enough to the average mass-luminosity relation. The group of the *W UMa* stars on the contrary in many cases presents overluminous companions; in two cases

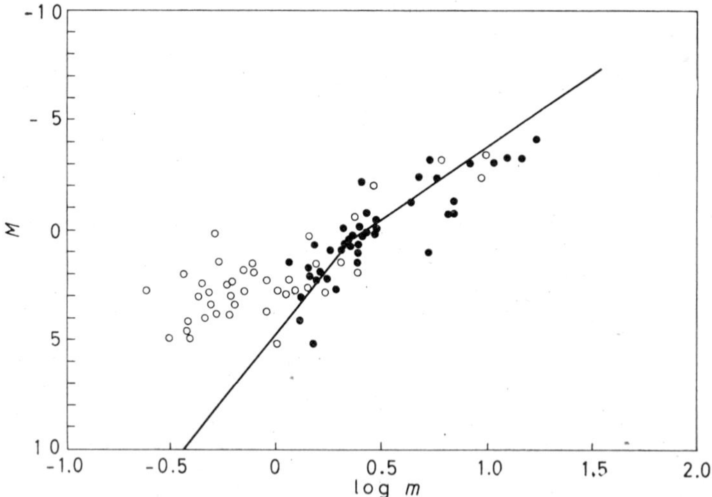

Fig. 5. – Mass-luminosity relation for *sd*-systems (● pr., ○ sec.).

both the components are overluminous; in four cases the primary is underluminous (Fig. 6).

The *R CMa* stars have as a rule both components overluminous. No mass-

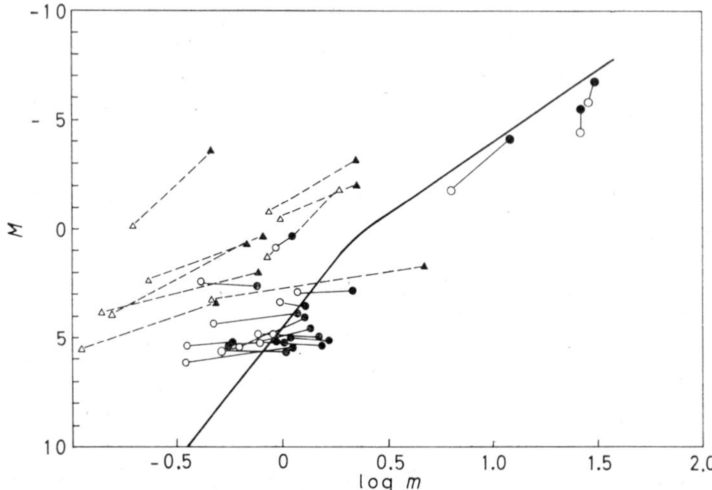

Fig. 6. – Mass-luminosity relation for *W UMa* systems (● pr., ○ sec.) and *R CMa* systems (▲ pr., △ sec.).

luminosity relation seems to exist for the *W UMa* stars and the *R CMa* stars, but the deviations are much stronger in the latter case (Fig. 6).

We shall see later what interpretation may be proposed for explaining such peculiar behavior.

Kopal also constructs the H-R diagram for the four groups. We observe that the detached systems have both components on the main sequence. The only exceptions are *MR Cyg* and *TV Cas*, which can be considered intermediate between *d* and *sd* systems (Fig. 7).

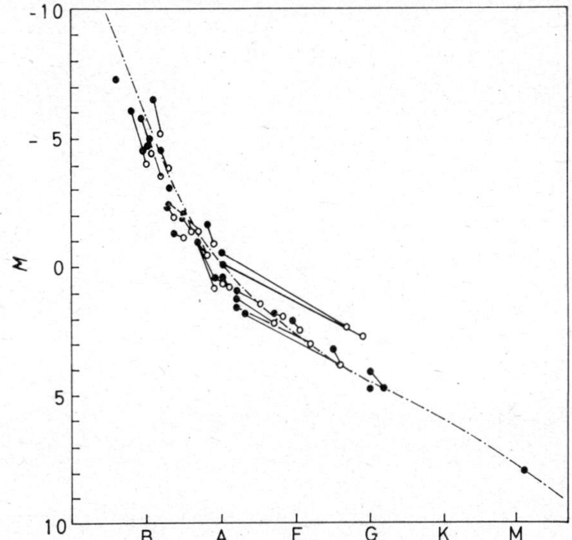

Fig. 7. – H-R diagram for *d*-systems (● pr., ○ sec.).

Sd systems as a rule have a subgiant for the companion; we point out that the main sequence primary of a secondary subgiant never is later than A8 (Fig. 8).

C systems of early spectral type and of late type (*W UMa*) are very close to the main sequence (Fig. 9).

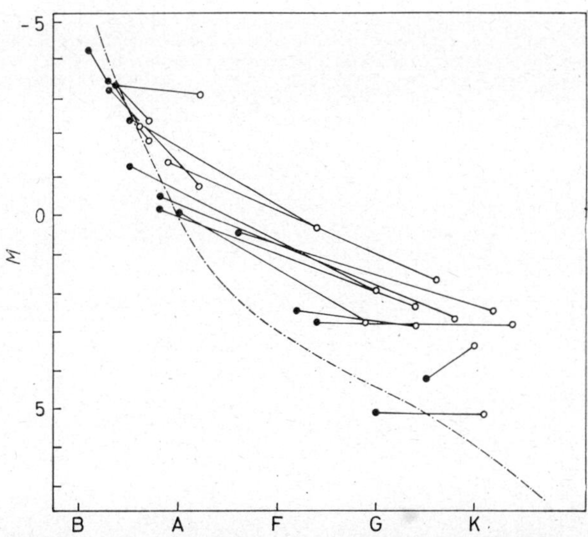

Fig. 8. – H-R diagram for *sd*-systems (● pr., ○ sec.).

The *R CMa* group regarding the position on the H-R diagram and the mass-luminosity relation, has a behavior similar to the *sd* group, in spite of the fact that the subgiant component is smaller in size than the maximum

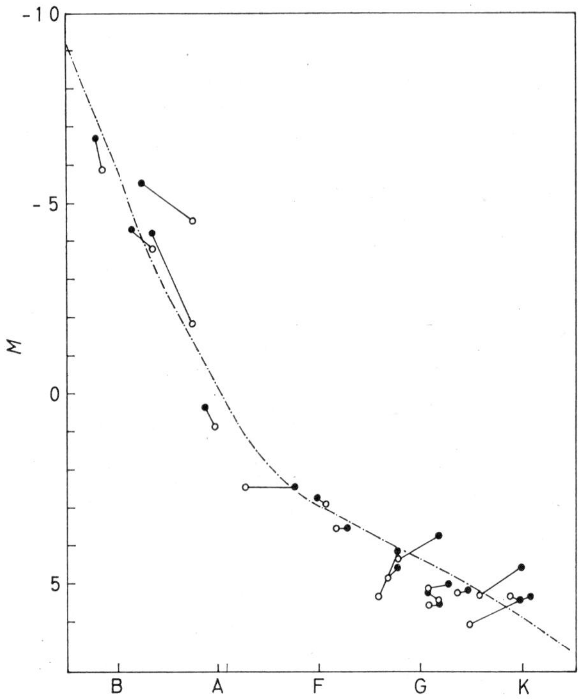

Fig. 9. – H-R diagram for early type *c*-systems and *UMa* systems (● pr., ○ sec.).

close equipotential surface, and therefore in this consideration these systems must be considered *d* systems (Fig. 10).

We summarize these results as follows:

d systems have both components on the main sequence and both components obey the mass-luminosity relation.

sd systems have the primary components on the main sequence (and earlier than F0) while the secondary is a red subgiant. The primary generally obeys the mass-luminosity law, while the companion is overluminous (except for the more massive and brighter systems which have a normal behavior).

c systems belong to the main sequence. The early type systems obey the mass-luminosity relation, but the *W UMa* do not, the companions being as a rule overluminous, and sometimes the primaries underluminous. The different behavior concerning the mass-luminosity law and the different average values of the mass ratios suggest that contact systems of early type and *W UMa* systems form two groups physically distinct.

R CMa systems have a main sequence star (earlier than F0) as primary, and a subgiant as secondary. The only exception is *R CMa* which has both components on the main sequence. Both components of the *R CMa* stars are overluminous (excep for *XZ Sgr*, with the primary underluminous).

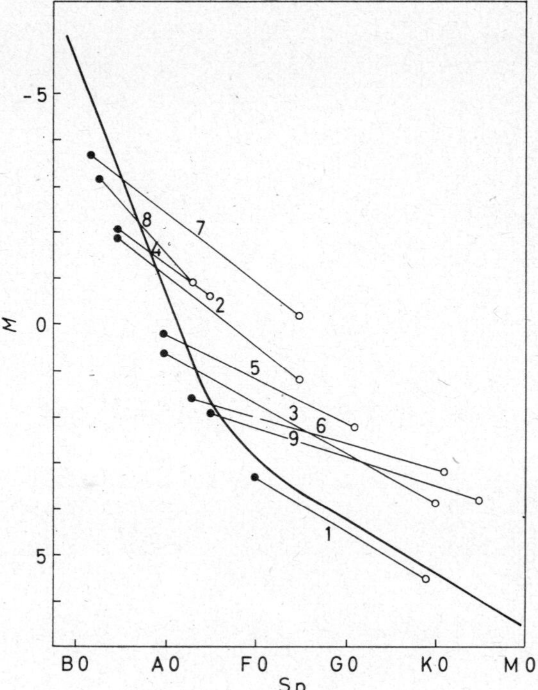

Fig. 10. – H-R diagram for *R CMa* systems (● pr., ○ sec.).

Other statistical relations among the close binary systems have been found using the data of Kopal. Fig. 11 gives the relation between the mass-ratio m_2/m_1 and the ratio $(R_1+R_2)/a$ which is an index of the closeness of the system. From this diagram there appears a sharp separation in groups corresponding to the five groups mentioned over (*d*, *sd*, *c* early type, *W UMa*, *R CMa*) and to a group which we shall call *sd-d*, composed of a main sequence primary

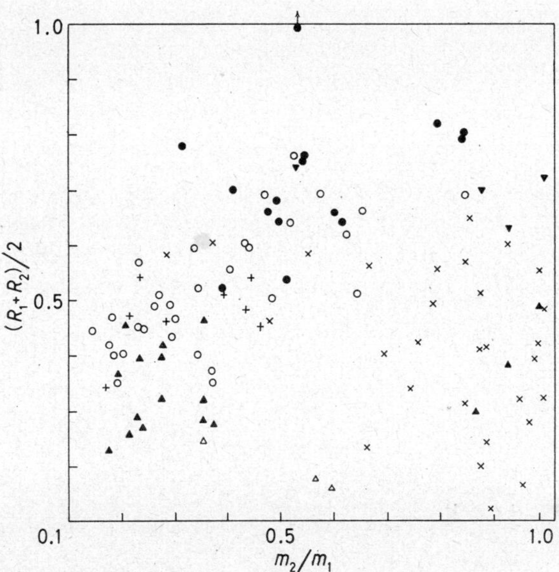

Fig. 11. – $(R_1+R_2)a$ *vs.* m_2/m_1. The symbols are: ▼ *c*-systems of early type; ● *W UMa* systems; ○ *sd*; × *d*; + *R CMa*; ▲ *sd-d*; △ ζ *Aur*. *32 Cyg* and *VV Cep*.

and a subgiant secondary which has not reached the Roche limit. We have:

	m_2/m_1	$(R_1+R_2)/a$
d	$\geqslant 0.65$	$\leqslant 0.60$
c (early type)	$\geqslant 0.70$	$\geqslant 0.70$
c (*W UMa*)	0.4 to 0.7	$\geqslant 0.70$
sd	$\leqslant 0.60$	0.35 to 0.70
sd-d	$\leqslant 0.40$	0.25 to 0.45
R CMa	$\leqslant 0.40$	0.30 to 0.55

We have derived for each star the mass and radius which are normal for its spectral type and luminosity. These data are collected in Table I. Concerning the radii we find that almost all the stars, primaries and secondaries have radii which differ in a negligible way from the normal value. In 208 stars only 9 have $0.60 \leqslant R_{obs}/R_{HR} \leqslant 0.74$ (where R_{HR} is the radius which is normal for the position of the star in the H-R diagram) and one, β *Per A*, has $R_{obs}/R_{HR} = 2.6$. This is the only one showing a strong difference between observed and normal H-R value of the radius. However β *Per* has a completely normal B8 V spectrum.

The comparison of observed and normal values of the masses confirms the results already shown by the mass-luminosity curves. Plotting the mass-ratio m_2/m_1 versus m/m_{mL} (where m is the observed mass and m_{mL} is the normal value according the mass-luminosity relation) (Fig. 12) we observe the grouping of *R CMa* star (primaries and secondaries), *sd-d* and *sd* secondaries in the region of low values of m/m_{mL} and of m_2/m_1. Note that two stars are strongly overmassive: β *Per A* and *XZ Sgr A*. However their spectra are normal. Generally all the *R CMa* stars, in spite of their strong overluminosity show normal spectra. Hence we have several stars which have radii normal for their luminosity and spectrum, but are strongly under-massive. For example, *RZ Sct* is a B2 star with absolute magnitude -3.7, located on the main sequence. Its mass is $0.47 m_\odot$ instead of the expected $10 m_\odot$! This star appears to be a normal B2 star, but using the words of Kopal it « is largely empty inside, containing only about one-tenth of the mass which a main sequence star of its absolute magnitude would be expected to possess, and maintaining, therefore, its extravagant external appearances on manifestly shaky grounds. And of the secondary component of *RZ Sct* the same is true in an even more exaggerated manner. The abnormally small mass of both stars is made manifest by the weak gravitational bond between them as revealed by the length of the orbital period and the low value of the orbital velocity; but if both components were travelling singly through space no obvious sign would warn us that such outwardly normal stars are in reality but over-distended "empty wind bags" posing on the main sequence (or above it) under false pretences! ».

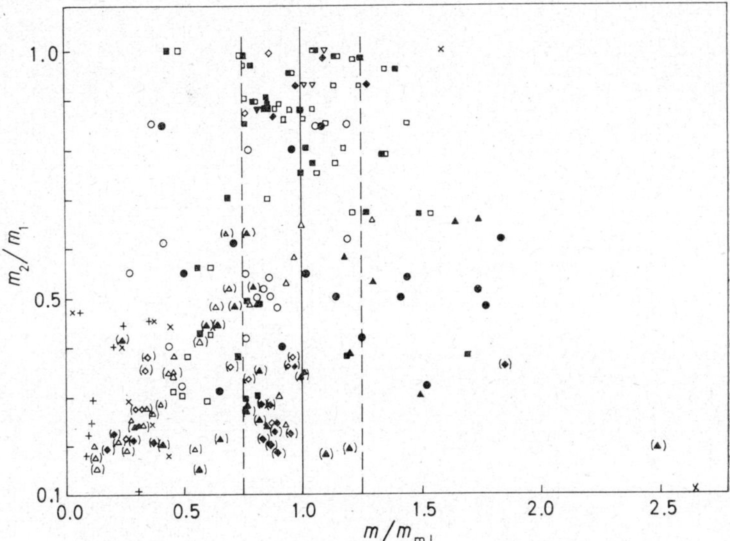

Fig. 12. – (m_2/m_1) *vs* m/m_{mL}. The symbols are:

Primaries	Secondaries	
▼	▽	*c* (early type)
●	○	*c* (*W UMa*)
▲	△	*sd*
■	□	*d*
◆	◇	*sd-d*
×	+	*R CMa*

A star « largely empty inside » is of course unexplicable by the theories on stellar structure. It seems much better to explain those observational facts assuming that the undermassive stars are small hot stars with a spectrum essentially continuous surrounded by a shell excited by the radiation of the star itself. This shell would be opaque and thick enough to simulate a star of radius and spectrum like those observed. High rotational velocity of the shell might mask the characteristics of dilution of radiation which must be expected in a shell spectrum, and simulate a main sequence spectrum. A proof for the existence of a shell in such types of stars is given by STRUVE'S (1950) observations of *RZ Sct*, and the more recent observations of the same star by HANSEN and MCNAMARA (1959). Struve found a rotational velocity of 50 km/s as given by the hydrogen lines, and 125 km/s as given by the helium lines. Hansen

TABL

Star	m_1	m_1/m_{mL}	m_2	m_2/m_{mL}	R_1	R_1/R_{HR}	R_2	R_2/R_{HR}
RT And	1.5	1.65	1.0	1.0	0.76	0.95	1.36	0.84
AB And	1.7	1.85	1.1	1.2	0.9	0.9	0.9	0.9
S Ant	0.76	0.51	0.42	0.29	1.4	0.94	1.1	1.0
V 599 *Aql*	12	0.86	6.4	1.28	7.8	0.98	4.4	0.88
σ Aql	6.8	1.35	5.4	1.35	4.2	0.84	3.3	0.94
SX Aur	10.5	1.30	5.6	0.94	4.8	0.96	4.3	0.86
TT Aur	6.7	1.02	5.3	1.18	3.3	0.74	3.2	1.06
WW Aur	1.82	0.96	1.75	0.96	1.9	0.95	1.9	0.96
AR Aur	2.6	0.87	2.3	0.96	1.8	0.90	1.80	0.9
EO Aur	27	1.1	27	1.6	13	1.0	16	1.06
β Aur	2.4	0.8	2.3	0.76	2.49	0.90	2.28	0.95
i Boo	1.07	1.42	0.54	0.72	0.66	1.0	0.64	1.0
SZ Cam	21	0.76	6.1	0.61	9.8	0.98	4.3	0.86
TX Cnc	1.5	1.75	0.8	0.84	0.74	0.92	0.78	0.98
RS CVn	1.9	1.27	1.7	1.0	1.8	0.96	5.1	1.02
TV Cas	1.7	0.57	1.0	0.6	2.6	1.3	2.4	1.2
AO Cas	30	1	28	1.05	14	1.4	10	1
CC Cas	21	0.78	10	0.83	9.3	0.93	4.5	1.12
U Cep	2.9	0.83	1.4	0.78	2.4	1.1	3.9	0.98
VW Cep	1.1	1.35	0.35	0.5	0.97	0.79	0.58	0.93
WX Cep	1.0	0.45	1.0	0.5	2.0	1.0	2.0	1.0
AH Cep	16.5	1.03	14.2	0.94	6.1	1.2	6.1	1.2
CW Cep	10.0	1.25	9.8	1.22	4.5	0.98	4.0	0.89
RZ Cam	1.6	1.78	0.8	0.9	0.9	0.74	0.9	0.76
U CrB	6.4	1.7	2.4	1.2	1.9	1.35	3.6	1.2
Y Cyg	17.4	1.15	17.2	1.15	5.9	1.18	5.9	1.18
GO Cyg	1.12	0.42	0.94	0.38	1.8	0.9	1.6	0.89
MR Cyg	3.0	0.86	2.6	1.45	3.4	0.86	2.6	1.05
V 453 *Cyg*	17.8	1.05	13.8	1.15	7.6	0.94	6.3	0.97
V 470 *Cyg*	12.5	0.83	11.1	0.92	7	1.0	6	0.75
V 477 *Cyg*	2.3	1.28	1.6	1.22	1.5	0.94	1.2	0.96
V 478 *Cyg*	14.4	0.76	14.2	0.75	7.6	1.02	7.6	1.02
TW Dra	2.2	0.78	0.62	0.34	3.5	0.95	5.1	1.02
YY Eri	0.97	1.08	0.57	0.71	0.89	0.89	0.81	0.89
YY Gem	0.64	1.06	0.64	1.07	0.62	0.62	0.62	0.62
Z Her	1.5	0.9	1.3	0.76	1.8	1.0	2.8	1.0
RX Her	2.1	0.81	1.9	0.82	2.2	1.0	1.8	0.9
TX Her	2.1	1.1	1.8	1.06	1.8	1.0	1.5	1.25
DI Her	3.7	0.86	3.3	0.78	2.1	0.92	2.4	0.96
U Her	7.9	0.99	2.8	0.47	4.5	0.9	4.3	0.78
VZ Hya	1.2	0.86	1.1	0.92	1.5	0.86	1.2	1.0
SW Lac	1.08	1.2	0.93	0.98	0.82	0.9	0.89	0.93
AR Lac	1.32	1.1	1.31	0.88	1.54	0.98	2.86	0.96
CM Lac	2.0	1.0	1.5	1.07	1.8	0.9	1.3	1.05
UV Leo	1.36	1.15	1.25	1.25	1.21	0.99	1.22	1.22
IM Mon	9.0	1.5	6.0	1.55	3.8	0.84	2.7	0.84
U Oph	5.30	1.05	4.65	1.0	3.4	0.85	3.1	0.9
WZ Oph	1.4	1.4	1.35	1.35	0.88	0.92	0.86	0.90
V 502 *Oph*	1.2	0.92	0.49	0.45	1.51	1.0	1.04	1.04
ER Ori	0.58	0.72	0.35	0.43	0.72	0.9	0.72	0.90
δ Ori	26	0.74	10	0.53	17	1.05	10	1.1
U Peg	1.25	0.96	1.00	0.78	1.18	1.1	1.18	1.0

Sp_1	Sp_2	M_1	M_2	P	a	$(R_1+R_2)/a$	m_2/m_1	Type
G0	K1	5.1	5.1	0.629	4.2	0.51	0.65	*sd*
G5	G4	5.1	5.0	0.332	2.80	0.64	0.62	*c*
A8	A3	2.6	2.5	0.648	3.32	0.76	0.55	*c*
B4	B8	−4.2	−1.8	1.849	16.4	0.74	0.53	*c*
B8	B9	−1.7	−0.9	1.950	15.2	0.49	0.79	*d*
B3.5	B6	−3.2	−2.4	1.210	12.0	0.76	0.53	*sd*
B3	B7	−2.5	−1.4	1.333	11.7	0.55	0.80	*d*
A7	A8	1.8	1.9	2.525	11.9	0.32	0.96	*d*
B9	A0	0.4	0.7	4.135	18.6	0.195	0.88	*d*
B3	B8	−5.5	−4.5	4.066	40	0.72	1.00	*c*
A0	A0	0.0	0.2	3.960	17.5	0.275	0.97	*d*
dG2	dG1	5.6	5.6	0.268	2.05	0.64	0.50	*c*
O9.5	(B2)	−5.9	−3.4	2.698	24.1	0.58	0.29	*d*
F8	F7	5.0	4.8	0.383	2.8	0.54	0.52	*c*
F4	gK4	2.7	2.8	4.798	18.3	0.38	0.93	*sd-d*
A0	(gF9)	−0.1	2.7	1.813	8.61	0.58	0.56	*d*
O8.5	O9	−6.7	−5.9	3.523	38	0.63	0.93	*c*
O8	B0	−6.0	−4.0	3.369	30	0.46	0.49	*d*
B8	gG8	−0.5	2.6	2.493	12.6	0.50	0.49	*sd*
dK1	G5	5.5	6.1	0.278	2.01	0.78	0.32	*c*
A2	A5	0.9	1.4	3.378	8.4	0.48	1.00	*d*
B0	B0.5	−4.6	−4.4	1.775	18.7	0.65	0.86	*d*
B3	B3.5	−3.2	−2.9	2.729	22.0	0.39	0.98	*d*
K0	G9	5.4	5.3	0.339	2.72	0.66	0.48	*c*
B5	(gG0)	−0.8	1.9	3.452	12.9	0.35	0.38	sd
O9.5	O9.5	−4.6	−4.6	2.996	28.5	0.42	0.99	*d*
B9	A0	0.4	0.9	0.718	4.3	0.80	0.85	*c*
A0	F7	−0.6	2.2	1.677	10.5	0.57	0.85	*d*
B2	B3	−4.6	−3.9	3.890	32.9	0.42	0.77	*d*
B2	B4	−4.3	−3.6	1.874	18.5	0.70	0.88	*c*
A3	F5	1.8	3.7	2.347	11.8	0.23	0.67	*d*
B0.5	B0.5	−5.0	−5.0	2.881	27.9	0.55	0.99	*d*
A6	K2	0.4	2.4	2.807	16.7	0.51	(0.28)	*sd*
G5	G4	5.2	5.3	0.321	2.26	0.75	0.55	*c*
dM1	dM1	7.8	7.8	0.814	3.96	0.32	1.00	*d*
F2	dG4	2.4	2.7	3.993	15.1	0.30	0.87	*sd-d*
A0	A1	0.4	0.8	1.779	9.7	0.41	0.89	*d*
A5	A7	1.6	2.2	2.060	10.6	0.31	0.85	*d*
B4	B5	−1.3	−1.3	10.550	38.7	0.117	0.90	*d*
B3	B7	−3.2	−2.2	2.051	15.0	0.59	0.34	*sd*
F5	F6	3.2	3.8	2.904	13.5	0.24	0.89	*d*
G3	G1	5.2	4.9	0.321	2.49	0.69	0.85	*c*
G5	K0	4.1	3.3	1.983	9.13	0.49	0.99	*sd-d*
A2	F2	1.1	3.2	1.605	8.8	0.34	0.75	*d*
G0	G2	4.1	4.2	0.600	4.1	0.60	0.93	*d*
B5	B8	−2.3	−0.8	1.190	11.6	0.56	0.67	*d*
B5	B6	−2.1	−1.7	1.677	12.8	0.51	0.88	*d*
G0	G0	4.8	4.8	4.183	10.9	0.16	0.96	*d*
G2	F8	3.8	4.3	0.453	2.97	0.52	0.40	*c*
G1	G1	5.3	5.4	0.423	2.16	0.66	0.61	*c*
B1	B2	−6.5	−5.2	5.733	45	0.60	0.38	*d*
F3	F2	3.6	3.4	0.375	2.93	0.82	0.80	*c*

TABLE I (*continued*)

Star	m_1	m_1/m_{mL}	m_2	m_2/m_{mL}	R_1	R_1/R_{HR}	R_2	R_2/R_{HR}
AG Per	5.01	0.84	4.47	0.9	3.1	1.0	2.8	0.88
ζ Pho	2.8	0.7	2.0	0.87	2.5	0.84	1.5	0.84
V Pup	16.6	1.18	9.8	0.98	6.0	1.05	5.3	0.94
U Sge	6.7	1.5	2.0	0.91	4.1	0.86	5.4	1.1
V 356 Sgr	12.1	1.2	4.7	0.47	4.9	0.88	12.7	0.92
μ' Sco	14.0	1.75	9.2	1.3	4.8	0.96	5.3	0.88
RZ Tau	2.2	1.45	1.2	0.86	1.4	1.0	1.3	0.93
W UMa	1.29	1.15	0.65	0.82	1.1	1.1	0.61	1.0
TX UMa	2.9	0.82	0.9	0.5	2.16	0.86	3.79	1.15
AH Vir	1.36	1.25	0.57	0.72	1.3	0.76	0.75	0.82
Z Vul	5.25	0.58	2.34	0.62	4.7	0.94	4.2	0.98
RS Vul	4.6	0.66	1.4	0.47	4.2	0.9	5.4	0.90
TW And	2.4	1.2	0.46	0.27	2.4	0.97	3.7	0.92
RZ Cas	1.8	0.82	0.63	0.45	1.5	0.84	1.8	1.1
W Del	2.1	0.66	0.44	0.24	2.5	0.90	4.0	0.90
Z Dra	1.4	0.78	0.38	0.38	1.5	0.94	1.6	1.0
RX Hya	1.5	0.83	0.38	0.29	1.6	1.2	2.3	0.86
SX Hya	1.7	1.0	0.59	0.45	1.3	0.93	3.2	0.97
Y Leo	1.1	0.58	0.46	0.25	1.8	1.0	2.1	0.88
δ Lib	2.6	0.65	1.1	0.61	3.5	0.88	3.5	1.1
TY Peg	1.5	0.58	0.23	0.145	2.3	0.88	2.4	1.1
RT Per	1.3	0.86	0.31	0.31	1.4	1.0	1.1	1.0
RY Per	2.5	0.42	0.49	0.14	2.7	0.68	5.4	0.84
ST Per	2.4	1.1	0.43	0.15	2.2	0.92	2.6	0.97
β Per	5.2	2.4	1.0	0.56	3.6	2.6	3.8	0.96
V 505 Sgr	2.3	0.80	1.2	0.70	2.3	0.82	2.3	1.1
X Tri	1.4	0.78	0.9	0.69	1.6	0.94	1.8	1.0
WU Mi	2.4	0.73	1.1	0.65	3.5	0.92	2.8	1.1
UW Vir	1.6	0.94	0.37	0.31	1.5	1.07	2.0	1.0
KO Aql	2.9	0.88	0.58	0.39	2.6	0.87	2.1	1.1
QY Aql	2.7	0.84	0.75	0.31	4.4	0.88	5.7	0.96
Y Cam	2.2	0.85	0.48	0.30	3.0	0.92	3.0	0.92
S Cnc	6.8	1.85	2.4	0.78	3.4	0.86	7.6	1.05
RS Cep	2.0	0.96	0.75	0.36	2.2	0.88	6.8	1.02
SW Cyg	2.5	0.86	0.70	0.41	2.5	0.83	4.4	1.05
VW Cyg	2.5	0.86	0.70	0.33	2.8	0.93	5.7	0.90
WW Cyg	4.3	0.96	1.5	0.72	3.5	0.78	4.4	1.05
RX Gem	2.4	0.96	0.53	0.22	2.4	0.86	6.0	0.92
RY Gem	2.7	0.90	0.62	0.28	2.6	0.87	6.3	1.0
TT Hya	2.0	0.96	0.71	0.36	2.1	1.0	4.8	0.96
AQ Peg	2.5	0.86	0.59	0.33	2.6	0.96	4.7	0.94
RW Per	2.0	0.91	0.36	0.18	2.2	0.92	5.1	0.92
Y Psc	2.1	0.88	0.51	0.37	2.3	0.92	3.3	0.82
R CMa	0.49	0.38	0.11	0.12	1.1	1.0	1.0	0.92
RW Gem	1.9	0.38	0.85	0.37	2.8	0.80	3.6	1.05
TL Mi	0.69	0.26	0.15	0.11	1.8	0.86	2.1	0.84
TU Mon	2.3	0.46	1.0	0.26	3.0	0.76	4.9	0.92
UU Oph	0.84	0.28	0.24	0.13	2.2	0.87	3.1	1.03
XZ Sgr	4.8	2.65	0.48	0.32	1.6	0.98	3.2	0.98
RZ Sct	0.47	0.05	0.20	0.055	3.4	0.62	6.9	0.90
λ Tau	2.3	0.25	0.9	0.22	3.4	0.70	4.8	0.96
S Vel	0.80	0.45	0.14	0.10	1.5	0.92	3.1	0.94

Sp_1	Sp_2	M_1	M_2	P	a	$(R_1+R_2)/a$	m_2/m_1	Type
B3	B4	−2.4	−1.9	2.029	14.2	0.41	0.88	*d*
B7	B9	−0.9	0.8	1.670	10	0.40	0.70	*d*
B1	B3	−4.3	−3.5	1.454	16.2	0.69	0.58	*sd*
B9	gG6	−1.4	1.6	3.381	19.4	0.49	0.30	*sd*
B3	A2	−3.4	−3.2	8.896	47	0.37	0.38	*sd*
B3	B7	−3.3	−2.5	1.446	15.4	0.66	0.66	*sd*
F0	F1	2.8	3.0	0.416	2.2	1.2	0.54	*c*
dF8	dF6	4.2	5.4	0.334	2.52	0.68	0.50	*c*
B8	gG4	−0.3	2.3	3.063	13.7	0.34	0.30	*sd*
dK0	dG6	4.6	5.5	0.408	2.9	0.70	0.42	*c*
B3	A2	−3.3	−0.7	2.455	15.1	0.59	0.45	*sd*
B5	(F4)	−2.5	0.3	4.478	20.9	0.46	0.31	*sd*
F0	gK1	1.5	2.9	4.123	15.4	0.40	0.19	*sd*
A0	gG1	0.9	3.4	1.195	6.4	0.52	0.35	*sd*
A0	gK0	−0.1	2.5	4.806	16.2	0.40	0.21	*sd*
A5	gK2	2.0	4.9	1.357	6.2	0.49	0.27	*sd*
A8	gK3	2.2	4.2	2.282	9.0	0.44	0.25	*sd*
A3	gK6	2.0	3.9	2.896	11.1	0.40	0.35	*sd*
A3	gK0	1.4	4.0	1.686	7.0	0.56	0.41	*sd*
A0	gG2	−0.8	2.2	2.327	11.6	0.60	0.44	*sd*
A2	gG1	0.6	2.9	3.092	10.7	0.44	0.15	*sd*
F2	gG4	3.0	4.9	0.849	4.4	0.57	0.24	*sd*
B4	F2	−2.3	−0.1	6.864	23	0.35	0.20	*sd*
A3	gG5	1.0	3.0	2.648	11.3	0.42	0.18	*sd*
B8	gK0	1.0	2.7	2.867	15.7	0.47	0.19	*sd*
A1	gF8	0.3	2.7	1.183	7.2	0.64	0.54	*sd*
A3	gG4	1.7	3.7	0.972	5.5	0.62	0.63	*sd*
A3	gG5	−0.1	2.9	1.701	9.1	0.69	0.48	*sd*
A2	gK	1.8	4.4	1.811	7.8	0.45	0.24	*sd*
A0	gF8	−0.1	3.0	2.864	12.8	0.37	0.20	*sd-d*
F0	gG9	0.2	1.6	7.230	23.6	0.42	0.28	*sd-d*
A7	gK1	0.7	3.4	3.306	13.0	0.46	0.21	*sd-d*
A0	gG5	−0.8	0.7	9.485	39.5	0.28	0.36	*sd-d*
A5	gK3	1.1	1.9	12.420	31.8	0.28	0.38	*sd-d*
A2	gK1.5	0.3	2.8	4.573	17.1	0.40	0.28	*sd-d*
A3	gK0	0.4	1.8	8.430	26.5	0.32	0.28	*sd-d*
B8	gG2	−1.4	1.6	3.318	16.9	0.47	0.36	*sd-d*
A4	gG8	0.7	1.4	12.209	32.1	0.26	0.22	*sd-d*
A2	gK1	0.2	1.8	9.301	30.5	0.29	0.23	*sd-d*
A3	gG7	1.1	1.8	6.953	21.3	0.32	0.36	*sd-d*
A2	gK2	0.3	2.5	5.548	19.3	0.275	0.24	*sd-d*
A5	gK0	1.1	2.1	13.198	31.4	0.23	0.18	*sd-d*
A3	gK8	0.8	3.9	3.766	14.1	0.40	0.24	*sd-d*
F0	gG9	3.3	5.5	1.136	3.9	0.54	0.24	*R CMa*
B5	F5	−1.9	1.2	2.865	11.9	0.54	0.45	*R CMa*
A0	gK0	0.6	3.9	3.020	8.3	0.47	0.22	*R CMa*
B5	A5	−2.1	−0.6	5.049	16.5	0.48	0.44	*R CMa*
A0	gG1	0.2	2.2	4.397	11.6	0.46	0.29	*R CMa*
A3	gK1	1.6	3.2	3.276	16.1	0.30	0.10	*R CMa*
B2	F5	−3.7	−0.2	15.190	22.7	0.45	0.47	*R CMa*
B3	A3	−3.2	−0.9	3.953	16	0.51	0.4	*R CMa*
A5	K5	1.9	3.8	5.934	13.6	0.34	0.175	*R CMa*

and McNamara studying high dispersion spectrograms of *RZ Sct* found that the H lines are double during eclipse phases. The weak component shows the same rotational disturbance as the He lines ($\sim$ 150 km/s) while the strong component shows a much smaller rotational disturbance.

This difference in rotational velocity suggests the existence of a stratum or shell of hydrogen high above the rest of the reversing layer of the brighter component.

8. – The rotation of spectroscopic binaries.

We can expect that nonevolved close binaries have synchronized rotational and revolutional motions, *i.e.* each star always exposes the same hemisphere to the other. This situation will change when one of the two stars begins to evolve and expand; then it must rotate more slowly in order to preserve its total angular momentum. As observational evidences of this, let us consider Fig. 13; PLAUT (1959) has collected the existing data on rotational velocities

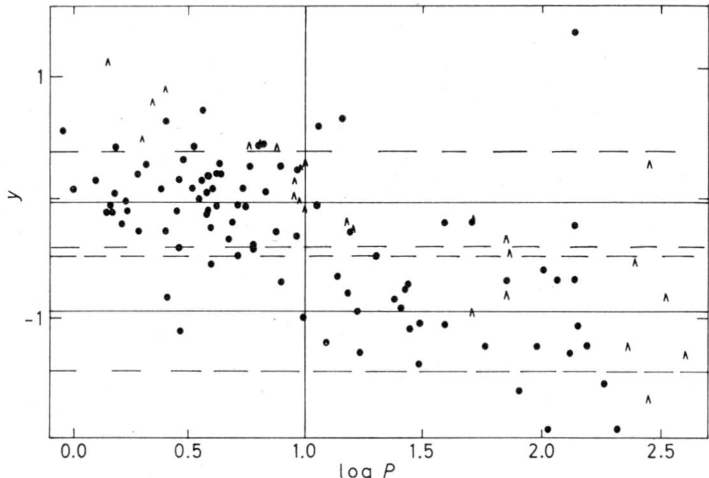

Fig. 13. – $\log (V_{rev}/V_{rot} \sin i)$ (indicated with y) *vs.* $\log P$.

of stars for deriving a relationship between the period of axial rotation and the period of orbital revolution for the brighter main sequence components of spectroscopic binaries. The ratio between $v_{rot} \sin i$ and $v_{rev} = 2\pi R/P$ (where v_{rev} is the rotational velocity in case the period of rotation equals the period of orbital revolution). For 57 stars for which $\log P < 1.0$ (P in days) the average value is $\log (v_{rev}/v_{rot} \sin i) = -0.03$. For $\log P > 1.0$ the average value of $\log (v_{rev}/v_{rot} \sin i)$ for 36 stars abruptly become less and equal to -0.97. This result suggests that synchronization is present only if the system is sufficiently close (and hence the period shorter than 10^d, otherwise the components rotate more rapidly than they would if there were synchronism between the two periods.

Among the evolved stars, take for example *β Lyrae.* It is probable that when the system was detached both components had a rotation period equal to the period of revolution. When one of the stars evolved and expanded it began to rotate more slowly. We observe today that the blue component of *β Lyrae* has a rotational equatorial velocity of 45 km/s. If this star would rotate in exact synchronization with its 12.9 orbital period the rotational velocity would be about 280 km/s. Such a high rotational velocity following synchronization with the period of revolution, combined with the tidal effects due to the presence of a close companion might favor strong loss of mass, at least in certain periods of the life of the star, and hence accelerate the evolutionary time scale.

II. – Interpretation of the Observed Facts.

1. – Introduction.

In the first lecture we underlined some of the more important observational facts about close binaries. How these facts can be interpreted, and what information they can give on the evolution of these systems is our next concern.

The questions on the evolution of a close binary can be summarized under one heading: what are the evolutionary effects of the loss and transfer of mass? In particular we may ask the following:

1) Do close binaries evolve along different tracks as compared with single stars and wide binaries?

2) Are the evolutionary time scales significantly different in close binaries from those in single stars and wide binaries?

3) What is the reason for the underluminosity which is common to the secondaries of *sd*, *sd-d*, *W UMa* types and to the primaries and secondaries or *R CMa* stars?

4) What is the solution to the problem of the subgiants? In other words: nonevolved systems such as *d* systems, are composed of two normal population-I stars and hence have normal internal constitution. Therefore we would expect that the most massive star evolves first, expanding and becomes a

subgiant while the secondary is still on the main sequence. However the observational evidence is the contrary: the most massive star is still on the main sequence, and the secondary has evolved on the red giant branch. This is one of the several puzzles presented by the close binaries.

5) Are the different physical groups already evolved, like *sd*, *sd-d*, *R CMa* and maybe some of the *c* systems, different stages of a same evolutionary path, or do they represent different evolutionary paths of groups differing initially in the mass-ratio m_2/m_1, or in the degree of closeness of the system?

All that we can safely assume is that the evolution of the components of a close binary system must depend upon the single masses of the components (as for the single stars), upon the mass ratio and upon the closeness of the system; *i.e.* upon

$$m_1\,, \qquad m_2\,, \qquad m_1/m_2\,, \qquad (R_1+R_2)/a\,.$$

2. – Evolutionary hypotheses.

We consider, on the basis of previous results, that *d*, *sd*, *c* of early type, *c W UMa* and *R CMa* stars are physical groups; let us now try to explain the evolutionary stage of each group. It is probably reasonable to assume that the stars which are farther from normality are also more advanced in evolution. We observe that detached systems have both components which are normal main sequence stars; hence they are probably nonevolved pairs of stars. Several *d* systems have mass-ratios very close to one. It is very reasonable to assume that early type *d* systems with $m_2/m_1 \sim 1$ evolve in early type contact systems, which have on the average about the same values for the mass-ratios and the periods as the early type *d* systems. For example, the *d* system *Y Cyg* (O 9.5 + O 9.5; $m_2/m_1 = 0.99$; $P = 3^{\mathrm{d}}.0$) might evolve in a *c* system like *AO Cas* (O 8.5 + O 9; $m_2/m_1 = 0.93$; $P = 3^{\mathrm{d}}.52$). In the list of eclipsing binaries for which all the photometric and spectroscopic data are available (Table I) there are 12 early type *d* systems with $m_2/m_1 \geqslant 0.70$, and 4 early type *c* systems with $m_2/m_1 \geqslant 0.70$. We find for them:

$$c \qquad \overline{m_2/m_1} = 0.91 \qquad \overline{P} = 2^{\mathrm{d}}.54 \qquad (\text{no. } 4)\,,$$

$$d \qquad \qquad = 0.87 \qquad \qquad = 3^{\mathrm{d}}.13 \qquad (\text{no. } 12)\,.$$

The total mass of the systems of both groups is on the average about the same but the density of the components of the *d* systems are sistematically higher than that of the components of the *c* systems of the same spectral type, proving that these last are already beginning to expand (see Table II and Fig. 14). It is not surprising that *c* early type systems are rarer than *d* systems because probably this phase of the life of the close systems lasts a short time as a consequence of the strong loss of mass which is apparent from spectrographic observations. It is possible that systems like β *Lyrae* represent a further stage than that of the *AO Cas* stars. Contact systems like β *Lyrae* are very rare objects. STRUVE (1958) computes that only one star having the unusual characteristics of β *Lyrae* exists within a distance of 4600 parsecs and therefore there are only 100 similar binary systems in the whole galaxy. Struve in asking why the β *Lyrae* systems are so rare suggests the following explanation. Let us suppose that β *Lyrae* is a star which during its evolutionary course undergoes a short stage (for example 10^4 years long) of copious mass loss. If the galaxy contains at any given time 100 β *Lyrae* stars, each of which lasts 10^4 years, then 100/10 000 β *Lyrae* stars per year lose their peculiarity and the same number of stars acquire these characteristics. In other words one β *Lyrae* star is created once in 100 years and one is lost in the same interval. Struve concludes that in the lifetime of the galaxy the number of ancestors and descendants of the β *Lyrae* stars must be of the order of 10^8.

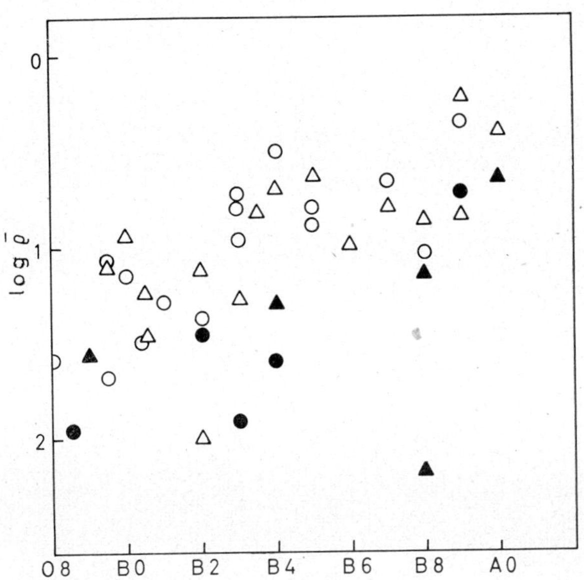

Fig. 14. – log ϱ *vs.* the spectral type for *c*- and *d*-systems of early type: ● *c* pr., ○ *c* sec.; ▲ *d* pr., △ *d* sec.

The descendants must be systems of small mass, larger periods and greater separations, but we cannot at the moment identify them. A possible suggestion is that they are *R CMa* stars. These systems have probably undergone a fast evolution, since both the components are strongly undermassive, and probably surrounded by thick shells. In Table I we have 9 *R CMa* systems; their average period is $4^d.98$ and the average total mass of this kind of system is of the order of two solar masses. Since the average total mass of a contact early type system is $34m_\odot$, the hypothetical evolution from *c* to *R CMa* requires a loss of mass of the order of $30m_\odot$; in 10^6 years this means an average loss of $3\cdot10^{-5}\,m_\odot$ per year. HUANG (1956) found that if a component of a binary star loses mass into space, the period will increase, but if it loses mass to its companion, the period will decrease. He finds $\delta P/P = -\delta m_1/m_1$. In our case the period is about $2^d.5$ for early type *c* systems, and 5 for the *R CMa* systems, and assuming $m_1 = 20$ we have $\delta m_1 = 10$ a value of the right order of magnitude.

Hence an evolution of this type is possible:
d early type systems (like *Y Cyg*) → *c* early type (like *AO Cas*) → *β Lyrae* type → *R CMa* type (?).

TABLE II.

c early type systems			*d* early type systems		
name	ϱ_1	ϱ_2	name	ϱ_1	ϱ_2
V 599 *Aql*	0.025	0.074	*σ Aql*	0.092	0.15
E0 *Aur*	0.0124	0.0066	*TT Aur*	0.185	0.162
A0 *Cas*	0.011	0.028	*AR Aur*	0.45	0.40
G0 *Cyg*	0.19	0.23	*SZ Cam*	0.022	0.076
V 470 *Cyg*	0.036	0.051	*CC Cas*	0.026	0.117
			AH Cep	0.074	0.064
			CW Cep	0.112	0.154
			Y Cyg	0.086	0.085
			V 453 *Cyg*	0.041	0.055
			V 478 *Cyg*	0.0325	0.0320
			DI Her	0.32	0.24
			IM Mon	0.16	0.14
			U Oph	0.135	0.105
			δ Ori	0.053	0.010
			AG Per	0.168	0.205
			ξ Pho	0.225	0.6

The contact systems *W UMa* type probably are nonevolved stars. The majority are late type main sequence stars, and hence they have not had enough time to evolve (provided that the loss of mass and the exchange of mass between the two components do not modify essentially the normal evolutionary path). They could be born in contact with each other and the overluminosity of the companion could be explained by heating due to radiative and corpuscular transfer of energy from the primary. The two stars have generally about the same spectrum; probably the secondary mimics the spectrum of the primary.

The *W UMa* stars have periods smaller than 1^d; they are characteristic for having rapid axial rotation (~ 100 km/s) while the main sequence stars of the same spectral type (F or K) have rotational velocities of less than 30 km/s. The two components are embedded in a common envelope through which mass exchange may take place. SHAPLEY (1948) has shown that the *W UMa* are at least 2 or 3 times as numerous per unit volume, as all other variables put together. STRUVE (1950) pointed out that Shapley's conclusions may simply indicate that the *W UMa* stars are more numerous than the early type binaries by about the same factor by which the main sequence single stars of late spectral type are more numerous than the early type single stars.

The idea that *W UMa* stars are young nonevolved objects is consistent with the result of a search for *W UMa* stars in clusters (SAHADE, 1959).

Sd, *sd-d* and *R CMa* are in this respect farther and farther from normality. We point out that all these systems with the secondary a subgiangt, have a main sequence primary whose spectral type is earlier than A8. We can make the speculation that *d* systems formed by two stars of appreciably different masses, and hence with appreciably different evolutionary time scales, become *sd* systems. But why in 99% of the cases is the less massive star the most evolved? The fact, observed independently by KOPAL (1954) and by CRAWFORD (1955) that almost all the *sd* systems have a secondary which fills the maximum closed volume it can in systems of given mass ratios, suggests that there is a particular reason for such a clustering of the sizes of the secondaries around the Roche limit. Therefore this may mean that the secondaries are secularly expanding and the growth of the star stops at its Roche limit. This observational fact excludes the possibility that *sd* systems are formed by a primary which has already arrived on the main sequence and by a secondary which is still on the contracting stage. In this last case, in fact, no clustering around the Roche limit is expected.

2·1. *Crawford's hypothesis.* – CRAWFORD (1955) suggests an ingenous explanation for the fact that almost all of the *sd* systems have a contact secondary and not a contact primary, as we might expect. He postulates that the contact secondaries where initially the more massive components and hence they evolved first and began to expand, reaching a point where they filled their maximum close equipotential surface. As they continued to expand, the matter began either to pass through the first lagrangian point to their companion, or was ejected from the system if violent mechanisms such as prominences predominated. Thus the actual secondaries became less massive and less luminous than the actual primaries. Kopal criticizes this hypothesis with the following arguments: 1) a complete lack of subgiants at their Roche limit having masses comparable to their companions. On Crawford's hypothesis such systems should be at least as frequent as those characterized by very unequal mass ratios; 2) the existence of subgiants having sizes considerably smaller than the inner contact surfaces; 3) the total amount of mass transfer demanded to transform the primary into the secondary is about 80% of the total. This would demand a removal of mass right down to the core of the primary, and the total amount of nuclear energy released in the core to do this would need a large amount of rapid helium burning which would be detectable. We can reply to the first objection that if the period of transfer of mass lasts a very short time, this would explain the lack of subgiants with masses comparable to those of their companions. To the second we can answer that perhaps the *sd-d* systems represent a more advanced stage than the *sd* systems and not the contrary. Concerning the third point the BURBIDGES (1958) observe that it is difficult to see how the conclusion of Kopal are reached. They observe that helium burning produces only 10 % of the energy released in hydrogen burning per gram of material. Furthermore helium burning would not lead to a sudden increase in luminosity. Kopal argued that in many cases there is not available sufficient nuclear energy to move material from the primary to the secondary but the Burbidges observe that the available nuclear energy exceeds the gravitational potential energy by several orders of magnitude.

2·2. *Kopal's hypothesis.* – On the strength of his objections, KOPAL (1956) proposed another explanation. The more massive star has evolved faster than the secondary. In the course of the expansion due to hydrogen shortage in the core the primary extends to its Roche limit in an interval of time which can be estimated of the order of 10^5 or 10^6 years. Hence less than one star in a hundred would be performing the expansion at any particular time, and in this way their scarcity could be explained. When the Roche limit is attained a loss of mass to the secondary will take place. This is probably limited only to the outer layers of the primary component, but the small transfer of mass from its surface should be sufficient to fill the Roche limit of the secondary. The mass ratio of the system would not be altered appreciably and so the secondary would remain the less massive star of the system, but its volume is increased at the Roche limit. Then, following the expansion, Kopal assumes that the primary collapses back to its original size with the same rapidity as it expanded, while the secondary would settle down at a much slower rate. The expansion of the primary will probably be repeated and when this occurs, the secondary will have already settled down, so as to make room for other gas expelled from the primary. The BURBIDGES (1958) object to this hypothesis observing that there are no arguments connected with the internal structure of the primary or of the secondary which suggest that they should rapidly contract again.

2·3. *Quantitative considerations concerning Crawford's hypothesis.* – MORTON (1960) computes the effect of mass loss on the radii of various stellar models and the effect of mass transfer on the size of the limiting Roche surface. He shows that only if the two stars are close enough and if the mass ratio is near to unity they can reach their limits before the end of their early evolutionary phases, but usually the primary components in most systems can reach their limits only in the Hertzsprung gap. Since the normal evolution in this region of the H-R diagram is so fast that very few single stars are discovered here it is not surprising that rarely binary systems have been found with a contact primary. Let be τ_N the nuclear time-scale, *i.e.* the number of years required for a star to burn 13% of its mass of hydrogen; we have

$$\log \tau_N = 10.1 + \log (m/m_\odot) - \log (L/L_\odot) .$$

When the luminosity of the star is not balanced by its nuclear energy generation the star will expand or contract until equilibrium is established. The time necessary to reach equilibrium is the Kelvin time-scale τ_K and is given in years by

$$\log \tau_K = 7.7 + \log (m/m)_\odot - \log (L/L_\odot) .$$

Since τ_K is shorter than τ_N by a factor of 250 we can expect to find less than 1% of *sd* binary systems with a contact primary; an expectation which is actually confirmed by observation. Morton's computations permit us to establish the change in radius with mass loss for a series of models in thermal equilibrium and to see the effect of mass transfer on the size of the primary lobe. Morton assumes that all the mass lost from the original primary is transferred to the original secondary. The results are summarized in Table III for two systems having the original mass ratio m_2/m_1 equal to 0.2 and to 1.

TABLE III.

m_1	R_1	Radius of the lobe for initial mass-ratio	
		$m_2/m_1=0.2$	$m_2/m_1=1.0$
$10m_\odot$	$5.5\,R_\odot$	$5.5\,R_\odot$	$5.5\,R_\odot$
3	5.5	3.1	3.8
2	2.8	2.8	3.3
1.8	2.55	2.7	3.2

Therefore the Roche limit for any initial mass ratio shrinks faster than the radius of the star and hence transfer of mass to the original secondary occurs steadily until the mass of the primary decreases from 10 to 2. At this stage so much matter has been transferred that the roles of the two stars are interchanged.

As consequence of the great mass loss the subgiants of close binary systems lack a hydrogen-rich envelope. In this case computations of stellar models having $Y = 0.42$ at their surface layers show that the two stars are two magnitudes brighter, their radii are larger by a factor of two and their effective temperatures higher by 0.05 in logarithm as compared with an unperturbed subgiant of the same mass and central chemical composition. This fact can explain the overluminosity of the secondary, if this actually was the original primary. An example of such hydrogen-poor components of a close binary system is υ *Sgr*. It can be expected that the new primary begins its evolutionary expansion possibly returning some of the matter to the companion.

We observe now that loss of mass from the system is probably a process at least as effective as the exchange of mass. If we assume the following evolutionary path: $d \rightarrow sd$, we observe that the average period for the d system is $2^{d}.8$ (average of 29 systems) and for the sd systems is $2^{d}.7$ (30 systems). According to computations by HUANG (1956), since the period is practically the same for the two groups, both the two processes, loss and transfer, occur. A further step in the evolutionary path might be represented by the sd-d systems. The average period for this group (for 17 systems) is $6^{d}.6$. When the original primary has become smaller than its Roche lobe the steady process of transfer of mass stops, but loss to outer space can occur if mass is violently ejected, under form of prominences.

In conclusion we suggest that d systems with mass-ratio appreciably less than one probably evolve according to the following path: $d \rightarrow sd \rightarrow sd - d$.

2·4. *Struve's hypothesis.* – The problem of the subgiants in close binaries has been examined by STRUVE (1954) which suggests another interesting possibility for explaining the fact that the secondary seems to evolve faster than the primary. Up to now we have accepted as obvious the hypothesis that the components of a close binary have the same age and the same chemical composition. But let us suppose, as Struve does, that a binary star which consists of two components of appreciably different mass, has the primary component of normal chemical composition, while the secondary is from the beginning deprived of a large amount of hydrogen. In this hypothesis the temperature of the secondary would be two or three times higher. For example a star of mass equal to $0.3\,m_{\odot}$ and average molecular weight such that $\mu_{\text{subg}}/\mu_{\text{norm}} \simeq 2$ would have an effective temperature of 6500 °K instead of 2500 °K, and its position of false subgiant would be explained. Struve says that this idea is suggested by analogy with the solar system, but remarks that we really do not know whether Jupiter is really more deficient in hydrogen than the sun, and that extrapolation from a mass ratio $m_2/m_1 \sim 0.001$ to a mass ratio of the order of 0.1 is dangerous.

3. – Overluminosity of the companion and interaction between the two stars.

We have often observed that in several cases the secondaries of *sd* systems are overluminous for their masses and that such anomaly is present and stronger in both stars of the *R CMa* systems. Such effects could be explained by assuming an anomalous structure of the star (as a thick shell mimicking the property of a normal atmosphere) or an anomalous chemical composition, or an interaction between the two stars, as for example corpuscular radiation from the primary to the secondary. A similar hypothesis has been advanced by STRUVE (1958) concerning *β Lyrae* and applied by HUANG (1959) to the spectroscopic binary *α Vir*.

Observations show that the period of *β-Lyrae* is becoming longer at a rate of 9 seconds per year (STRUVE, 1958). Huang has shown that if this increase is caused by a symmetrical escape of mass from the system we must have $dm_1/dt = -5.6 \cdot 10^{22}$ g s$^{-1} = 10^{-3}\, m_\odot$/year. Struve estimates that the loss of mass suggested by the spectroscopic observations is about 100 times less than the value computed by Huang from the change in period. Huang reconsidered the problem, assuming that the ejection is not symmetrical but occurs only from the rear side of the visible B8 component. In this case the loss of mass could perhaps be lowered by a factor of ten but not by one hundred. Hence Struve suggests that either the change in period is not of evolutionary character, or the loss of mass occurs also in the form of spectroscopically unobserved particles, such as cosmic rays or similar phenomena. This hypothesis has suggested to Huang that an ejection of corpuscolar radiation might also explain the observed overluminosity of the companion of several close binaries. If we assume with FESSENKOFF (1949) and MASSEVITCH (1949) that the ejection of corpuscolar radiation increases with luminosity, we may expect the excess of luminosity of the secondary to increase with the luminosity of the primary. A plot (Fig. 15) of the excess luminosity of the secondaries of close systems versus the bolometric magnitude of the primary is traced by HUANG (1958) and confirms these expectations. The sources from which the data are taken are not given. We have made a similar plot using the data of Kopal for a greater number of systems, but no correla-

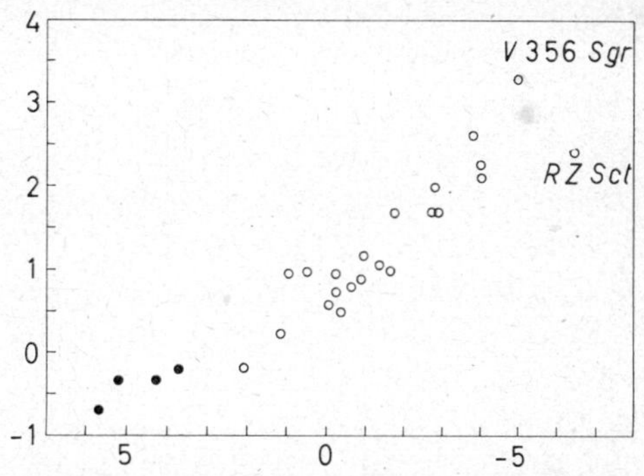

Fig. 15. – Relation between the excess of luminosity of the secondary and the absolute bolometric magnitude of the primary, according to Huang. ● *W UMa* systems; ○ *Algol* systems.

tion is found for the *sd* and *sd-d* systems, and a very scattered relation exists for the *W UMa* and the *R CMa* stars (Fig. 16 and 17). However there is no reason why a star should not dissipate some of the energy produced in its interior by corpuscolar radiation, and it is perfectly possible that phenomena similar to the solar flares (from which solar corpuscles mainly come) occur in stars other than the sun even on a much larger scale. It is therefore possible that at least part of the excess luminosity may be explained by corpuscular radiation. Concerning α *Vir* we report the results of HUANG (1959). The two components have visual magnitudes

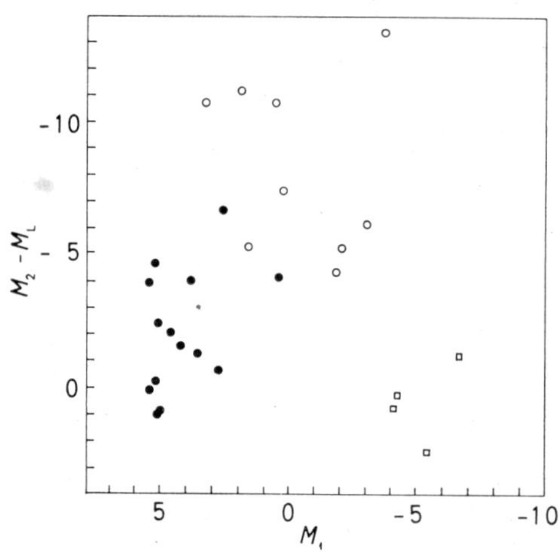

Fig. 16. – Relation between the excess of luminosity of the secondary, and the absolute magnitude of the primary, using Kopal's data. ○ *R CMa* systems, ● *W UMa* systems, □ *c* early type systems.

$$M_1 = -2.0 \qquad Sp_1 = \text{B2 V},$$

$$M_2 = -0.3 \qquad Sp_2 = \text{B3 V},$$

while according to the empirical mass-luminosity relation we should have

$$M_1 = -1.7 \qquad \text{and} \qquad M_2 = +0.4 .$$

Moreover the spectral types being B2 V and B3 V, they correspond to $M_1 = -2.6$ and $M_2 = -2.0$. Hence the secondary is slightly overluminous for its mass, but there is a strong discrepancy between its absolute magnitude and its spectral type. Huang suggested that the spectral type of the secondary should be actually B7 but that its atmosphere is heated by the primary whose spectrum it

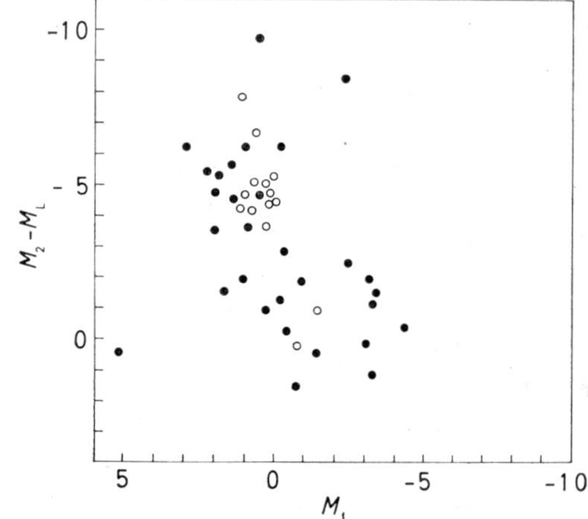

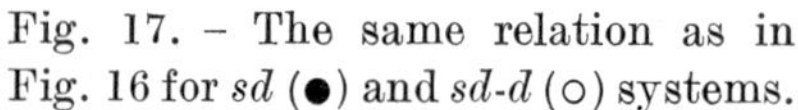
Fig. 17. – The same relation as in Fig. 16 for *sd* (●) and *sd-d* (○) systems.

matches. Huang remarks that the hypothesis of heating by the primary has been introduced in the belief that there is no systematic difference in internal structure between main sequence stars in visual binaries and in spectroscopic binaries.

What is the nature of the heating process? According to Huang there are three possibilities: transfer of energy through: 1) hydrodynamical circulation of stellar material between the two stars; 2) electromagnetic radiation; 3) high-energy corpuscular radiation.

Struve has satisfactorily explained the overluminosity of the secondary star of a *W UMa* system by the mass circulation which certainly occurs between the two components. However, this explanation is not acceptable for binary stars having a separation much larger than the sum of the radii, because when the size of the two components is smaller than the size of their respective lobes no exchange of matter is possible.

In the second case, heating of the secondary by electromagnetic radiation from the primary, the radiation which is falling on the secondary should be equal to the excess of luminosity that the secondary radiates away as a result of its rise in surface temperature. In the case of α *Vir* since the temperature of the secondary is raised from about 12 000 °K (B7) to about 18 000 °K (B3) we have for the excess luminosity:

$$\sigma(18\,000^4 - 12\,000^4)\ \text{erg cm}^{-2}\,s^{-1} = 4.6 \cdot 10^{12}\ \text{erg cm}^{-2}\,\text{s}^{-1}\,.$$

Now the energy sent by the primary is $\sigma T_1^4/(a/R)^2 = 10^{10}$ erg cm^{-2} s^{-1} (assuming $T_1 \sim 20\,000$ °K and $a/R \sim 30$). This value is more than two orders of magnitude smaller than required for explaining the luminosity excess of the secondary. However, for closer binaries where $a/R \sim 2$, the energy sent by electromagnetic radiation is sufficient for explaining the luminosity excess of the secondary.

Huang examines the third possibility, corpuscular radiation. He observed that at the distance of the earth from the sun, which is of the order of 200 solar radii, the earth receives appreciable corpuscular radiation from the sun. It is therefore reasonable to expect an active interchange of corpuscular radiation between the two components of close binary systems. We may imagine that they are embedded in an unique corona. If for example we consider, as in the case of α *Vir*, that the excess of energy is $4.6 \cdot 10^{12}$ erg cm^{-2} s^{-1} the net rate of energy received per cm^2 and per second by the secondary surface must be

$$\tfrac{1}{2}\,(N_1 m v_1^2 - N_2 m v_2^2) = 4.6 \cdot 10^{12}\ \text{erg cm}^{-2}\,\text{s}^{-1}\,.$$

Assuming that the velocity of ejection of high-velocity particles increases with the luminosity, v_2 can be neglected. Since the protons reaching the earth from the sun have $v \sim 3000$ km/s or more, Huang assumes $v_1 \sim 10^4$ km/s Therefore we have

$$N_1 = 4.7 \cdot 10^{18} \text{ cm}^{-2} \text{ s}^{-1} ,$$

for the number of protons reaching the secondary per cm^2 and per second. This is a reasonable value since it corresponds to a density of $5 \cdot 10^9$ protons cm^{-3} near the secondary surface. From solar corpuscular radiation we can infer that this density near the solar surface is of the order of $4 \cdot 10^7$ to $4 \cdot 10^9$ protons cm^{-3}. The charged particles will follow the magnetic lines of force of the magnetic field of the two stars, hence the corpuscular radiation from the primary will hit the secondary on both hemispheres.

4. – Binary systems of ζ *Aurigae* type.

This group of five components 31 *Cyg*, 32 *Cyg*, ζ *Aur*, *VV Cep*, and probably ε *Aur* (HACK, 1957, 1961, STRUVE, 1962) distinguishes itself among the other classes of eclipsing systems by having the primary which is a supergiant. These five systems are such that the separation of the components is one order of magnitude greater than the radius of the more extended companion. Hence the reciprocal perturbations are not negligible but probably they are not strong enough to alter the evolution of the single stars. The position of these five systems in the H-R diagram suggests that the more massive primary is already evolved off the main sequence, while the secondary is still on the main sequence as in the case of 31 and 32 *Cyg* and ζ *Aur* or is also evolving off the sequence as in the case of the more massive system *VV Cep* and perhaps of ε *Aur*. The positions of these five systems in the H-R diagram suggests that at least in these cases of very detached systems the loss of mass (the proof of which exists

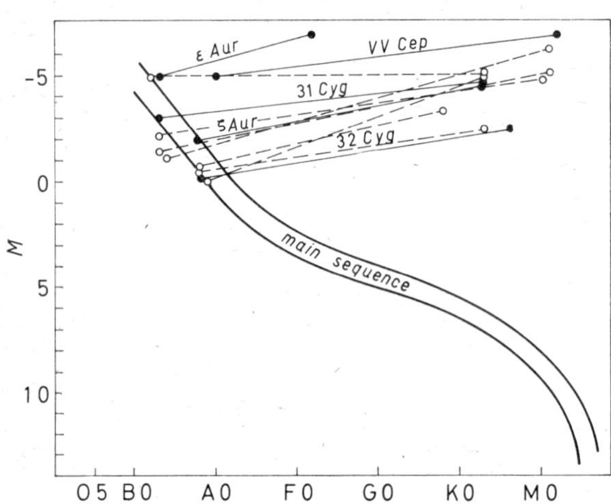

Fig. 18. – The position in the H-R diagram of the visual binaries having *B*-type secondary, (○ - - - ○) and of the 5 systems 31 and 32 *Cyg*, ε and ζ *Aur*, and *VV Cep* (● ——— ●).

from spectroscopic observations: extended atmospheres of 31, 32 *Cyg*, ζ *Aur*, *VV Cep*, and eclipsing ring or shell of ε *Aur*) is not important enough to change the evolutionary path.

It is interesting to observe the H-R diagram constructed by BIDELMAN (1958) for visual binaries having B type secondaries and to plot on it these 5 pairs of eclipsing binaries. We have a confirmation that the evolution of these systems follows the same path as that of the visual binaries (Fig. 18).

5. – Close binary systems at the end of the evolutionary path.

KRAFT (1958) has shown that probably the recurrent nova *T CrB* is a double line spectroscopic binary of period 227.6 days. It consists of a hot subdwarf secondary and of a red giant primary. Also *AE Aqr* and *SS Cyg* which are members of the *U Gem* class (or dwarf novae) are double stars; both *AE Aqr* and *SS Cyg* consist of a hot subdwarf secondary and of a red dwarf primary. In 1954 WALKER discovered that also Nova *DQ Her* is an eclipsing binary. It was suspected then that all novae and dwarf novae are members of close binary systems. Therefore a systematic investigation of spectra of novae and dwarf novae is now in progress (KRAFT, 1962). Of 19 dwarf novae that can be observed at M. Palomar, eight have been studied to date and all have been found to be binaries. The orbital periods range from 3.5 to almost 9 hours. In every case it appears that the red star supplies material for a ring surrounding the blue star. Of 17 novae in the range of the Mt. Palomar telescope, 6 have been studied so far: 5 are certainly binaries and one is suspected of being a binary. Hence it appears from these investigations that a class of close binaries exists consisting of a hot subdwarf secondary and a red primary. Here again we face the puzzle that the less massive star is in a more advanced evolutionary stage than the primary. In these last phases of the evolution two groups at least are well defined, one having a red giant as primary, such as *T CrB*, and the other having a red dwarf primary such as *SS Cyg*. In the case in which the primary is a giant it is reasonable to expect that it fills its lobe and hence it loses mass through the first lagrangian point; but in the case of the *U Gem* stars it seems improbable that the red dwarf companion fills its lobe, however there is the spectroscopic evidence of the gaseous ring coming from the primary and surrounding the secondary. It is possible that the *U Gem* primaries may be old stars which were red giants and are now evolving toward the left of the H-R diagram. In this case they could be surrounded by extended rarefied envelopes filling their lobes.

KRAFT (1962) suggests that the *U Gem* star may represent the evolution of *W UMa* stars. He observes that only the *W UMa* systems have periods short enough to be comparable with that of the *U Gem* stars. He gives the following Table IV, comparing the properties of the two classes of stars:

TABLE IV.

U Gem	*W UMa*
$P \sim 0^d.25$	$P \sim 0^d.37$
$m_1/(m_1 + m_2) \sim \frac{1}{2}$	$m_1/m_1/(m_1 + m_2) \sim \frac{1}{3}$
$m_1 + m_2 \sim 1.5$ to 2	$m_1 + m_2 \sim 1.2$ to 2.5
$M_v \sim +9.5$	$M_v \sim +4.5$
$z \sim 37$ parsec	$z \sim 40$ parsec
No. of stars: 6	10

where z represents the average height over the galactic plane.

6. – Conclusion.

We conclude by giving a summary of the evolutionary picture which seems most probable, on the basis of observational evidence.

If the system is enough detached to remain so, even when the two components evolve and expand, the evolution follows the normal path as single and wide double stars. We have an example given by the class of the ζ *Aur* systems.

All the problems which appear when there is a contact secondary must be correlated with phenomena of loss and exchange of mass, and these phenomena are probably very important in determining the evolution of the system.

The great percentage of close binaries discovered among the novae and the dwarf novae leads us to believe that this explosive phase of stellar life is encountered only by close binaries and not by single stars. An argument in favor of this idea is that if all the stars which have arrived at the white dwarf stage, previously went through the nova stage, the novae outbursts would be much more frequent than observed. We suppose therefore that the condition of close system favors the nova outburst.

The values of the single masses, m_1 and m_2 and the index of closeness of the system $(R_1 + R_2)/a$ are probably the main factors determining the rapidity of evolution of the system, while the mass ratio m_2/m_1 probably determines the type of evolutionary path.

We suggest tentatively the following subdivision according to the evolutionary stage of each system:

d main sequence system
W UMa contact systems } both components nonevolved

sd and *sd-d* systems:	one component already evolved and the other component a main sequence non-evolved star
c systems of early type *R CMa* systems Novae and dwarf novae	both components already evolved.

REFERENCES

BIEDELMAN, W. P.: 1958, *P. A. S. P.*, **70**, 158.
BURBIDGE, E. M. and G. R.: 1958, *Handbuch der Physik*, vol. **51**, p. 146.
CRAWFORD, J. A.: 1955, *Ap. J.*, **121**, 71.
FESSENKOV, V.: 1949, *Russ. A. J.*, **26**, 67.
GOULD, N. L.: 1957, *P. A. S. P.*, **69**, 541.
HACK, M.: 1957, *P. A. S. P.*, **69**, 389.
HACK, M.: 1961, *Mem. S. A. I.*, **32**, 351.
HANSEN, K. and McNAMARA, D. H.: 1959, *Ap. J.*, **130**, 791.
HARPER, W. E.: 1937, *Publ. Dom. Ap. Obs. Victoria*, **7**, 1.
HUANG, S. S.: 1956, *A. J.*, **61**, 49.
HUANG, S. S.: 1958, *P. A. S. P.*, **70**, 473.
HUANG, S. S.: 1959 *Ann. d'Ap.*, **22**, 527.
JASCHEK, C. and M.: 1957, *P. A. S. P.*: **69**, 546.
KOPAL, Z.: 1955*a*, *Les Particules Solides dans les Astres* (Cointe-Sclessin, Liège), p. 684.
KOPAL, Z.: 1955*b*, *Ann. d'Ap.*, **18**, 379.
KOPAL, Z.: 1956, *Ann. d'Ap.*, **19**, 298.
KRAFT, R. P.: 1958, *Ap. J.*, **127**, 625.
KRAFT, R. P.: 1959, *Ap. J.*, **130**, 110.
KRAFT, R. P.: 1962, *Ap. J.*, **135**, 408.
KRAT, T. V.: 1944, *Russ. A. J.*, **21**, 20.
KRZEMINSKY, W.: 1962, *P. A. S. P.*, **74**, 66.
KUIPER, G. P.: 1935, *P. A. S. P.*, **47**, 15.
KUIPER, G. P.: 1941, *Ap. J.*, **93**, 141.
MARTYNOV, D. YA.: 1957, *Non Stable Stars*, ed. G. H. HERBIG (Cambridge), p. 142.
MASSEVITCH, D. C.: 1949, *Russ. A. J.*, **26**, 207.
MERGENTALER, J.: 1950, *Contr. Wroclaw Astr. Obs.*, n. 4.
MORTON, D. C.: 1960, *Ap. J.*, **132**, 146.
PLASKETT, J. S. and PEARCE, J. A.: 1931, *Publ. Dom. Ap. Obs. Victoria*, **5**, 99.
PLAUT, L.: 1959, *P. A. S. P.*, **71**, 167.
PRENDERGAST, K. H.: 1960, *Ap. J.*, **132**, 162.
SAHADE, J.: 1959, *Modèles d'Etoiles et Evolution Stellaire* (Cointe-Sclessin, Liège), p. 170.
SAHADE, J.: 1960, *Stellar Atmospheres*, ed. J. L. GREENSTEIN (Chicago, Ill.), p. 467.
SAHADE, J. and STRUVE, O.: 1957, *Ap. J.*, **126**, 87.
SHAPLEY, H.: 1948, *Harv. Obs. Monography*, n. 7, p. 249.

STRUVE, O. 1946, *Ap. J.*, **103**, 76.
STRUVE, O.: 1950, *Stellar Evolution* (Princeton N.J.), p. 168.
STRUVE, O.: 1954, *Les Processus Nucléaires dans les Astres* (Cointe-Sclessin, Liège), p. 235.
STRUVE, O.: 1957, *Sky and Telescope*, **17**, 70.
STRUVE, O.: 1958*a*, *P. A. S. P.*, **70**, 38.
STRUVE, O.: 1958*b*, *Etoiles à Raies d'Emission* (Cointe-Sclessin, Liège), p. 377.
STRUVE, O.: 1962, *Sky and Telescope*, **23**, 127.
WOOD, F. B.: 1950, *Ap. J.*, **112**, 196.
WOOD, F. B.: 1957, *Non Stable Stars*, ed. G. H. HERBIG (Cambridge), p. 144.

Tipografia Compositori - Bologna - Italy